ENERGY POVERTY

ENERGY FLOWERS

Energy Poverty

Global Challenges and Local Solutions

Edited by
ANTOINE HALFF, BENJAMIN K. SOVACOOL,
AND JON ROZHON

OXFORD
UNIVERSITY PRESS

OXFORD
UNIVERSITY PRESS

Great Clarendon Street, Oxford, OX2 6DP,
United Kingdom

Oxford University Press is a department of the University of Oxford.
It furthers the University's objective of excellence in research, scholarship,
and education by publishing worldwide. Oxford is a registered trade mark of
Oxford University Press in the UK and in certain other countries

© Oxford University Press 2014

The moral rights of the authors have been asserted

First Edition published in 2014

Impression: 1

Published in the United States of America by Oxford University Press
198 Madison Avenue, New York, NY 10016, United States of America

British Library Cataloguing in Publication Data
Data available

Library of Congress Control Number: 2014939569

ISBN 978–0–19–968236–2

Printed and bound by
CPI Group (UK) Ltd, Croydon, CR0 4YY

Links to third party websites are provided by Oxford in good faith and
for information only. Oxford disclaims any responsibility for the materials
contained in any third party website referenced in this work

In memory of Amulya K.N. Reddy
and
Abeeku Brew-Hammond
two visionaries in the fight against energy poverty
men of both words and deeds
whose intellect, leadership and devotion to the people
touched many a life,
and will remain a lasting inspiration to us all

Contents

Part II: Lessons Learned

Part III: Challenges and Policy Options

List of Figures

List of Tables

Introduction: The End of Energy Poverty

Pathways to Development

Antoine Halff, Benjamin K. Sovacool, and Jon Rozhon

Access to electricity and other forms of energy has become so ubiquitous in Western society and other developed economies that it is easy to forget how recent it really is. Although we take it for granted, energy access is still, even in the most advanced economies, a relatively new development. How startling it is, when watching a 1962 documentary about Paris—Chris Marker's *Le Joli Mai*—to be reminded that as many as 20 per cent of Parisian apartments at that time still lacked access to electricity (and 10 per cent lacked water access!). Even in the United States, electrification prior to Franklin Roosevelt's New Deal was far from universal: only after the 1936 Rural Electrification Act and the establishment of the Rural Electrification Administration were virtually all Americans able to enjoy the benefits of electricity (Trombley 1954). A generation ago, 0.5 billion Chinese had no access to electricity whatsoever. Today, China is the world's factory and ranks as one of the top consumers of electricity. Given how new this all is, perhaps it should not come as a surprise that for so many of us access to modern energy services remains a distant dream.

The world has made giant strides in the last century, and especially in the last 30 years, in extending access to modern energy services to most of humanity. Yet however extraordinary that achievement may be, many of us still lag behind. In fact, the number of those of us still mired in energy poverty is staggering: depending on definitions, as many as one fourth or even one third of humanity. The geographical distribution of these energy-poor is strikingly lopsided. While residual pockets of energy poverty can be found in various parts of the globe, the vast majority of the energy-poor span two main regions: South Asia (especially India) and Sub-Saharan Africa. Not coincidentally, this is also where most of the world's poor can be found. From the standpoint of energy access, these regions are the 'last mile': delivering energy services to them is, for various reasons, trickier than elsewhere.

Until recently, few seemed to care. Even among those who made it their mission to end poverty in general, awareness of this problem was limited. Thankfully, that is changing. The world has come to recognize the central role of energy in human and economic development. The United Nations and others have identified universal energy access as a major policy goal, humanity's shared responsibility.

Yet while this goal has now moved to the front burner, it remains elusive. Sizing up the problem, and finding solutions to it, is what this book is about.

Universal energy access as a policy issue is an idea whose time has come—at long last. Less than 15 years ago, it was virtually nowhere to be seen on the global agenda. Pioneering, visionary third-world economists and scientists embraced the cause of energy access in their own countries and made it part of their life's mission. The late Amulya Kumar N. Reddy in India and Jose Goldemberg (co-author of one of this book's chapters) in Brazil are chiefs among those early prophets.[1] Outside of their country and an elite circle of international experts, however, their work remained all too hidden. Remarkably, energy access did not make the cut as one the UN Millennium Goals: health and education, yes; energy, no. Until very recently, there has been precious little mention of energy access in the vast economic development literature. Energy economists also ignored the subject—so much so that when International Energy Agency economist Fatih Birol (also a co-author of this book) called on them, as recently as 2007, to make 'a place for energy poverty' in the agenda, the call came with a question mark 'Energy Economics: A Place for Energy Poverty in the Agenda?' (Birol 2007).

Since then, the UN has been making up for lost time. Recognizing 'the importance of energy access for sustainable economic development and supporting achievement of the Millennium Development Goals', the General Assembly designated 2012 as the 'International Year of Sustainable Energy for All'. Secretary-General Ban Ki-moon's Advisory Group on Energy and Climate Change—a high-profile gathering of top business leaders and UN agency heads—has called for 'a major UN initiative' to achieve universal energy access by 2030. Other international institutions, such as the World Bank and the International Energy Agency, have more or less formally enshrined universal energy access in their mission and have designed high-profile programmes and activities to promote the cause. More and more articles get published, and conferences are convened, on the subject by the day. Until now, however, there had been no comprehensive, book-length treatment of the subject—a gap that this volume seeks to fill.

The benign neglect in which the economic-development and poverty-alleviation literatures have long held energy poverty is not hard to understand. This vast body of work is as old as economic theory itself—going back all the way to Adam Smith's 1776 classic, *An Inquiry into the Nature and Causes of the Wealth of Nations*. Since political economy, like economic development itself, predates

---

[1] See, among others, J. Goldemberg, T.B. Johansson, Amulya K.N. Reddy, and R.H. Williams, 'Energy Efficiency from the Perspective of Developing Countries', *Energy for Sustainable Development* 1(2), July 1994; Amulya K.N. Reddy, 'Poverty-Oriented Energy Strategies for Sustainable Development', paper presented to the International Workshop on Environment and Poverty, 22–24 July 1993, Dhaka (Bangladesh), organized by the Bangladesh Centre for Advanced Studies for the Global Forum on Environment and Poverty; Amulya K.N. Reddy, 'Energy Strategies for Sustainable Development in India', paper for presentation at the Conference on Global Collaboration on Sustainable Energy Development, Copenhagen, 25–28 April 1991; José Goldemberg, Thomas B. Johansson, Amulya K.N. Reddy, and Robert H. Williams, 'Energy for the New Millennium', *Ambio* 30(6), 2001, pp. 330–7; José Goldemberg, T.B. Johansson, Amulya K.N. Reddy, and Robert H. Williams, *Energy for a Sustainable World*, Washington, DC: World Resources Institute, 1987, 119 pp.; José Goldemberg, T.B. Johansson, Amulya K.N. Reddy, and Robert H. Williams, *Energy for a Sustainable World*, New Delhi: Wiley Eastern Limited, 1988, 517 pp.

modern energy services, it is not entirely surprising that the energy/development linkage initially fell outside of its common frame of references. As Douglas Barnes and his co-authors make clear in this book's chapter on energy access and development, while the idea of a linkage between the two is now broadly accepted, the science of understanding and measuring how the linkage exactly works is still in its infancy. For instance, they note, 'more research is necessary on all aspects of the various impacts of better household energy use and how this relates to poverty alleviation'.

Energy poverty—in contrast with income poverty—is still not included in most measures of human poverty. That is in part because, although there have been many attempts at quantifying the scope of the problem, a consensus has yet to emerge on how to measure energy poverty. As Barnes et al. write, 'most international organizations measure energy poverty as outputs...rather than outcomes.... Thus, unlike income poverty—which is usually based on measures of the minimum consumption of food and non-food items necessary to sustain life—energy poverty lacks a solid theoretical basis'. Yet it may be argued that access to energy services, or the lack thereof, is in fact a more revealing and relevant metric to capture poverty than traditional measures based on income or wealth deprivation. While income and wealth, in Amartya Sen's terminology, are a *resource*, energy access represents a *capacity*. Sen makes a compelling case for 'using the capability approach over the resource-centred concentration on income and wealth as the basis of evaluation [of economic advantage]'. As he explains, the idea of capability 'is linked with substantive freedom', and thus gives 'a central role to a person's *actual* ability to do the different things that she values doing'. Conversely, 'the identification of poverty with low income is well established, but there is, by now, quite a substantial literature on its inadequacies' (Sen 2010, pp. 253–4). In that sense, lack of access to electric lighting, clean cooking, heating or cooling, electricity-dependent health services, and other benefits of modern energy services may be more relevant for policymakers and other stakeholders than more established income or wealth measurements.

Because the field of energy and energy access is so inherently dynamic, this book makes no claim to being the definitive work on the matter. Rather, its goal is to offer the first authoritative resource documenting all aspects of energy poverty, a state-of-the-art reference tool for energy entrepreneurs, policymakers, NGOs, development economists and all others interested in the field. It is divided into three main sections.

Part I sets the scene by presenting the most up-to-date statistics concerning energy poverty and seeks to size up the problem and take stock of its large-scale implications. The section deals with generic issues of definitions and methodology, sums up the literature on the linkage between energy and development, and provides an in-depth analysis of the public health implications of energy access, including focus studies at both the household level (examining more specifically the health costs of traditional stoves and the benefits of clean-cooking facilities) and that of the community (considering the case of energy access for health-care facilities). Also addressed in that section are the gender dimension of energy poverty, a discussion of the global cost of ending energy poverty, and a look at the energy–water nexus and how it affects energy access.

Part II focuses on case and country studies and lessons learned, including a more detailed look at the specific policies deployed around the world by various countries—sometimes with mixed results—to remove energy poverty. Regions and countries under review include Africa, Asia-Pacific, China, India, the Middle East and North Africa, and Brazil. These country reviews are provided by writers who come from, or work in, the places they describe and thus in addressing the issues bring an insider's perspective to bear.

Last, Part III is focused on extracting lessons learned and policy options from the global experience in attempting to eradicate energy poverty. This discussion includes issues of scale, access to capital and financial engineering, and the role of the various agents and partners in energy access, including government, the private sector, international diplomacy, the World Bank, and other development aid organizations, philanthropies, and NGOs.

While this book is premised on the recognition that energy access is a necessary condition for economic development and must be part of any plan to promote development and combat poverty, we are under no illusion that the provision of modern energy services is sufficient, by itself, to guarantee development. The history of international aid is rife with examples of energy projects that went nowhere and failed to lift local residents out of the 'poverty trap'. William Easterly, an impassioned and articulate critic of foreign aid as a tool of development, evokes several cases of energy projects that failed to deliver on their promise: in Pakistan, he visits an impoverished village that 'just got electricity' but whose basic needs were all otherwise unfulfilled: 'no telephone, no running water, no doctor, no sewage, no roads' (Easterly 2002, p. 7). In Ghana, Easterly evokes the case of the Akosombo hydroelectric dam built in the 1960s by Ghanaian leader Kwame Nkrumah on the Volta River with help from the UK, the US, and the World Bank. The dam created Lake Volta, the world's largest man-made lake. Easterly quotes the research of a Ghanaian PhD student that compared the dam's performance 20 years later:

> to the high hopes held by Nkrumah and his foreign and domestic advisors for industrialisation, transport, agriculture, and overall economic development. Lake Volta was there, an electric generator was there, and an aluminum smelter was there. Production of aluminium in the smelter had fluctuated up and down, but did grow on average about 1.5 per cent a year from 1060 to 1992. But that was it for the project's benefits.... People living next to the lake, including the 80,000 whose old homes had been submerged, suffered from waterborne illnesses like river blindness, hookworm, malaria, and schistosomiasis. The large-scale irrigation projects that the planners had envisioned never worked. The lake transport...had 'ended up in complete failure'.

Research on energy poverty makes it clear that electrification and the provision of modern energy services are not in and of themselves a panacea. For instance, as Barnes et al. note in the chapter cited earlier, 'the effect of rural electrification on small businesses and farm income is determined by the nature of the local community, complementary programs, and the ability of rural entrepreneurs'. Writing from an altogether different perspective, Amulya Reddy holds as a necessary condition for the success of energy projects that they first and foremost benefit their host community. In his view, development projects in general 'must start with the people at the project sites and then radiate outwards; otherwise, these people at the epicentre become the victims of development' (Reddy 1993, p. 8). Energy projects are not a cure for failures of governance, corruption, red tape, and other woes; rather, those issues must often first be addressed, and the right incentives must be in place, in order for their benefits to bear fruit.

Just as the effectiveness of energy access projects depends at least in part on the broader economic, political, and social context in which they are developed, so too must energy projects be assessed in terms of their overall impact on society, including both costs and benefits, rather than their energy upside alone. The environmental cost of energy access projects is an obvious example of potential unintended consequences which ought to be taken into consideration. This is an area where more work needs to be conducted. Allan Hoffman in Chapter 9 touches on the water requirements of energy access projects. Dealing with the Chinese experience, former president Jiang Zemin had emphasized that 'the solution to China's energy issues . . . involves addressing the ecological and environmental concerns raised by large-scale energy use' (Jiang 2008). Building on those remarks, Han Wenke, lead author of this book's chapter on the Chinese experience, cautioned elsewhere that 'in the process of economic development, resource and environmental preservation must be viewed as highly important. For developing countries and regions, development should be the top priority . . . However . . . preserving energy resources and protecting the environment are also very critical. Otherwise, the overall economic and developmental achievements might be killed and buried' (Han 2010, p. 46). Critics of the US energy access projects of the 1930s have emphasized their toll on the nation's hydrological network (Reisner 1993). China's Three Gorges Dam hydroelectric facility, the world's largest hydropower project, which was begun in 2003 and completed in 2012, has attracted its own share of controversy, and Beijing itself has openly voiced concern about the emissions from its many coal-fired electric power plants and publicly stated its goal of reducing them.

The thought that energy access projects necessarily carry a heavy price tag in terms of incremental greenhouse gas emissions or other adverse environmental impacts could not be farther from the truth, however. In its eleventh and twelfth Five-Year Plans, China set both pollution control and energy access as targets. Those policy goals are not incompatible. Hard pollution limits were added to the eleventh Five-Year Plan for the first time, and Beijing took those policies further in the twelfth Plan through energy efficiency and environmental protection measures and 'optimisation of the energy structure'. Many of the Chinese energy-access programmes described in this book, such as mini-hydro projects and other renewable energy developments, entail no significant environmental cost. As IEA Chief Economist Fatih Birol emphasizes in the opening chapter of this book, energy access for all may not 'cost the earth' anything. Traditional energy use, such as the burning of traditional biomass, carries its own—not insignificant— environmental cost, including high particulate matter and other emissions. Transitioning to cleaner forms of modern energy services can alleviate those costs even as it improves both the reach and the quality of modern energy access.

While this book seeks to identify best practices and broadly applicable recipes for energy access and development, it also recognizes that there may not be a 'one-size-fits-all' solution to the problem of energy poverty. While every country's experience in dealing with energy poverty may carry lessons for the others, successful programmes may not always be directly replicable and transferable from one country to the next without significant adjustments. For example, China's extraordinary success in alleviating energy poverty at home is partly based on the unique traits of its political system. Even the most proven policy measures must be tailored to the specific social, economic, and political conditions of the countries in which they are applied. With this in mind, we have refrained from the temptation of smoothing out differences in tone and voice among the book's chapters. While this may give the book a somewhat heterogeneous and fragmented feeling, we feel that this diversity is very much part of the message.

Yet there are also insights from recent trends in energy access developments that may resonate on a global scale. There is a degree of tension between the policy target of extending access to sustainable energy for all and rising concerns about resource scarcity that had recently gained currency in many circles. Although the recent surge in North American energy production from unconventional sources (shale gas, US light tight oil, Canadian oil sands) has since pushed worries about energy scarcity to the back burner, there are many reasons why production constraints may resurface in the future as a policy issue, reigniting a policy drive to rein in, rather than expand, energy consumption. Even without any disruption in energy supply, the rising cost of energy fuels a conservation agenda that could be construed as conflicting with the goal of energy access. In this context, the innovative nature of many energy access projects is worth noting.

Rather than extending traditional energy infrastructure models over 'the last mile' at considerable cost, many of the latest, more promising energy-access projects entail innovative models based on distributed generation, renewable energy, and gains in energy efficiency. Renewable energy projects whose requirements might have seemed prohibitive in a conventional market context look attractive and economical in this setting. Thus the goal of 'energy for all' may foster the deployment of projects that might otherwise have been seen as uneconomical, deployments that could ultimately cause frontier technologies to mature and costs to fall. In that sense 'energy for all' might be less the 'last mile' of traditional energy access than a laboratory of sorts, enabling a much-needed energy transition for emerging markets and industrialized economies.

REFERENCES

Birol, F. (2007). 'Energy Economics: A Place for Energy Poverty in the Agenda?', *The Energy Journal* 28(3).

Easterly, W. (2002). *The Elusive Quest for Growth, Economists' Adventures and Misadventures in the Tropics*. Cambridge, MA: MIT Press.

Han, W. (2010). 'China's Experience with Energy Access and Development', *Geopolitics of Energy* 32(10–11).

Jiang, Z. (2008). 'Reflections on Energy Issues in China', *Journal of Shanghai Jiaotong University* 13(3): 257–74.

Reddy, A. (1993). 'The Making of a Socially-Concerned Scientist: Personal Reflections of a Maverick' [online document], <http://www.amulya-reddy.org.in/Publication/1993_07_SEMINAR2.pdf> (accessed 24 October 2013).

Reisner, M. (1993). *Cadillac Desert: The American West and its Disappearing Water*. London: Penguin Books.

Sen, A. (2010). *The Idea of Justice*. London: Penguin Books.

Trombley, K. (1954). *The Life and Times of a Happy Liberal: A Biography of Morris Llewellyn Cooke*. New York: Harper & Brothers Publishers.

Part I

Taking Stock of Energy Poverty

1

Achieving Energy for All Will Not Cost the Earth

Fatih Birol

1.1. ASSESSING THE ISSUES

History, it is often said, is written by the victors. The defeated, because they are defeated, often have no voice in the narrative. Likewise, it might be argued that economic theory and statistics focus on the haves, at the expense of the have-nots. The poor, because of their very poverty, don't have much of an economic footprint. All too often they fall below the radar and fail to register in international data. For far too long such has been the case of the energy-poor in energy economics and policy circles. However pressing their needs or devastating the consequences of their deprivation not only for them but for the world at large, policymakers have long ignored their condition, if only for lack of decent data. The battle to eradicate energy poverty is first and foremost a statistical battle.

Measuring the size of the energy-poor population, mapping its distribution, tracking global progress in eradicating energy poverty, assessing the cost of universal energy access have therefore been long-standing priorities of the International Energy Agency (IEA). We all know that the first step towards managing or overcoming any challenge is to measure it, because in life what gets measured gets managed. With that in mind, over ten years ago, the IEA's *World Energy Outlook* (2002) did a first-ever assessment of energy and poverty and found that at that time 1.6 billion people had no access to electricity. For the very first time, the entire world understood the size and nature of the challenge of universal energy access.

Since then, rapid economic development in several developing countries, increasing urbanization, and ongoing energy access programmes have led to some progress. Despite this, today nearly 1.3 billion people—or around four times the population of the United States—still do not have access to electricity. Twice as many, around 2.6 billion people, rely on the traditional use of biomass for cooking and heating.

Sub-Saharan Africa, which accounts for only 12 per cent of the global population but almost 45 per cent of those without access to electricity, has made some progress in recent years, particularly in Ethiopia, Angola, Ivory Coast, and Senegal. However, it is also the region of most concern. Today Sub-Saharan

Africa's electricity consumption is comparable with that of New York State, yet its population is almost 45 times greater. Despite pockets of progress, the number of people without access to electricity in the region is actually expected to rise in decades to come as the rate of population growth outpaces the rate of connections.

This situation is intolerable and needs to change. Energy is a critical enabler for all forms of development. In addition to its own tangible benefits, the positive multiplier effects of access to modern energy are huge. Modern energy services enhance the life of the poor in countless ways. Electric light extends the day, providing extra hours for reading and work. Modern cookstoves save women and children from daily exposure to noxious cooking fumes. Refrigeration allows local clinics to keep needed medicines on hand. And modern energy can directly reduce poverty by raising a low-income country's productivity and extending the quality and range of its products—thereby putting more wages into the pockets of the deprived. In short, ensuring universal access to modern energy services is a moral imperative that we can no longer afford to ignore.

There is no universally agreed and universally adopted definition of modern energy access. For the purpose of the IEA's analysis, energy access refers to a situation in which households enjoy the benefits of 'reliable and affordable access to clean cooking facilities, a first connection to electricity and then an increasing level of electricity consumption over time'. Defining access to modern energy services at the household level necessarily excludes other relevant categories such as electricity access to businesses and public buildings that are crucial to economic and social development, that is schools and hospitals. Access to electricity involves a first connection to electricity for a household and consumption of a specified minimum level of electricity, the amount varying based on whether the household is in a rural or an urban area. People are assumed to start with a level of 500 kWh per household per year in urban areas and 250 kWh per household per year in rural areas. The consumption level of people who receive electricity access over time increases to reach the regional average level.

The IEA's definition of energy access also includes provision of cooking facilities which can be used without harm to the health of those in the household and which are more environmentally sustainable and energy efficient than traditional biomass cookstoves used in many developing countries. This definition refers primarily to biogas systems, liquefied petroleum gas (LPG) stoves, and advanced biomass cookstoves that have considerably lower emissions and higher efficiencies than traditional three-stone fires for cooking. LPG stoves and advanced biomass cookstoves are assumed to require replacement every five years, while a biogas digester is assumed to last 20 years. Related infrastructure, distribution, and fuel costs are not included in investment cost estimates.

Based on this definition, energy poverty is overwhelmingly concentrated in Asia and Sub-Saharan Africa, though there are sizeable pockets of energy poverty in the Middle East and Latin America. As shown in Figure 1.1, Sub-Saharan Africa ranks first in the number of people without electricity. In India, however, the population without clean cooking facilities is even larger than in Africa. In Asia as a whole, more than 1.8 billion people have no access to clean cooking facilities, compared to nearly 700 million people in Africa. The size of this population defines the scope of the policy challenge.

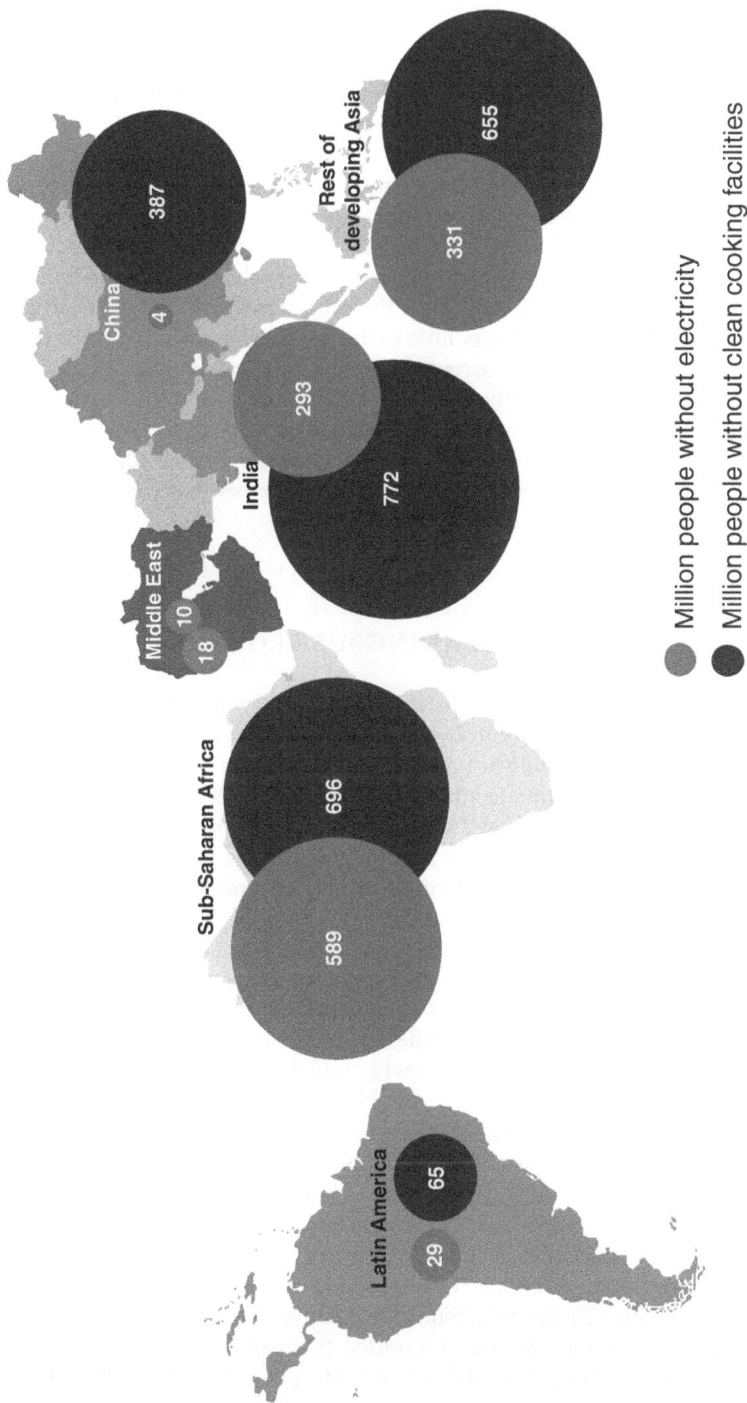

China 4 387

Rest of developing Asia 331 655

India 293 772

Middle East 18 10

Sub-Saharan Africa 589 696

Latin America 29 65

● Million people without electricity

● Million people without clean cooking facilities

This document and any map included herein are without prejudice to the status of or sovereignty over any territory, to the delimitation of international frontiers and boundaries and to the name of any territory, city or area.

Figure 1.1. Number of people without modern energy access by selected region, 2010

When measuring against the backdrop of climate change, the plight of the energy-poor is even more daunting. That is because the energy-poor also happen to count among the population most vulnerable to the adverse environmental impacts of global warming. That is not an abstract consideration. Despite efforts to mitigate climate change, we recently passed a grim milestone with the concentration of carbon dioxide in the atmosphere, topping 400 parts per million at the Mauna Loa Observatory in Hawaii. This is uncharted territory in human history. Normally, the body of climate evidence should provoke a sense of urgency, but there seems to be worryingly few signs that this is matched by decisive action. Amid concerns over global economic pressures, climate change has quite frankly slipped down the policy agenda. And while the countries where energy poverty is a problem obviously contributed very little to the historical build-up of emissions, many of them are particularly susceptible to the physical impacts of climate change. This means that the difficulties that communities in Africa, Asia, and Latin America are currently facing due to lack of access to modern energy services are set to be compounded by the prospect of dislocation of human settlements and changes to rainfall patterns, drought, flood, and heat-waves that would severely affect food production, human disease, and mortality.

1.2. MAKING MEANINGFUL PROGRESS

Part of the statistical challenge of combating energy poverty is to assess the cost of poverty alleviation and eradication. Ironically, those of us working in this field know that universal access would not cost very much, yet it is not at all clear that it will be achieved. We have estimated that it could be met by 2030 with investment of $49 billion per year in the period 2011–2030. The additional investment would provide electricity connections for almost 50 million people per year on average, and clean cooking facilities to 135 million people per year on average. That might sound like a lot—and is more than five times the level of investment observed in recent years. Yet it amounts to just 3 per cent of the total investment that is expected to be made in the world's energy supply infrastructure over the period. In other words, to ensure that those of us who already have access to modern energy services enjoy uninterrupted supplies, we are going to spend over 30 times more money than would be needed to provide initial access to the population at the bottom of the energy ladder.

Encouragingly, the level of attention given to improving modern energy access has recently risen, as has the level of expectation about the ultimate result. This started with the United Nations designation of 2012 as the Year of Sustainable Energy for All, to be followed by a Decade of Sustainable Energy for All from 2014, coupled with the decision by the UN Secretary-General to include universal access to modern energy within his Sustainable Energy for All initiative (SE4All) (UN 2013). At the Rio+20 Summit, countries then recognized the critical role of energy in the development process, committed themselves to measures to improve energy access, and emphasized the need for further action. They stated their determination to act to make sustainable energy for all a reality. Financial commitments to improve modern energy access have been lodged under the

SE4All initiative by multilateral development banks, direct government sources, and the private sector. These still amount to less than 5 per cent of the nearly $1 trillion that we estimate will be required to achieve universal modern energy access by 2030, so significant additional funding and policy action will be necessary. But the initiative is having a positive impact in mobilizing awareness and a greater unity of purpose to tackle this issue.

The recent Power Africa initiative launched by US president Obama in 2013 is also very promising (US Government 2013). The initiative aims to double electricity access in Sub-Saharan Africa, where more than two-thirds of the population lack access. An initial set of countries including Ethiopia, Ghana, Kenya, Liberia, Nigeria, and Tanzania are partners of the initiative and have already set ambitious goals for electricity access. Power Africa intends to help countries develop their own resources for power generation and transmission, and to expand mini-grid and off-grid solutions. The initiative would initially provide electricity access to at least 20 million new households and commercial entitites. The United States is committing more than $7 billion over five years, and private companies are adding an additional $9 billion.

For anyone concerned that bringing electricity to 1.3 billion people, and clean cooking facilities to 2.6 billion people, would further dent our chances of meeting ambitious climate goals or of enhancing energy security, our analysis has some reassuring news. It shows that achieving the goal by 2030 would increase global electricity generation by only 2.5 per cent, demand for fossil fuels by less than 1 per cent, and carbon dioxide emissions by 0.6 per cent. The small size of these increases is linked to the low level of energy consumed per capita by the people provided with modern energy access and to the relatively high proportion of renewable solutions that would be adopted, particularly in rural and peri-urban households.

While international support is of course important in solving the problem, real progress will also require a scaling up at the local and regional levels of actual programmes on the ground to supply households with access to modern energy services. In terms of electricity access, the unit cost of electricity delivered through an established grid is typically less than the unit cost through mini-grids or off-grid systems. But in remote areas, where distance may render it too costly to connect communities to the national or regional grid, decentralized projects will be needed.

Many developing countries subsidize electricity to households, on the grounds that the benefits subsidies provide are judged to exceed the long-term costs to government. If not well designed, these subsidies can result in significant losses of economic efficiency, wasteful habits on the part of consumers, and adverse environmental effects. These risks were highlighted by our recent analysis in the *World Energy Outlook* (IEA 2011, IEA 2012), which found that only 8 per cent of spending on fossil-fuel subsidies makes its way to the poorest 20 per cent of the population. One way to improve this situation is to have governments pay part of the capital cost of connection, or have utility companies spread the connection charge out over several months. Another approach is the so called 'lifeline rate', a special subsidy for poor families—with 'poverty' defined by both household income and electricity use. The lifeline-rate system avoids a number of the pitfalls of other forms of subsidies, although it is still hard to design in such a way that it does not benefit the rich even more than the poor.

1.3. MEASURING SUCCESS

Although the magnitude of the challenge can seem daunting, we can take inspiration from what has already been achieved. China, for example, secured electricity access for well over 0.5 billion people between 1980 and 2000, enabling it to achieve near universal access by 2010. The electrification goal was part of China's poverty alleviation campaign in the mid-1980s. A key factor in its success was the central government's determination and its ability to mobilize contributions at the local level. The electrification programme was backed with subsidies and low-interest loans. The programme also benefited from the country's plentiful coal resources and its low-cost domestic production of elements ranging from hydro generators down to light bulbs. China also avoided a trap into which many other nations have fallen: most Chinese customers pay their bills on time. If they do not, their connections are cut off.

Brazil is another success story that can provide others with important insights into strategies for overcoming energy poverty. The Luz Para Todos programme, which was launched in 2003 when only 84 per cent of the rural population had access to electricity, aims to achieve universal access to electricity by 2014 and is well on its way to meeting the goal, with the country now boasting an electrification rate of around 99 per cent (Ministry of Mines and Energy 2012). The programme provides an electricity connection free of charge, together with three lamps and the installation of two outlets in each home. In Brazil's case, those people who remain without electricity represent a particular challenge, as they live mostly in the Amazon, where the population is thinly spread (about four inhabitants per square kilometre) and where extension of the power grid is difficult. Recognizing this, new guidance is being given on how to set up decentralized renewable energy systems. It is estimated that the programme has generated nearly 300,000 new jobs and an increase in income of more than one-third for those households that have gained access. It has also played an important role in stimulating social programmes providing health services, education, water supply, and sanitation.

An essential part of any successful initiative to achieve universal modern energy access will be to have the means to track progress, so as to be able to inform governments and other stakeholders of what is being achieved and what more needs to be done. Since 2004, the IEA has published an Energy Development Index (EDI), which is devised as a composite measure of a country's progress in transitioning to modern fuels and modern energy services, as a means to help better understand the role that energy plays in human development (see Figure 1.2). Since 2012, the new EDI has included important additional indicators with strong explanatory value for energy access—such as productive use of energy—presented in the form of country-level results. A substantial amount of data has been collected, allowing the IEA to present EDI results for 80 countries and to compare EDI results over time. Such an indicator can help ensure that policy and financing commitments achieve maximum impact.

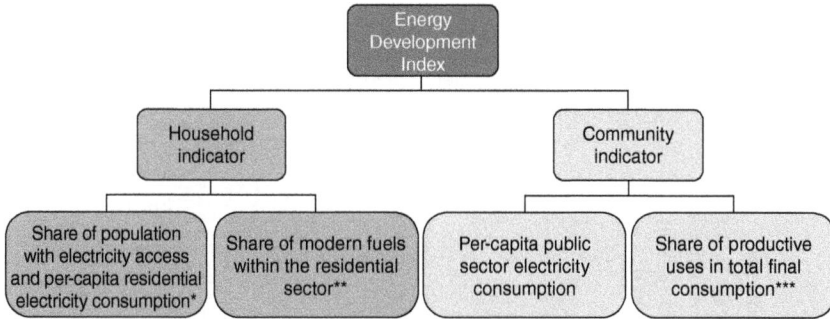

Figure 1.2. Components of the Energy Development Index

* The geometric mean of the two variables is taken. ** Excludes electricity to avoid double counting. *** Includes industry, agriculture, services, transport, and other non-specified energy use.

The ambition for the EDI is to develop a multidimensional indicator that tracks energy development country by country, distinguishing between developments at the household level and at the community level. In the former, we focus on two key dimensions: access to electricity and access to clean cooking facilities. The categories are necessarily broader when looking at community-level access. In the case of public services, the focus is on the use of modern energy in schools, hospitals and clinics, water and sanitation, street lighting, and other communal institutions or services. In the case of productive use, the focus is on modern energy use as part of economic activity. An additional aspect of modern energy use, captured, to an extent, within productive use, is transport. This is particularly important in the early stages of economic development because a significant share of energy consumed in the transport sector is used for productive economic purposes.

Figure 1.3 ranks 80 countries according to their overall EDI score. It also shows the relative contribution of each of the constituent indicators discussed above and, where available, shows a country's EDI score in 2002 for comparison. Many of the countries with the highest EDI score are in the Middle East, North Africa, or Latin America. Countries in Sub-Saharan Africa represent a significant share of those in the lower half of the EDI country scores. Those countries with a low overall EDI ranking tend to have a low result on the clean cooking and public services indicators. For the countries for which the IEA had both 2002 and 2010 data a general improvement over time is observed. Looking across regions as a whole suggests that, on average, the biggest improvements have taken place in East Asia and North Africa.

We can also take encouragement that universal access can be achieved by looking at the progress that has been made in other critical areas of human development. The UN Report of the High-Level Panel of Eminent Persons on the Post-2015 Development Agenda (UN 2013) highlighted that the 13 years since the millennium have seen the fastest reduction in poverty in human history: there are 0.5 billion fewer people living below an international poverty line of $1.25 a day. Child death rates have fallen by more than 30 per cent, with about

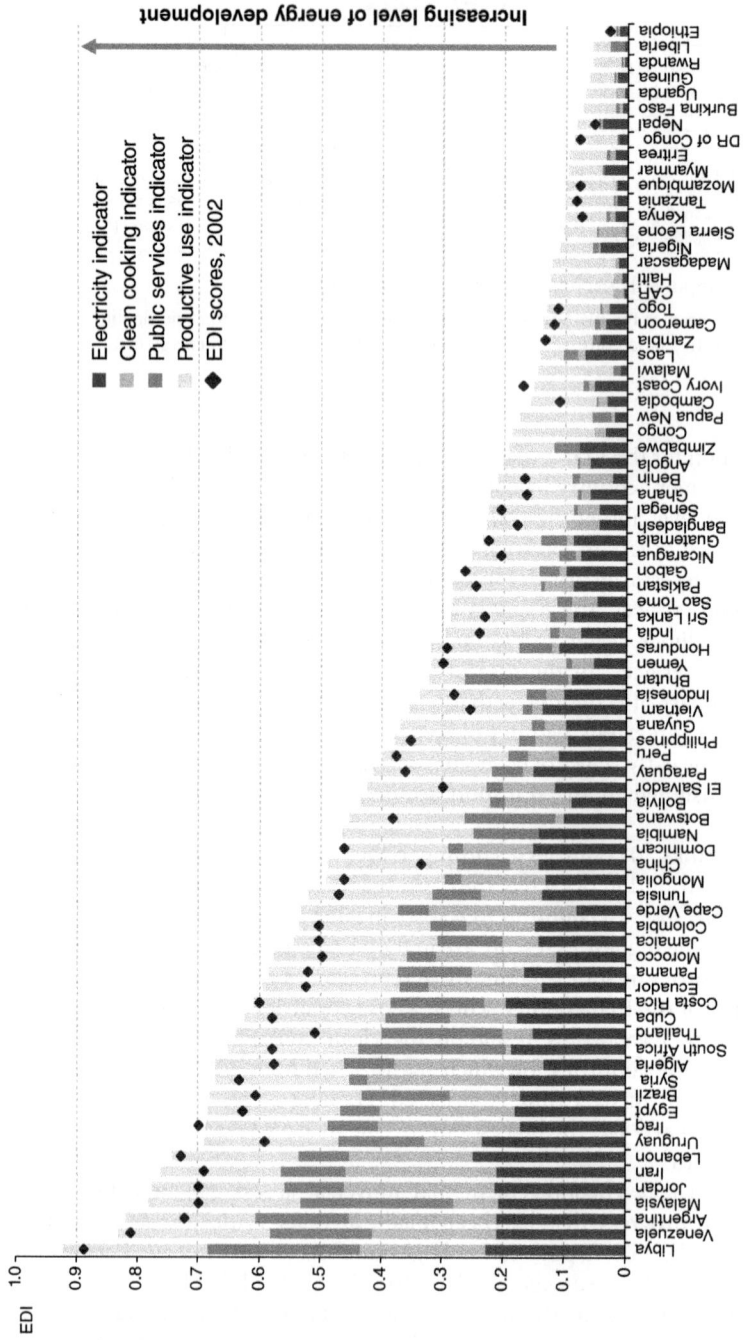

Figure 1.3. Energy Development Index country results, 2010 (also showing 2002 results where available)

Increasing level of energy development

Electricity indicator
Clean cooking indicator
Public services indicator
Productive use indicator
♦ EDI scores, 2002

EDI

1.0
0.9
0.8
0.7
0.6
0.5
0.4
0.3
0.2
0.1
0

Libya
Venezuela
Argentina
Malaysia
Jordan
Iran
Lebanon
Uruguay
Iraq
Egypt
Brazil
Syria
Algeria
South Africa
Thailand
Cuba
Costa Rica
Ecuador
Panama
Morocco
Jamaica
Colombia
Cape Verde
Tunisia
Mongolia
China
Dominican
Namibia
Botswana
Bolivia
El Salvador
Paraguay
Peru
Philippines
Guyana
Vietnam
Indonesia
Bhutan
Yemen
Honduras
India
Sri Lanka
Sao Tome
Pakistan
Gabon
Nicaragua
Guatemala
Bangladesh
Senegal
Ghana
Benin
Angola
Zimbabwe
Congo
Papua New
Cambodia
Ivory Coast
Malawi
Laos
Zambia
Cameroon
Togo
CAR
Haiti
Madagascar
Nigeria
Sierra Leone
Kenya
Tanzania
Mozambique
Myanmar
Eritrea
DR of Congo
Nepal
Burkina Faso
Uganda
Guinea
Rwanda
Liberia
Ethiopia

3 million children's lives saved each year compared to 2000. Deaths from malaria have fallen by one-quarter.

This remarkable progress puts past efforts to improve energy access to shame. But there are some important lessons we can take away from it. The success has been attributed to a combination of economic growth, better policies, and the global commitment to the Millennium Development Goals, which set out an inspirational rallying cry for the whole world. Many of these same elements now exist in terms of the battle being about universal access. Thanks to the *World Energy Outlook* reports, we have a much improved picture of the global energy access situation and of the role played by electricity and clean cooking facilities in eradicating the worst effects of poverty and putting poor communities on the path to development. This has contributed to recent international recognition that the inadequate energy access situation cannot continue and there must be a commitment to effect the necessary change. Targets have been set, as have indicators to monitor progress. We also now know that positive experiences exist, showing that real progress is possible and giving others insights into how it can be achieved. And, importantly, there are signs that additional investment in universal access is being mobilized and that programmes are being put in place to dramatically scale up access programmes on the ground.

Of paramount importance now is to ensure that the building momentum acts as a catalyst for even greater future action by governments, industry, the private sector, and financial institutions. For this to happen, all of the players who have recently become active in this area—as well those of us who have been looking at it for much longer—must persevere with the issue until universal access is achieved. As energy is the source of all life, so modern energy can be the source of a better life for all.

Box 1.1. Projections for energy access

The energy access projections presented in this chapter come from the IEA's World Energy Model, which integrates trends in demography, economy, technology, and policy. This kind of integrated analysis offers valuable insights into the globe's energy trajectory and what will have to be done to attain universal access to modern energy services by 2030. The projections for access to electricity and to modern cooking solutions are based on separate econometric panel models that regress the electrification rates and rates of reliance on biomass for different countries over many variables to test their level of significance. In the case of electrification, the variables that were determined to be statistically significant and thus included in the equations are per capita income, demographic growth, urbanization level, fuel prices, level of subsidies for electricity consumption, technological advances, electricity consumption, and electrification programmes. In the case of cooking solutions, variables that were determined to be statistically significant and consequently included in the equations are per capita income, demographic growth, urbanization level, prices of alternative modern fuels, subsidies to alternative modern fuel consumption, technological advances, and government programmes to promote modern cooking.

REFERENCES

Ministry of Mines and Energy, Brazil. (2012, December). *Programa Luz Para Todos*, <http:// luzparatodos.mme.gov.br/luzparatodos/Asp/informativos.asp> (accessed July 2013).

IEA (International Energy Agency). (2002). *World Energy Outlook 2002*. Paris: OECD/IEA.

IEA (2011). *World Energy Outlook 2011*. Paris: OECD/IEA.

IEA (2012). *World Energy Outlook 2012*. Paris: OECD/IEA.

UN (United Nations). (2013). *A New Global Partnership: Eradicate Poverty and Transform Economies through Sustainable Development—The Report of the UN High Level Panel of Eminent Persons on the Post-2015 Development Agenda*. New York: UN.

US Government. (2013). *Fact Sheet: Power Africa*, <http://www.whitehouse.gov/the-press-office/2013/06/30/fact-sheet-power-africa> (accessed 30 June 2013).

2

Defining, Measuring, and Tackling Energy Poverty

Benjamin K. Sovacool

2.1. INTRODUCTION

For roughly half of the global population, existence—and energy consumption—is remarkably distinct from the lifestyles most people in industrialized countries have become accustomed to. Imagine a daily ritual without consistently hot showers or baths, no indoor lighting at night, poorly cooked food, and debilitating health problems associated with indoor air pollution. Think about life with no steady pumping of water for drinking and irrigation, few televisions, mobile phones, or computers, and limited access to the fruits of modern civilization. For those in the developing world, the search for energy fuels and services is an arduous, continual, exhausting battle. Women and children spend hours each day, time they could otherwise utilize on productive work or education, carrying fuel and water loads often in excess of their own weight, with calamitous consequences on their health, their natural environment, and their community.

Yet many, if not most, developing countries still lack the capacity and technology to shift to more sustainable and affordable supplies of energy without external assistance. One survey of the 24 least developed countries in the world found that 22 of them each had less than 1 per cent of their region's total energy resources (UNESCAP 2008: 185). With scarce energy reserves of their own, these countries must rely either on the global trading system or on development assistance from benevolent middle- and upper-income countries, both outside of their control (Ruggie 2006).

Expanding modern energy access for rural and increasingly poor communities, therefore, is a daunting task. Those without electricity or dependent on traditional fuels tend to have income levels, purchasing power, and consumption levels far below what private companies and electric utilities typically deem profitable, and the reluctance of such companies is increased by the inaccessibility of these communities to national electricity grids. Public officials, like their private counterparts, prioritize investments in urban infrastructure where most of their constituents reside, and they often subsidize grid electricity to existing customers instead of expanding access to rural ones or incorporating off-grid technologies.

How large is the problem? What business models and technologies can best address energy poverty? What barriers remain, and where does new research need to take us? This chapter addresses each of these questions in turn. Section 2.2 defines and measures the prevalence of energy poverty using the most up-to-date statistics from the International Energy Agency and other organizations. Section 2.3 discusses an array of grid, micro-grid, and off-grid technologies that can expand energy access and looks at the partnership models that businesses and the international community utilize to disseminate these technologies. Section 2.4 suggests corrections to nine common misconceptions related to energy poverty and provides ideas for improvement. The chapter concludes by calling for a new paradigm of energy access and development.

2.2. DEFINING AND MEASURING ENERGY POVERTY

This section of the chapter discusses the invisibility of the energy-poor, summarizes rural energy needs and the concept of the 'energy ladder', and presents updated measurements of energy poverty (within the uncertainties of the data involved).

2.2.1. The invisibility of the energy-poor

For all too long the energy-poor have been invisible in the energy and development literature. Energy scholars Vaclav Smil and William E. Knowland (1980) declared more than three decades ago that the 'real energy crisis' involved not the Organization of Petroleum Exporting Countries and oil suppliers in the Middle East but lack of access to energy in the developing world. Such a crisis has simmered for the past three decades; it is thus a crisis that never really dissipates, despite the emergence of other 'crises' such as rising prices, surging global demand, and increasingly heated competition for scarce energy resources.

Close to a fourth of humanity still lives without electricity or other modern forms of energy, while as much as a third of the world's population still relies at least in part on solid fuel such as cow dung or firewood. This is a major issue in the world today, and until recently an under-unacknowledged obstacle in the way of poverty eradication and development efforts around the world, especially—though not exclusively—in the 'bottom billion' economies of Sub-Saharan Africa and South Asia.

Despite its importance, the topic of energy poverty has been largely neglected in energy planning discussions and energy publications in an OECD context, at least until recently. Daniel Kammen and Michael R. Dove (1997) wrote that advanced and modern technologies related to electricity and motorized transport (think 'nuclear reactors' and 'electric vehicles') were highly favoured topics of energy policy discussion. However, 'mundane' technologies—such as cookstoves, biogas units, heating and cooling systems, and other less 'state-of-the-art' constructions—were minimally discussed, even though these technologies affected the greatest number of people and had the most substantial impact on the environment in everyday life.

Almost ten years later, Fatih Birol (2007), the Chief Economist for the International Energy Agency, argued that, 'unfortunately, the energy-economics community has given far less attention to the challenge of energy poverty among the world's poorest people'. And most recently, a series of content analyses of the top energy journals noted that only 3 per cent of authors came from least developed countries and only 8 per cent of papers addressed topics even loosely related to energy poverty and energy development (D'Agostino et al. 2011; Sovacool et al. 2011).

Remarkably, even the UN Millennium Development Goals of 2000 did not say a word about energy; anybody with experience on the ground knows that development efforts cannot succeed without a robust energy component. This is not just a problem for the poor, but a global concern. Energy deprivation is a leading contributor to disease epidemics, underdevelopment, unemployment, social discontent, political unrest, and instability—it threatens the energy haves as well as the have-nots.

Only in recent years has energy poverty moved up the global agenda. In October 2011, the International Energy Agency, jointly with branches of the United Nations, devoted a section of its World Energy Outlook to energy poverty. The UN declared 2012 the 'International Year for Sustainable Energy for All'. Thus, over the past few years different scholars and various organizations have designed and refined methodologies and processes to make the energy-poor population visible—to assess its size and measure progress in alleviating energy poverty and extending access. The next section of the chapter summarizes their efforts.

2.2.2. Defining energy poverty and the energy ladder

To begin to understand the stark energy needs of the world's underprivileged population, it is necessary to explore the concepts of energy poverty, the energy ladder, and energy equity.

As there is no simple definition of poverty, conceptualizing 'energy poverty' is not a simple task. Recent work, including the United Nations Development Programme's Human Development Report, has noted that poverty is not a static or fixed state, but instead a multidimensional concept encompassing caloric intake, life expectancy, housing quality, literacy, access to energy, and a variety of other factors (United Nations Development Programme 2010). Such poverty is frequently expressed from an income perspective: to be 'poor' is to earn less than $2 per day when adjusted for the purchasing power parity of countries. Under this definition, a shocking 40 per cent of the global population is poor (Sovacool 2012). Sticking with the UNDP's multidimensional notion of poverty, non-income dimensions of poverty such as health, education, and living conditions can be just as important as sources of employment or wages. Within this list of non-income dimensions, two energy indicators are found: electricity (having no electricity constitutes poverty) and cooking fuels (relying on wood, charcoal, and/or dung for cooking constitutes poverty). Such a conception of energy poverty has been confirmed by the International Energy Agency and other multilateral organizations, which state that energy poverty is comprised of a lack of access to electricity and reliance on traditional biomass fuels for cooking (International Energy Agency et al. 2010; Jones 2010; IIASA 2012).

Thus, the UNDP explicitly defines energy poverty as the 'inability to cook with modern cooking fuels and the lack of a bare minimum of electric lighting to read or for other household and productive activities at sunset' (Gaye 2007: 4). The Asian Development Bank takes a slightly broader approach to defining energy poverty and tells us that it is 'the absence of sufficient choice in accessing adequate, affordable, reliable, high-quality, safe and environmentally benign energy services to support economic and human development' (Masud et al. 2007: 47).

Several ways of measuring such energy poverty exist. One method is to track the minimum amount of physical or animate energy needed for basic needs such as cooking and lighting, often including the minimum amount of nutritious food necessary for a healthy life. Another is to look at the poorest people in a given country, say households in the lowest-income quintile, and then to detail the types and amounts of energy they use. Yet another is to measure how much income is spent on energy services; typically a family that spends more than 10 to 15 per cent of their earnings on energy services per month or year is considered 'energy-poor' or classified as in 'fuel poverty' (Dutta 2011). Still another, taken by Sanchez (2010), is to declare that a person is 'energy-poor' if they have access to less than 35 kilograms of liquefied petroleum gas for cooking per year (from liquid or gaseous fuels) and access to less than 120 kWh of electricity per year.

The most common concept illustrating energy poverty involves 'energy ladders' for services such as heating and cooking. One study defines the energy ladder as 'the percentage of population among the spectrum running from simple biomass fuels (dung, crop residues, wood, charcoal) and coal (or soft coke) to liquid and gaseous fossil fuels (kerosene, liquefied petroleum gas, and natural gas) to electricity' (Holdren and Smith 2000). The idea implies that the primary types of energy used in rural areas or developing countries can be arranged on a 'ladder' with the 'simplest' or most 'traditional' fuels and sources, such as animal power, candles, and wood, at the bottom and the more 'advanced' or 'modern' fuels such as electricity or refined gasoline at the top. The ladder is often described in terms of efficiencies, with the more efficient fuels or sources higher on the ladder. For example, kerosene is three to five times more efficient than wood for cooking, and liquefied petroleum gas is five to ten times more efficient than crop residues and dung (Barnes and Floor 1996). Table 2.1 depicts the energy ladder as synthesized from a variety of academic studies.

These rungs of the energy ladder suggest a meaningful difference between how the rich and poor consume energy—with implications on equity and affordability. As Figure 2.1 shows, there can be a significant difference in dependence on solid fuels for cooking between the poorest and richest quintiles. In Ethiopia, the poorest in society spend about 10 per cent of their income on energy services, compared to less than 7 per cent for the wealthiest. In India, the poorest spend more than 8 per cent, the wealthiest less than 5 per cent. In South Africa, the gap between rich and poor expenditures is about 3 per cent; in Uganda, the difference is 60 per cent (World Health Organization 2006). Studies have found that the highest income quintiles of households consumed 3 to 21 times more energy than the lowest income quintiles—which means access to energy, or lack of it, can both

Table 2.1. The energy ladder for household energy use

Energy service	Developing countries			Developed countries
	Low-income households	Middle-income households	High-income households	
Cooking	Wood (including wood chips, straw, shrubs, grasses, and bark), charcoal, agricultural residues, dung	Wood, agricultural residues, coal, kerosene, biogas	Wood, kerosene, biogas, liquefied petroleum gas, natural gas, electricity	Electricity, natural gas
Lighting	Candles, kerosene (sometimes none)	Kerosene, electricity	Electricity	Electricity
Space heating	Wood, agricultural residues, dung (often none)	Wood, agricultural residues	Wood, coal, electricity	Oil, natural gas, electricity
Other appliances	None	Electricity, batteries	Electricity	Electricity

reflect and worsen social inequality (United Nations Development Programme 2009; Modi et al. 2005). Some of these assessments have noted that in all countries analysed, the poor consumed less energy than other strata of households, but spent more of their income on it. The price per unit of energy for these poor users was usually higher, since they have difficulty accessing electricity grids or liquid fuels with higher energy densities.

While commonly applied in the energy poverty literature, however, the concepts of energy ladders and energy equity are imperfect and have been criticized for not fully capturing the intricacies of how households consume energy. People can be on several rungs at the same time; the movement is not all unidirectional, up or down. In many countries the upper quintile hold on to traditional biomass in various aspects of cooking. Yet at its core, the model assumes that consumers will always seek to move up the ladder, that dung, crop residues, and wood are the 'energy of the poor', and that price differences have little impact on how homes choose their energy fuels or technologies.

By contrast, one survey in Botswana found that fuelwood is chosen by households across the entire spectrum of users, with even richer homes relying on wood due to its low cost and availability (Hiemstra-van der Horst and Hovorka 2008). The study noted that even though electricity and other sources of energy such as kerosene were available, wood remained the most popular, and that the quantity of wood consumed did not vary (as expected) with the income level of households. Such trends are not limited to Africa. A separate study of 34 cities in a sample of developing countries in Africa, Asia, the Caribbean, Latin America, and the Middle East concluded that the transition to other parts of the energy ladder are rarely linear nor predetermined; instead, many homes rely on multiple types of fuels at the same time and will switch between them based on price and availability (Barnes et al. 2004). Indeed, such studies imply that household decisions about energy are shaped not only by price and affordability, but also by tradition, social expectations, and availability.

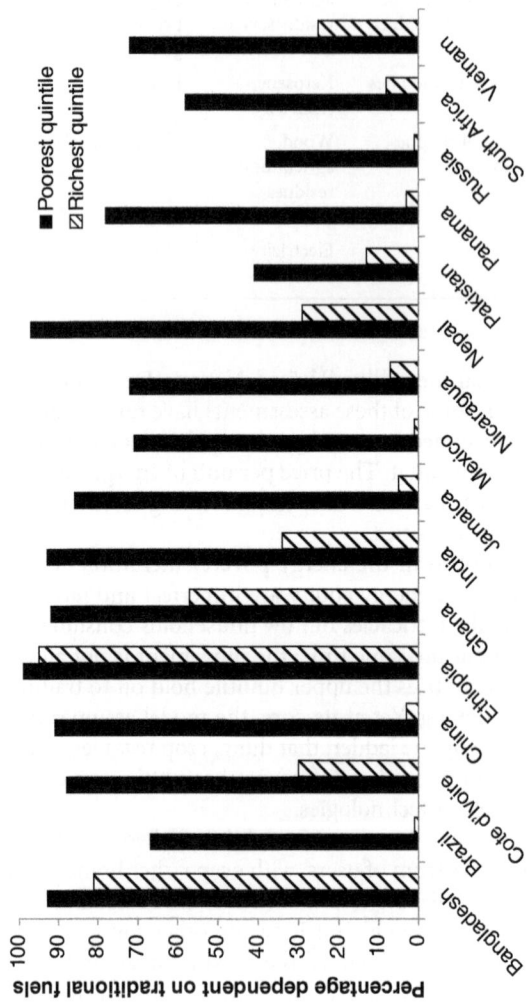

Figure 2.1. Dependence on solid fuels according to poorest and richest quintiles

2.2.3. Tracking energy poverty

Notwithstanding the complexities involved in defining energy poverty, organizations such as the International Energy Agency, the World Health Organization, and various United Nations agencies have done a remarkable job compiling statistics on those without access to electricity as well as those dependent on traditional fuels.

There is broad consensus on two key criteria: electricity access and clean cooking. Under these definitions, according to the most recent data available from the International Energy Agency (2013), in 2010 approximately 1.3 billion people did not have access to electricity, and 2.6 billion people relied on the traditional use of biomass for cooking—numbers reflected in Tables 2.2 and 2.3. This analysis focuses on the traditional use of biomass for cooking, but there are

Table 2.2. People without access to electricity by region, 2010

Region	Population without electricity	Electrification rate	Urban electrification rate	Rural electrification rate
	millions	%	%	%
Developing countries	1,265	76.1	92.1	63.7
Africa	590	43	72	24
North Africa	1	99	100	99
Sub-Saharan Africa	589	32	64	13
Developing Asia	628	83	96	74
China & East Asia	157	92	98	88
South Asia	471	70	92	61
Latin America	29	94	98	76
Middle East	18	91	99	75
Transition economies and OECD	2	99.8	100	99.5
World	1,267	81.5	94.7	68

Source: IEA 2013

Table 2.3. People relying on traditional use of biomass for cooking by region, 2010

Region	Population relying on traditional use of biomass	Percentage of population relying on traditional use of biomass	Percentage of urban population	Percentage of rural population
	millions	%	%	%
Africa	698	68	44	83
North Africa	3	2	1	2
Sub-Saharan Africa	696	81	56	95
Developing Asia	1,814	51	17	72
China and East Asia	716	36	12	56
South Asia	1,098	69	27	87
Latin America	65	14	5	50
Middle East	10	5	1	14
World	2,588	37.8	12.5	63.7

Source: IEA 2013

also about 400 million people (not included in the table) that rely on coal for cooking and heating purposes, which causes air pollution and has serious potential health implications when used in traditional stoves. These people are mainly in China, but there are also significant numbers in South Africa and India.

Figures 2.2 and 2.3 (taken from the latest IEA data) indicate that 13 countries account for a majority of the world's energy-poor: Bangladesh, China, the

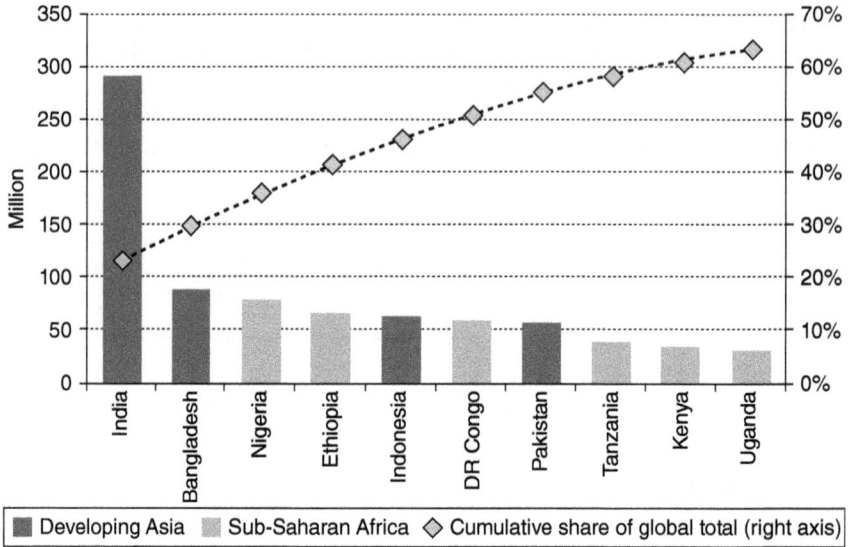

Figure 2.2. Countries with the largest population without access to electricity, 2010

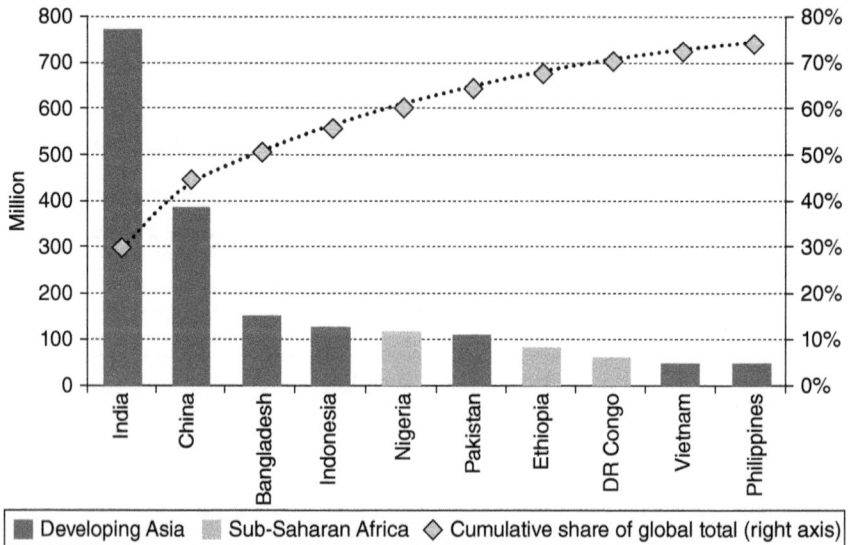

Figure 2.3. Countries with the largest population relying on traditional use of biomass for cooking, 2010

Democratic Republic of the Congo, Ethiopia, India, Indonesia, Kenya, Nigeria, Pakistan, Philippines, Tanzania, Uganda, and Vietnam.

Tracking progress and refining the measurements are of course critical to addressing energy poverty on the ground. But when presenting these figures it should be noted that there is still some level of ambiguity—due to politics, constrained institutional capacity, and the very disenfranchised nature of the poor—that contribute to inconsistent definitions and methodologies. The numbers from the IEA, in other words, should be bracketed as 'best-guess' estimates that contain elements of uncertainty.

2.3. MODELS AND TECHNOLOGIES FOR ERADICATING ENERGY POVERTY

Though estimating the global extent of energy poverty with perfect precision is impossible, that should not distract us from the fact that it is a serious problem. In that light, this section of the chapter introduces technologies and business models for addressing energy poverty.

2.3.1. Technologies for eradicating energy poverty

Table 2.4 presents the most common way of classifying technologies for eradicating energy poverty: extensions of the electricity grid, mini- or micro-grids, and off-grid technologies.

Table 2.4. Summary of technological options that expand energy access

	Conventional electricity grids	Micro-grids	Off-grid technology
Scale	National, regional, even international	Community	Household
Geographic radius	More than 50 square kilometres	1 to 49 kilometres	<1 kilometre
Number of customers	Thousands to millions	Dozen to hundreds	Usually a dozen or fewer
Installed capacity	More than 10 MW	20 kW to 10 MW	<20 kW
Technologies involved	Large-scale, centralized, capital-intensive	Medium-scale and small-scale	Very small-scale
Investment required	Billions of dollars	Millions of dollars to hundreds of thousands	Thousands of dollars
Examples	North China grid, Electricité de France grid, New England Independent System Operator (NEISO) grid	Community-scale solar PV systems in Bangladesh, micro-hydro networks in Nepal and Sri Lanka	Individual solar home systems, pico-hydro units, biogas digesters, cookstoves, residential wind turbines

Grid electrification

Lack of electricity from the national grid limits the productive hours of the day for business owners and heads of households, and it also inhibits the types of business opportunities available. Grid electrification, combined with appropriate government and with financial and technical training, can make a variety of income-generating activities possible, including mechanical power for milling grain, illumination for factories and shops, heat for processing crops, and refrigeration for preserving products (Eric and Kartha 2000). In the Philippines, for instance, investments in electrification are largely justified on the grounds that households typically see income gains of $81 to $150 per month when they become connected to the grid, as Table 2.5 reveals—representing an increase in income of 10 to 100 per cent.

In Papua New Guinea, household surveys concerning electrification conducted by the World Bank (2004) found that:

- in all cases, lighting is considered the most important and immediate benefit;
- knowledge of the outside world and entertainment opportunities offered by TV and VCR (today it would be DVDs or MP3s) are viewed as key benefits, especially by men;
- women report time savings of three to four hours per day;
- lighting and TV are said to have contributed positively to children's education through extended study hours and informal learning;
- electrified homes have incomes higher than non-electrified homes by 27 to 100 per cent;
- increases in assets are attributed mainly to the acquisition of electricity-producing equipment and appliances (World Bank 2004).

Similarly, in Vanuatu, household surveys indicate that people living in homes without electricity desperately want it for services such as lighting, water pumping, cold storage of fish and meat, TVs and DVD players, and mobile phone charging (Fischer and Pigneri 2011). Improved electricity provision can also deliver power to schools and small hospitals and can serve local industries such as sawmills, crop processing, workshops, and other emerging forms of micro-entrepreneurship. Recent data from the World Bank (2013), furthermore, indicates that villages in Africa that lack electricity are less likely to have a secondary school or a post office, two key institutions for participating effectively in the global economy.

Table 2.5. Summary of rural household benefits from electricity in the Philippines

Benefit category	Benefit value (USD)	Unit (per month)
Less expensive and expanded use of lighting	36.75	Household
Less expensive and expanded use of radio and television	19.60	Household
Improved returns on education and wage income	37.07	Household wage earner
Time savings for household chores	24.50	Household
Improved productivity of home business	34.00 (current business), 75.00 (new business)	Business

Source: World Bank 2002a

Micro-grids and mini-grids

A 'mini-grid' refers to a localized or isolated grouping of electricity generation, distribution, storage, and consumption within a confined geographic space (Kaplan and Sissine 2009). While in some instances mini-grids can be interconnected to national electricity networks, in most cases they operate autonomously and at lower loads and voltages. Though definitions vary, mini-grids are often locally managed, they involve less than 10 MW of installed capacity, they serve small household loads, and they possess a radius of 50 kilometres or less. 'Micro-grids' are even smaller, typically operate with less than 100 kW of capacity and at even lower voltage levels, and possess a 3- to 8-kilometre radius (Mukwedeya 2011). Mini- and micro-grids can be powered by fossil fuels, such as diesel generators or fuel cells, or by renewable energy sources such as micro-hydro dams, solar PV plants, biomass combustion, or wind turbines. When configured properly, such mini- and micro-grids can operate more cost-effectively than centralized generation from a power grid (Thiam 2010; Szabo et al. 2011; Wachenfeld 2012; Casillas and Kammen 2012).

The International Energy Agency expects micro- and mini-grids to play an instrumental role in global electrification efforts over the next decade, especially in the Asia-Pacific. The IEA's 2011 numbers suggest that national grid extension is the most suitable option for all urban areas and for around 30 per cent of rural areas, but it is not a cost-effective option in more remote rural areas. Therefore, out of a total electrification requirement of 838 TWh, 56 per cent (or 470 TWh) is expected to be provided via mini-grid and isolated off-grid technology—see Figure 2.4.

Off-grid or isolated units

Rural and poor households throughout the world need not be served only by the grid or micro- and mini-grids. They can also receive electricity through isolated micro-hydro dams, solar lanterns, and photovoltaic solar home systems, among other options discussed in this section.

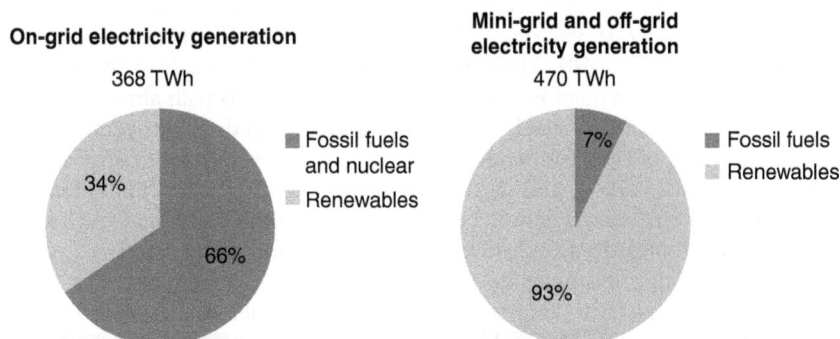

Figure 2.4. Contributions of grids, micro-grids, and off-grid technologies to universal energy access

Source: International Energy Agency, Energy for All: Financing Access for the Poor © OECD/IEA 2011

Improved cookstoves (ICS)

The poorest households in the developing world tend to use simple three-stone fires for cooking, burning wood, agricultural waste, and dung with high moisture content. This results in low efficiencies and high amounts of smoke, with many cookstoves in the developing world averaging 10 to 12 per cent in terms of their efficiency, meaning as much as 90 per cent of the energy content of the wood or charcoal used in them is wasted (Jones 2010). In some cases, existing cookstoves can be drastically improved by something as simple as adding a chimney or more insulation around the stove to retain heat; in other cases, older stoves may be replaced with new stoves with drastic increases in efficiency.

Though the term 'improved' is certainly subjective, modern stoves can take a variety of forms. Sanchez (2010) argues that an 'improved' cooking source is one that requires less than 4 person-hours per week per household to collect fuel, has a conversion efficiency above 25 per cent, and meets WHO guidelines for air quality. ICSs therefore frequently require a switch away from charcoal or polluted wood to 'healthier' fuels such soft biomass, crop residues, and firewood; they have a grate and an improved combustion chamber; and they almost always have a chimney. They utilize high temperature ceramics, fire-resistant material and longer-lasting metals, and they possess more insulation and a better frame that guides hot gases closer to cooking pots. They can cook more food at once and many have coils around the combustion chamber to heat water while cooking is in progress. Some improved stoves are connected to radiators or space heaters so that heat may be recycled and/or vented to other rooms, and some stoves send heat through pipes directly into a brick platform that occupants sleep on at night (Brown and Sovacool 2011).

Newer designs of cookstoves, especially those made in the past decade, only multiply such benefits. Thermoelectric generators (TEG) are becoming cost-competitive and enable stoves to generate both heat and a small amount of electricity; they use this to power a fan, which increases the efficiency of combustion, or to charge small devices such as mobile phones (Roth 2011). These TEG stoves have been shown to reduce harmful pollutants at 10 to 20 times the rate of ordinary improved stoves. Natural-draft rocket stoves reduce emissions of key health pollutants by two to three times compared to an ordinary cookstove, and natural-draft gasifier stoves can reduce those pollutants by five or six times (Venkataraman et al. 2010). Micro-gasifiers, those small enough to fit under a cooking pot at a 'convenient height', can cleanly burn biogas with almost smoke-free combustion, provide a steady flame with no waiting, and can be operated over extended periods of time (no tending of a fire). Moreover, micro-gasifier stoves reduce soot, black carbon, and particulate matter. They need less total biomass fuel due to their efficiency. Figure 2.5 shows some of the environmental and health advantages from improved cookstoves, rocket stoves, and gasifier stoves.

As with our other technologies, ICSs do have some shortcomings. The most obvious is that the term 'improved' changes over time. An improved stove installed ten years ago is probably no longer an improvement over existing models, and vendors have been known to call stoves 'improved' even when they are not. Some women have expressed a concern that ICSs cook food too quickly; that is, they had grown accustomed to the fuel amounts and timing associated

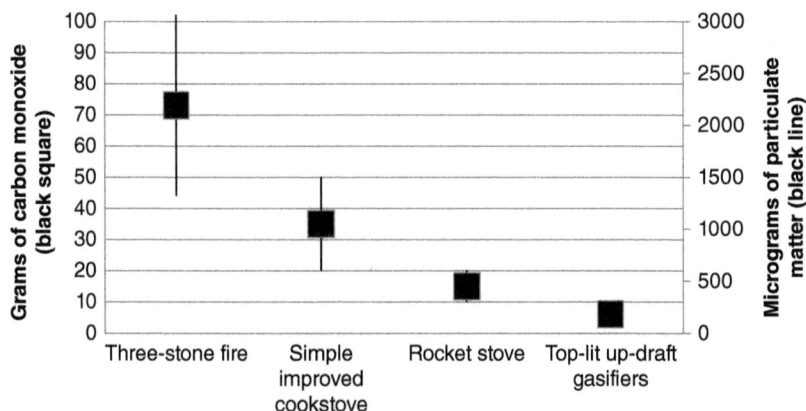

Figure 2.5. Emissions from four types of cookstoves (five litre water-boiling test)
Source: Roth 2011

with an older stove and became quickly frustrated when the new stove 'ruined' their meals. ICSs may not meet a family's entire cooking needs—families may wish to boil, bake, and broil with other cooking devices—and they still depend primarily on fuelwood, meaning that they contribute to some of the burdens associated with its collection and use (though these are substantially less for a truly 'improved' stove).

Solar home systems (SHS)

The typical solar home system consists of a solar photovoltaic module, battery, charge controller, and lamp. Customers in off-grid and rural areas can often choose from a variety of systems and technologies from a 10 watt-peak (Wp) unit for the poorest customers to a 150 Wp unit for wealthier clients. Larger systems often have the capacity to connect televisions, radios, and other electric appliances.

SHSs offer a very cost-effective way for rural communities around the world to acquire energy services without relying on expensive fossil fuels (such as kerosene) or capital-intensive efforts to extend national electricity grids. Costs of battery charging often range from $6 to $15—depending on the type of battery, price of energy, and location. Moreover, overall system costs vary greatly between countries. One assessment of SHSs in China, Kenya, Indonesia, Philippines, Sri Lanka, Brazil, the Dominican Republic, and Mexico found that complete systems cost as little as $10 per installed Wp in China and as much as $100 per Wp in Brazil, differences reflecting a multiple of ten (Miller and Hope 2000).

Globally, the World Bank (2007) has calculated that 50 Wp and 300 Wp SHSs are already cost-competitive in many areas with diesel and gasoline distributed generators and will see their costs decline, and their advantages over fossil fuelled systems increase, even further by 2015. SHSs also have immense safety and health benefits—they can displace the use of combusting kerosene, coal, and fuelwood indoors which can lead to higher rates of morbidity and mortality among women and children—and can be easily installed and maintained with minimal amounts of training. Thus, done properly, SHS programmes can enhance quality of life, provide a new source of skilled employment for rural technicians, and enhance energy security and reliability due to their decentralized nature.

Still, SHSs do have some drawbacks. They do not work well at higher latitudes with less sunlight, in tropical areas prone to frequent storms, or on foggy mountaintops. Due to the capital expense of purchasing one, they tend to be utilized only by middle- and higher-income rural homes, and they may lead to luxurious use (powering things such as soap operas on a television) rather than meeting subsistence needs. In some cases they serve only as a 'gateway' to electricity coming from the national grid, meaning they do not truly substitute for fossil fuels and conventional electricity systems. Their high value has made them prone to theft and sabotage in some areas—one villager told the author that they are stolen at the same rate as 'automobiles are in rich countries'—and owners and operators must be assiduous in their charging practices for batteries and system maintenance. A final limitation is that these systems generally only provide lighting and small amounts of electricity rather than energy for heating or cooking (Karekezi and Kithyoma 2002).

Nonetheless, more than 40 national SHS programmes exist around the world with more than 1.3 million systems installed at a collective cost greater than $700 million (Magradze et al. 2007). SHSs thus represent a vital technology employed by multilateral financial institutions in their efforts to curb energy poverty through off-grid electrification.

Solar lanterns

Solar lanterns, sometimes referred to as 'pico-PV systems' and 'solar LEDs', are very small solar units, often independent solar flashlights or lanterns, that can use light-emitting diodes (LEDs) or other lighting devices. These small systems, less than 10 Wp with a voltage up to 12 V, have advantages over SHSs because they are often less capital-intensive and more versatile. When costs of equipment are amortized over three years inclusive of fuel, energy, replacement lamps, wicks, mantles, and batteries, it has been estimated that the cheapest option by far is a 1 W solar LED (Mills 2005). Put another way, solar lanterns can pay for themselves in one month to two years compared to kerosene lamps and have lower costs of lifetime ownership than almost any other system on the market, as Figure 2.6 shows. Even countries with high electrification rates have market potential for solar lanterns, which can be used in remote communities or areas where grid electricity is unreliable or unaffordable, or for those who work or roam far from electricity grids, such as herders and fishers.

Biogas digesters and biomass gasifiers

Biogas is a clean fuel produced through anaerobic digestion of animal, agricultural, and domestic wastes. These three forms of organic waste and water typically enter a vessel where they are left to ferment and decompose, producing both biogas and digested slurry that can be turned into an organic fertilizer (Gautama et al. 2009). Smaller-scale, two-to-three-cubic-metre biogas plants tend to be used in homes and communities, suitable for providing gas and heat for cooking three meals a day for an average-sized family. Commercial-scale systems exist as well, with these larger units offering enough gas to meet the energy needs of

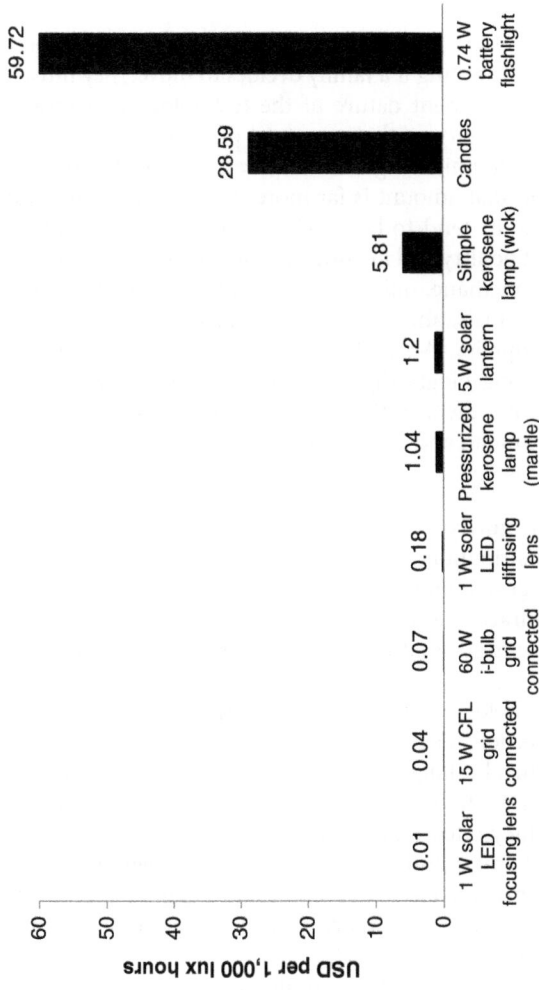

Figure 2.6. Cost of illumination services for various lighting sources

Source: Mills 2005. Lux-hours describe the perceived luminous intensity of light by the human eye, multiplied by the number of functional hours

neighbourhoods, restaurants, tea stalls, and bakeries. These larger systems, installed near large farms, poultry suppliers, and livestock ranches, can supply enough gas for up to 1,000 families.

By relying on biogas, these units minimize reliance on traditional forms of biomass, animal dung, and charcoal (with attendant negative environmental and social impacts), and also protect communities from disease by enhancing sanitation. The plants harness gas obtained from livestock and, yes, even human excrement. Biogas systems quite literally have people using their own waste to meet their energy needs.

That said, biogas digesters face some challenges. Once built, they become fixed to a house or structure, meaning if a family decides to move, they must leave their system behind. This permanent nature of the technology also means that if a family or community defaults on their payment plan, banks cannot easily disassemble and repossess the units. Digesters require a sufficient supply of waste to operate optimally, and that amount is far more than a single family can provide (which is why most units tend to be purchased by wealthier families that own significant numbers of poultry and livestock). When built improperly or damaged, biogas units can leak methane, quickly erasing any gains made from displacing more carbon-intensive fuels, and the untreated slurry waste has been shown to pollute some water supplies. Anaerobic digestion occurs much more slowly at higher altitudes, meaning biogas digesters work poorly at high elevations, and some have refused to use them in the belief they are 'unclean' since they are, in essence, powered by human and animal excrement.

Micro-hydro dams

Unlike their larger counterparts which require reservoirs, micro-hydro dams utilize low-voltage distribution systems and simpler designs that often have a natural river intake, de-sanding basin, masonry-lined canal, forebay, penstock, powerhouse, short tailrace, and electronic load controller. By 'micro' we refer to what is commonly discussed as either 'mini', 'micro', and 'small' hydro units from 5 kW to 10 MW.

Micro-hydro units hold a distinct set of advantages compared to the other technologies in this section. They can provide not only electricity but also mechanical energy for milling, husking, grinding, carpentry, spinning, and pump irrigation. They are much easier to operate and are cleaner, safer, and cheaper than the diesel generator sets they often replace; moreover, local people can be trained to manage them without any technical background in engineering or maintenance. These units can also provide electricity in remote mountain areas unsuited for biogas (because fermentation takes more time at higher altitudes) or SHSs (because of consistent fog and cloud cover).

They are, however, not perfect. To work properly, micro-hydro systems need continuous, dedicated maintenance. Their multi-functionality, the fact that they can perform multiple energy services at once, can become a 'curse' when they break down or need refurbishment, leaving communities suddenly without a vital technology for lighting, agricultural production, education, and so on. Larger micro-hydro units require upstream and downstream communities to cooperate and consent to water rights of way—meaning conflicts between communities can prevent effective deployment—and they can disrupt river flows and degrade fisheries when built improperly. Lastly, the energy produced by micro-hydro units is not always equitably distributed within communities and villages (Sovacool and Drupady 2012).

2.3.2. Financing and business models

Technologies are only one piece of the puzzle; appropriate financing and business models are also needed. Based on an extensive four-year assessment of 1,156 energy access and development programmes being implemented throughout the Asia-Pacific region, the eight approaches presented in Table 2.6 are the most widespread: (1) a technology improvement model, (2) a microfinance model, (3) a project finance model, (4) a cooperative model, (5) a community fund model, (6) a fee-for-service or ESCO model, (7) a cross-subsidization model, and (8) a hybrid model.

A 'technology improvement' model attempts to 'push' the supply of a given technology by improving its performance, often through research subsidies, product guarantees, warranties, technical standards, or improved manufacturing techniques. It has been largely used for solar home systems.

A 'credit model' or 'microfinance model' operates when local dealers sell their products to rural clients on credit against collateral or personal guarantees. It is commonly applied to SHSs, biogas units, and improved cookstoves. Payment is made in installments, and this type of partnership has high installation expenses due to the transaction costs associated with acquiring credit and high- to medium-quality products. This model also excludes poor families without the ability to provide collateral.

A 'project finance model' supports small- and medium-scale projects with loans and financial assistance from commercial banks. These projects are often on a commercial or village scale and involve micro-grids or sales of electricity back to the national grid.

A 'cooperative' model operates when households or investors band together to create their own cooperative which contributes all or some of the installation or operation of energy equipment. It is usually used for larger systems, such as solar micro-grids, commercial-scale biogas units, or micro-hydro dams.

A 'community fund' model directs efforts not at technology per se but at building the capacity of public institutions, private companies, or energy end-users themselves.

A 'fee for service' model is one where renewable energy technology is owned, operated, and maintained by a supplying company, but the customer pays regular fees for using it. It, too, has been utilized for all types of renewable energy with varying degrees of quality and installation cost.

A 'cross-subsidization' model operates when wealthier homes, or particular subclasses of electricity or energy customers, pay higher energy rates to produce money that then offsets the cost of expanding access to energy for poorer customers. It is usually applied to grid-extension efforts and occasionally to solar home systems.

Many approaches use 'hybrid' models that involve one or many of these approaches integrated together.

Table 2.6. Successful business models for expanding energy access

Model	Description	Example	Primary partners	Technology	Cost (USD)	Accomplishments
Technology improvement and market development	A 'supply push' structure where the partnership develops a renewable energy technology to reduce costs	China's Renewable Energy Development Programme, 2002–2007	World Bank, Global Environment Facility, National Development and Reform Commission, local solar manufacturers	Solar home systems	$316 million	Distributed more than 400,000 units in 5 years
End-user microfinance	A 'demand pull' which gives loans to energy users so that they can purchase renewable energy equipment	Grameen Shakti in Bangladesh, 1996–2010	International Finance Corporation, Infrastructure Development Company Limited, Grameen Bank	Solar home systems, biogas digesters, and improved cookstoves	—	Installed almost half a million solar home systems, 132,000 cookstoves, and 13,300 biogas plants among 3.1 million beneficiaries
Project finance	Small- and medium-scale projects are supported with loans and financial assistance from commercial banks	Energy Services Delivery Project in Sri Lanka, 1997–2002	World Bank, GEF, Ceylon Electricity Board, national banks	Solar home systems, grid-connected hydro, off-grid village hydro	$55.3 million	Installed 21,000 solar home systems and 350 kW of installed village hydro capacity in rural Sri Lanka, in addition to 31 MW of grid-connected mini-hydro capacity
Cooperative	Communities own renewable energy systems themselves	Cinta Mekar Micro-hydro Project in Indonesia, 2004–2011	Yayasan Ibeka, Hidropiranti Inti Bakti Swadaya, Directorate General of Energy Electricity Utilization, PLN, UNESCAP, Cinta Mekar Cooperative	Micro-hydro	$225,000	Constructed a 120 kW micro-hydro scheme that has electrified thousands of homes and creates thousands of dollars of monthly revenue funnelled back to the village
Community mobilization fund	Revenues from renewable electricity or energy production are invested back into local communities	Micro-hydro Village Electrification Scheme in Nepal, 2004–2011	World Bank, Government of Nepal, United Nations Development Program, Nepal Alternative Energy Promotion Center, district development communities, village development communities, micro-hydro functional groups	Micro-hydro	$5.5 million (original proposal)	Distributed 250 units benefiting 50,000 households in less than 10 years

Energy services company (ESCO) 'fee-for-service'	Private sector enterprises purchase technology and then charge consumers only for the renewable energy 'service' that results	Zambia's PV-ESCO Project, 1999–2009	Ministry of Energy, Stockholm Environmental Institute, Swedish International Development Authority	Solar home systems	—	Three ESCOs currently lease the services of 400 solar panels and have hundreds of clients wait-listed
Cross-subsidization	Tariffs on one type of electricity are funnelled into a fund to support renewable energy	Rural Electrification Project in Lao PDR, 2006–2009	Electricité du Lao PDR, Ministry of Energy and Mines, World Bank, Global Environment Facility, provincial electrification service companies	Solar home systems and grid-connected hydroelectricity	$13.75 million	Electrified 36,700 previously off-grid homes and disbursed more than 9,000 solar home systems
Hybrid (end-user microfinance and ESCO 'fee-for-service')	Private sector enterprises purchase technology and then charge consumers only for the renewable energy 'service' that results	India's Solar Lantern Project, 2005–2012	Small-Scale Sustainable Infrastructure Fund, Solar Electric Light Company, local banks and entrepreneurs	Solar lanterns	—	Distributed 80,000 units across 25 separate cities

Source: Sovacool 2013

As the application of these eight distinct models implies, the potential markets for energy access can become quite large. According to data from the World Resources Institute (Bairiganjan et al. 2010), providing energy services to the 'bottom of the pyramid', in India, or the 114 million households that earn only $1 to $8 (adjusted for PPP) per day represented a potential untapped market of more than $2 billion per year through small hydropower and biomass waste systems.

2.4. THE COMPLEXITY OF ENERGY POVERTY

This section of the chapter highlights the complexity of energy poverty and corrects nine common misconceptions.

2.4.1. Africa isn't everything

Many public officials intuitively think that the greatest pockets of energy poverty on a map are located in Africa or Sub-Saharan Africa. However, this is not the case. As Figure 2.7 illustrates, more than half of those without access to electricity reside in the Asia-Pacific, as do more than three-quarters of those wholly or partially dependent on traditional fuels for their cooking and heating needs. Indeed, energy use per capita for the countries comprising South Asia—India, Pakistan, Sri Lanka, Bangladesh, Nepal, and Bhutan—is *less* than per capita consumption for all of Africa. There are many pockets of energy poverty elsewhere, often in the hardest to reach places such as the Andean mountains or the rainforests of the Amazon. Making matters more complicated, the global map of energy poverty is perpetually changing as international actors make progress in combating it and as standards of definition and measurement evolve.

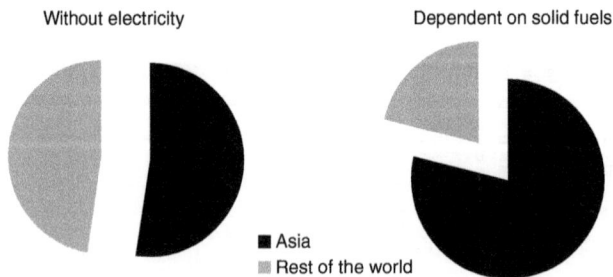

Figure 2.7. Asian proportion of those in 'energy poverty'

Source: United Nations Development Programme (2013). Achieving Sustainable Energy for All in the Asia-Pacific. Bangkok: UNDP APRC

2.4.2. Urban areas matter

A second common misconception is that energy poverty is exclusively a rural problem. This is untrue. One recent investigation of 34 cities in Bolivia, Botswana, Burkina Faso, Cape Verde, Haiti, India, Indonesia, Mauritania, Philippines, Thailand, Yemen, Zambia, and Zimbabwe concludes that 'the poor in urban areas of developing countries face special problems in meeting their basic energy needs'. The report notes that most urban poor are migrants that continue to rely on traditional fuels they collect on the periphery of urban areas; these people also pay higher prices for usable energy because of the inefficiency of stoves and lamps (Barnes et al. 2004).

2.4.3. The extremely poor need help

Another misconception is that energy access programmes often benefit the poorest. Instead, the very poor 'fall through the cracks'; they are too politically distant and it is economically too costly to provide them with energy services, even under many international programmes. It may be that energy access is a higher development goal, not a lower one; as such, attainment of energy security and reductions in energy poverty will happen only when more basic needs, such as the repayment of debt, financing of education, and satisfaction of community responsibilities are accomplished. In addition, many multilateral financial institutions such as the Asian Development Bank and World Bank must demonstrate positive cost-benefit ratios for all of their projects, since they are indeed giving *loans* rather than *grants*, and many energy access projects have timelines that are too risky for these development partners. Thus, to ensure equal development and access for all, there is a need for specific programming to reach the poorest at the bottom of the ladder who are not served by commercial energy providers or large-scale energy projects that demand positive profit margins from an early point.

The most recent projections from the International Energy Agency (IEA) subtly, but clearly, underscore that many of the poor are not likely to reach the goals of SE4ALL anytime soon. In projecting the future in their latest World Energy Outlook, the IEA (2012) estimate that almost 1 billion people will still be without electricity by 2030 and that 2.6 billion people will still be without clean cooking facilities. That same year, the number of people without clean cooking technologies in India will amount to *twice* the population of the United States, and overall the IEA forecast that 39 per cent of people in the Asia-Pacific region would lack access to modern cooking. The IEA also estimates that about $76 billion would be required to achieve universal access to clean cooking by 2030 (an average of $3.8 billion per year) and $1 trillion would be needed for universal access to energy and electricity (an average of $50 billion per year). As of 2012, however, only 3 per cent of this needed investment has been committed. It is the poorest households unlikely to be served by the private sector, government programmes, or financial institutions—the energy-poor that even the IEA projects will not provide with access to modern energy by 2030—that development partners will need to consider serving.

2.4.4. Energy access won't trash the planet

A fourth logical misconception is that if global society were to suddenly add billions of people to the grid, and provide them with modern cooking and heating energy services, it would have alarming impacts on both global energy consumption (which would dramatically increase) and the earth's climate (with greenhouse gas emissions rising astronomically). Instead, because modern energy services tend to be much more efficient and less polluting than traditional technologies and practices, the IEA (2012) projects that achieving universal access would only increase world energy demand by a meagre 1.1 per cent by 2030 and would increase world carbon dioxide emissions by 0.7 per cent—numbers illustrated in Figure 2.8.

2.4.5. Energy poverty involves more than heat and light

Another major misconception is that only lighting and cooking matter. As discussed, most of the literature on energy access has an explicit (or implicit) focus on provision of light and heat. However, this misses at least two instrumental energy services: mechanical or productive energy, and mobility (Sovacool et al. 2012).

Mechanical or productive energy services have great potential to tremendously reduce time spent on fuelwood gathering, to improve air quality in homes, and to raise household and community incomes. Some of the most fundamental services required for reducing poverty and promoting human development involve mechanical energy and increasing the productivity of human labour. Mechanical power enables activities such as pumping, transporting, and lifting water, irrigating fields, processing crops, small-scale manufacturing, and natural resource extraction. As Bates et al. (2009) forcefully argue:

> Experiences show that mechanical power helps alleviate drudgery, increase work rate and substantially reduce the level of human strength needed to achieve an outcome,

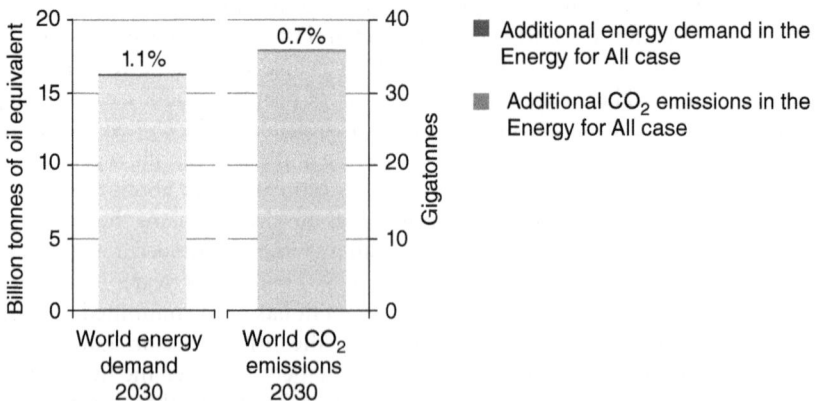

Figure 2.8. Global energy and climate impacts to universal energy access in 2030
Source: International Energy Agency, World Energy Outlook 2012 © OECD/IEA 2012

thus increasing efficiency and output productivity, producing a wider range of improved products, and saving time and production costs... In this regard, financing of
mechanical power is often one of the most cost effective ways to support poor people.

Mechanical power is not merely a derivative of other forms of energy such as
electricity, but an instrumental energy service in its own right.

Another key energy service is also generally lacking from the global discussion
of energy poverty: mobility. Though precise numbers are difficult to obtain, as it is
less studied, a significant proportion of the world population has transportation
choices constrained by lack of infrastructure, fuel scarcity, the distances or time
involved with travel, expense, or a combination of them all (Woodcock 2007).
Most of the world's poor do not have access to mass transportation and cannot
afford private motorized vehicles. Many rely on non-motorized transport, which
includes walking, bicycles, rickshaws, handcarts, and animal-drawn transport
(Booth et al. 2000; World Bank 2002b).

For both the rural and urban poor, low mobility—regardless of the technology
or mode of transport involved—stifles the attainment of better living standards. It
reduces the ability to earn income, strains economic resources, and limits access to
education and health services and markets. Many poor communities depend on
shared or hired motorized transport for daily mobility, disproportionally straining
the budgets of poor families. For instance, in New Delhi, poor households spend
approximately 20 to 25 per cent of their income attempting to meet daily mobility
needs (Kaltheier 2002). Low mobility is particularly burdensome for the rural
poor because it consumes a significant amount of their non-economic assets, such
as physical capital and time. For instance, in Bhutan, 96 per cent of the country's
poor live in rural areas and must walk hours or even days to reach the nearest road
(IFAD 2011). In rural Brazil, poor children often walk for more than an hour to
arrive to school or to the school bus location (Carvalho et al. 2010). While improved
cookstoves, access to mechanical services, and electricity can certainly alleviate some
of this burden, improving mobility is equally important. Providing poor populations with greater access to mobility has, among other things, been linked with an
increased use of social services and higher involvement in the political process
(Gannon et al. 2002). Mobility provides opportunities for the poor to increase their
earning potential, to visit hospitals, shop for cheaper goods, and attain educational
and employment goals.

2.4.6. There is no 'one-size-fits-all' solution

Though many countries can be grouped according to some of their common
energy poverty needs, each one is unique—and that specificity must be acknowledged when designing particular policy and technological interventions. In other
words, countries differ in terms of the large-scale investments, reforms, and energy
solutions needed. In light of this, planners must fully recognize the specificity of
each country and region in a world that is vastly diverse.

It could be that the best immediate energy development strategy for countries
such as Bhutan, Cambodia, Laos, and Nepal is encouraging cookstoves and
biomass gasifiers and digesters fed from manure and animal waste for household

needs, and to harness micro-hydro and larger-scale hydroelectric dams to displace fossil fuels on the national grid. Yet in India, small hydropower and electricity from biomass and waste may represent the cheapest technological solutions, supplemented with the use of solar home systems in areas remote from the grid where they can compete favourably with diesel prices. Brazil, China, and South Africa have provided a significant proportion of their population with access to electricity, but have lagged behind in providing them access to cleaner and modern forms of cooking. In Sub-Saharan Africa, the best approach could be off-grid solar systems coupled with solar-diesel-powered micro-grids.

One way to reflect these differences is through an energy access cost curve, a curve that shows household effectiveness for yearly investments in energy access. Figure 2.9 presents such a curve for Vanuatu, and it illustrates that the total number of people that could be provided with cost-effective energy access per year is 19,000. However, the figure reveals that the composition, configuration, cost, and number of people served by these measures vary, though it does suggest that the least costly four are cookstoves, solar lanterns, community woodlots, and diesel micro-grids.

These examples, and the particular cost curve for Vanuatu, suggest that no 'one-size-fits-all' approach exists for achieving energy access goals in developing countries.

2.4.7. Failure is as salient as success

Another implicit misconception is that we can learn the most from 'best practices' or case studies of success. While certainly important, examples of success are only half of the equation; we can learn just as much, maybe even more, from project failures. Admittedly, the notion of each is difficult to pin down, since the terms are charged and politicized. As former US President John F. Kennedy said, 'victory has a thousand fathers, but defeat is an orphan'.

Yet the energy policy community and perhaps the development community as a whole need to better understand the dynamics of failure alongside the better-known reasons for success. One World Bank official has attested that 'in all likelihood, there are more failures than successes—you just never hear about them. It is easier to fail than to succeed, but all too often failures are swept under the carpet' (quoted in Sovacool and Drupady 2012). An appreciation of past experiences and the barriers encountered from such failures can accelerate feedback and lead to new programmes that actually meet some or all of their targets.

But this lesson goes deeper. As one energy development expert in Papua New Guinea has stated:

> It annoys me to see organizations put positive spin on projects that more rightly deserve to be described as mediocre or failures. Sure, we can't always have wins, but too often outcomes are smoothed over with soft rhetoric meaning the reasons behind failed projects don't make it far enough back up the food chain to make a difference to how future projects are designed and implemented . . . I suppose learning about Papua New Guinea culture and politics, and past failures, might require more effort than [project] managers were willing to provide (quoted in Sovacool and Drupady 2012).

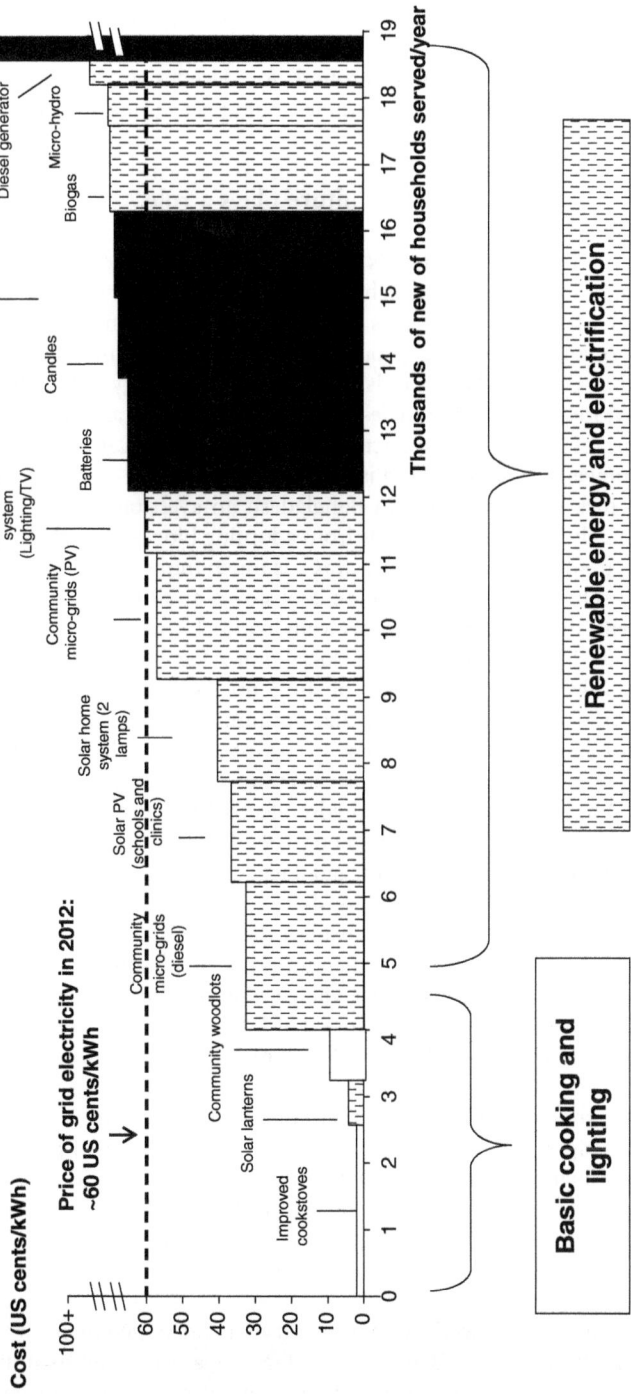

Figure 2.9. Energy access cost curve for Vanuatu

Note: The vertical axis of the figure shows the cost for various options in US cents per kWh, with some (below the dotted line) costing less than grid electricity already. Others (above the dotted line) cost more than the grid, but are *all* cheaper than grid extension, which is cost-prohibitive at more than $1.20 per kWh. The horizontal axis of the figure shows the number of people served by each option (i.e. the larger the size of the square, the wider the bar, the more people served)—with improved cookstoves (about 2,500 people), batteries (1,900 people), and kerosene lamps (1,200 people) having the most potential, but only one of them—cookstoves—low costs of 3 to 4 cents per kWh, whereas solar lanterns cost about 5 cents per kWh, micro-grids vary from 40 to 58 cents per kWh, and biogas and micro-hydro units cost about 70 cents per kWh. Figure 2.9 does not show energy efficiency on the chart, only because efficiency programmes are at a nascent stage of development in Vanuatu; if fully developed, they would probably fall competitively with the lowest cost energy access solutions.

Source: United Nations Development Programme 2013

There is nothing wrong with idealism and taking comfort in programmatic successes, but examples of unsuccessful case studies can motivate energy and development professionals to remain critical and self-reflective. Moreover, many examples of success have been 'around' for decades in the literature on programme design and development studies, but appear to have been forgotten.

2.4.8. Culture plays a role

A more subtle misconception is that technology and economics are all that matter—that if one perfects a technology, and gets the price signals right, energy access can occur. History shows instead that well-designed programmes with sufficient financial incentives promoting high-quality technology have nonetheless failed to convince households to enhance their energy access due to cultural reasons. Moreover, as technologies get rolled out to more and more remote areas, they invariably come into contact with more isolated local communities, where the very powerful role of culture is evident.

To list just a few examples, in Bangladesh an aversion to pigs has prevented predominately Muslim households from adopting biogas units that would run on pig waste, despite the fact that such waste is much more efficient than cattle dung. Other households refuse to purchase cookstoves at all because they are uncomfortable with the idea of piping in gas from livestock and human excrement, which they see as 'impure'. In Nepal, a social norm against collecting revenue for electricity inhibits the profitability of some micro-hydro schemes. Some believe hydroelectric facilities should serve the community for free, and that poor families should not have to pay for electricity. As one villager has remarked, 'almost everyone in rural Nepal is poor, so a cultural stigma exists against charging rural households tariffs for micro-hydro electricity that match costs' (quoted in Sovacool and Drupady 2012). In Papua New Guinea, SHSs have been prone to unusually high rates of vandalism, sabotage, and theft. Under a *wantok* system rooted in tribal traditions, clans there share resources. Solar panels, which benefit a particular house or individual instead of the community, assault this system of *wantok*. Tribal communities have therefore smashed hundreds of solar panels or, worse, threatened their owners. One village elder mentioned that he would 'never' want to purchase a solar panel because if 'if I did put one in my village, but not all of the surrounding villages, they would kill me' (quoted in Sovacool and Drupady 2012).

In short, these examples reveal that cultural attitudes and social expectations can play as significant a role as price signals, poorly designed programmes, and improperly aligned regulations in impeding the use of off-grid renewable energy applications. They also imply that no matter how well developed or perfected a given energy technology or energy system becomes, it could have little to no impact without systematic and scientific efforts to ensure such technologies are culturally compatible.

Culture, moreover, need not always serve as an impediment; it can catalyse programmatic success. In Bangladesh, a taboo against letting men inside homes during the day enabled Grameen Shakti to empower women as entrepreneurs and technicians, and in Sri Lanka, a culture of *shramadana* convinced communities to give their own time or materials for the civil work and construction of micro-hydro units. These examples, and others, imply that culture can either impede or accelerate the adoption of modern energy technologies.

2.4.9. Metrics need improving

What these examples and misconceptions reveal is that we may need new standards of evaluation. Energy poverty is a multidimensional phenomenon irreducible to merely one region (Africa), one area (urban), one group (the poor), two services or two sets of technology (heat and light), factors for success (failure is important), or cultural homogeneity. One more nuanced way to conceptualize energy poverty would be to conceive of energy services as existing in a matrix. The United Nations Secretary-General's Advisory Group on Energy and Climate Change (UN-Energy 2010), an intergovernmental body composed of representatives from businesses, the United Nations, and research institutes, has recently suggested that energy access ought to be categorized into an incremental matrix shown in Table 2.7.

In this matrix, first come basic human needs that can be met with electricity consumption of 50–100 kWh per person per year, 50–100 kg of oil equivalent or modern fuel per person per year, and the ownership of an improved cookstove. Second are productive uses, such as access to mechanical energy for agriculture or irrigation, commercial energy, or liquid transport fuels. Consumption here rises to 500–1,000 kWh per person per year and 150 kg of oil equivalent. Third are modern needs, which include the use of domestic appliances, cooling and space heating, hot and cold water, and private transportation, which in the aggregate result in the consumption of about 2,000 kWh per person per year and 250–450 kg of oil equivalent. The matrix includes all four key energy services (lighting, heating, mechanical power, and mobility).

If one accepts this finding, then new ways of quantifying and measuring energy poverty, and collecting data about it, are needed. The World Bank, International

Table 2.7. Energy services and access levels

Level	Electricity use	kWh per person per year	Solid fuel use	Mobility	Kilograms of oil equivalent per person per year
Basic human needs	Lighting, health, education, communication	50 to 100	Significant	None, walking or bicycling	50 to 100
Productive uses	Agriculture, water pumping for irrigation, fertilizer, mechanized tilling, processing	500 to 1,000	Minimal	Mass transit, motorcycle, or scooter	150
Modern society needs	Domestic appliances, cooling, heating	2,000	Minimal	Private transportation	250 to 450

Source: UN 2010

Energy Agency, World Health Organization, United Nations, and other organizations could be directed to start collecting information on mobility and mechanical power. All four energy services discussed here—lighting, heating and cooking, mechanical power, and mobility—should be integrated into national and international development and energy targets and strategies. The awareness and capacity of communities and national governments need to be built up so that they themselves can inform policymakers about their needs and local solutions to energy poverty. Put another way, policymakers need to build awareness so that they can make themselves aware.

2.5. CONCLUSION

The arguments advanced in this chapter challenge conventional definitions and approaches to energy poverty in a few meaningful ways.

First, they remind us that any attempts to elevate energy access within national and international energy policy should learn from the modern shift in development theory and avoid adopting overly technical approaches. Development practitioners warn that the kind of 'mass customization of technologies and delivery models' that Bazilian et al. (2010: 5410) advocate risks effectively depoliticizing the underlying patterns of injustice that create and maintain inequities in the delivery of energy services (Sesan, 2011). Without careful consideration of cultural and demographic sensitivities now commonplace among development practitioners, reframing energy poverty as an essential element in national and international energy planning could merely reify the modes of domination that deeply influence the lives of the planet's energy-poor.

Second, one way to avoid abandoning altogether the evolution of development theory is for energy planners and development practitioners to strive to see energy poverty as a services-oriented issue, or a fundamental human rights concern, rather than a fuel or technological issue. As Amory Lovins (1976: 65) ruminated many decades ago, 'people do not want electricity or oil, nor such economic abstractions as "residential services", but rather comfortable rooms, light, vehicular motion, food, tables, and other real things'. This means we ought to be moving towards 'mobility security' and 'light security' rather than 'oil security' and 'electrification'.

Third, the relationship between energy services and energy poverty is synergistic. The fact that the majority of the world's development experts continue to measure energy poverty as two-dimensional ('heat' and 'light', or 'cookstoves' and 'electricity') tells us just how far we need to go in pushing mechanical power and mobility as rightfully desirable attributes. Moreover, lack of access to any of the four services can inhibit or exacerbate the use of the others. For example, lack of access to mobility can influence the availability or price of kerosene (lighting), fuelwood collection times (impacting heating and cooking), and transporting processed products to market (mechanical power). This means that energy poverty and energy services have an interactive relationship and should not be viewed independently of each other.

Table 2.8. Three paradigms of energy access and development

	Donor gift paradigm (1970s–1990s)	Market creation paradigm (1990s and 2000s)	New 'sustainable programme paradigm' (2010s–?)
Actors	One, usually a government or just one development donor	Multiple government agencies and/or multilateral donors	Multiple public, private, and community stakeholders
Primary goal	Technology diffusion	Market and economic viability	Environmental and social sustainability
Focus	Equipment, often single systems	Multiple fuels (e.g. 'electricity' or 'fuelwood')	Energy services, income generation, institutional and social needs
Standardization	Little standardization between projects	Some standardization	Harmonized with certificates, testing regimes, and national standards
Implementation	One-time disbursement	Project evaluation at beginning and end	Continuous evaluation and monitoring
After-sales service and maintenance	Limited	Moderate	Extensive
Ownership	Given away	Sold to consumers	Cost-sharing and in-kind community contributions
Awareness raising	Technical demonstrations	Demonstrations of business models	Demonstrations of business, financing, institutional, and social models

Fourth, we may need to adopt a new paradigm of energy development and access. As Table 2.8 shows, the 'classic' approach to energy development assistance in the 1970s and 1980s focused exclusively on single fuels, with projects implemented by a central agency and involving a single financer or borrower. Such projects predominately favoured large systems, with little local participation, and strong subsidies for fuels or capital equipment. They overemphasized energy production and installed capacity, and focused on universal access, regardless of the level of development (Barnes and Floor 1996; Martinot et al. 2002). Central to this approach was the notion that developed countries should 'give' technology and assistance away to developing countries out of a sense of moral obligation.

A new paradigm arose in the 1990s and 2000s, emphasizing the need for multiple fuels, a market approach supported by technical assistance and training, and multiple borrowers pushing smaller, high-return projects. Greater involvement with other donors was incentivized, and projects became more oriented toward consumer demands, integration with broader development efforts, and higher levels of local involvement and investment. This approach worked under the assumption that if prices of technology could be brought down beyond a certain point, or local manufacturing established up to a particular threshold, the adoption of modern energy systems would become self-sustaining (Magradze et al. 2007). Or, as one consultant put it, 'companies mistakenly believed that if they provided a great product, or delivered a great service, customers would purchase it—nothing else was needed' (Sovacool and Drupady 2012).

The newest paradigm, however, is fundamentally different. It maintains a focus on polycentrism, or the involvement of multiple actors from multiple spheres. Programmes extend beyond technological diffusion and market viability to encompass goals such as environmental sustainability, the reduction of greenhouse gas emissions, and local job creation. Their focus is on energy services and income generation rather than fuels or equipment, but they still recognize the necessity of high-quality, standardized, and certified technology. Evaluation and monitoring are continuous, after-sales service and maintenance are extensive, and communities share costs and in-kind contributions to projects. These projects recognize that the definition of energy affordability varies according to market segments, relative to incomes, market applications, and geography, and that broader social and political factors must be promoted alongside technology and market development.

For if we are right about the complex multidimensionality of energy poverty, then we functionally need to be more sensitive to those that suffer from it and to better adapt our paradigms to understand it. Only then will we be able to effectively lift the world's poorest out of their persistent state of energy deprivation.

REFERENCES

Bairiganjan, Sreyamsa, et al. (2010). *Power to the People: Investing in Clean Energy for the Base of the Pyramid in India*. Washington, DC: World Resources Institute.

Barnes, Douglas F. and Willem M. Floor. (1996). 'Rural Energy in Developing Countries: A Challenge for Economic Development', *Annual Review of Energy and Environment* 21, 497–530.

Barnes, Douglas F., Kerry Krutilla, and William Hyde. (2004). *The Urban Household Energy Transition: Energy, Poverty, and the Environment in the Developing World*. Washington, DC: Resources for the Future.

Bates, Liz, et al. (2009). 'Expanding Energy Access in Developing Countries: The Role of Mechanical Power'. Practical Action Report. Washington, DC: UNDP.

Bazilian, Morgan, Ambuj Sagar, Reid Detchon, and Kandeh Yumkella. (2010). 'More Heat and Light', *Energy Policy* 38, 5409–12.

Birol, Fatih. (2007). 'Energy Economics: A Place for Energy Poverty on the Agenda?' *The Energy Journal* 28(3), 1–6.

Booth, D., et al. (2000). *Poverty and Transport*. World Bank. Available at <http://www.odi.org.uk/resources/download/2689.pdf>.

Brown, M.A. and B.K. Sovacool. (2011). *Climate Change and Global Energy Security: Technology and Policy Options*. Cambridge, MA: MIT Press.

Carvalho, W.L., et al. (2010). 'Rural School Transportation in Emerging Countries: The Brazilian Case', *Research in Transportation Economics* 29, 401–9.

Casillas, Christian E. and Daniel M. Kammen. (2012). 'The Challenge of Making Reliable Carbon Abatement Estimates: the case of Diesel Micro-grids', *Sapiens* 5(1), 1–9.

D'Agostino, A.L., B.K. Sovacool, K. Trott, C.R. Ramos, S. Saleem, and Y. Ong. (2011). 'What's the State of Energy Studies Research? A Content Analysis of Three Leading Journals from 1999–2008', *Energy* 36(1), January, 508–19.

Dutta, S. (2011). *Sustainable Energy Development in the Asia Pacific: A Discussion Note*. Bangkok: UNESCAP.

Eric, D.L. and S. Kartha. (2000). 'Expanding Roles for Modernized Biomass Energy', *Energy for Sustainable Development* 4(3), 15–25.

Fischer, Barry, and Attilio Pigneri. (2011). 'Potential for electrification from biomass gasification in Vanuatu', *Energy* 36, 1640–51.

Gannon, C., et al. (2002). 'Transport'. In Jeni Klugman, ed., *A Sourcebook for Poverty Reduction Strategies*, Washington, DC: The World Bank, pp. 326–67.

Gautama, R., S. Baralb, and S. Heart. (2009). 'Biogas as a Sustainable Energy Source in Nepal: Present Status and Future Challenges', *Renewable and Sustainable Energy Reviews* 13, 248–52.

Gaye, Amie. (2007). 'Access to Energy and Human Development', *Human Development Report 2007/2008*, United Nations Development Program Human Development Report Office Occasional Paper.

Hiemstra-van der Horst, Greg, and Alice J. Hovorka. (2008). 'Reassessing the "Energy Ladder": Household Energy Use in Maun, Botswana', *Energy Policy* 36, 3333–44.

Holdren, John P. and Kirk R. Smith. (2000). 'Energy, the Environment, and Health', in Tord Kjellstrom, David Streets, and Xiadong Wang, eds., *World Energy Assessment: Energy and the Challenge of Sustainability*, New York: United Nations Development Program, pp. 61–110.

IIASA. (2012). *Global Energy Assessment*. Laxenberg: IIASA. Available at: <http://www.iiasa.ac.at/Research/ENE/GEA/>.

International Energy Agency, United Nations Development Program, United Nations Industrial Development Organization. (2010). *Energy Poverty: How to Make Modern Energy Access Universal?* Paris: OECD.

International Energy Agency. (2011). *Energy for All: Financing Access for the Poor*. Paris: OECD.

International Energy Agency. (2012). *World Energy Outlook 2012*. Paris: OECD.

International Energy Agency. (2013). *Global Status of Modern Energy Access*. Paris: OECD.

International Fund for Agricultural Development. (2011). 'Rural Poverty in Bhutan'. Available at <http://www.ruralpovertyportal.org/web/guest/country/home/tags/Bhutan>.

Jones, Richard. (2010). 'Energy Poverty: How to Make Modern Energy Access Universal?' Special early excerpt of the World Energy Outlook 2010 for the UN General Assembly on the Millennium Development Goals. Paris: International Energy Agency/OECD.

Kaltheier, R.M. (2002). 'Urban Transport and Poverty in Developing Countries'. Division 44 Environmental Management, Water, Energy, Transport. Available at <http://www.gtkp.com/assets/uploads/20091127-182046-6236-en-urban-transport-and-poverty.pdf>.

Kammen, Daniel M. and Michael R. Dove. (1997). 'The Virtues of Mundane Science', *Environment* 39(6), July/August, 10–41.

Kaplan, Stan Mark, and Fred Sissine, eds. (2009). *Smart Grid: Modernizing Electric Power Transmission and Distribution*. New York: The Capitol Net Publishing.

Karekezi, S. and W. Kithyoma. (2002). 'Renewable Energy Strategies for Rural Africa: Is a PV-led Renewable Energy Strategy the Right Approach for Providing Modern Energy to the Rural Poor of Sub-Saharan Africa? *Energy Policy* 30, 1071–86.

Lovins, Amory B. (1976). 'Energy Strategy: The Road Not Taken', *Foreign Affairs* 55(1), October, 65.

Magradze, N., A. Miller, and H. Simpson. (2007). *Selling Solar: Lessons from More Than a Decade of Experience*. Washington, DC: Global Environment Facility/International Finance Corporation.

Martinot, E., et al. (2002). 'Renewable Energy Markets in Developing Countries', *Annual Review of Energy and the Environment* 27, 309–48.

Masud, Jamil, Diwesh Sharan, and Bindu N. Lohani. (2007). *Energy for All: Addressing the Energy, Environment, and Poverty Nexus in Asia*. Manila: Asian Development Bank.

Miller, D. and C. Hope. (2000). 'Learning to Lend for Off-Grid Solar Power: Policy Lessons from World Bank Loans to India, Indonesia, and Sri Lanka', *Energy Policy* 28, 87–105.

Mills, Evan. (2005). 'The Specter of Fuel-Based Lighting', *Science* 308, 27 May, 1263–4.

Modi, Vijay, Susan McDade, Dominique Lallement, and Jamal Saghir. (2005). *Energy Services for the Millennium Development Goals*. Washington and New York: The International Bank for Reconstruction and Development/The World Bank and the United Nations Development Program.

Mukwedeya, Brian Moyo. (2011). *Off Grid Systems: Micro and Mini-grids For Rural and Marginalised Communities*. Zimbabwe International Energy and Power Corporation, Harare, 27–29 September.

Roth, Christa. (2011). *Micro-gasification: Cooking with Gas from Biomass*. Berlin: GIZ.

Ruggie, John. (2006). *Interim Report of the Special Representative of the Secretary-General on the Issue of Human Rights and Transnational Corporations and Other Business Enterprises*. UN Doc E/CN.4/2006/97. Geneva: United Nations.

Sanchez, T. (2010). *The Hidden Energy Crisis, 2010*. Practical Action Publishing.

Sesan, T.A. (2011). *What's Cooking? Participatory and Market Approaches to Stove Development in Nigeria and Kenya*. Thesis submitted for the degree of Doctor of Philosophy, University of Nottingham, UK.

Smil, Vaclav, and William E. Knowland, eds. (1980). *Energy in the Developing World: The Real Energy Crisis*. Oxford: Oxford University Press, 1980.

Sovacool, B.K. (2012). 'The Political Economy of Energy Poverty: A Review of Key Challenges', *Energy for Sustainable Development* 16(3), September, 272–82.

Sovacool, B.K. (2013). 'Expanding Renewable Energy Access with Pro-Poor Public–Private Partnerships in the Developing World', *Energy Strategy Reviews* 1(3), March, 181–92.

Sovacool, B.K. and I.M. Drupady. (2012). *Energy Access, Poverty, and Development: The Governance of Small-Scale Renewable Energy in Developing Asia*. New York: Ashgate.

Sovacool, B.K., S. Saleem, A.L. D'Agostino, C.R. Ramos, K. Trott, and Y. Ong. (2011). 'What About Social Science and Interdisciplinarity? A 10-year Content Analysis of Energy Policy', in D.L. Goldblatt et al., eds., *Tackling Long-Term Global Energy Problems: The Contribution of Social Sciences*, New York: Springer.

Sovacool, B.K., C. Cooper, M. Bazilian, K. Johnson, D. Zoppo, S. Clarke, J. Eidsness, M. Crafton, T. Velumail, and H.A. Raza. (2012). 'What Moves and Works: Broadening the Consideration of Energy Poverty,' *Energy Policy* 42 (March), 715–19.

Szabo, S., K. Bodis, T. Huld, and M. Moner-Girona. (2011). 'Energy Solutions in Rural Africa: Mapping Electrification Costs of Distributed Solar and Diesel Generation versus Grid Extension', *Environmental Research Letters* 6, 1–9.

Thiam, D.-R. (2010). 'Renewable Decentralized in Developing Countries: Appraisal from Micro-grids Project in Senegal', *Renewable Energy* 35(8), 1615–23.

UN-Energy (2010). *Energy for a Sustainable Future: The Secretary-General's Advisory Group on Energy and Climate Change Summary Report and Recommendations*. New York: UN, 28 April.

United Nations Development Programme. (2009). *Contribution of Energy Services to the Millennium Development Goals and to Poverty Alleviation in Latin America and the Caribbean*. Santiago, Chile: United Nations, October.

United Nations Development Programme. (2010). *Human Development Report 2010*. New York: UNDP.

United Nations Development Programme. (2013). *Achieving Sustainable Energy for All in the Asia-Pacific*. Bangkok: United Nations Development Programme-Asia Pacific Research Centre, August.

United Nations Economic and Social Commission for Asia and the Pacific [UNESCAP]. (2008). *Energy Security and Sustainable Development in Asia and the Pacific*. Geneva: UNESCAP, ST/ESCAP/2494, April.

Venkataraman, C., A.D. Sagar, G. Habib, N. Lam, and K.R. Smith. (2010). 'The Indian National Initiative for Advanced Biomass Cookstoves: The Benefits of Clean Combustion', *Energy for Sustainable Development* 14, 63–72.

Wachenfeld, V. (2012). 'PV-Hybrid Diesel Systems: Why it makes Sense to Combine Diesels Systems with PV', presentation to Intersolar Europe, 13–15 June.

Woodcock, J. (2007). 'Energy and Transport'. *The Lancet* 9592(370), 1078–88.

World Bank. (2002a). *Rural Electrification and Development in the Philippines: Measuring the Social and Economic Benefits.* Washington, DC: World Bank ESMAP Program, May.

World Bank. (2002b). *Cities on the Move: A World Bank Urban Transport Strategy Review.* Washington, DC: World Bank.

World Bank. (2004). *Papua New Guinea Energy Sector and Rural Electrification Background*, Note 1, March.

World Bank. (2007). 'Technical and Economic Assessment of Off-Grid, Mini-Grid and Grid Electrification Technologies'. ESMAP Technical Paper 121/07.

World Bank. (2013). Data, accessed 30 May 2013, available at <http://web.worldbank.org/WBSITE/EXTERNAL/EXTOED/EXTRURELECT/>.

World Health Organization (WHO). (2006). *Fuel for Life: Household Energy and Health.* Geneva: WHO.

3

The Development Impact of Energy Access

Douglas F. Barnes, Hussain Samad, and
Sudeshna Ghosh Banerjee

The growing evidence of the socio-economic benefits of electricity, along with the growing acceptance of the importance of alleviating indoor air pollution through clean cooking solutions, reinforces the notion that programmes to expand the use of more efficient and cleaner forms of energy are of critical importance for developing countries. The evidence is now becoming overwhelming that without modern forms of energy, development is very difficult, if not impossible, for today's poorest countries.

Intuitively, one can easily understand that in households with electricity people are better able to undertake activities that require higher levels of lighting, such as reading and studying (Barnes, Peskin, and Fitzgerald 2003). They can also listen to the radio, watch television, and attend to more household chores (World Bank 2002b, 2004). In contrast, the kerosene lanterns and candles in households without electricity emit a dull light inadequate for reading or close work (Nieuwenhout et al. 1998; Van der Plas and de Graaff 1988). In households with no electricity, family members retire early after a fairly unproductive evening.

Improved access to modern cooking solutions contributes to improved health and reduces premature mortality, especially among women. In households that rely on traditional methods of cooking with biomass fuel, women are exposed to small particulates present in smoke that can reach 20 times maximum recommended levels. Indoor smoke pollution from inefficient use of biomass for cooking is estimated to cause more than 3 million premature deaths annually in developing countries. Also, the time that women and children spend collecting firewood takes away from more desirable activities; collecting fuel can also lead to neck and back injuries.

The poverty line is a common measure used by the development community for assessing development impacts. There is a large body of literature on how to measure income poverty and expenditure poverty and on the reliability of alternate measures (Ravallion 1998; Ravallion and Bidani 1994; Pradhan and Ravallion 1998; Haughton and Khandker 2009). The concept of an energy poverty line, however, does not exist. No international or government agency tracks energy poverty and its relationship to income poverty because consensus has not yet been reached on the methodological and conceptual hurdles involved in defining and measuring such a relationship. However, in recent years, there has been progress in understanding the relationship between energy and development, and this bodes well for further understanding the nature of energy poverty.

This chapter is a selective review of the literature on the relationship of energy and development, focusing on household-level outcomes. We will first examine the economic and social impacts of rural electrification before turning our attention to the benefits of adopting better fuels and stoves for cooking. Both topics are now receiving greater attention thanks to recent international initiatives such as Sustainable Energy for All and the Global Alliance for Clean Cooking, both of which stress the importance of modern household energy for health, productivity, and social well-being.

3.1. THE ECONOMIC AND SOCIAL BENEFITS OF RURAL ELECTRIFICATION

After a period of doubt several decades ago, the evidence has been building that electrification stimulates local economies, which in turn contributes to higher levels of development. One exception to this is a recent study on productive uses of electricity in Africa, but many of the case studies in that report suffered from a lack of complementary conditions for promoting development, a lack common in Africa (Mayer-Tasch et al. 2013). The relationship between energy use and economic development has been understood for decades (Figure 3.1), but many questioned whether the observed effects reflected a causal relation rather than

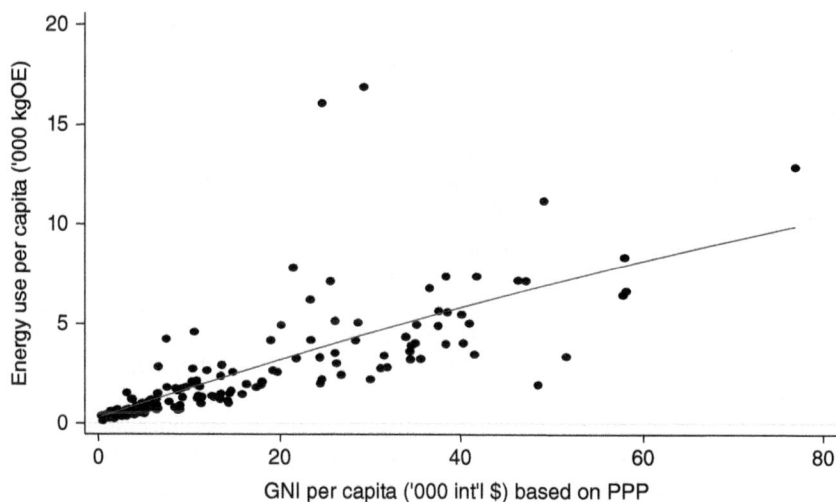

Figure 3.1. Conventional energy use per capita, by economic development level (GNI), 2010

Note: Horizontal axis is gross national income converted to current international dollars based on purchasing power parity; vertical axis is primary energy use in kilograms of oil equivalent per capita before transformation to end-use form.

Source: Figure prepared for this report using World Bank databank of World Development Indicators 2013

simple correlation. More recently, studies examining the relationship of access to electricity and income confirm that under the right conditions electricity indeed plays a role in raising levels of economic development (Khandker, Barnes, and Samad 2013; Dinkelman 2011; Van de Walle et al. 2013). For example, an empirical study of farm productivity in India indicates that, even after factoring in moderate price increases to improve the reliability of power supply, small-scale farmers could increase their income by about 50 per cent, while medium- and large-scale farmers, who have higher levels of investment, could expect increases of about 15 per cent (Monari and Mostefai 2001). A study in Kenya found that income from some small-scale businesses doubles after the introduction of a decentralized community electricity system (Kirubi et al. 2009). In Bangladesh, having access to electricity had a cumulative impact on rural household income of as much as 21 per cent, resulting in a corresponding drop in the poverty rate of about 13 percentage points (Barnes, Khandker, and Samad 2011; Khandker, Barnes, and Samad 2012).

Electricity is widely recognized as indispensable for raising households' standard of living and broader economic development. Once households connect to the electricity grid, they get an immediate benefit from better household lighting. With brighter light in the home, children spend more hours studying, adults have more flexible hours for completing chores and reading books, and home-based businesses remain open longer in the evenings, producing more items for sale. After rural families connect to the grid, television sets, fans, and an array of other household appliances gradually become more affordable. Rural electrification can also raise productivity and income when farmers switch from manual to grid-powered irrigation and small industries begin using electric tools and machinery. Indeed, a large body of literature discusses the many direct and indirect benefits of electricity (for example, Khandker 1996; Filmer and Pritchett 1998; Roddis 2000; World Bank 2002a, 2004, 2008; Barnes, Peskin, and Fitzgerald 2003; Kulkarni and Barnes 2004; Cabraal, Barnes, and Agarwal 2005; World Bank 2011a; Kirubi et al. 2009; Tanguy 2012; Barakat et al. 2002). Given electricity's substantial benefits, access to it and to other sources of modern energy has been identified as essential to fulfilling the Millennium Development Goals adopted by the United Nations (UNDP 2005) and is being actively considered as a possible Sustainable Development Goal in the post-2015 development framework.

Given the multiple interconnections among a broad array of appliances, outputs, and intermediate outcomes, the task of sorting out the direction of causality of electrification impacts is understandably complex (Figure 3.2). Once connected to the grid, consumers purchase a variety of appliances, often beginning with electric lamps, followed by radios, televisions, space coolers/heaters, cooking devices, and small machines. These appliances, in turn, produce such short-term or immediate outputs as a greater quantity of higher quality lighting, access to knowledge and information, greater comfort, better food preservation, productive motive power, and more efficient cooking—all benefits that are not possible without electricity (World Bank 2008). These outputs, in turn, can lead to such intermediate outcomes as extended study time, longer hours of home-business operation, greater exposure to business knowledge and information, better health, and more efficient production activities. Over time, these intermediate outputs can lead to more comprehensive development outcomes that might

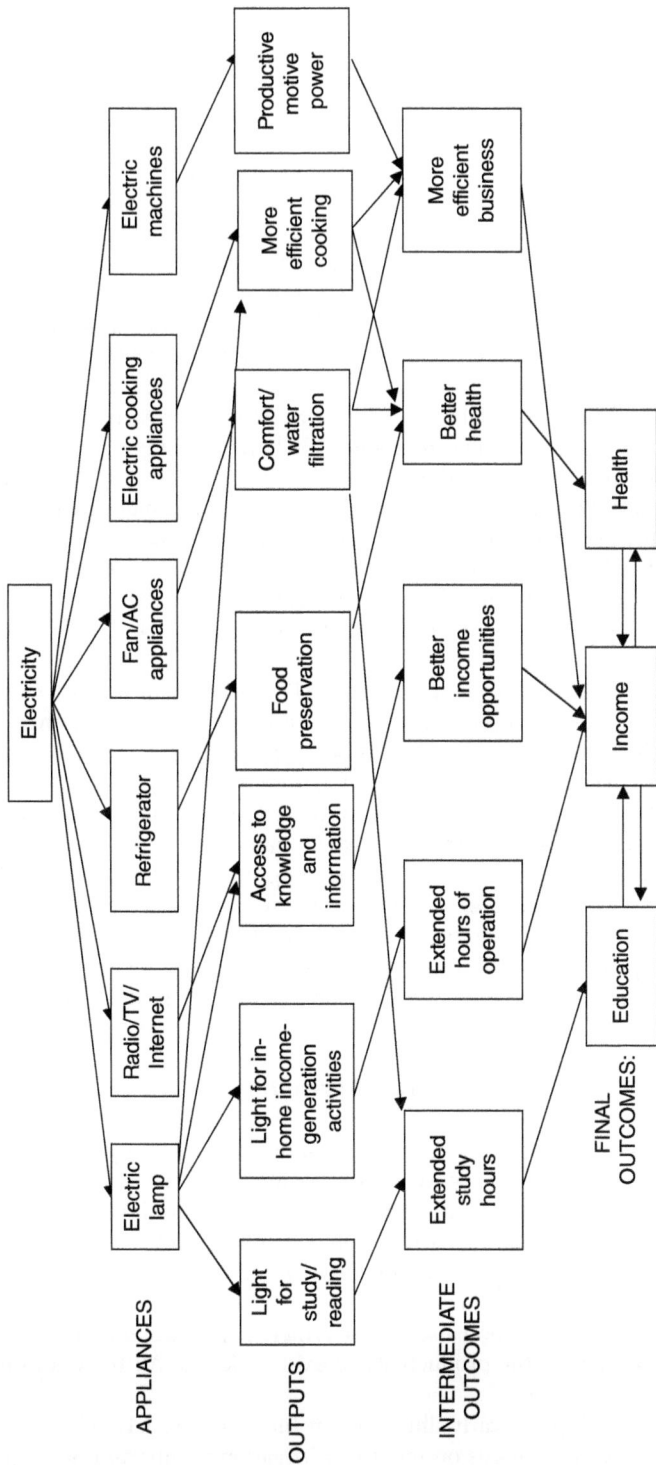

Figure 3.2. Social and economic impacts of electricity on households

Source: Khandker, Barnes, and Samad 2013

include improvements in education, income, and health. Since there are multiple channels through which electrification can lead to development outcomes, the accumulated benefits can be quite high.

3.1.1. Household lighting

After adopting electricity, most households immediately begin using electric lamps. This higher quality electric lighting enables household members to read and study during evening hours, increases productivity, and raises incomes and quality of life. Compared to candles or kerosene lamps, which households without electricity commonly use, electricity converts energy into lighting more efficiently and with much brighter results. A candle or kerosene wick lamp emits about 12 lumens (a measure of brightness), and a hurricane kerosene lamp 32 lumens, whereas a 60-watt light bulb can produce 730 lumens. Using a single 60-watt bulb four hours a day, a household uses 260 kilo-lumen hours of light per month. By contrast, burning a hurricane kerosene lamp four hours a night yields only 4 kilo-lumen hours per month; depending on prices, this may cost between one-quarter as much and the same as electric lighting (Nieuwenhout et al. 1998). One study in Peru (Meier et al. 2010) found that for an average family the price of lighting per kilolumen hour was about 90 US cents for kerosene and only 2 cents for electricity.

The energy-poor in Africa spend about $17 billion a year on costly, inefficient, fuel-based lighting sources such as kerosene lamps, which not only give poor-quality light (as mentioned) but also pose fire hazards and pollute the indoor environment (Lighting Africa 2008). According to a 1998 survey in the Philippines, where the price of electricity was fairly high, the cost of switching to electric lights was only marginally higher than using kerosene for all income classes (Figure 3.3). The amount of spending on lighting increases whether a household uses kerosene or electricity, but the spending on electricity is only slightly higher than the amount that would have been spent on kerosene, and yet a household gains about 100 times the amount of light. This light in turn allows household members to read, study, socialize, and develop new businesses.

3.1.2. Agriculture

The impact of rural electrification on agricultural development is achieved through productive motive power. For those areas that use gravity-fed irrigation, the impact of electricity is much lower than in those with individual agricultural pump sets (Asaduzzaman, Barnes, and Khandker 2009; World Bank 2002a, 2002b). With the advent of electricity it is easier for farmers to irrigate their fields, as electric pumps require low maintenance and are more efficient compared to diesel alternatives. Irrigation also allows farmers to produce multiple crops in one year and to improve the productivity of existing farms. All these lead to higher crop yield and income.

This relationship is clearly illustrated in India (Figure 3.4), where historically there has been an emphasis on the use of irrigation pumps and new agricultural

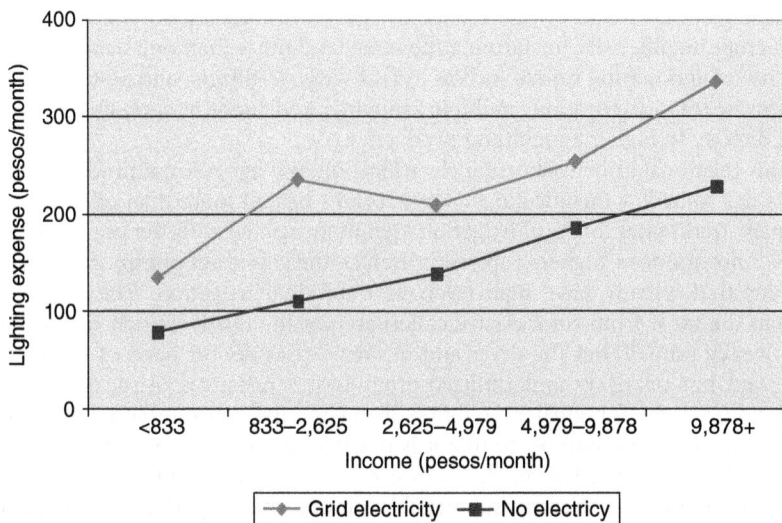

Figure 3.3. Household lighting expenses by grid electrification status in the Philippines, 2002

Source: World Bank 2002a

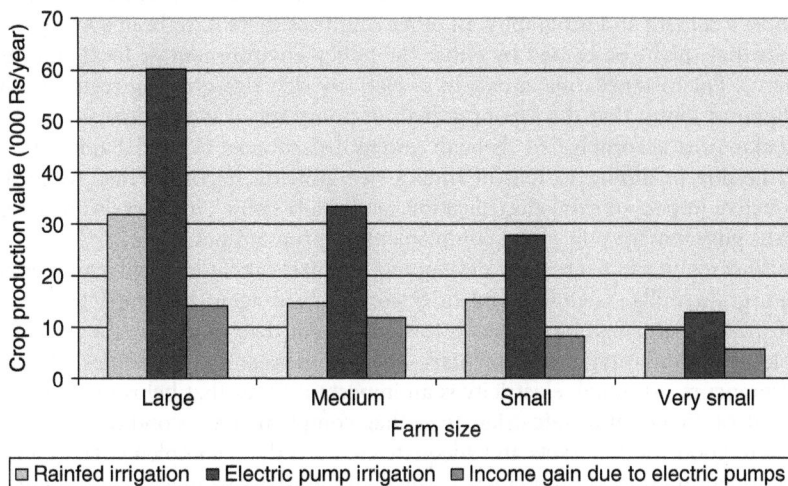

Figure 3.4. The relationships between farm size, electric pump irrigation, and crop value in India, 1998

Source: Barnes, Fitzgerald, and Peskin 2003

technologies to improve agricultural productivity (World Bank 2002b; Barnes, Peskin, and Fitzgerald 2003). In India the price of electricity is subsidized heavily, reflecting the government's commitment to productive uses of electricity. In villages connected to the central grid, farmers have over time replaced diesel engines or purchased new electric-powered pump sets, resulting in a higher percentage of farmland being irrigated. Irrigation in turn is associated with both high cropping intensity (or farming the same land more than one season a year), and use of agricultural inputs such as hybrid seeds, fertilizers, and pesticides. The improvements in irrigation, multiple cropping, and modern agricultural inputs lead directly to higher agricultural productivity.[1]

This relationship between rural electrification and agricultural productivity in India has some key qualifications. Some regions benefit more than others. Interestingly, traditional forms of irrigation are not associated with the use of agricultural innovations or higher crop yields. In fact, the growth of pumps is slowest in regions that already have high levels of traditional irrigation. The areas that benefit the most from rural electrification are not the relatively rich rice regions with heavy rainfall, but the dryer and in many cases poorer parts of India with sufficient but relatively underutilized groundwater resources. Also, in terms of irrigation, except for greater pollution diesel pumps generate about the same outcomes as electric pumps, so that it is possible that many of the same benefits could be accomplished through the use of diesel. The exception is that deep tubewells require an electric submersible pump, and irrigation would not be possible without electricity under such circumstances.

While India's effort to improve rural development through electrification has been relatively successful, the caveat is that due to high electricity subsidies the power companies have come under financial pressure. However, even without the subsidies, the impact on agriculture in India would be significant due to the nature of India's climate and geography. In other countries there have been some mixed results that might be caused by either the policy environment or local land-use patterns. For instance, one survey in a relatively rich rice-growing region in the Philippines found that the rate of growth of pump sets was very low since most irrigation was accomplished through gravity-fed sources (World Bank 2002a). This finding is similar to that of India's rice-growing regions. Thus, while the productive impact of rural electrification can be substantial, it depends on factors such as government policy and complementary programmes.

Businesses in rural areas of developing countries include small commercial shops, grain mills, sawmills, and brickworks. Once again the impact of rural electrification on small businesses is to some extent determined by the nature of the local community, complementary programmes, and the abilities of rural entrepreneurs. Although electricity is an important input that helps in the development of small rural industries, the other complementary conditions include access to good rural markets and adequate credit. If these complementary conditions are not present, the anticipated growth of industries from electrification may be somewhat slow in some areas (Zomers 2001). But areas without electricity have an even worse record of business development. Thus the conclusion is that electrification is an important condition for the development of rural businesses, but it cannot produce an explosion of industry and commerce in the absence of other complementary development programmes.

---

[1] In one interesting case in Kenya, farmers increased their use of tractors alongside the access to electric welding facilities necessary for tractor repair (Kirubi 2009).

3.1.3. Household income

Businesses in rural areas of developing countries include home businesses, small commercial shops, grain mills, sawmills, coffee and tea processing, and brick kilns (for a review, see Cabraal et al. 2005). The effect of rural electrification on small businesses and farm income is determined by the nature of the local community, complementary programmes, and the ability of rural entrepreneurs. In South Africa, Dinkelman (2011) examined the impacts of the massive electrification rollout programme between 1996 and 2001 on employment growth among rural communities and found using several data sources that female employment grew up to 35 per cent as a result of electrification. Apparently, electric labour-saving effects in the home allowed women to participate in the labour market and microenterprise activities.

Similar impacts on women's employment were also found by Grogan and Sadanand (2012) in their study of electrification in Nicaragua using nationally representative LSMS data of 2005. They found that electrification increased women's employment by 23 per cent, with no significant effect on male employment. Kumar and Rauniyar (2011) also found a very large income effect in their study of rural electrification in Bhutan—a rise of 60–70 per cent in non-farm income as a result of rural electrification, which materialized through greater productivity in microenterprises and home-based businesses.

Recently several studies have revealed that household income increases as a consequence of the adoption of electricity (Table 3.1). But again, the impact of electricity varies depending on the local situation and sources of electricity. Four of the most recent studies demonstrate that impacts on both farm and non-farm income can vary significantly due to the way electricity is used by households. In all four countries, there is an increase in non-farm income for households that have adopted electricity. All these studies used econometric techniques that control for the proclivity of households with higher incomes to adopt electricity. In a study that examined existing businesses before and after having electricity in Kenya, it was found in one village that for many small businesses gross revenues increased between 25 per cent and 100 per cent (Kirubi et al. 2009).

Improved business income for households with electricity actually may have been caused by the tendency of households to establish businesses once electricity became available in their region (Figure 3.5). For instance, in a study in the

Table 3.1. Rural electrification impacts: improvements in household income due to electricity (per cent change)

Country and type of electricity	Farm income	Non-farm income	Total income
Bangladesh (grid)[1]	31.3	35.3	21.2
India (grid)[2]	0	68.8	38.6
Vietnam (grid)[3]	0	27.5	28.0
Nepal (micro-hydro)[4]	0	11.2	0

[1]Khandker, Barnes, and Samad 2012; [2]Khandker et al. 2012; [3]Khandker, Barnes, and Samad 2013; [4]Banerjee, Singh, and Samad 2011.

Note: Figures show per cent changes in income due to household access to electricity. Zero impact means that the calculated benefits are not statistically significant.

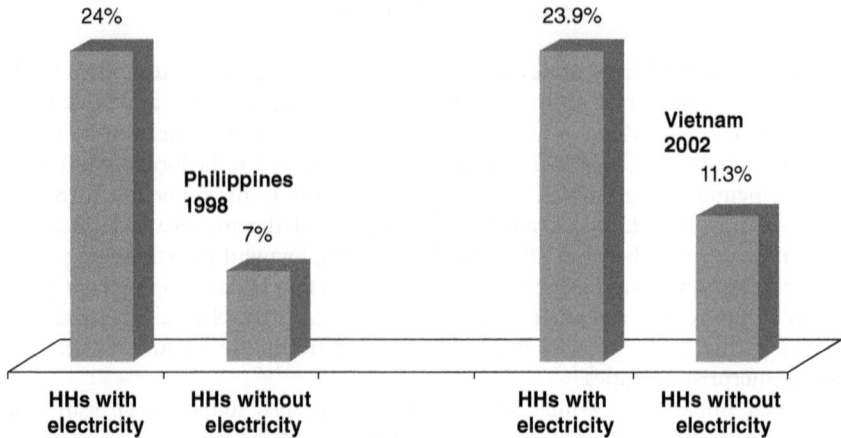

Figure 3.5. Percentage of households (HHs) with business income for households with and without electricity: Philippines 1998 and Vietnam 2002

Source: World Bank 2002a; IOS 2005

Philippines, small businesses were found to be more active in areas with electricity, contributing to family income (World Bank 2002a). The majority of these businesses are small general stores for food and other necessities, often run out of the home.

A similar finding was also discovered in Vietnam, where previously established household income improved with the adoption of electricity. Although not directly related to income, households with electricity in Vietnam are also more likely to establish a home business (Khandker, Barnes, and Samad 2013). The majority of households in both countries still do not have a business, but there may be other paths through which electricity improves household income, as indicated in Figure 3.2.

3.1.4. Energy poverty

About ten years ago energy poverty was measured by access to electricity. More recently the United Nations and the United Kingdom's Department for International Development (DFID) have broadened definitions of energy poverty to encompass multiple indicators, using somewhat arbitrary weights. The International Energy Agency (IEA) has never actually defined energy poverty (other than noting its relation to lack of access to modern energy) but has advocated that better ways of using biomass energy for cooking should be an important policy objective for household energy. Also, most international organizations measure energy poverty indicators as outputs (for example, lack of electricity connections) rather than outcomes (welfare gains from electricity consumption). Thus, unlike income poverty—which is usually based on measures of the minimum consumption of food and non-food items necessary to sustain life—energy poverty lacks a solid theoretical basis.

Table 3.2. Rural electrification impacts: improvements in expenditures (per cent) and poverty rates (percentage points)

Country and type of electricity	Expenditure	Moderate poverty
Bangladesh (grid)[1]	11.3	13.3
India (grid)[2]	18.0	13.3
Vietnam (grid)[3]	22.7	—
Nepal (micro-hydro)[4]	0	—
Bangladesh (SHS)[5]	0	—

[1]Khandker, Barnes, and Samad 2012; [2]Khandker et al. 2012; [3]Khandker, Barnes, and Samad 2013; [4]Banerjee, Singh, and Samad 2011; [5]Samad et al. 2013

Note: Figures show changes in expenditure (per cent) and poverty (percentage points) due to household access to electricity. Zero impact means that the calculated benefits are not statistically significant. The dashes mean that impact estimation was not carried out due to limited data. SHS = solar home system.

In this section we examine the relationship between rural electrification and expenditure/income poverty. We look at recent studies in Bangladesh, India, Vietnam, and Nepal. A recently conducted impact evaluation of solar home systems (SHS) programme in rural Bangladesh operated by the Infrastructure Development Company Limited (IDCOL) is also reported. This programme was responsible for promoting 1.5 million solar home systems in the last decade. Two of the studies are for off-grid electricity provision and three are for grid electricity.

All three rural grid-electrification programmes had an impact on expenditures (a proxy for income), but the two off-grid programmes had no impact (Table 3.2). In addition, the programmes in Bangladesh and India reduced poverty rates by 13 percentage points. These findings have interesting policy implications because it appears that the smaller off-grid electricity systems may have a significant impact on social outcomes, but are not as important for stimulating improvements in income (as measured by expenditure). The reason for this might be that grid electricity provides higher levels of power with a diverse set of uses, compared to the limited electricity services offered by micro-hydro or solar home systems.

At this stage, the finding that grid electricity is more important than off-grid systems for increasing incomes is only suggestive, and more research is necessary before reaching any conclusions. But this at least identifies the importance of measuring the socio-economic impact of off-grid electricity systems. Today, there are many claims that off-grid electricity might be an adequate substitute for grid service, but more research is needed to substantiate such claims.

3.1.5. Education

Most of the early studies of the relationship between rural electrification and education simply indicate that there is a positive association between the two (Madigan et al. 1976; Saunders et al. 1975: 111). In the past the problem was determining if electricity was prompting higher education levels, or if households with higher education and higher incomes were the ones adopting electricity. In addition, rural schools might possibly be located in communities with electricity rather than without electricity. Beginning in the 1980s studies in India, Colombia, and Indonesia (Barnes 1988) began to show that having electricity was related to literacy and school attendance even after controlling for household income and distance to schools. In Colombia, a study indicated that the education of heads of households was higher for families with electricity even after controlling for the level of family income (Velez et al. 1983). Whether by providing lights for reading or by providing a conducive environment for locating community schools, rural electrification and education were associated even after controlling for several important intervening variables.

The evidence is beginning to accumulate regarding the strong positive relationship between rural electrification and education. A household survey conducted in 2010 in Bhutan found that rural electrification increased children's study time in the evening by 10 minutes and grade completion by three-quarters of a year (Kumar and Rauniyar 2011). A recent study of county-level electrification (made possible by a hydropower dam) in Brazil suggests that counties achieving full electrification saw a drop of 22 per cent in illiteracy, a 19 per cent reduction in the population with less than four years of education, and an increase of 1.2 years in schooling completion (Lipscomb, Mobarak, and Barham 2013). Further, in the case of the Philippines (World Bank 2002a), in households with electricity women were found to spend more hours reading (Figure 3.6).

In households without electricity, women simply do not read—at any level of household income. In households with electricity, even poor women read more than do women in wealthier households without electricity. The evidence is clear that as income rises the level of reading also rises. These results are compelling because many women at lower levels of income cannot read, which brings down the average reading hours for low-income households with electricity.

Building on this evidence, some recent studies using advanced statistical techniques have established the causality between electrification and educational outcomes by controlling for the possibility that those with higher levels of education are more likely to adopt electricity (Table 3.3). In Bangladesh both boys and girls in households with electricity spend more time studying and had higher levels of grades completed in school. Also, it does not appear that the source of electricity mattered, as children from households with decentralized

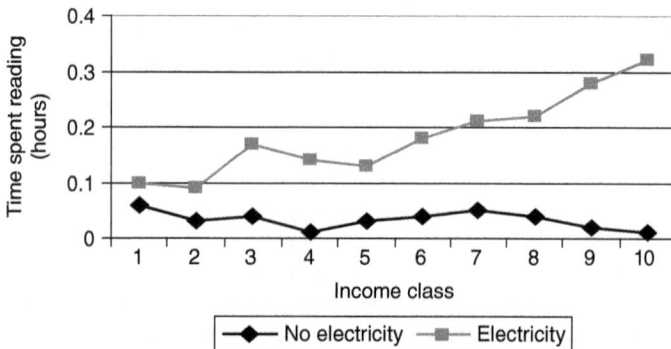

Figure 3.6. Impact of electricity on women's time spent reading: Philippines, 1998

Source: World Bank 2002a

Table 3.3. Rural electrification impacts on education outcomes

Country and type of electricity	Study time in the evening (minutes/day)		School enrollment (per cent)		Grade completion (years)	
	Boys	Girls	Boys	Girls	Boys	Girls
Bangladesh (grid)[1]	21.9	12.3	—	—	0.23	0.16
India (grid)[2]	11.6	13.5	6.0	7.4	0.28	0.49
Vietnam (grid)[3]	—	—	6.3	9.0	0.12	0
Nepal (micro-hydro)[4]	7.7	12.0	—	—	0	9.24
Bangladesh (SHS)[5]	7.0	8.2	—	—	—	—

[1]Khandker, Barnes, and Samad 2012; [2]Khandker et al. 2013; [3]Khandker, Barnes, and Samad 2014; [4]Banerjee, Singh, and Samad 2011; [5]Samad et al. 2013

Note: Figures show changes in outcomes for boys and girls of 5–18 due to household access to electricity. SHS = solar home system.

sources of electricity in Bangladesh (solar home systems) and Nepal (community micro-hydro systems) also studied longer than those from households without any form of electricity. The implication is that electrification indeed contributes to better education.

In conclusion, a growing number of studies confirm that having electricity from either the grid or decentralized systems has a long-term impact on educational levels in developing countries. Of course, it is necessary to have complementary investments in schools, teachers, and teaching materials to make this impact even greater (World Bank 1999; Tanguy 2012).

3.1.6. Equity impact

The average impacts reaped for all households with electricity (see Tables 3.2 and 3.3) may not indicate whether poor households benefit more or less from electrification compared to more wealthy ones. This may be critical from a policy-making perspective because critics may argue that rural electrification projects do not benefit the poor very much, and so funds might be better allocated to other projects with higher impacts on the poor. In a further analysis for the same countries, rural electrification impacts for various percentiles (quantiles) of income and expenditure have been examined (Table 3.4). Although the magnitude of the impact varies across the countries, the overall trend is obvious: richer households tend to benefit more than poorer ones. Such findings are not surprising since richer households consume electricity in multiple ways (lighting, extended business operations, productive tools, TVs, computers, and so on), while the electricity consumption of the poor is limited mainly to lighting.

Policymakers might need to revisit the details of electrification intervention to ensure more equitable benefits. One way to accomplish this might be to make credit and appliances available in poor areas, or to find ways to provide incentives to poor households to take more advantage of the benefits of electricity. But it should be kept in mind that poor households definitely benefit from having electricity, though—considering the expense—purchasing appliances may reap less benefits than those experienced by their wealthier counterparts.

Table 3.4. Rural electrification impacts: improvements in income and expenditure by quantile (per cent)

Income/expenditure quantile	Bangladesh[1]		India[2]		Vietnam[3]	
	Income	Expenditure	Income	Expenditure	Income	Expenditure
15th	12.4	0	25.9	0	0	0
25th	10.5	9.1	29.7	0	0	0
50th	15.1	13.6	36.1	16.2	29.8	0
75th	21.5	13.8	40.4	25.3	36.1	38.1
85th	23.9	13.5	45.7	30.2	40.6	0

[1]Khandker, Barnes and Samad 2012; [2]Khandker et al. 2012; [3]Khandker, Barnes and Samad 2013

Note: Figures show per cent changes in income and expenditure due to household access to electricity. Zero impact means that the calculated benefits are not statistically significant.

3.2. BENEFITS OF CLEAN COOKING AND HEATING

Open fires and primitive stoves have been used for cooking since the beginning of human history. They have come in various sizes and styles, having been adapted to myriad cultures and food preparation methods. As society has progressed, more sophisticated stove models have been developed. Today's modern kitchens reflect many types of standardized and specialized cooking devices available, from coffee- and teapots to toasters and gas cooktops.

In many developing countries, the poor still burn biomass energy to meet their basic household cooking needs. These open fires are fairly inefficient at converting energy into heat for cooking; the amount of biomass fuel needed each year for basic cooking can reach up to 2 tons per family. In addition, collecting this fuel can take an hour a day on average. Furthermore, these open fires and primitive cookstoves emit a significant amount of smoke which fills the home; this indoor cooking smoke has been associated with a number of diseases, the most serious of which are chronic and acute respiratory illnesses such as bronchitis and pneumonia. We will later revisit the issues related to cooking with biomass.

Recent impact studies on improving cooking and heating practices have had a narrow focus on health benefits. In the 1990s there was more research on social or economic benefits, but recently that interest has levelled off with more focus on health benefits. Actually, much more research is necessary on all aspects of the various impacts of better household energy use and how this relates to poverty alleviation.

3.2.1. Health

Each year, indoor air pollution, caused mainly by smoke emitted by traditional biomass cooking stoves, accounts for an estimated 3.7 premature million deaths and 4.6 per cent of the burden of disease (Smith et al. 2014), making it the second leading cause of disease behind smoking. A regional breakdown of fatalities from indoor air pollution indicates that the situation is extremely bad in East Asia and South Asia. Women and children are disproportionately affected; their exposure to small particulates is 10 to 20 times higher than the maximum acceptable levels recommended by the WHO (WHO 2006). The WHO has estimated the benefits (including time savings and avoided health costs) of having one half of traditional fuel-using populations switch to LPG (liquefied petroleum gas) or improved stoves between now and 2015: the annual benefits are estimated at about $90 billion for the switch to LPG and $105 billion for improved stoves. The benefit–costs ratios are generally above 3 for LPG and about 10 times higher for improved stoves due to their lower costs (Hutton et al. 2006).

This pervasive use of solid fuels—including wood, coal, straw, and dung—and traditional cookstoves results in high levels of household air pollution, the extensive daily drudgery required to collect fuels, and serious health impacts. In a recent study of the health benefits of improved stoves in rural Madagascar, Dasgupta, Martin, and Samad (2013) found that use of ethanol stoves lowers household members' exposure to harmful $PM_{2.5}$ concentration by over 50 per cent and to CO by 90 per cent. The same study also showed that improved wood stoves reduce concentration of CO by 69.1 per cent over traditional charcoal stoves. Typically, LPG, kerosene, electricity, and biomass energy used in less polluting stoves also result in lower levels of pollution, drudgery, and time spent collecting fuel for cooking or heating. The main studies addressing such issues involve household fuel use and its impact for gender, poverty, and the environment.

3.2.2. Energy policies and household fuel use

Energy policies have a significant impact on the fuels that households use for cooking and heating. Energy transition is an ongoing process that involves movement from the inefficient use of traditional energy to the efficient use of modern fuels for cooking, heating, lighting, and other purposes. It can be said with some confidence that in the long term people will switch away from the inefficient ways of using biomass fuels and will use energy for a much wider range of services than they do now. But the long-term solution to these problems cannot be forced indiscriminately on to countries or cities regardless of their stage in energy transition. These problems include stress on wood resources around some urban areas, low standards of energy services, high prices of wood, and poor markets in modern fuels.

Rather than switching completely to new fuels and devices, households prefer to 'stack' them for increased flexibility (Figure 3.7). This is seen in Mexico both before and after rural households adopted *patsari* stoves. LPG is preferred for reheating food and cooking meals that require little time, *patsari* cookstoves are preferred for tortilla making and other cooking tasks, and open fires continue to be used for water heating (Wang et al. 2012).

While subsidies can assist the poor to transition to more modern forms of cooking and heating, the situation is complicated by household decisions regarding their energy use. Subsidies can cause positive outcomes, but at other times there can be negative consequences (see Table 3.5). Often, not only the poor but also the middle-class and the rich get the subsidy. In Indonesia, for example, in the past the subsidization of kerosene did help the poor, but it was an unnecessary boon to higher-income consumers and kept the middle class from switching to

Diversification of cooking tasks

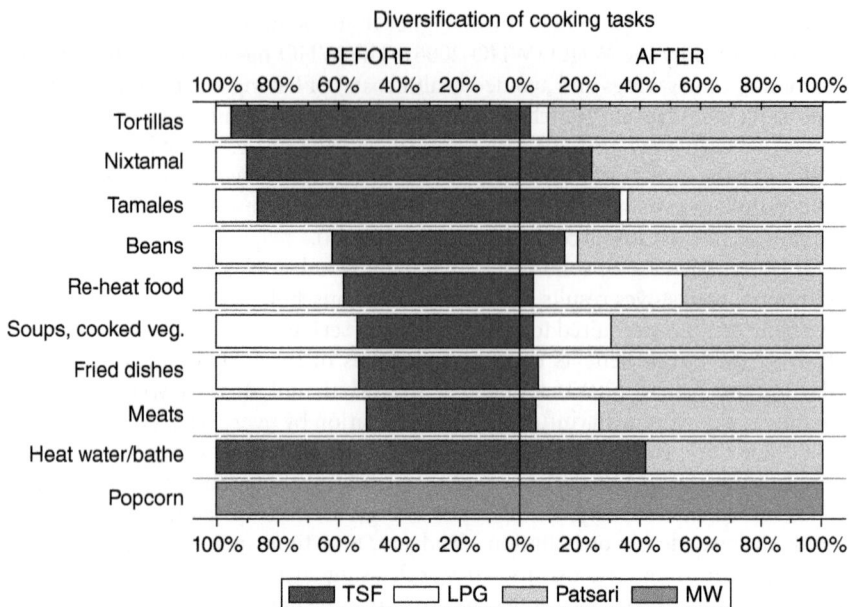

Figure 3.7. Fuels and appliances used before and after adopting *patsari* stoves in Mexico

Note: TSF = three-stone fire, LPG = liquefied petroleum gas, MW = microwave.

Source: World Bank 2012

Table 3.5. Impact of energy pricing and supply policies on rural communities and the urban poor

Supply policy	Energy pricing policy		
	Subsidized prices	Market prices	Fuel taxation
Limited or targeted supply	Subsidy is redirected away from poor to other groups.	Higher-income groups are served first.	Traditional fuel prices are unaffected by those of alternatives.
	Rural and poor people lack access to fuel.	Rural and poor people lack access to fuel.	Rural and poor people lack access to fuel.
Unlimited or untargeted supply	Modern fuel subsidies mean lower prices for traditional fuels.	Traditional fuel prices are capped at price of alternatives.	First costs of service, along with fuel costs, constrain poor from purchasing fuel.
	Rural and poor people can access service, but it is fiscally unsustainable.	Rural and poor people can access service, and it is fiscally sustainable.	Rural and poor people can access service, but it is expensive.
	Poor people benefit from lower fuel prices, but other income groups benefit more.	First costs of service constrain poor people from purchasing modern fuels.	Traditional fuel prices are often high because of higher-priced alternative fuels.

Source: Barnes and Toman 2004

superior fuels such as LPG. Subsidies to household fuels such as kerosene often wind up being diverted to other markets, including transportation. In Pakistan the government had subsidized kerosene to assist the poor, but much of the subsidized kerosene was diverted away from households to the transportation sector.

The main point is that the energy policies to encourage switching to cleaner methods of cooking and heating sometimes involve unintended consequences. More research is necessary to understand these issues, to achieve the right balance between incentives to promote cleaner cooking and heating, and to achieve long-term use of both stoves and fuels such as LPG. This may involve a combination of trying new programmes in conjunction with supportive policies that enable poor people to afford better methods of cooking and heating.

3.2.3. Gender dimensions

The use of solid fuels and traditional cooking methods has a disproportionate impact on the health and drudgery of women; women are affected in two ways. First, the burden of fuelwood collection most often falls to women (and to a lesser extent to children). This sometimes entails walking long distances and carrying heavy loads, thereby enduring some safety hazards. The time spent collecting fuel can be as high as one hour per day (World Bank 2002b; Kumar and Hotchkiss 1988), time which could be better used for more productive activities. For instance, in a national survey of over 40,000 households in India, it was found that both women and men collect firewood, mostly in rural areas. For those who collect from their own village, women spend about 18 hours per month collecting fuelwood, and men pitch in another 8 hours for a total of about 26 hours (Figure 3.8). The transition from traditional cookstoves to improved ones can reduce the drudgery, as shown in Figure 3.8. Continuous collection of biomass this way can also lead to a gradual deterioration of the local environment and depletion of biomass supplies, meaning even longer walks and greater drudgery. Second, women, again along with children, spend many hours in indoor cooking areas, inhaling smoke from the incomplete burning of biomass fuels in inefficient stoves. As a result, they are more vulnerable to the health risks associated with indoor air pollution.

Although women have cooked with traditional biomass stoves for thousands of years, it is only in recent years that the accompanying health risks and work burden have been fully understood. The past 'invisibility' of this issue to policymakers explains, in part, why governments and international organizations previously failed to assign this issue the priority it deserves. Much of the time and energy women spend on domestic tasks remains invisible to policymakers since nonmarket productive work is not counted in economic statistics or national accounts (Charmes 2006).

Even at present, women's time often is not considered a priority by policymakers, who are focused on large-scale energy projects, in part because women are at the lowest level of paid and unpaid work. In formulating policies and investment priorities, energy policymakers should take into account women's nonmarket productive work. Clearly, the lives of women—especially poor women—and children stand to benefit significantly from energy programmes that account for the output

Douglas F. Barnes et al.

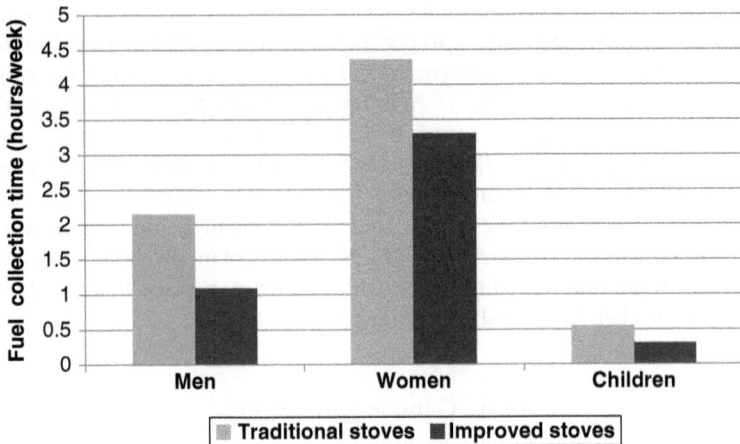

Figure 3.8. Fuel collection time of men, women, and children by stove type in rural India, 2005

Source: India Human Development Survey 2005

produced by women and the burdens they bear in the production process. But it must be acknowledged that the relationship between interventions and changed exposure to cooking smoke is complex. For instance, in China it was found that in cold-climate regions the type of stove had more of an impact on exposure to pollution than the programmes that attempted to change cooking behaviour (Baris and Ezzati 2006). The conclusion was that, to have an impact on lower exposure to pollution levels, training in fire-tending techniques needed to be coupled with the use of different stoves. At the close of India's improved stoves programme, results of a 2000 research project showed that satisfaction among women users in Maharashtra state (see Figure 3.9) was largely attributable to the initiative and sustained efforts of the Appropriate Rural Technology Institute (ARTI), the organization that provided technical support for the national programme. Women users appreciated that the stoves were developed by traditional potters. About half of users—most of whom would have otherwise purchased fuelwood for lack of crop residues—reported fuel savings as a major benefit. Users in households where the kitchen was the innermost room valued the efficient smoke removal. Many viewed time savings as a benefit, since two pots could be used at once. They also valued the cleaner cooking environment, especially in villages where sugarcane root was the primary cooking fuel. Some recognized the link between less indoor cooking smoke and better health. Women users also perceived drawbacks to using the improved stoves. The problems reported centred mainly on the chimney (leakage and need of frequent cleaning), inappropriate pothole size, inconvenient grates, and more fuel.

Solutions for reducing the burden on women are likely to involve smaller, demand-side interventions and investments different from the more visible large-scale energy projects. The strategies for small-scale electricity goods and services such as solar home systems and lamps are similar to those that involve the development and implementation of clean cooking solutions. Depending on the

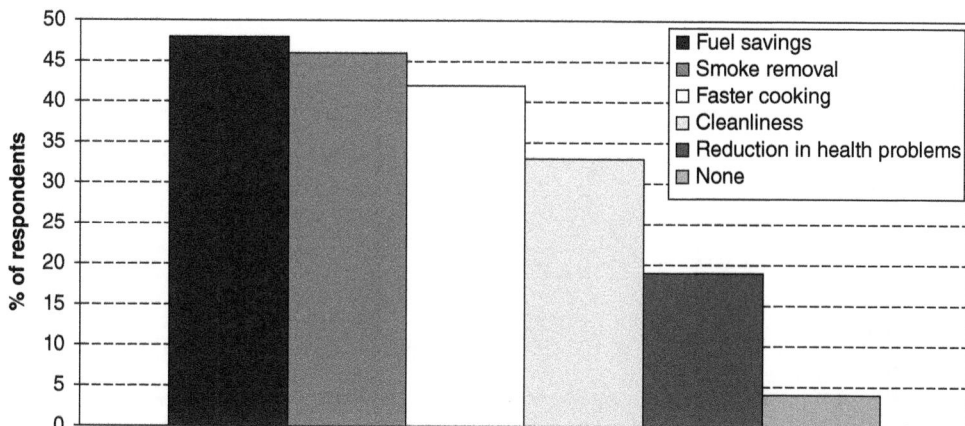

Figure 3.9. Perceived user benefits of improved stoves in Maharashtra, India, 2000 (per cent)

Source: Barnes, Kumar, and Openshaw 2012

possibilities within countries, it may be possible to merge these similar agendas to achieve the scale necessary to have larger and more visible investments.

3.2.4. Poverty

Overall, the use of solid fuel and biomass is closely intertwined with poverty. Based on the cross-sectional data for 2010, the general pattern is that the use of biomass fuels declines at a far lower rate than the increase in income levels (Figure 3.10). More specifically, the proportion of households using biomass for cooking declines by approximately 1.6 per cent for every 10 per cent of income growth. This means that those using biomass energy generally are poor, and the question is how having access to more modern stoves or more modern forms of energy might have an impact on poverty. Research evidence on this topic (see Barnes, Krutilla, and Hyde 2005) suggests that energy policies contribute extensively to the transition from traditional to modern ways of cooking. As a consequence, there are ways beyond improving income to impact significantly the shape of energy transitions, and this would include changing pricing policies, developing better stoves, encouraging market development, and providing incentives for technological innovation in energy appliances for the poor.

3.2.5. Environment

The environmental consequences of biomass use, first put before the international community several decades ago as the 'other energy crisis', involve indoor air pollution and degradation of local commons (Eckholm 1975). Fuel collection leads to a deterioration of the local environment and depletion of biomass, meaning ever-longer walks to collect fuels. In India, the time spent collecting

Douglas F. Barnes et al.

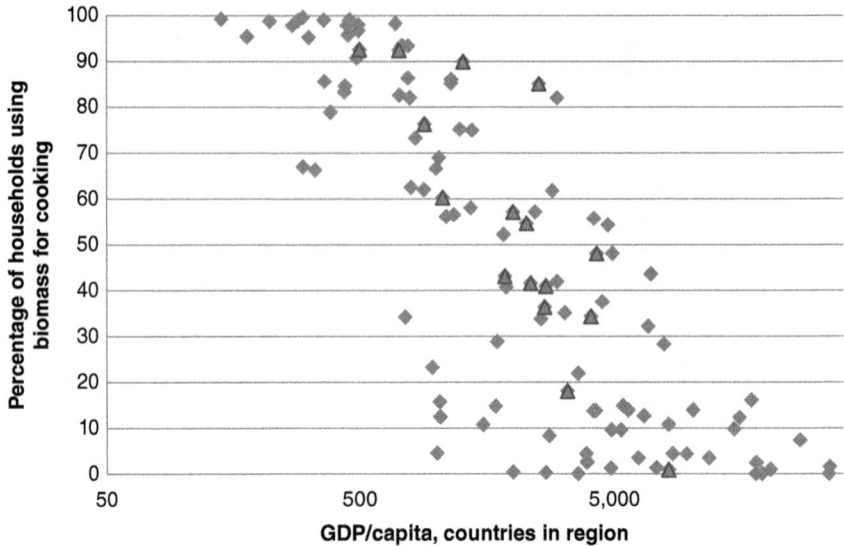

Figure 3.10. Household biomass energy use vs. GDP in developing countries, 2007 (triangles = countries in East Asia and Pacific; diamonds = all other countries)

Source: WHO and UNDP 2009

fuel per household is estimated at nearly one hour per day (World Bank 2004; Heltberg, Arndt, and Sekhar 2000). In Haiti, the overall decline in forested areas resulting from charcoal production for urban use is well documented (Lewis and Coffey 1985; Stevenson 1989). There also is evidence that biomass fuels burned in traditional ways contribute to a buildup of greenhouse gases (Venkataraman et al. 2010), as well as other climate risk forcers including black carbon, in the atmosphere (Ramanathan and Carmichael 2008; World Bank 2011b). Consequently, although more research is necessary on this important topic, it appears that household use of modern energy and efficient appliances for heating homes and cooking food can have significant benefits for the environment.

3.3. CONCLUSION

Once a better understanding of energy poverty is achieved, pro-poor policies that influence energy access and pricing of modern energy services can be implemented to reduce energy poverty. Although income is a key factor, energy policies and access to higher quality energy services also matter, especially for poor households (O'Sullivan and Barnes 2006). The impact of electricity on development outcomes recently has taken a step forward with more sophisticated studies that address such issues as impact on education, income, and business development. But more work is needed on how these development impacts affect the level of energy poverty in developing countries.

To summarize, several possible channels exist through which increased higher-quality energy and better use of existing fuels could disproportionately affect economic development (Barnes and Toman 2004):

- Reallocation of household time (especially that of women) away from energy provision and toward improved education, income generation, and greater specialization of economic functions.
- Economies of scale in more industrial-type energy provision.
- Greater flexibility in time allocation through the day and evening.
- Enhanced productivity of businesses and farms.
- Improvements in education.
- With more flexible and reliable as well as more plentiful energy, greater ability to use a more efficient capital stock and take advantage of new technologies.
- Lower transportation and communication costs—greater market size and access, more access to information (combined result of energy and other infrastructure).
- Health-related benefits: reduced smoke exposure, clean water, refrigeration (direct benefits and higher productivity).

More efficient use of traditional energy for cooking and heating seems as important in moving people out of energy poverty as the use of electricity, but unfortunately has garnered less attention. Although many health studies have been completed, the more general impacts of both using better stoves and switching to cleaner cooking practices such as adopting LPG—including avoided household drudgery, less time collecting fuel, shorter meal preparation times, and impacts on women and children—have not been the focus of many studies. More research on the impacts is necessary to examine the whole range of benefits of better stoves and the use of more modern fuels such as LPG for cooking. While improved and equitable access to electricity is necessary, policies for alleviating energy poverty should be based on understanding their actual development outcomes.

REFERENCES

Asaduzzaman, M., Barnes, D., and Khandker, S. (2009). *Restoring Balance: Bangladesh's Rural Energy Realities*. Energy Sector Management Assistance Programme, Special Report 006/09. Washington, DC: World Bank.

Banerjee, S., Singh, A., and Samad, H. (2011). *Power and People: The Benefits of Renewable Energy in Nepal*. Washington, DC: World Bank.

Barakat, A., et al. (2002). 'Economic and Social Impact Evaluation Study of the Rural Electrification Program in Bangladesh'. Report submitted to the National Rural Electric Cooperative Association International, Dhaka.

Baris, E. and Ezzati, M. (2006). *Sustainable and Efficient Energy Use to Alleviate Indoor Air Pollution in Poor Rural Areas in China*. Washington, DC: World Bank.

Barnes, D. (1988). *Electric Power for Rural Growth: How Electricity Affects Rural Life in Developing Countries.* Boulder, CO: Westview.

Barnes, D., Khandker, S., and Samad, H. (2011). 'Energy Poverty in Rural Bangladesh', *Energy Policy* 39: 894–904.

Barnes, D., Krutilla, K., and Hyde, W. (2005). *The Urban Household Energy Transition: Social and Environmental Impacts in the Developing World.* Washington, DC: RFF Press.

Barnes, D., Kumar, P., and Openshaw, K. (2012). *Cleaner Hearths, Better Homes—New Stoves for India and the Developing World.* New Delhi: Oxford University Press.

Barnes, D., Peskin, H., and Fitzgerald, K. (2003). 'The Benefits of Rural Electrification in India: Implications for Education, Household Lighting, and Irrigation'. Draft paper prepared for South Asia Energy and Infrastructure. Washington, DC: World Bank.

Barnes, D. and Toman, M. (2004). 'Energy, Equity and International Development'. Working paper, Initiative for Policy Dialogue. New York: Columbia University.

Cabraal, R., Barnes, D., and Agarwal, S. (2005). 'Productive Uses of Energy for Rural Development', *Annual Review of Environment and Resources* 30: 117–44.

Charmes, J. (2006). 'A Review of Empirical Evidence on Time Use in Africa from UN-Sponsored Surveys', in C.M. Blackden and Q. Wodon, eds, *Time Use and Poverty in Sub-Saharan Africa*, Washington, DC: World Bank, pp. 39–72.

Dasgupta, S., Martin, P., and Samad, H. (2013). 'Addressing Household Air Pollution: A Case Study in Rural Madagascar', Policy Research Working Paper, Washington, DC: World Bank.

Dinkelman, T. (2011). 'The Effects of Rural Electrification on Employment: New Evidence from South Africa', *American Economic Review* 101(7): 3078–108.

Eckholm, E. (1975). 'The Other Energy Crisis: Firewood', Worldwatch Paper 1, Washington, DC: Worldwatch Institute.

Filmer, D. and Pritchett, L. (1998). *The Effect of Household Wealth on Educational Attainment around the World: Demographic and Health Survey Evidence.* Washington, DC: World Bank.

Grogan, L. and Sadanand, A. (2012). 'Rural Electrification and Employment in Poor Countries: Evidence from Nicaragua', *World Development* 43: 252–65.

Haughton, J. and Khandker, S. (2009). *Handbook on Poverty and Equality.* Washington, DC: World Bank.

Heltberg, R., Arndt, T., and Sekhar, U. (2000). 'Fuelwood Consumption and Forest Degradation: A Household Model for Domestic Energy Substitution in Rural India', *Land Economics* 76(2): 213–32.

Hutton, G., Rehfeus, E., Tediosi, F., and Weiss, S. (2006). *Evaluation of the Costs and Benefits of Household Energy and Health Interventions at Global and Regional Levels.* Geneva: World Health Organization.

IOS (Institute of Sociology). (2005). 'Rural Electrification Impact Study and Analysis'. Photocopy. Hanoi: Vietnam Academy of Social Sciences.

Khandker, S. (1996). 'Education Achievements and School Efficiency in Rural Bangladesh', World Bank Discussion Paper no. 319, Washington, DC: World Bank.

Khandker, S., Barnes, D., and Samad, H. (2012). 'The Welfare Impacts of Rural Electrification in Bangladesh', *Energy Journal* 33(1): 199–218.

Khandker, S., Barnes, D., and Samad, H. (2013). 'Welfare Impacts of Rural Electrification: A Panel Data Analysis from Vietnam', *Economic Development and Cultural Change* 61 (3): 659–92.

Khandker, S. H. Samad, R. Ali, D. Barnes. (2012). Who Benefits Most from Rural Electrification. Policy Research Working Paper No. 6095, World Bank Washington DC.

Kirubi, C., et al. (2009). 'Community-Based Electric Micro-Grids can Contribute to Rural Development: Evidence from Kenya', *World Development* 37(7): 1208–21.

Kulkarni, V. and Barnes, D. (2004). 'The Impact of Electrification on School Participation in Rural Nicaragua', Working Paper, College Park, MD: University of Maryland.

Kumar, S. and Hotchkiss, D. (1988). 'Consequences of Deforestation for Women's Time Allocation, Agricultural Production, and Nutrition in Hill Areas of Nepal', International Food Policy Research Institute Research Report 69, Washington, DC.

Kumar, S. and Rauniyar, G. (2011). 'Is Electrification Welfare Improving? Non-Experimental Evidence from Rural Bhutan', MPRA Paper No. 31482, Asian Development Bank.

Lewis, L. and Coffey, W.J. (1985). 'The Continuing Deforestation of Haiti', *Ambio* 14(3): 158–60.

Lighting Africa [website]. (2008). <http://www.lightingafrica.org/> (accessed September 25, 2013).

Lipscomb, M., Mobarak, M., and Barham, T. (2013). 'Development Effects of Electrification: Evidence from the Topographic Placement of Hydropower Plants in Brazil', *American Economic Journal: Applied Economics* 5(2): 200–31.

Madigan, F., Herrin, A., and Mulcahy, W. (1976). *Evaluation Study of the MISAMIS Oriental Rural Electric Service Cooperative (MORESCO)*. Report. Washington, DC: US Agency for International Development.

Mayer-Tasch, L., Mukherjee, M., and Reiche, K. (2013). *Productive Uses of Energy—PRODUSE*. Eschborn, Germany: GIZ. Also published by ESMAP, World Bank, Washington, DC.

Meier, P., Tuntivate, V., Barnes, D., Bogach, S., and Farchy, D. (2010). *Peru: National Survey of Rural Energy Use*. ESMAP Energy and Poverty Special Report, World Bank, Washington, DC.

Monari, L. and Mostefai, D. (2001). 'India: Power Supply to Agriculture Summary Report', Energy Sector Unit, South Asia Regional Office. New Delhi: World Bank.

Nieuwenhout, F., van de Rijt, P., and Wiggelinkhuizen, E. (1998). *Rural Lighting Services*. Petten: Netherlands Energy Research Foundation.

O'Sullivan, K. and Barnes, D. (2006). *Energy Policies and Multitopic Surveys: Guidelines for Questionnaire Design in Living Standards Measurement Studies*, World Bank Working Paper No. 90, Washington, DC: World Bank.

Pradhan, M. and Ravallion, M. (1998). 'Measuring Poverty Using Qualitative Perceptions of Welfare', Policy Research Working Paper No. 2011. Washington, DC: World Bank.

Ramanathan, V. and Carmichael, G. (2008). 'Global and Regional Climate Changes Due to Black Carbon', *Nature Geoscience* 1: 221–7.

Ravallion, M. (1998). 'Poverty Lines in Theory and Practice', Living Standard Measurement Study Working Paper 133. Washington, DC: World Bank.

Ravallion, M. and Bidani, B. (1994). 'How Robust Is a Poverty Profile?' *The World Bank Economic Review* 8(1): 75–102.

Roddis, S. (2000). 'Poverty Reduction and Energy: The Links between Electricity and Education'. Photocopy. Washington, DC: World Bank.

Samad, H., Khandker, S., Asaduzzaman M., and Yunus, M. (2013). 'The Benefits of Solar Home Systems: An Analysis from Bangladesh', Policy Research Working Paper 6724. Washington, DC: World Bank.

Saunders, J., Davis, M., Moses, G., and Ross, J. (1975). *Rural Electrification and Development: Social and Economic Impact in Costa Rica and Colombia*. Boulder, CO: Westview.

Smith, K., et al. (2014). 'Millions Dead: How Do We Know and What Does It Mean? Methods used in the Comparative Risk Assessment of Household Air Pollution'. Draft paper. Berkeley: University of California.

Stevenson, G. (1989). 'The Production, Distribution and Consumption of Fuelwood in Haiti', *Journal of Developing Areas* 24: 59–76.

Tanguy, B. (2012). 'Impact Analysis of Rural Electrification Projects in Sub-Saharan Africa', *World Bank Research Observer* 27(1).

UNDP (United Nations Development Programme). (2005). *United Nations Millennium Development Goals.* Washington, DC: World Bank. <http://www.un.org/millenniumgoals> (accessed 16 October 2013).

Van de Walle, D., Ravallion, M., Mendiratta, V., and Koolwal, G. (2013). 'Education Achievements and School Efficiency in Rural Bangladesh', Discussion Paper 319, Washington, DC: World Bank.

Van der Plas, R. and de Graaff, A. (1988). *A Comparison of Lamps for Domestic Lighting in Developing Countries.* World Bank Industry and Energy Department Working Paper (Energy Series) No. 6. Washington, DC: World Bank.

Velez, E., Becerra, C., and Carrasquilla, A. (1983). *Rural Electrification in Colombia.* Report by Instituto SER de Investigacion for Resources for the Future and the U.S. Agency for International Development. Washington, DC: Resources for the Future.

Venkataraman, C., Sagar, A., Habib, G., and Smith, K. (2010). 'The National Initiative for Advanced Biomass Cookstoves: The Benefits of Clean Combustion', *Energy for Sustainable Development* 14(2): 63–72.

Wang, X., et al. (2012). *What Have We Learned about Household Biomass Cooking in Central America?* ESMAP, Washington, DC: World Bank.

WHO (World Health Organisation). (2006). *Fuel for Life: Household Energy and Health.* Geneva: WHO.

WHO and UNDP. (2009). *The Energy Access Situation in Developing Countries: A Review Focusing on the Least Developed Countries and Sub-Saharan Africa.* New York: UNDP.

World Bank. (1999). *Poverty and Social Developments in Peru, 1994–1997.* World Bank Country Study. Washington, DC: World Bank.

World Bank. (2002a). *Rural Electrification and Development in the Philippines: Measuring the Social and Economic Benefits.* ESMAP Report no. 255/02. Washington, DC: World Bank.

World Bank. (2002b). *Energy Strategies for Rural India, Evidence from Six States.* ESMAP, Washington, DC: World Bank.

World Bank. (2004). *The Impact of Energy on Women's Lives in Rural India.* ESMAP Report no. 215/05. Washington, DC: World Bank.

World Bank. (2008). *The Welfare Impact of Rural Electrification: A Reassessment of the Costs and Benefits. Independent Evaluation Group Impact Evaluation.* Washington, DC: World Bank.

World Bank. (2011a). 'Vietnam's Rural Electrification Story: State and People, Central and Local, Working Together', Asia Sustainable and Alternative Energy Program (ASTAE), Washington, DC.

World Bank. (2011b). *Household Cookstoves, Environment, Health, and Climate Change: A New Look at an Old Problem.* Washington, DC: World Bank Environment Department.

Zomers, A. (2001). *Rural Electrification: Utilities' Chafe or Challenge?* Enschede, Netherlands: Twente University Press.

4

The World Bank's Perspective on Energy Access

Sudeshna Ghosh Banerjee, Mikul Bhatia, Elisa Portale,
Ruchi Soni, and Nicolina Angelou

Energy access is one of the most important development challenges of the twenty-first century. A large swathe of the world's population still remains in the dark, relying on kerosene and candles on a day-to-day basis or cooking with environmentally harmful wood fuels, excluded from the income-generating and human development outcomes achievable with energy access. There is now unambiguous evidence of the benefits of electricity provision. At the minimum, the quality of electrical lighting (measured in lumens) is many times better than that of kerosene—the most common alternative lighting fuel. In addition, students are able to study for longer hours, instances of respiratory and gastrointestinal diseases are reduced, the safety and security of women is heightened, and women's role in the household decision-making process is increased. The importance of clean cooking has even greater effects on women, who bear a disproportionate burden of traditional methods of fuel collection and transport. The smoke from inefficient cookstoves is a major contributor to indoor air pollution, and is responsible for 3 to 5 per cent of the global disease burden and about 1.3 million premature deaths annually around the world (Lim et al. 2012).[1]

The World Bank has supported energy access initiatives through lending (including grants), technical assistance activities, and making global expertise available to its clients. This commitment is now enshrined in the Energy Sector Directions Paper, which sets out to secure the affordable, reliable, and sustainable energy supply needed to support the World Bank Group's twin goals of ending extreme poverty and building shared prosperity. These overarching objectives are aligned with the Sustainable Energy for All (SE4ALL) initiative, which includes universal access to modern energy services by 2030 as one of its three goals. SE4ALL was launched in 2011 by the United Nations secretary-general. Recognizing the urgent action required to addressing global energy challenges; the World Bank Group has joined and is now co-leading the initiative.

---

[1] <http://www.who.int/healthinfo/global_burden_disease/GlobalHealthRisks_report_full.pdf>.

With the International Energy Agency (IEA), the World Bank recently co-led the production of a Global Tracking Framework (GTF) that measured the status of the three SE4ALL objectives at their starting point in 2010, against which future improvements can be measured. For the universal access goal, the report not only traces the historical evolution of access and presents its current status, but also proposes a new multi-tier framework for defining and measuring access to electricity and modern cooking solutions. Thus far, access to electricity has been typically measured using a binary metric (either having a connection or using electric lighting), while access to modern cooking solutions is measured based on the use of non-solid fuels.[2]

This chapter discusses these access trends in Section 4.1 and the multi-tier metric in Section 4.5. In addition, the chapter summarizes the lessons learned from the Bank's engagements in electrification and household energy projects.

4.1. GLOBAL TRENDS IN ACCESS

About 1.2 billion people or 17 per cent of the global population lived without electricity in 2010. The access deficit is overwhelmingly rural and located in Sub-Saharan Africa and the South Asian regions. In fact, Sub-Saharan Africa alone constitutes half of the access deficit. India is home to the largest unelectrified national population of more than 300 million people, followed far behind by Nigeria and Bangladesh. In fact, the access deficit is localized in a select group of countries. The top 20 countries with the largest unelectrified population contribute 74 per cent of the total access deficit.

Among the regions, the access rate is lowest in Oceania (where one of four people has electricity) and Sub-Saharan Africa (where one of three people has electricity). Electrification rates in urban areas are far ahead; only in Oceania and Sub-Saharan Africa do they remain lower than 90 per cent. Comparatively, in rural areas, the challenge is still stark in most of the developing world, except the Caucasus and Central and East Asia.

About 2.8 billion people or 41 per cent of the global population relied primarily on solid fuels in 2010. About 78 per cent of that population lived in rural areas, and 96 per cent were geographically concentrated in Sub-Saharan Africa, East Asia, South Asia, and South-east Asia. The situation is particularly dismal in rural areas—especially in Sub-Saharan Africa, where 94 per cent of the rural population relies on solid fuels. India and China have by far the largest populations using solid fuels, constituting close to half of the total access deficit. The top 20 countries with the largest solid-fuel-using populations contribute 85 per cent of the 2.8 billion people reliant on solid fuels for cooking.

---

[2] *Non-solid fuels* include: (i) liquid fuels such as kerosene, ethanol, or other biofuels; (ii) gaseous fuels such as natural gas, liquefied petroleum gas [LPG], and biogas; and (iii) electricity. *Solid fuels* include (i) traditional biomass, for example, wood, charcoal, agricultural residues, and dung; (ii) processed biomass, such as pellets and briquettes; and (iii) other solid fuels, such as coal and lignite.

Within the developing world, the rate of access to non-solid fuel varies from 19 per cent in Sub-Saharan Africa to about 95 per cent in West Asia[3] and 100 per cent in North Africa. The access rate in rural areas is particularly stark—for instance, in Sub-Saharan Africa 94 per cent of the rural population still relies on solid fuels. In urban areas more than 70 per cent of the population has access to non-solid fuel, except in Sub-Saharan Africa (42 per cent) (see Figure 4.1).

Energy access increased rapidly in the 20 years to 2010 but is now static. Between 1990 and 2010 electricity access rose from 76 per cent to 83 per cent, underpinned

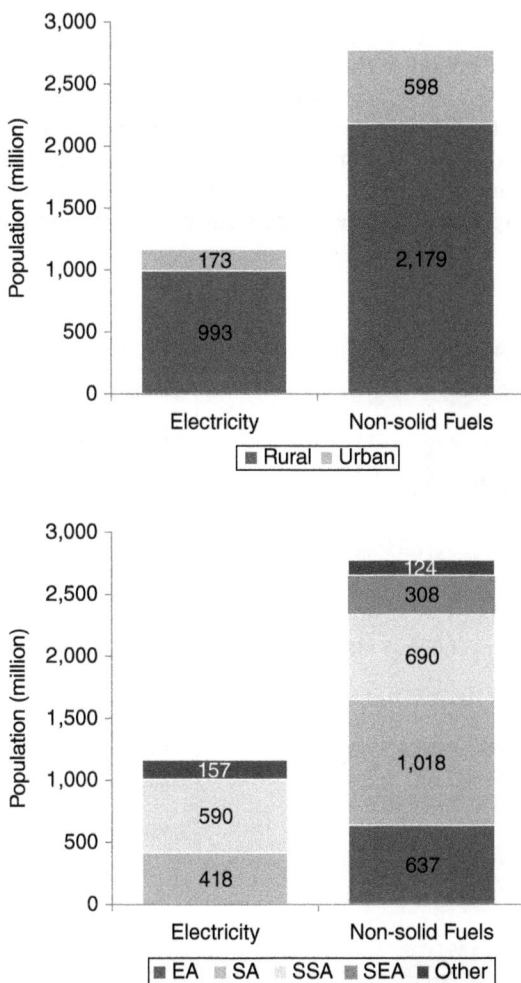

Figure 4.1. Composition of the access deficit in electricity and non-solid fuels, 2010 (EA = East Asia; SA = South Asia; SSA = Sub-Saharan Africa; SEA = South-east Asia)

[3] High-income countries with a gross national income (GNI) of more than $12,276 per capita (World Bank, <http://data.worldbank.org/about/country-classifications>) and countries in the Developed Country group according to the UN aggregation.

by growth in rural access, which increased from a lower base of 61 per cent to 70 per cent. Urban access remained stable, at 94 to 95 per cent. This overall increase resulted in a total of 1.7 billion people gaining access to electricity, a step ahead of the population increase of 1.6 billion (see Figure 4.2). Of these 1.7 billion people, 1.2 billion were in urban areas, suggesting that most of the electrification was localized here. For cooking purposes, the global access to non-solid fuels rose from 47 per cent in 1990 to 59 per cent in 2010. Both rural and urban access also grew. This growth in access to non-solid fuels fell slightly behind the population growth of 1.6 billion. As in the case of electricity, a greater part of this increase (1.2 billion people) occurred in urban areas.

The growth in access to modern energy has been only around 1 per cent on average. The upper bound of the access growth rate is witnessed in the United Arab Emirates (UAE) and Qatar for electricity—3.5 per cent for both electricity and non-solid fuels. A select group of countries made dramatic progress—14 countries increased access to primary non-solid fuels by more than 2.5 per cent. The largest absolute increase of electrified population was achieved in India, where 24 million annually gained access since 1990, followed by China and

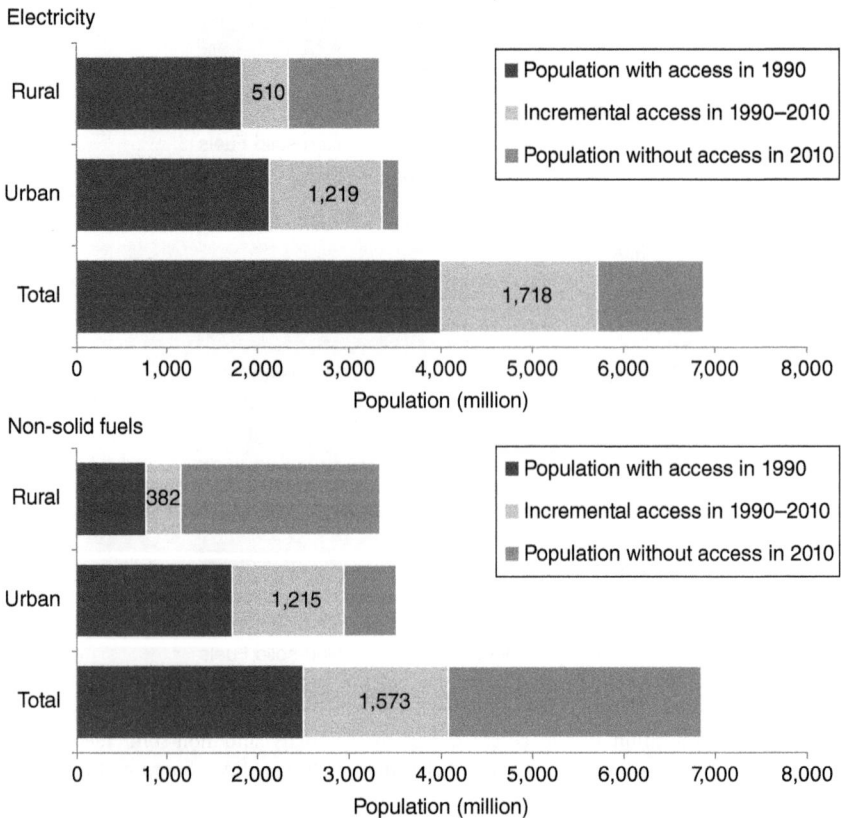

Figure 4.2. Incremental increase in energy access, 1990–2010

Source: Banerjee et al. 2013

Indonesia. India charted a remarkable trajectory in access to non-solid fuels as well (20 million people a year gained access), followed by China and Brazil.

Looking ahead, the access profile in 2030 will remain as alarming as today. Under the World Energy Outlook's New Policies Scenario (NPS), in which existing and announced policy commitments are forecast to continue, the global access deficit for electricity will be cut by 2030 to about 990 million, around 12 per cent of the global population at that time. Only Sub-Saharan Africa will witness an increase in unelectrified population in rural areas. To achieve universal access to electricity by 2030, some 50 million more people will have to gain access to electricity each year than under the NPS. It is estimated that universal access to electricity by 2030 will require an investment of around $890 billion over the period (2010 dollars), of which around $288 billion is projected to be forthcoming under the NPS. An additional $602 billion would be required to provide universal access to electricity by 2030—an average of $30 billion per year.

For non-solid fuels, the global access deficit will remain the same in 2030 as in 2010—that is, more than 30 per cent of the projected global population in that year (see Figure 4.3). Sub-Saharan Africa will experience an increase in people without primary access to non-solid fuels in both urban and rural areas. To achieve universal access, modern cooking solutions will need to be provided to an additional 135 million people per year on average, over and above those gaining access under the NPS. It is estimated that universal access to modern cooking solutions by 2030 would require an investment of about $89 billion over the period (in 2010 dollars), of which about $13 billion is projected to be forthcoming under the NPS, meaning that an additional $76 billion ($3.8 billion per year, 2011–30) would be required to provide universal access to modern cooking solutions by 2030.

4.2. THE WORLD BANK'S CONTRIBUTION TO ELECTRICITY ACCESS

The Bank's engagements in client countries in the developing world encompass the sector value chain—from generation to transmission and distribution (T&D) to direct connections to consumers. As indicated in Tables 4A.1–4A.4 in the Appendix to this chapter, these interventions have spanned different sectors and involved different profiles. The results of these interventions are summarized in three indicators in the Bank's corporate scorecard on energy, and are publicly available as part of a suite of indicators across many sectors. These indicators—on generation capacity, T&D lines, and people provided with access—provide a cumulative snapshot of results from projects approved between 2000 and 2013. At the upstream end of the sector value chain, Bank support has resulted in 13,490 megawatts of generation capacity and 98,323 circuit kilometres of T&D lines. Further downstream, Bank-supported programmes have resulted in an increase in the number of people provided with access to electricity.

Electricity

Non-solid fuels

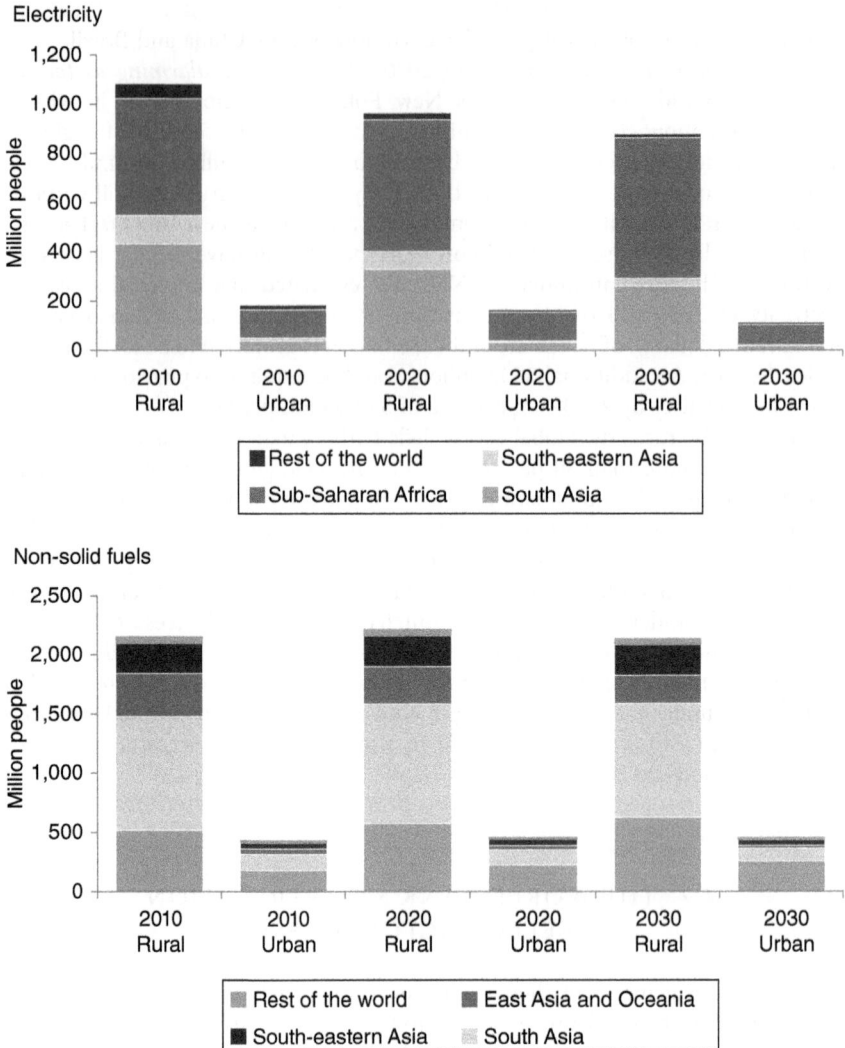

Figure 4.3. Projected energy access profile in 2030 (the new policies scenario)
Source: Banerjee et al. 2013

Bank-supported programmes[4] have provided 42 million people with new access to electricity between 2000 and 2013. New beneficiaries from electricity provision may have emerged from both direct and inferred access. Direct access is measured as the number of people that benefited from new-grid or off-grid connections—a total of *17.5 million people*. Inferred access is measured as the number of people that benefited from Bank-funded generation capacity, a proportion of whose output is reasonably estimated to be powering new household connections—a

---

[4] Includes both closed and active projects.

total of *24.4 million*. Inferred access is a conservative estimate of an indicative number of new connections that could have been directed to access expansion in select client countries. It is important to estimate the Bank's contribution to both direct and inferred access. In many countries, governments typically finance the last-mile connections themselves or through consumer contributions rather than by borrowings. The Bank's efforts have often involved upstream investments to ensure that capacity is available for access expansion.

Direct access emerged from Bank programmes in 26 countries. Only a few countries were responsible for direct access, most notably Kenya and Bangladesh. In Bangladesh, a sustained Bank engagement in both rural grid and off-grid solar home systems has resulted in 9.7 million people gaining from direct access. Kenya, Rwanda, Mali, and Cambodia are also countries where together more than 5 million new beneficiaries have been added through Bank-supported projects (see Appendix A, Table 4A.1). In the remaining 21 countries the scale of engagement has been smaller, and has collectively provided electrification to 2.7 million people.

Inferred access is conservatively estimated to have materialized in 19 low-access countries,[5] for a total of 24.4 million people, through both national and regional programmes. The beneficiaries are in South Asia and Sub-Saharan Africa as well as in Nicaragua in Latin America and Timor-Leste in East Asia. A proportion of the Bank's generation capacity expansion may have contributed to the increase of 15.2 million in Asia and 9.2 million in Sub-Saharan Africa. In Asia, India has added 12.1 million people to connection and Bangladesh 1.7 million people. In Sub-Saharan Africa, the Democratic Republic of Congo, Uganda, and Botswana lead the way, with 3.4 million, 1.3 million, and 1.2 million people connected respectively.

4.3. LESSONS LEARNED FROM WORLD BANK INTERVENTIONS IN ELECTRIFICATION

Traditionally, most countries in the world have mainstreamed the grid-based route in their endeavour to reach universal access. Until recently, off-grid interventions constituted a minor proportion of total investments in access expansion, usually on an experimental pilot basis and on a small scale. The options are now multifaceted—countries are choosing among a variety of options ranging from the solely grid-based to innovative mini-grid and stand-alone home systems. Bank investments also reflect this transition. A group of 14 energy access projects implemented between 2000 and 2013[6] suggests that a little more than 78 per cent of the new connections generated are grid. An overwhelming majority of connections has emerged from Vietnam and South Asia, the former from grid and the latter from both grid and off-grid, specifically from an innovative solar home system (SHS) programme in Bangladesh. Most of the new off-grid connections in

---

[5] Lower than the global average access rate of 83 per cent.
[6] Includes projects that were approved before fiscal year 2000.

the analysed projects are SHS and solar photovoltaic (PV) installations. In Lao People's Democratic Republic (Lao PDR), off-grid connections are mini-/micro-hydro and biomass resources.

Together, the 14 projects have provided energy access to 7 million new connections or 33 million people, with a total financing envelope of $1.3 billion (see Figure 4.4). About 94 per cent of the total financing of $1.3 billion came from the concessional arm of the World Bank—the International Development Agency—either in the form of grants or credits. Of the total financing, $624 million was devoted to South Asia, followed by $286 million to East Asia, and $300 million to Sub-Saharan Africa. Bangladesh is also an example of good value for money (number of connections realized compared with total financing), with $280 spent for each connection.

There have been innovative institutional arrangements in grid-based access expansion projects supported by the World Bank. Two recent interventions stand out. First, in Lao PDR a project incorporated a 'Power to Poor' (P2P) programme based on the finding of a socio-economic survey that approximately 20 to 40 per cent of households were still not connected to the grid two years after grid arrival in the village. This programme had a unique gender dimension—interest-free credit was provided to the poorest rural households and to rural households headed by women that could not afford the up-front charges for connection to the grid. These families could pay back the credit in monthly installments (based on their affordability) into a revolving P2P fund, which would be used to support other disadvantaged families. Second, Rwanda's ongoing Energy Access Rollout Project, which began in 2009, showcases the sector-wide approach (SWAp)

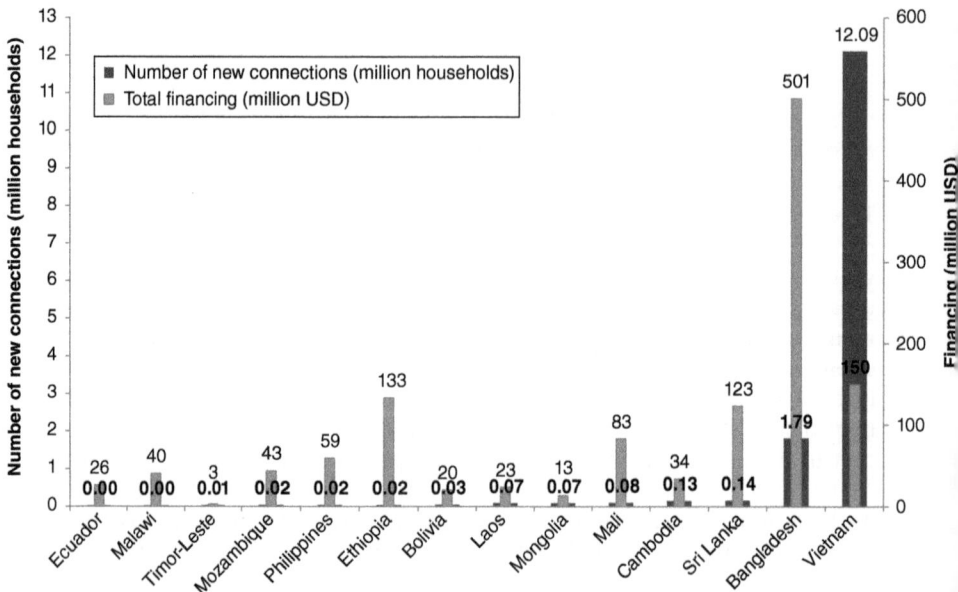

Figure 4.4. Number of connections and total financing of 14 closed energy access projects
Source: Authors' compilation

that pools together donor resources in a country-led, results-oriented programme. The programme (with a total financing of $378 million) aimed to increase the access rate from 6 per cent in 2009 to 16 per cent in 2013. This was among the first attempts at implementing a SWAp approach in the energy sector; previously, it had been a mainstay of social infrastructure projects. The Rwandan SWAp is underpinned by a sector investment prospectus that lays out the five-year funding requirements, as well as detailed geospatial planning that projects electrification efforts until 2020. The SWAp is implemented by a sector working group in which the World Bank is a co-chair. As a result of these focused efforts, Rwanda has managed to provide 60,000 connections annually compared to only 1,000 connections before the programme began in 2009.[7]

Among off-grid projects, the solar PV engagements in Bangladesh and Mongolia are particularly noteworthy. In Bangladesh, the focus on SHSs for rural consumers transformed the market. The programme involved dissemination of SHSs through direct sales by private companies, in partnership with microfinance institutions (MFIs) and nongovernmental organizations (NGOs). The benefits accrued to all stakeholders—the consumers benefited from the large-scale distribution of SHSs and the better service provided at lower costs and new grid connections; the private companies, MFIs, and NGOs benefited through SHS sales and new business opportunities. The engagement also leveraged resources from several other donors contributing to off-grid solar electrification and led to the creation of a renewable energy industry that not only made a substantial dent in the access deficit in Bangladesh but also set a precedent for many countries. In Mongolia, portable SHSs transformed herder communities in remote areas and contributed substantially to the Government of Mongolia's National 100,000 Solar Ger Electrification Program. Because of this engagement, the government could competitively procure large numbers of SHSs at bulk prices to be sold to herders through public distribution channels and private dealers. This is an example of a unique public–private partnership (PPP) that finally resulted in lower prices for consumers. The success of this programme ensured availability of electricity to 60 to 70 per cent of the herder population of Mongolia (Jayawardena, Rivera, and Ratnayake 2013).

The lessons presented below draw from the Implementation Completion and Results Reports (ICRs; World Bank 2002–2009) for these projects, as well as two recent documents that evaluate and summarize Bank engagements in electrification projects (World Bank 2010 and IEG 2010).

No 'best-practice' model or standardized institutional option exists. One of the key lessons to be taken away from these different projects is that one size does not fit all: there have been successful cases based on public, private, and cooperative models and schemes, as well as rural electrification agencies. There is no one institutional model for electrification that has worked and could work everywhere. Countries have been able to succeed in their electrification efforts using diverse institutional approaches, provided their programmes and strategies include institutional, technical, economic, and financial design and implementation

---

[7] <http://www.esmap.org/sites/esmap.org/files/ESMAP_Energy_Access_RwandaSWAp_KS013-12_Optimized.pdf>.

arrangements ensuring their efficient execution and their financial and operational sustainability.

An effective implementing agency is the backbone of any implementation programme. The management of rural electrification programmes requires the leadership of a strong entity (either a distribution utility or a specially designated agency) with an efficient administration and the technical capacity to support the supply chain of contractors and small service providers. An effective implementing agency is one of the most basic requirements to manage the relatively complex business of large-scale grid-based electrification. A variety of approaches have been successful: a separate rural electrification authority (Bangladesh), the setting up of rural electric cooperatives (Costa Rica), allocating rural electrification to a department of the national distribution company (Thailand), and delegating access expansion to the regional offices of the utility (Tunisia). There are common factors behind the models that have worked well: (i) the leadership of the programme is vested in a single strong entity with a stable and long-tenure management backed by an institutionally capable team, (ii) there are clearly established roles and responsibilities, and (iii) there is adequate technical capacity to support the supply chain of contractors and small service providers.

Indigenization is the key, and customized design standards and low-cost technologies should be incorporated. Many countries have been successful in reducing construction costs using technical standards adapted to rural demand patterns. In Costa Rica, the Philippines, and Bangladesh, adoption of the well-proven single-phase distribution system[8] used in the US rural electrification programme of the 1930s brought major savings. Since the main use of electricity is expected to be for lighting and small appliances (typical of many rural areas), there is no reason to apply the design standards used for much more heavily loaded urban systems. Although consumption does grow, this is usually at a slower pace and (provided the necessary design provisions are made) systems can be upgraded relatively cheaply later.

The long-term sustainability of electricity service is essential. Sustainable access means providing not just the village infrastructure but also the connections. Electricity distribution also needs to be financially and operationally sustainable to be able to attract investors and meet the growing demand of customers. A robust system of tariffs and subsidies is required to complement—but not replace—the limited contribution of low-income consumers and to ensure the sustainability of the service. A tariff/subsidy policy that recognizes the full cost-recovery of an efficient service is important. Every country must formulate its own 'best-fit' strategy (including what priority to allocate to electrification and the type and level of subsidies provided) on the basis of social, economic, and political conditions.

Productive use of electricity should be considered in the bundle of services to be provided. The planning process should assess the potential for productive uses of electricity and include measures for their promotion. The experience of several

---

[8] Single-phase electric power refers to the distribution of alternating current electric power using a system in which all the voltages of the supply vary in unison. Single-phase distribution is used when loads are mostly lighting and heating, with few large electric motors.

countries (including Bangladesh, Peru, and Thailand) suggests that promotion of and capacity building for productive uses of electricity in rural areas can increase the productivity of rural businesses, enable a more efficient use of the supply infrastructure, and improve the financial viability of distribution companies, thereby enhancing the economics of electrification. The complementary services identified in many studies as facilitating productive uses of electricity and business development are: access to information and knowledge (education, training, business development services, dissemination campaigns, available and qualified human resources), access to markets (access to raw materials, sufficient demand for the product or service, complementary infrastructure such as roads or information and communication technology), increased opportunities for local entrepreneurs to generate income by modernizing production methods and raising the value of production, and availability of equipment and tools (sufficient stock).

In Peru, examples include the installation of electric motors to grind grains and to process coffee in agricultural activities or the use of electric pumps to irrigate the land and improve growing conditions and yields. In Senegal the Electricity Services in Rural Areas Project, funded by the World Bank, also tested and implemented multi-sectoral approaches to the promotion of productive uses of electricity. Experiences from the Mali project have shown that rural micro-entrepreneurs are highly interested in expanding economic activities when electricity is available. However, such productive energy uses are not spontaneously induced when energy services start to be delivered in an off-grid area. Customized support for developing, scaling up, and sustaining productive uses is highly likely to make a significant impact on the feasibility and viability of rural electrification investments. In Cambodia, the Rural Electrification Fund (REF) included a specific Technical Assistance budget for the development of additional productive uses of electricity. This combination successfully built the necessary consumer base to make the network investments viable.

Grid extension and off-grid options are not mutually exclusive and could be implemented in parallel and, under specific conditions, in sequence. The rural electrification programme needs to be guided by a transparent long-term multi-year vision aimed at coordinating grid/off-grid efforts and should be supported by studies on the optimization of technology options, a grid/off-grid comparative economic analysis, and publicly disclosed market studies on the ability and willingness to pay. In the Lao PDR project, the Renewable Energy Master Plan was able to provide real guidance on electrification planning since, along with the database, it was updated to keep pace with electrification.

A few lessons emerge from recent Bank interventions in off-grid electrification, particularly in solar PV.

Flexibility to adapt to the changing environment is important. The project should be sufficiently flexible to adapt easily to changes in market conditions and consumer needs. It should maintain flexibility during the execution phase and be able to change based on evidence on the ground. In Bangladesh, consumer buy-back schemes of purchasing Solar Home Systems (SHS) directly, mostly in conjunction with microcredit hire-purchase agreements, reduced the risk perception and increased the uptake of SHS. As the project evolved, technology advancements (such as the introduction of LED bulbs) reduced the cost and increased the efficiency of SHS.

Flexibility of design is important to address any midcourse corrections. In the Philippines the dealer-based model of selling solar home systems fell short of achieving the access targets set by the government, and a fee-for-service mode was adopted during the course of the project. The same was the case in Cambodia. The initial pace of SHS installation was very sluggish, primarily because rural households could not afford the up-front payments to the suppliers. The model was then changed to a 'hire-and-purchase model', with the Rural Electrification Fund carrying out bulk purchases of SHSs and the private sector providing installation and other services. Also, households were allowed to pay in installments (over as long as 48 months), which sparked interest and enabled them to reach the installation target in less than six months. In Mongolia the project planned for a network of private dealers that would self-finance the purchases of SHSs and then sell them to herders. But there was a midcourse correction because the ability of the private sector to buy in bulk was overestimated, as was their ability to reach widely dispersed herders. Instead, the project adopted the government's parallel programme of buying the SHSs and then distributing them to herders through far-reaching government channels. Similarly, in Sri Lanka, the grid expansion went faster than expected; a few years into the project, it was clear that sales target could not be met. Therefore, cheaper but good-quality systems were imported and there was a shift to cash sales and a modification of the sales services networks. The targets were also revised downwards to reflect the prevailing market situation.

Quality control matters. It is crucial to arrange for quality assurance of product performance at the beginning of a project to establish credibility and consumer confidence. Qualifications criteria should be strict enough to attract only genuine bidders and enhance product testing. The need for quality assurance for SHSs was determined early in the Bangladesh project. Procurement of the SHSs was the responsibility of the MFIs and NGOs, who were to follow established commercial practices. Stringent quality standards were set, including a five-year warranty for batteries, and these quality standards were strongly enforced. In Mongolia the design included adoption of international standards for SHSs as well as robust after-sales service and warranties. As a result, the credibility of these products in the eyes of the consumers was enhanced. This also created a service infrastructure that was able to buy (in small quantities), distribute, sell, and service SHSs and meet the scattered herder population's electrification needs.

Sustainability of the programme must be thought through at the start and is the key for any success story. While designing off-grid electrification business models, it is advisable to think carefully about sustainability up front and to set up strategy beyond the project life. In designing SHS programmes for rural areas in Cambodia, the choice of appropriate system sizes based on robust up-front analysis, suitable delivery approaches, and post-installation operation and maintenance arrangements had a significant bearing on the efficiency, cost-effectiveness, and sustainability of those programmes. In Laos the focus was on the implementing agency carrying the principal responsibility to ensure the sustainability of the SHS programme, particularly when installation, operation, and maintenance of the SHS were outsourced to the private sector. Any such programme requires an appropriate monitoring and evaluation (M&E) system that ensures clearly defined terms of reference and establishes appropriate compensation schemes.

Cost sharing with consumers is the key to sustainability. The successful experiences in Bangladesh and Mongolia epitomize that poor households are willing to pay for energy services. The Bangladesh project (through a minimal subsidy per SHS and the leveraging of MFI services) demonstrated that even low-income rural households are willing and able to pay for SHSs to have access to improved lighting services. The Mongolia project illustrated that while the affordability for the herder population was limited, there was still a strong willingness to pay for good-quality and reliable products and services if the consumers were well informed and after-sale services were accessible to a dispersed population. The Mongolian experience also showed that payment for SHSs created a sense of ownership compared to the distributing of grants.

M&E, beneficiary assessments, and impact evaluation surveys can be a strong feedback mechanism. All World Bank projects focus on developing strong M&E systems and data-collection tools. In Bangladesh the data collected through project M&E had a strong impact on improving project implementation. In particular, in the case of SHSs, feedback from the field helped the project team and the Infrastructure Development Company Limited incorporate new technical specifications and technologies to improve service delivery to poor households. Similarly, feedback from the project teams also proved crucial for the establishment of improved SHS-testing facilities and improved service provisions for MFIs and NGOs. In Mongolia a beneficiary assessment—'A Survey of Herder Electrification'—was carried out in October 2012 to evaluate the impact of the dissemination of SHSs to the herder population. The preliminary analysis of the data collected resulted in a quantitative assessment of the user satisfaction and socio-economic impacts linked with the installation of SHSs. Similarly, in Bangladesh, an impact evaluation of SHSs was carried out.

Testing various delivery and institutional models is important. The Philippines project highlights the importance of testing various delivery and institutional models for off-grid electrification investments and of setting conservative targets. Generally, adoption of new technologies and business models in off-grid areas advanced at a gradual pace in the initial years, and then picked up pace once immediate implementation issues were sorted out and institutional capacity established. The Philippines project also offers a cautionary tale of thinking through context-specific options, rather than imposing successful experiences elsewhere to adapt to a particular situation.

4.4. LESSONS LEARNED FROM WORLD BANK INTERVENTIONS IN MODERN COOKING SOLUTIONS

World Bank funding of Projects on Clean Cooking Solutions helped leverage and contributed to approximately $1.2 billion in the past 20 years. Ekouevi and Tuntivate (2011; Ekouevi 2013) have reviewed a total of about 70 World Bank Group projects covering the past 20 years. The selected projects (see Appendix A, Table 4A.3) had the objective of improving household cooking and heating energy access through fuelwood management or improved stoves. The total cost of these projects was $1.2 billion and the World Bank's contribution was $698 million, of which $161 million was devoted specifically to household fuels. All the projects were in Sub-Saharan African countries, with the exception of the Mongolia Urban Stove Improvement Project, which was financed by the Global Environment Facility (GEF).

During the period of the review, the Bank funded four biogas projects for household cooking and lighting in China and Nepal. The total cost of these projects was $1 billion, to which the Bank contributed $365 million, with 70 per cent allocated to household energy access components. Similarly, the Bank financed eight household energy access projects on natural gas for cooking and heating—mostly in European and Central Asian countries, and one project in Colombia (see Appendix A, Table 4A.4). The total project cost was $203 million, to which the Bank contributed $126 million. The total cost of specific components on household access to natural gas was $142 million.

Figure 4.5 provides a summary of the funding allocated to fuelwood and stove, biogas, and natural gas programmes. It appears that relatively few resources are allocated to fuelwood and stove programmes.

Providing clean and efficient stoves and fuels to poor households in developing countries is a complex challenge. Providing clean cooking solutions is an issue cutting across many disciplines (such as environment, forestry, energy, health, and household economics), and linked to contextual social and cultural considerations. The projects undertaken by the World Bank present a mixed story of successes and challenges. The complexity and cross-sectoral nature of the

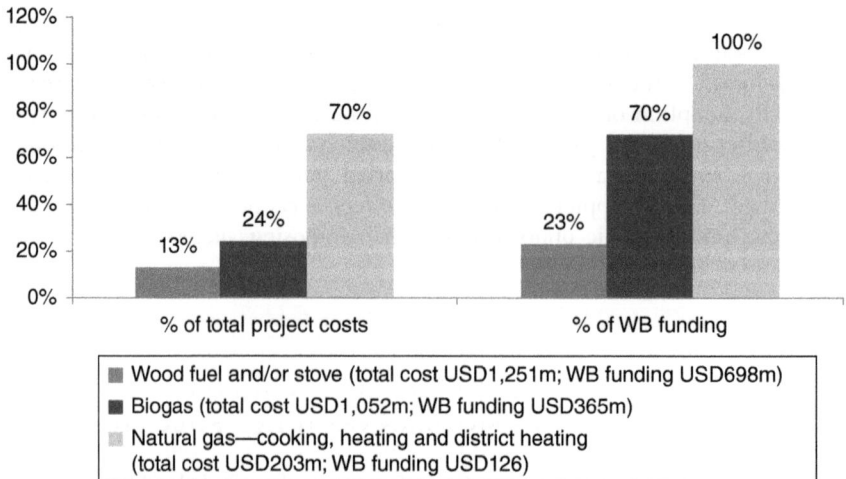

Figure 4.5. Household access component as a percentage of total project cost and of total World Bank funding

Note: WB = World Bank.
Source: Ekouevi, Koffi and Voravate Tuntivate. 2012. Household Energy Access for Cooking and Heating: Lessons Learned and the Way Forward. Washington, D.C.: World Bank. DOI: 10.1596/978-0-8213-9604-9. License: Creative Commons Attribution CC BY 3.0

challenge is reflected in the mixed results that have been obtained in the field over the years. The lessons learned from these experiences are summarized below.

A holistic approach to household energy issues is necessary. Successful programmes are designed with a holistic approach on how household energy access can contribute to the global agenda of social transformation and poverty reduction. It is important that the proposed programme prioritize (i) supply-side interventions to ensure that the fuelwood supply is sustainable; (ii) demand-side and interfuel substitution with the introduction and dissemination of improved stoves and alternative household fuels, such as kerosene and liquefied petroleum gas (LPG); and (iii) the capacity to develop and strengthen institutions to create the regulatory incentives for the sustainable production of fuelwood and for the facilitation of fuel switching.

Developing market-based clean cooking solutions are important. All World Bank studies point to the importance of developing market-based clean cooking solutions. Incentives are needed for local entrepreneurs to design, manufacture, and market safe cookstoves that are tailored to the country or region, made with local materials, and adapted to local cooking practices.

Public awareness campaigns are prerequisites for successful interventions. From past experience, it has been observed that successful programmes often pay particular attention to public awareness, education, and information campaigns. Households need to be sensitized to the risks they incur by cooking with inefficient stoves. Programmes that have assumed that households would spontaneously adopt improved stoves or participate in forest management initiatives have often failed. Households need to perceive and be convinced about the direct and indirect benefits associated with these interventions.

Local participation is fundamental. Experience indicates that the active participation of communities, governments, NGOs, and the private sector is fundamental for household energy access projects to be successful and sustainable. For example, local communities need to be involved at an early stage to ensure that they own supply-side forest management initiatives. A clear rule of engagement should be discussed for communities to know their rights and responsibilities, the prerogatives of the national forest service, and the role of NGOs and local associations.

Consumer fuel subsidies are not a good way of helping the poor. Experience has shown that across the board consumer fuel subsidies are not the best way of helping the poor. Affluent households tend to benefit the most from prevailing fuel subsidies, given that in most cases, energy consumption increases in parallel with income. For governments these subsidies may result in heavy fiscal deficits, diverting direct public expenditures away from productive and social sectors. Alternative options could include social protection programmes.

Both market-based support and public support are relevant in the commercialization of improved stoves. A market-based approach in the commercialization of improved stoves is often viewed as the best way to ensure the sustainability of programmes. Evidence indicates, however, that a certain level of public funding is necessary at the initial programme stages for improved stoves programmes to take off. This is particularly true in settings where the business environment is not well developed. Funding is usually needed to support research and development, marketing, quality control, training related to stove design and maintenance, and

M&E. The development of stove standards and certification protocols relies on the availability of public funding. Without this initial support, small enterprises find it difficult to participate in improved stoves programmes, and scaling up is unrealistic. A challenge is to determine what level of public funding is adequate and the right time to transition to a fully market-based business model.

The needs and preferences of improved stove users should be given priority. Successful programmes pay attention to the needs and preferences of the users of improved stoves. Targeting households susceptible to buying and using these improved stoves, and working with them to supply a suitable stove that responds to their needs, is critical. By first focusing on households that can afford to adopt an improved stove, the programme can subsequently capitalize on the benefits of the demonstration effects produced. Successful, improved stove programmes are also designed bearing in mind the preferences of the users. Experience has shown that when these factors are ignored, stove dissemination rates are low and programmes are not sustainable.

Experience from pilot projects in Central America shows that fuelwood users respond well when improved cookstoves meet the needs of a specific circumstance. This was seen in cases when fuelwood is purchased and is becoming increasingly expensive (particularly in the case of former urban LPG users); when health issues are clearly understood by the whole family (as in the case of Honduras); when incentives are provided to lower the up-front costs of stoves (but are not seen as a gift); when improved cookstoves are tailored to local cooking practices, resulting in tangible fuel and time savings; and when the stoves do not involve major changes in use of fuelwood or in cooking habits and where they appeal to the 'modernity' aspirations of users.

Strengthening the supply chain is essential. A persistent challenge is that clean cooking remains a 'poor person's problem.' When short-term incentives have prompted business people to try to build a market for improved cookstoves, their efforts have often foundered. The genuinely safer advanced cookstoves were not affordable, not adapted to local needs, or not locally made and thus in short supply. A number of players in this space—including local small and medium enterprises, international social enterprises, domestic conglomerates, and multinational corporations—now hold promise for scaling up, given the right conditions. Adequate guidance to them when they start off and initial hand-holding of public sector partners or developmental partners (to help prioritize design appeal and product quality, market the stoves, and spread the word about their benefits) is very beneficial. This is a win-win solution for everyone.

Durability of improved stoves is important for their successful dissemination. For households that can afford an improved stove, the decision to adopt one or not depends on their perception of the durability of the stove. The durability depends on the quality of the materials used in the production of the stove, the resistance of the stove in the climatic context where it is used, how it is used, and the maintenance that it needs. In addition to technical considerations such as heat-transfer efficiency and combustion efficiency, it is important to account for durability issues in the design and construction of improved stoves.

Microfinance can be a good tool to help the poor gradually afford an improved stove. Programmes that have included microfinance options to help households afford the stoves tend to be more successful. The poor need to have a time horizon to pay gradually for the improved stoves. For example, in Bangladesh, Grameen Shakti has been working with international donors to provide cookstoves as part of its microfinance activities. This dimension is very important. Having an improved stove is not perceived as a first priority by the poor, but integrating the adoption of an improved stove in a broader programme and creating opportunities to generate income is a different proposition.

The challenge is to put these lessons into practice. Addressing these challenges requires a multipronged approach across four key drivers that have been identi-fied through reviewing successful programmes: awareness raising, markets and preferences, technologies and standards, and innovative financing.

World Bank Group experts in household energy are working with clients through the Clean Stove Initiative in East Asia, as well as in Central America. Efforts are also under way through the Africa Clean Cooking Energy Solutions Initiative in Sub-Saharan Africa. The Bank Group is also a partner in the Global Alliance for Clean Cookstoves, a PPP that seeks to create a thriving global market for clean and efficient household fuels and cookstoves. The World Bank's operational pro-grammes are either global or regional in focus; each is associated with a different business model to improve access to clean cookstoves and fuel in an effort to scale up (see Table 4.1).

4.5. A COMPREHENSIVE DEFINITION AND MEASUREMENT OF ENERGY ACCESS

Measuring energy access has proved to be a complex task, as multidimensional issues need to be taken into consideration. Convenient binary metrics, as are currently available, fail to capture several important aspects of the problem such as multiple access solutions, supply problems, and a differentiation between access to electricity supply versus electricity services. Therefore, a multi-tier metric has been proposed in the Global Tracking Framework (Banerjee et al. 2013) that addresses these definitional and measurement issues.

4.5.1. Multiple access solutions

Energy can be accessed through multiple solutions. Although a connection to the national grid has often been considered essential to access electricity, isolated mini-grid and off-grid options (for example, solar lanterns or stand-alone home systems) are increasingly emerging as interim or long-term solutions in many developing countries, particularly in remote rural areas.[9] It is, therefore, important to capture expansion of access through mini-grid and off-grid solutions along with main grid connections. The performance of each solution in terms of quantity and quality of

---

[9] The IEA has projected that about 60 per cent of households not connected to the main grid at present are likely to obtain electricity through such solutions by 2030.

Table 4.1. World Bank programmes and initiatives that focus on clean cooking

Programme or initiative	Geographic focus	Implementation period (years)	Activities	Partners	Business model	Impact
Energy Sector Management Assistance Programme	World	Established in 1983. Operates with five-year strategic business plans. Currently implementing its 2008–13 business plan.	Analytical work to support technical assistance and lending operations; knowledge sharing.	GACC, Sustainable Energy for All, 13 bilateral donors.	Provide funding to regional teams working on clean cooking initiatives.	Awareness raising within and outside the Bank and facilitation of lending operations.
Africa Clean Cooking Energy Solutions	Sub-Saharan Africa (pilot countries: Democratic Republic of Congo, Senegal, Uganda)	2012–14	Enterprise development and support to energy-for-cooking lending operations.	GACC	Enterprise-based platform to promote clean fuels and technologies.	Awareness raising in Sub-Saharan Africa and mobilization of the private sector and NGOs.
Clean Stove Initiative	East Asia and Pacific (pilot countries: China, Indonesia, Mongolia, Lao PDR)	2012–15	Capacity building, policy development, knowledge sharing, institutional strengthening.	AusAid, ASTAE, GERES, China Alliance for Clean Stoves.	Result-based financing.	Awareness raising in East Asia and the Pacific. Mobilization of private sector operators and partner organizations.
Household Biomass Cooking Review	Central America (pilot countries: El Salvador, Guatemala, Honduras, Nicaragua)	2011–13	Review of country and regional programmes.	SICA, OLADE, CEPAL, GACC.	Knowledge forum.	Regional ownership of clean cooking solutions.
South Asia Household Energy	South Asia (pilot country: Bangladesh)	2013–18	Bangladesh (disseminate 1 million cookstoves and 20,000 biogas units).	IDCOL, GIZ, USAID.	Social mobilization, market facilitation, and enterprise development.	Awareness raising and mobilization of partner organizations.

Note: ASTAE = Asia Sustainable and Alternative Energy Program; CEPAL = Comisión Económica para América Latina; GACC = Global Alliance for Clean Cookstoves; GERES = Groupe Energies Renouvelables, Environnement Et Solidarités; GIZ = Gesellschaft für Internationale Zusammenarbeit; IDCOL = Infrastructure Development Company Limited; NGO = nongovernmental organization; OLADE = Latin American Energy Organization; SICA = Sistema de la Integracion Centroamericana; USAID = US Agency for International Development.
Source: Ekouevi 2013

the electricity supplied can vary widely, and hence needs to be assessed accordingly.

Similarly, there are multiple types of cooking solutions, using not only different fuels but also various types of cookstoves. Although non-solid fuels have so far been considered as modern, new types of cookstoves—often called *improved* or *advanced*—may be equally clean and/or efficient despite using solid fuels. The emergence of these cookstoves is significant, since it is projected that a large part of the developing world will continue to rely on solid fuels for cooking despite increasing use of non-solid fuels. The performance of each cooking solution can vary widely and needs to be assessed accordingly. Additionally, many households in the developing world use multiple fuels and cookstoves in parallel, a phenomenon known as *stacking*. Even those households that have adopted a modern fuel or an advanced cookstove may continue to use a secondary or tertiary cooking solution that delivers an inferior performance on some parameters—and thus impede the achievement of complete access. There could be several reasons behind such practice. For instance, secondary solutions may better fit with local culinary preferences, or may serve other purposes such as space heating, food drying, and so on. Households may also be forced to use secondary solutions due to irregular availability of the primary fuel or affordability issues.

4.5.2. Supply problems

An energy supply has multiple attributes or characteristics that could vary widely. In many developing countries, grid electricity suffers from irregular supply, frequent breakdowns, and problems of low or fluctuating voltage. Often, electricity is provided during odd hours when the need is minimal, such as midnight or midday. Connection costs and electricity charges also constrain the use of energy among poor households. In many countries, illegal and secondary connections serve a significant proportion of the population. This leads to substantial revenue loss and undermines safety.

In the case of cooking, the performance of the cooking solution could vary quite widely in terms of parameters (such as pollution, efficiency, safety, and so on) depending on the cookstove and fuel used.[10] It is usually not possible to evaluate these attributes through simple observation, and a laboratory test using a standardized protocol is required. In many instances, the performance observed during the laboratory test may not be achieved in practice because of user behaviour, cooking practices, and site conditions. Lack of proper maintenance of the cookstove or disuse of required accessories such as chimneys, hoods, or pot skirts could result in lower performance. Besides the technical performance, time and effort invested in collecting the fuel and preparing the cookstove are also important factors to consider. As cooking is highly gendered in most societies, time and effort invested by women and children in cookstove preparation, cleaning, and cooking are important dimensions that need to be taken into consideration.

---

[10] But some fuels—such as electricity, natural gas, LPG, and biogas—are 'stove-independent' and are always considered of high technical performance.

4.5.3. Electricity supply versus electricity services

Electricity supply is useful only if it allows desired energy services to be run adequately. The actual use of electricity services depends not only on the availability of electricity supply, but also on the ownership of appropriate electrical appliances. But the ownership of appliances and actual use of electricity services depends upon a number of factors such as income, spending priorities, and cultural preferences. Energy providers cannot be held accountable for access to electricity services, but only for access to adequate electricity supply. Thus, it is important to distinguish between the two.

4.5.4. Multi-tier framework for measuring access to household electricity

The multi-tier framework for measuring access to household electricity consists of two distinct yet intertwined electricity measurements: access to electricity supply and use of electricity services (see Table 4.2). The framework is technology-neutral and accommodates grid, mini-grid, and off-grid solutions to reflect a wide range of electricity access levels.

The first metric—access to electricity supply—is based on increasing levels of attributes across tiers (such as quantity (peak available capacity), duration, evening supply, affordability, legality, and quality), whereby more and more electricity services become feasible. The second metric—access to electricity services—is based on the ownership of appliances requiring an increasing level of electricity supply to meaningfully use the said services (Table 4.2). For example, a household availing 16 hours of electricity a day but only 2 hours in the evening through a 100-watt

Table 4.2. Multi-tier framework for measuring access to household electricity

Access to electricity supply

Attributes	Tier 0	Tier 1	Tier 2	Tier 3	Tier 4	Tier 5
Peak available capacity (W)	–	>1	>50	>500	>2,000	>2,000
Duration (hours)	–	≥4	≥4	≥8	≥16	≥22
Evening supply (hrs)	–	≥2	≥2	≥2	≥4	≥4
Affordability	–	–	√	√	√	√
Legality	–	–	–	√	√	√
Quality (voltage)	–	–	–	√	√	√

Use of electricity services

Tier 0	Tier 1	Tier 2	Tier 3	Tier 4	Tier 5
–	Task lighting AND phone charging (OR radio)	General lighting AND television AND fan (if needed)	Tier 2 AND any low-power appliances	Tier 3 AND any medium-power appliances	Tier 4 AND any high-power appliances

Source: Banerjee et al. 2013

SHS will be classified as Tier 2 on access to electricity supply. But if the same household uses electricity only for lighting and phone charging, then it would be classified as Tier 1 on access to electricity services.

As the above example shows, a higher level of electricity supply does not automatically result in greater access to electricity services. Likewise, poor electricity supply does not constrain affluent households from availing energy services through the use of backup solutions (such as generators and invertors). Nevertheless, for the less affluent, energy services typically lag behind improvements in supply, as consumers tend to acquire electrical appliances gradually over a period. Accurate measurement and understanding of gaps between access to electricity supply and the actual use of electricity services is important to devise interventions to improve energy access.

The multi-tier framework evaluates both the extent of access (how many households have access) and the intensity of that access (the level of access that households have). It allows for an aggregated measure of access to electricity supply as well as use of electricity services, using two separate indices that can be calculated for any geographical area (Box 4.1).[11] In parallel, a disaggregated analysis of different supply attributes may also be conducted to obtain deeper insights into the underlying phenomenon. For example, the share of households receiving fewer than four hours of electricity per day, or the share of households facing affordability issues, or the share of households using televisions, fans, and so on, can be easily determined. The progress toward universal access can be tracked by comparing the value of the indices over time.

Box 4.1. Indices of access to household electricity

Index of access to electricity supply =
$\sum(P_T \times T)$
P_T = % of households at tier T of electricity supply
T = tier number of electricity supply
$\{0,1,\ldots,5\}$

Index of access to electricity services =
$\sum(P_T \times T)$
P_T = % of households at tier T of electricity services
T = tier number of electricity supply
$\{0,1,\ldots,5\}$

Source: Banerjee et al. 2013

4.5.5. Multi-tier framework for cooking

The multi-tier framework measuring access to cooking solutions first evaluates the technical performance of the primary cooking solution (including the fuel and the cookstove), and then adjusts the access level by considering how this solution meets the specific cooking needs of the household. As in the case of electricity, the methodology is fuel-neutral and reflects a wide range of access levels (Table 4.3). The technical performance is evaluated based on four attributes: efficiency, indoor pollution, overall pollution, and safety. The cooking solutions are categorized into

---

[11] Such as a village, a city, a district, a province, a country, a continent, or the whole world.

Table 4.3. Multi-tier framework for measuring access to cooking solutions

	Step 1: Technical performance				
Attributes	Grade-E	Grade-D	Grade-C	Grade-B	Grade-A
Efficiency Indoor pollution		Certified non-BLEN-manufactured cookstoves			
Overall pollution Safety	Self-made cookstoves or equivalent	Uncertified non-BLEN-manufactured cookstoves			BLEN cookstoves or equivalent

Step 2: Actual use					
Level 0	Level 1	Level 2	Level 3	Level 4	Level 5
				Grade-A without CCA	with CCA
			Grade-B without CCA	with CCA	
		Grade-C without CCA	with CCA		
	Grade-D without CCA	with CCA			
Grade-E without CCA	with CCA				

Note: BLEN = biogas–LPG–electricity–natural gas; CCA = conformity, convenience, and adequacy.
Source: Banerjee et al. 2013

five grades based on the fuel and cookstove used. Self-made cookstoves are assigned the lowest grade, while biogas–LPG–electricity–natural gas (BLEN) cookstoves receive the highest grade. Depending on whether the manufactured non-BLEN[12] cookstove has been tested (and therefore its technical performance is known), the corresponding grade is assigned.[13]

Beyond the technical performance, the framework also considers three additional attributes, conformity, convenience, and adequacy (CCA), to reflect the overall user experience and to capture the impact of cooking on the daily lives of users. Conformity refers to the household's adherence to the instructions that come with the specific cooking solution: for example, use of required accessories and the performance of scheduled maintenance. If suggested instructions are not followed, the cooking solution is not in full conformity. The time taken for the household to collect the fuel and prepare the cookstove is used as a criterion to evaluate convenience. Convenience refers to whether the household spends less than 12 hours per week on fuel collection, and less than 15 minutes per meal on stove

---

[12] Non-BLEN cookstoves include all cookstoves using solid or liquid fuels, that is, stove-dependent fuels.

[13] It is not possible to know the technical performance of a non-BLEN cookstove if no laboratory testing has been made. Such cookstoves are placed at Grade D by default.

preparation. A convenient cooking solution should also offer an easy cooking experience.[14] Adequacy refers to whether the household uses a secondary cooking solution or not. If yes, for what reasons? Adequacy is not achieved if the household uses a second solution because the primary fuel is not always available or too expensive, or the primary stove does not have the required number of burners, or it does not satisfy the culinary preferences.[15] The use of a primary cooking solution is deemed insufficient if its use is constrained by any of these factors.

The technical grade of a cooking solution is adjusted to account for these three attributes and thus to show the final level of access. If all three CCA attributes are satisfied, the technical grade is raised to a higher level; if not, the technical grade remains unchanged at the lower level.

As in the case of electricity, the multi-tier framework evaluates both the extent of access and the intensity of that access. An index of access to cooking solutions can be computed for any geographical area[16] (Box 4.2). In parallel, a disaggregated analysis may also be conducted either by type of fuel or by performance level of cookstoves. For instance, the share of households using a self-made cookstove or a tested manufactured non-BLEN cookstove; or the share of households using solid fuels with a higher-than-grade-D cookstove; or the share of households using a secondary solution, facing convenience issues, or affordability issues can be easily determined. The progress made toward universal access to modern cooking solutions can be tracked by comparing the value of the index over time.

In summary, the multi-tier framework offers a unique opportunity to accurately measure access to energy and offers the choice to a country to set its own targets. At present, multi-tier frameworks are constrained by limited data from the omnibus household surveys. Dedicated household energy surveys are necessary to track energy access on a regular basis. The utility of such surveys extends beyond multi-tier tracking. The data obtained could be a gold-mine for multiple stakeholders including governments, regulators, utilities, project developers, civil society organizations, developmental agencies, financial institutions, appliance manufacturers, international programmes, and academia. These stakeholders could apply the available data toward tracking progress, project selection, developing

---

Box 4.2. Index of access to cooking solutions

Index of access to cooking solutions = $\sum (P_T \times T)$
P_T = % of households at tier T
T = tier number $\{0,1,\ldots,5\}$
Source: Banerjee et al. 2013

---

[14] Easy cooking experience refers to whether the user is satisfied with the speed of cooking and level of attendance required.

[15] If the secondary cooking solution is of equal or higher grade compared to the primary one, adequacy is achieved, independent from the underlying reason. Higher-grade secondary solutions are considered as aspirational or transitional, and therefore do not adversely impact the overall level of access.

[16] Such as a village, a city, a district, a province, a country, a continent, or the whole world.

markets, policy design, regulatory processes, market research, and academic analysis. This would be a useful step not only in solving the energy access challenge but also in clearly establishing the linkages between energy access and socio-economic development.

The adoption of multi-tier metrics would require large amounts of data. Therefore, their implementation is not possible without robust enhancements to the existing data collection instruments. Existing omnibus surveys, although valuable, do not go beyond basic questions related to electricity connections and use of different fuels. Therefore, new energy-focused household surveys need to be administered globally. Energy-focused surveys can be conducted either at the national or project level. The contribution of energy projects (such as electricity generation, T&D, off-grid solutions, tariff reform, and energy efficiency) can be evaluated and tracked over time using the multi-tier metrics. The impact of these interventions can be determined in terms of the number of users moved to a higher tier of access by improving levels of energy supply attributes and the intensity of such a movement.

4.6. LOOKING AHEAD TO UNIVERSAL ACCESS TO ELECTRICITY BY 2030

Achieving universal access to modern energy services by 2030 will cost more than $49 billion a year. Country governments cannot do this alone. A development institution such as the World Bank can support only a handful of programmes. This transformative process will require the aggressive and ambitious action of numerous stakeholders including governments, the private sector, NGOs, and civil society aiming together to achieve universal access by 2030. So far, a total of about 80 countries have opted into SE4ALL, and these countries account for about half of the access deficit. Achieving universal energy access is a global effort, and within SE4ALL, the World Bank Group and the UN are joined by many partners, including the opt-in countries; donors such as Norway, Denmark, the United States, and Iceland; and a growing number of private companies and civil society organizations (including the World Energy Council). Worldwide consensus behind the SE4ALL goals is critical to achieve the level of investment required to meet the access challenge.

While this is a daunting challenge, it is also an opportunity to unleash innovation in both public and private sectors to expand access and make service available in an affordable, reliable, sustainable manner to the wide swathe of unserved people. There are new and tested ways that can be adopted and customized to move this agenda forward. The World Bank, with its mandate as a knowledge bank, is contributing not only with finances but also by disseminating implementable lessons, supporting demonstration projects, and making its convening power available for building consensus on these complex topics.

APPENDIX A

Table 4A.1. Solar home system (SHS) programmes supported by the World Bank in selected countries

Country	Project name	Type of model	Implementing agency
Bangladesh	Rural Electrification and Renewable Energy Development	To implement the off-grid component, the REB (through the PBSs) was tasked with extending a fee-for-service SHS programme, whereby the systems would be installed and owned by the REB and consumers would pay a monthly fixed fee for using the systems. The systems were sold to consumers using a microfinance scheme through NGOs/MFIs (called 'partner organizations', POs).	Rural Electrification Board (REB) Infrastructure Development Company Limited (IDCOL)
Cambodia	Rural Electrification and Transmission Project	A new hire-and-purchase delivery model (through bulk purchase) was implemented for the SHS programme of the REF, since the original output-based approach model failed in delivering the targeted outputs.	Electricité du Cambodge (EDC) Electricity Authority of Cambodia (EAC) Ministry of Industry, Mines and Energy (MIME) Rural Electrification Fund (REF)
Laos	Rural Electrification Phase I	Hire and purchase through a bulk purchase model.	Electricité du Laos (EdL) Ministry of Energy and Mines (MEM)
Mali	Household Energy and Universal Access Project	Fee-for-service model.	Malian Agency for Domestic Energy and Rural Electrification (AMADER)
Mongolia	Renewable Energy and Rural Electricity Access Project	Given the success of the government's parallel programme, whereby SHSs were prefinanced and purchased in bulk and then promptly distributed through far-reaching government channels, a decision was taken to modify the SHS component to emulate the government model in order to strike a balance between cost-recovery and affordability.	Ministry of Mineral Resources and Energy (MMRE) supported by the Energy Authority
Philippines	Rural Power Project	Lease-to-own model ECs would be in charge of installing PV systems for households and public facilities in remote *barangays* (small administrative districts) within their franchise areas. The households would make periodic repayments to cover the ECs' transaction costs and build a revolving fund to cover the procurement of more SHSs over time. In the end, system ownership would be transferred to the households. Fee-for-service model was piloted under the so-called 'PV mainstreaming' approach. Six ECs were tasked with procuring, installing, operating, and maintaining the SHSs, which would be owned by the ECs in exchange for a monthly fee from the households they serve.	Development Bank of the Philippines Department of Energy

Note: EC = Electric Cooperative; MFI = microfinance institutions; NGO = nongovernmental organization; PBS = Palli Bidyut Samity; PV = photovoltaic; SHS = solar home systems.
Source: Authors' compilation from World Bank ICRs (Implementation completion and results reports)

Table 4A.2. Profiles of electrification projects

Project ID	Project name	Country	Approval fiscal year	Closing fiscal year	Restructuring (if any)	Additional financing
P074040	Renewable Energy Development	Bangladesh	2002	2012	Yes, twice	Yes, twice
P073367	Decent Infrastructure for Rural Transformation	Bolivia	2003	2009	Yes, three times	No
P064844	Rural Electrification and Transmission	Cambodia	2004	2012	Yes, once	No
P063644	Power and Communication Sectors Modernization and Rural Services	Ecuador	2002	2008	No	No
P077380	Energy Access	Ethiopia	2006	2012	Yes, twice	No
P075531	LA-Rural Electrification Phase I	Laos	2006	2012	Yes, twice	Yes
P057761	Infrastructure Services SIM	Malawi	2006	2012	Yes	No
P073036	Household Energy and Universal Access	Mali	2004	2012	None	No
P099321	Renewable Energy for Rural Access	Mongolia	2007	2012	Yes, twice	No
P071942	Energy Reform and Access Programme	Mozambique	2004	2011	Yes, twice	No
P113159	Rural Power Project	Philippines	2004	2012	Yes, twice	Yes
P077761	Renewable Energy for Rural Economic Development	Sri Lanka	2002	2011	Yes, twice	Yes
P066396	System Efficiency Improvement, Equitization and Renewables	Vietnam	2002	2012	Yes, five times	Yes

Source: Authors' compilation from World Bank ICRs (Implementation completion and results reports)

Table 4A.3. Projects funded by the World Bank promoting access to fuelwood and/or stove component (USD millions)

	Project	Year	Total project cost	IBRD, IDA, GEF, GPOBA	HH energy access component	% of total project costs	Project closing date
1	Niger: Energy Project	1989	65.90	30.40	16.20	25	31 December 1996
2	Mali: Household Energy	1995	11.20	11.20	11.20	100	31 December 2000
3	Madagascar: Energy Sector Development	1996	102.60	44.20	2.90	3	31 December 2005
4	Senegal: Sustainable and Participatory Energy Management (PRODEGE I)	1997	19.93	19.93	19.93	100	31 December 2004
5	Chad: Household Energy	1998	6.30	5.27	6.30	100	30 June 2004
6	Mongolia: Urban Stove Improvement (GEF)	2001	0.75	0.75	0.75	100	31 March 2007
7	Ethiopia: Energy Access Project	2002	199.12	132.70	5.44	3	30 June 2013
8	Mali: Household Energy and Universal Access	2003	53.35	35.65	13.47	25	30 June 2012
9	Madagascar: Environment Programme	2004	148.90	40.00	2.50	2	30 June 2011
10	Senegal: Electricity Services for Rural Areas	2004	71.70	29.90	4.60	6	31 December 2012
11	Benin: Energy Services Delivery	2004	95.70	45.00	6.20	6	31 December 2011
12	Rwanda: Urgent Electricity Rehabilitation	2004	31.30	25.00	0.90	3	30 April 2011
13	Chad: Community-based Ecosystem Management	2005	94.45	39.76	2.50	3	30 March 2011
14	Benin: Forests and Adjacent Lands Management (GEF)	2006	22.35	22.35	22.35	100	30 November 2011
15	Burkina Faso: Energy Access	2008	41.00	41.00	6.70	16	30 April 2013
16	Benin: Increase Access to Modern Energy	2009	178.50	72.00	5.50	3	30 June 2015
17	Rwanda: Sustainable Energy Development (GEF)	2009	8.30	8.30	8.30	100	n.a.
18	Mozambique: APL for Energy Development and Access	2010	80.00	80.00	6.30	8	30 June 2015
19	Senegal: Second Sustainable and Participatory Energy Management (PRODEGE II)	2010	19.37	15.00	19.37	100	30 November 2016
	Total		**1,250.72**	**698.41**	**161.41**		
	Average loan/credit		65.83	36.76	8.50		

Note: GEF = Global Environment Facility; GPOBA = Global Partnership on Output-based Aid; HH = household; IBRD = International Bank for Reconstruction and Development; IDA = International Development Association.
Source: Ekouevi 2013

Table 4A.4. Projects funded by the World Bank promoting household access to natural gas for cooking and heating (USD millions)

Project	Year	Total project cost	IBRD, IDA, GEF, GPOBA	HH energy access component	% of total project costs	Project closing date
1 Bosnia and Herzegovina: Emergency District Heating Reconstruction Project	1996	44.50	20.00	44.50	100	31 March 1999
2 Moldova: Energy Project	1996	12.63	9.08	12.63	100	31 December 2001
3 Bosnia and Herzegovina: Emergency Natural Gas System Reconstruction Pro	1997	40.53	10.00	40.53	100	31 July 1999
4 Armenia: Urban Heating Project	2005	21.95	15.00	13.90	63	31 December 2009
5 Armenia: Access to Gas and Heat Supply for Poor Urban Households (GPOBA)	2006	3.09	3.09	3.09	100	31 December 2009
6 Belarus: Chernobyl Recovery	2006	60.90	50.00	8.50	100	31 December 2011
7 Colombia: Natural Gas Distribution for Low Income Families in the Caribbean Coast Project (GPOBA)	2006	5.10	5.10	5.10	100	13 March 2008
8 Tajikistan: Energy Emergency	2008	13.90	13.90	13.90	100	31 December 2012
Total		**202.60**	**126.17**	**194.55**	**96**	
Average loan/credit		25.33	15.77	24.32		

Note: GEF = Global Environment Facility; GPOBA = Global Partnership on Output-based Aid; HH = household; IBRD = International Bank for Reconstruction and Development; IDA = International Development Association.
Source: Ekouevi 2013

REFERENCES

Banerjee, Sudeshna Ghosh; Bhatia, Mikul; Azuela, Gabriela Elizondo; Jaques, Ivan; Sarkar, Ashok; Portale, Elisa; Bushueva, Irina; Angelou, Nicolina; and Inon, Javier Gustavo, (2013). *Global Tracking Framework*. Vol. 3 of *Global Tracking Framework: Sustainable Energy for All*. Washington, DC: The World Bank. <http://documents.worldbank.org/curated/en/2013/05/17765643/global-tracking-framework-vol-3-3-main-report>.

Ekouevi, K., (2013). *Scaling up Clean Cooking Solutions: The Context, Status, Barriers and Key Drivers*, Washington, DC: The World Bank.

Ekouevi, K. and Tuntivate, V., (2011). *Household Energy Access for Cooking and Heating: Lessons Learned and the Way Forward*, Washington, DC: The World Bank.

Jayawardena, Migara, Rivera, A. Salvador, and Ratnayake, Chrisantha, (2013). *Capturing the Sun in the Land of Blue Sky: Providing Portable Solar Power to Nomadic Herders in Mongolia*. Washington, DC: The World Bank.

IEA (International Energy Agency), (2012). *World Energy Outlook 2012*. Paris: IEA.

IEG (Independent Evaluation Group), (2010). *The Welfare Impact of Rural Electrification: A Reassessment of the Costs and Benefits. An IEG Impact Evaluation*. Washington, DC: The World Bank.

Lim, S., et al., (2012). 'A comparative risk assessment of burden of disease and injury attributable to 67 risk factors and risk factor clusters in 21 regions, 1990–2010: A systematic analysis for the Global Burden of Disease Study 2010'. *The Lancet* 380(9859): 2224–60.

World Bank, (2002a). *Implementation Completion and Results Report (ICR): Power and Communications Sectors Modernization and Rural Services Project (PROMEC) in Ecuador*. Project ID P063644. Washington, DC: The World Bank.

World Bank, (2002b). *Implementation Completion and Results Report (ICR): Renewable Energy Development in Bangladesh. Project ID P074040. Washington, DC: The World Bank*.

World Bank, (2002c). *Implementation Completion and Results Report (ICR): Renewable Energy for Rural Economic Development in Sri Lanka*. Project ID P077761. Washington, DC: The World Bank.

World Bank, (2003). *Implementation Completion and Results Report (ICR): BO Decent Infrastructure for Rural Transformation in Bolivia*. Project ID P073367. Washington, DC: The World Bank.

World Bank, (2004a). *Implementation Completion and Results Report (ICR): Energy Reform and Access Program in Mozambique*. Project ID P071942. Washington, DC: The World Bank.

World Bank, (2004b). *Implementation Completion and Results Report (ICR): Household Energy and Universal Access in Mali*. Project ID P073036. Washington, DC: The World Bank.

World Bank, (2004c). *Implementation Completion and Results Report (ICR): Rural Electrification and Transmission in Cambodia*. Project ID P064844. Washington, DC: The World Bank.

World Bank, (2006a). *Implementation Completion and Results Report (ICR): Energy Access Project in Ethiopia*. Project ID P077380. Washington, DC: The World Bank.

World Bank, (2006b). *Implementation Completion and Results Report (ICR): Infrastructure Services in Malawi*. Project ID P057761. Washington, DC: The World Bank.

World Bank, (2006c). *Implementation Completion and Results Report (ICR): Rural Electrification Phase I in Laos*. Project ID P075531. Washington, DC: The World Bank.

World Bank, (2007a). *Implementation Completion and Results Report (ICR): Energy Services Delivery Project in Timor-Leste*. Project ID P095593. Washington, DC: The World Bank.

World Bank, (2007b). *Implementation Completion and Results Report (ICR): Renewable Energy for Rural Access in Mongolia*. Project ID P099321. Washington, DC: The World Bank.

World Bank, (2007c). *Implementation Completion and Results Report (ICR): Rural Energy Project in Vietnam*. Washington, DC: The World Bank.

World Bank, (2009). *Implementation Completion and Results Report (ICR): Additional Financing for Rural Power in the Philippines*. Project ID P113159. Washington, DC: The World Bank.

World Bank, (2010). *Addressing the Electricty Access Gap. Background Paper for the World Bank Group Energy Sector Strategy*. Washington, DC: The World Bank.

5

Health Benefits from Energy Access in LMICs

Mechanisms, Impacts, and Policy Opportunities

Nigel Bruce and Chen Ding

5.1. INTRODUCTION

5.1.1. The importance of energy to health and development

Access to energy is central to our daily lives, providing the means to cook food and stay warm (or cool in hot conditions), lighting, and many other activities and services that are essential to human health and economic activity. And yet, well into the second decade of the twenty-first century, billions of people simply lack these services altogether or have to make do with intermittent and inadequate supply, as well as traditional energy sources which are inefficient, polluting, and unsafe and which have changed little since prehistoric times (IEA 2012; Bonjour et al. 2013).

The aim of this chapter is to provide an overview of how current energy access for households and communities impacts on health and to examine the potential that greater awareness of health benefits can have in advancing universal access to modern energy as quickly as possible. The focus is on households and local communities in low- and middle-income countries (LMICs).[1] Aspects of energy access relating to the potential benefits to health and society through power generation, transport, and industry are outside the scope of this chapter. In considering the benefits of access to electricity, however, it is important to keep in mind that although this energy source is clean at the point of use, depending on how the electricity is generated, air pollution (e.g. from coal, oil, etc.) may have negative impacts on the health of exposed populations.

---

[1] The World Bank classifies low-income countries as those with an annual gross national product (GNP) per capita equivalent to $103; lower-middle-income countries have an annual GNP per capita of between $1,036 to $4,085; upper-middle-income countries must have an annual GNP per capita of between $4,086 to $12,615; <http://data.worldbank.org/about/country-classifications/country-and-lending-groups#Low_income>.

Adequate access to energy and the technologies used to harness energy for useful purposes in and around the home, or lack of these, impacts health in myriad ways. Some of these effects on health can be considered to be direct, while others act through more indirect routes which are less well described but are also of considerable importance.

The clearest direct effects include those resulting from air pollution both inside the home and in outdoor air, and through safety issues such as burns, scalds, and fires, or injuries and sexual violence incurred by women and children walking far from their homes to collect fuelwood. The lack of energy services (notably electricity) in a substantial proportion of health facilities in developing countries can also be considered as impacting directly on the health of the population served, due to the restrictions this places on delivering high-quality health care.

Indirect impacts on health may be seen through the contributions that adequate energy can make to education by provision of power for lighting and ICT in homes and schools, processing food for consumption and sale, pumping clean water, irrigation for agriculture, and in many small- and large-scale commercial activities which support economic development and help families escape from poverty.

A socio-environmental perspective on the wider determinants of health as described by Dahlgren and Whitehead (1991) provides a useful framework for understanding how important energy is for health and how it acts through a variety of direct and indirect mechanisms (Figure 5.1).

Energy impacts on almost all of the health determinants illustrated in Figure 5.1. These impacts can be viewed in both positive and negative ways, largely depending on how the energy is supplied and consumed. The air pollution referred to above lies in the outer area of environmental conditions for ambient air

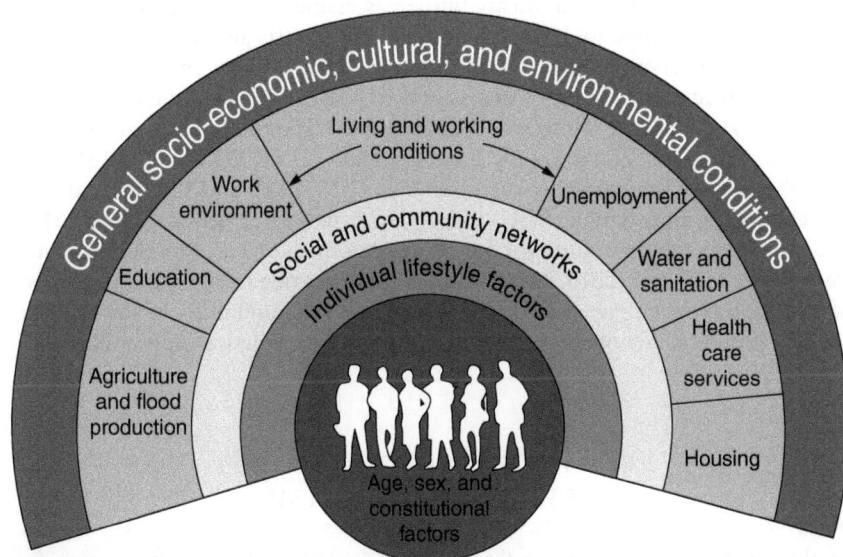

Figure 5.1. The wider determinants of health

Source: Dahlgren G, Whitehead M. 1991. Policies and Strategies to Promote Social Equity in Health. Stockholm, Sweden: Institute for Futures Studies

pollution and impacts on the climate, and in the housing segment for indoor or household air pollution and safety issues such as burns and house fires. These negative effects can be mitigated or eliminated by the supply of clean energy and efficient and safe use within the home or service setting. On the positive side, energy is important for the provision of quality health-care services, for clean water, for developing employment and income-generating opportunities, and for education and agriculture, and through communication and other means energy has important impacts on social and community networks.

This chapter aims to provide an overview of the importance of energy to health through consideration of its influence on these many aspects of the environment, living conditions, services, and society. Only some parts of this picture have been quantified, leaving others to be discussed mainly in terms of their potential for positively or negatively impacting on health. The energy access situation across these various determinants is summarized in Section 5.2, and the ways in which these affect health together with the role that energy plays are further elaborated in Section 5.3. In Section 5.4 we consider how policy could make more of the interrelationships between energy, health, environment, and development, and we examine current efforts to strengthen multi-sectoral action that includes the health sector.

5.1.2. Evidence review methods

The scope of this chapter is very broad, and a full, up-to-date, systematic review of all of the constituent topics would not have been possible. For some aspects, for example the health impacts of household air pollution (HAP), recent outputs from systematic reviews and disease burden assessments are available (Lim et al. 2012). For other topics, we have reviewed existing reports and data on household and community energy access and their health impacts. These include data from the WHO Household Energy Database (WHO 2013), the International Energy Agency's World Energy Outlook (IEA 2012), the WHO/UNICEF Joint Monitoring Programme for Water Supply and Sanitation (UNICEF/WHO 2012), UNESCO's database on energy access in schools (UN 2014), and the International Telecommunications Union (ITU) website for data on ICT access (Sanou 2013). We also frequently cite findings and sources from the NGO Practical Action's *Poor People's Outlook* (2010, 2012, 2013), as these provide comprehensive reviews of key aspects of energy use for households and communities in low-income settings (Practical Action 2010; Practical Action 2012; Practical Action 2013).

5.2. OVERVIEW OF THE ENERGY ACCESS SITUATION IN LMICS

In this section, we summarize the situation in terms of access to energy for households and community services in low- and middle-income settings. This is incomplete, not only in terms of what is available for at least some of the topics, but also qualitatively, for example in respect of the reliability of energy/fuel supply, or the type and effectiveness of technologies including stoves and lighting used in people's homes. Nevertheless, this does provide a fair overview of the level of access and some of the key gaps, and emphasizes how divided the world still is when it comes to the availability of these most basic services that—apart from during periodic 'oil crises' and large-scale electric power failures from storms—developed countries take for granted.

5.2.1. Household energy

Cooking

Data on the primary fuels used for cooking is collected through nationally representative surveys and compiled in the WHO household energy database (WHO 2013). A key indicator of lack of access to modern cooking fuels and of HAP is the percentage of homes relying primarily on solid fuels, which in LMICs are typically burned in open fires or in simple, poorly vented stoves. A recent report by Bonjour et al. describes trends for solid fuel use (SFU) over the period 1980–2010 compiled from a total of 586 national survey data points for 155 countries, including 97 per cent of all low- and middle-income countries (defined here as having less than USD12,276 per capita in 2011–2012) (Bonjour et al. 2013). The percentage of households relying primarily on solid fuels for cooking fell from 62 per cent (95 per cent CI: 58, 66 per cent) to 41 per cent (95 per cent CI: 37, 44 per cent) between 1980 and 2010. The actual number of persons exposed, however, remained stable at around 2.8 billion during three decades due to population increase. Solid fuel use is most prevalent in Africa and South-east Asia, where more than 60 per cent of households cook with solid fuels (Figure 5.2). Although most regions have shown relative declines in SFU, Africa shows the slowest rate of change (Figure 5.3).

In the past, it has been generally assumed that those not using solid fuels were cooking mainly on clean fuels. While this was a reasonable assumption for liquefied petroleum gas (LPG), natural gas, and electricity, evidence on the emissions and possible health effects of kerosene use suggest that this fuel should not be considered in the 'clean' category (Lam et al. 2012). An estimated 14 per cent of households in urban Africa and 7 per cent across South-east Asia use kerosene for cooking (WHO 2013), although the very high proportion (more than 50 per cent of urban homes) cooking with kerosene in Indonesia has been markedly reduced by the large-scale conversion to LPG within that country (Budya and Arofat 2011).

While these data provide a reasonable picture of the fuels used for cooking across LMICs, it must also be recognized that multiple fuel and technology use is more or less the norm. For example, many homes listed as primary solid-fuel users may have LPG for some uses (e.g. preparing hot beverages, cooking when in a hurry), and vice versa, so the implications for health are somewhat less clear-cut than the simplistic view obtained from the primary fuel data would suggest. The cookstove (or stoves) may also be used for a range of other purposes, including

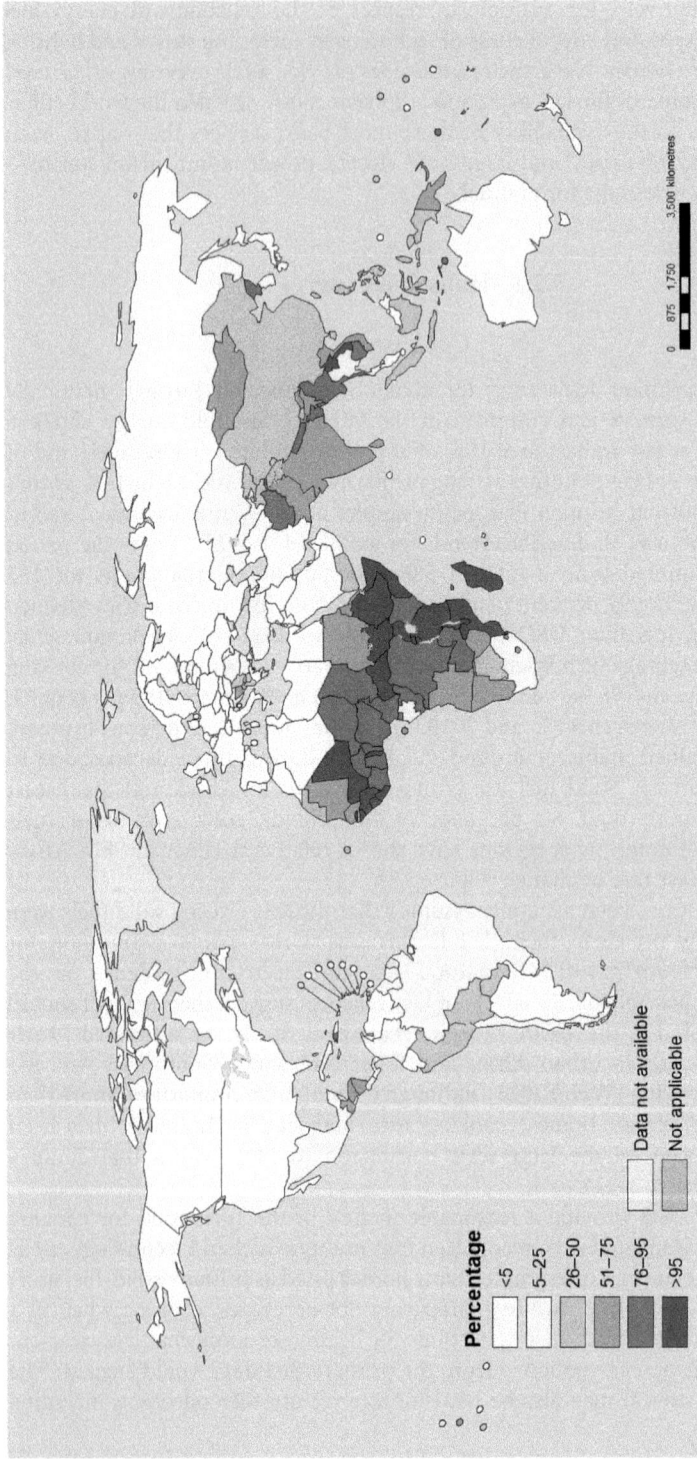

Figure 5.2. Population using solid fuels (wood, dung, crop wastes, charcoal, coal) as their primary cooking fuel, 2012

Source: Reproduced, with the permission of the publisher from Household Energy Database. World Health Organization, 2012 <http://www.who.int/indoorair/health_impacts/he_database/en/index.html>

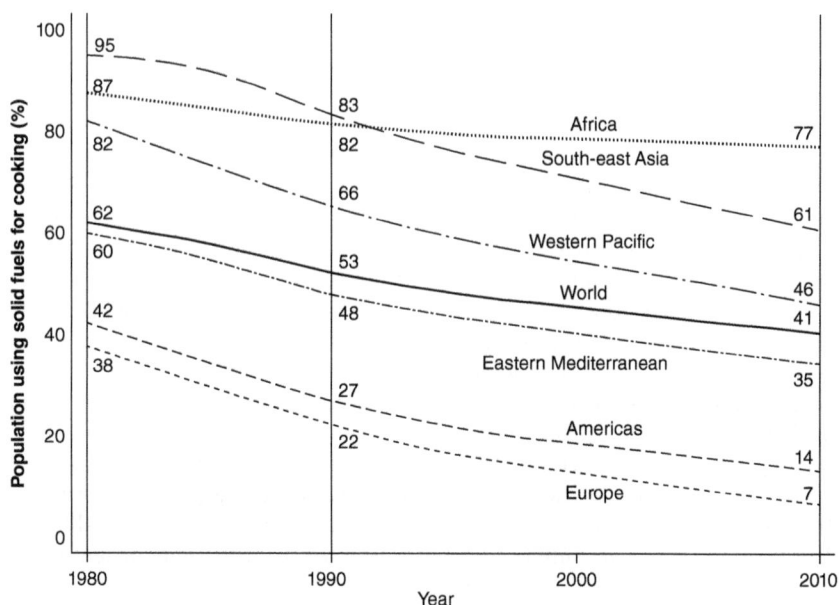

Figure 5.3. Trends in population using solid fuels (wood, dung, crop wastes, charcoal, coal) as their primary cooking fuel, from 1980 to 2010, by WHO region

Source: Sophie Bonjour et al., Solid Fuel Use for Household Cooking: Country and Regional Estimates for 1980–2010, Environmental Health Perspectives DOI: 10.1289/ehp.1205987, <http://ehp.niehs.nih.gov/1205987/#tab3>

food drying and preserving, preparation of food for sale, and making alcoholic beverages, in addition to providing warmth and lighting for some.

Heating and cooling

Low-income households in many higher-altitude and temperate areas of the world need energy for warmth, and households in other regions need energy for drying clothes and other possessions during rainy seasons. Data on heating needs is generally far less complete than for cooking and again derives from the WHO database (WHO 2013). For many of those relying on solid fuels and traditional stoves for cooking, the fire serves both purposes and may also provide lighting. For others, a separate solid fuel stove (biomass or coal) is used for heating, and this may or may not have a flue (chimney). Use of heating stoves varies widely: for example, around 25 per cent of homes in Bhutan use a coal stove for heating; in the Democratic Republic of Korea around 90 per cent use solid fuels for heating, and many also do in China, especially in rural areas. Kerosene is used extensively as a heating fuel in some regions, notably the Middle East, with an estimated 90 per cent of homes in Iran and 85 per cent in Jordan using either kerosene or gas (the data source does not distinguish these fuels).

Cooling in very hot regions of the world may be provided by fans, where electricity is available. Air conditioning also requires electrical power and is very energy intensive, so it is even less available to low-income households.

Lighting

Electrical power provides the most effective and clean energy source for lighting at the point of use. Since lighting is one of the first and main applications of electricity used by households, the rate of connection to electrical power is a useful, if incomplete, indicator of access to electric light. It is incomplete since many consumers in LMICs experience unreliable supply during many hours of the day; in addition, solar PV lamps are becoming increasingly available, with (for example) estimates of between 1 per cent and 3 per cent of homes stating this as their primary source of lighting in some African countries (WHO 2013).

The IEA estimates that in 2010, 1.27 billion people lacked access to electricity (IEA 2012), although many more will have unreliable supplies. Access varies greatly within and between regions and countries: thus, in 2010, 68 per cent of the population of Sub-Saharan Africa lacked access compared to 18 per cent in developing Asia (India 25 per cent) and 6 per cent in Latin America. Between 80 and 90 per cent of those without access in all regions were in rural areas.

For those without electricity (and many without reliable power), lighting is obtained from the open fire and from candles, kerosene lamps, and flashlights. Most kerosene lamps are simple wick types, which burn the fuel inefficiently and emit surprisingly high levels of health-damaging pollutants (Lam et al. 2012). As is the case for heating, data on primary (let alone secondary) energy sources for lighting are very incomplete. Drawing on the WHO database, it was found that in Africa, for 11 out of 21 countries with data, an average of 50 per cent of homes relied on kerosene for lighting, but more than 80 per cent did so in some countries; surveys from five countries across South-east Asia found between 8 and 40 per cent of homes relying on kerosene for lighting. In general, a much higher percentage of homes in Latin America have access to electricity (WHO 2013).

5.2.2. Other household-related uses of energy

In addition to the basic needs for cooking, heating, and lighting, energy either is required for or can enhance the efficiency of a whole range of other household activities including income-generating activities and small-scale agriculture, accessing safe water supplies, and information and communications technologies (ICTs).

Income generation and small-scale agriculture

Households may engage in a wide range of income-generating activities, many of which require energy in some form. Preparation and processing of food, crafts, and other activities conducted at home typically require heat from the stove, light for working in dark interiors or at night, and electrical power for some equipment; access to these forms of energy is discussed in Section 5.2.1. Another very important activity, both for income generation and for contributing to the nutrition and food security of the family, that has additional energy needs is agriculture.

Approximately 2.5 billion people or 45 per cent of the developing world's population depend on agriculture and agri-based activities (Utz 2011). Poor people in these regions are primarily smallholder farmers or farm labourers and rely on three major farm-power systems, namely human work, animal work, and application of renewable and fossil-based technologies (Utz 2011). Basic human work does not require energy inputs (other than adequate nutrition and sufficient levels of health) and is used for tilling, harvesting and processing, and rain-fed irrigation. Animal work such as draught animal power (DAP) is the next level and requires indirect energy inputs (such as the production of animal feed) (Utz 2011). The most advanced stage is the use of renewable, fossil-fuel-based, and hybrid applications such as wind pumps, water wheels, and diesel generators (Utz 2011).

Though recent cross-country data are limited, past FAO studies show that developing countries relied almost equally on hand power, DAP, and tractors between 1997 and 1999 (FAO 2003). As a percentage of land area cultivated, 35 per cent was prepared by hand power, 30 per cent by DAP, and 35 per cent by tractors (FAO 2003). However, the proportions varied significantly across regions, with Africa relying on hand power and DAP by 80 per cent (FAO 2003). A more recent study of 14 farming communities across Sub-Saharan Africa found hand power to be the predominant source of agricultural power, representing on average 30 per cent of the households in a community (and as high as 70 per cent in a community studied in Malawi) (FAO 2005).

Water supply

Energy plays an important role in provision of adequate supplies of water for a range of purposes including clean water for human consumption (the energy–water linkage will be further described in Section 5.3.2). Data on water supply access is comprehensive at the country level, thanks to the WHO/UNICEF Joint Monitoring Programme for Water Supply and Sanitation which is tasked to provide comparable estimates across countries using data from national governments. As of 2010, over 780 million people were still without access to improved sources[2] of drinking water, which is critical to health (UNICEF/WHO 2012). Amongst those without access, 39 per cent were in Sub-Saharan Africa, compared to only 10 per cent in Latin America, North Africa, and large parts of Asia (UNICEF/WHO 2012). The difference is also striking within LMICs: improved water coverage for countries classified as 'least developed'[3] is only 63 per cent compared to 83 per cent overall in the developing world (UNICEF/WHO 2012). These numbers are considered optimistic, given that even improved sources of water may be difficult to protect from contamination (UNICEF/WHO 2012).

In Sub-Saharan Africa, where access to improved water sources is the lowest globally, 13 per cent of the population relies on surface water from rivers, dams,

---

[2] According to the WHO/UNICEF JMP, 'improved' water sources include the use of piped water, public taps, boreholes, protected springs and dug wells, and uncontaminated rainwater. 'Unimproved' sources include unprotected dug wells, unprotected springs, untreated surface water, and water transported by tankers and carts.

[3] Regional groupings based on the WHO/UNICEF JMP 2012 report; countries in the 'least developed' group include Afghanistan, Angola, Bangladesh, Central African Republic, Chad, and Yemen. For a complete list, refer to the annex of the report. UNICEF/WHO (2012). Progress on Drinking Water and Sanitation: 2012 Update. New York, UNICEF and World Health Organization.

Table 5.1. Coverage and energy requirement of commonly used irrigation technologies

Regions	Technology	Description	Irrigated area	Energy requirement
Developing	Manual	Buckets or watering cans	< 0.5 ha	Low
	Surface/ gravity fed	Water moves over land depending on gravity flow; water in ditches can be pumped or lifted by human or animal power	Unlimited	Low
Developed	Sprinkler	Piping water to one or more central locations within the field and distributing it by high pressure sprinklers or guns	Unlimited	High
	Drip/ micro- irrigation	Delivers water directly at or near root zone of plants drop by drop	Unlimited	Medium

Source: Adapted from Practical Action 2012

lakes, and ponds, which is often contaminated (UNICEF/WHO 2012). Though groundwater is less likely to be contaminated[4] and is usually closer to the source of demand, successfully drilling for groundwater requires energy-driven drilling technologies and expertise, which are often lacking in Sub-Saharan Africa (IAH 2006).

Energy inputs are also important for improving irrigation methods. Increasing energy access helps households and communities engaged in agriculture to transition from manual and surface/gravity fed methods with low energy requirements to sprinkler and drip/micro-irrigation methods with higher energy requirements (Utz 2011), Table 5.1.

The type and capacity of irrigation pump technology also depends on energy availability. For example, in a study cited by GIZ and conducted by Winrock International, Empowering Agriculture, and USAID, basic manual hand and treadle pumps are only able to irrigate around 2 hectares of land, while motorized pumps greatly improve the land under irrigation to more than 4 hectares (Utz 2011).

In LMICs, poor farmers still rely primarily on human and animal power, and this is reflected in figures of low total area in production under irrigation of only 4 per cent in Sub-Saharan Africa, compared to 29 per cent in East Asia (Utz 2011).

ICT including mobile phones

Recent years have seen the adoption of ICT growing more rapidly in developing than in developed countries. Between 2006 and 2013, mobile-cellular subscriptions grew by on average 20 per cent in the developing world compared to 6 per cent in the developed world, while mobile-broadband subscription grew by 74 per cent and 27 per cent respectively (Sanou 2013). The number of fixed-broadband

---

[4] Groundwater, depending on depth and local mineral deposits, can become contaminated; toxic contamination with arsenic in Bangladesh and elsewhere provides a graphic illustration of this.

users is also increasing rapidly in the developing world, reaching an average of 19 per cent growth in the same period versus only 8 per cent in the developed world (Sanou 2013). The use of these technologies of course relies on electricity for powering equipment and charging batteries.

While ICT adoption has been has been rapid, it is still evident that LMICs with the lowest electricity penetration rates have the lowest populations of ICT users. According to ITU estimates, in 2013, 90 per cent of the 1.1 billion households not connected to the Internet were in the developing world, primarily in Africa and the Asia-Pacific region (Sanou 2013). Africa's Internet penetration rate in 2013 was still only 16 per cent, or half of that of the Asia-Pacific; this can be understood mainly as a result of poor electricity coverage (Sanou 2013).

5.2.3. Health facilities

Energy is critical for the delivery of high-quality health care, including for vaccine storage, powering diagnostic and surgical equipment, incubators, the provision of adequate light, and much more. Recent evidence shows that many facilities in LMICs lack electrical power completely or that the supply is inadequate and unreliable.

Although international institutions such as WHO and USAID have recently begun to systemically gather energy access data in the health sector through country surveys, cross-regional data are still rare. Globally, the *Poor People's Energy Outlook* 2013 estimates that approximately 1 billion people in the world rely on health facilities with no access to electricity (Practical Action 2013). The most in-depth data available for health facilities relates to Sub-Saharan Africa: according to a recent study based on 13 health facility surveys assessing electricity access since 2000 among clinics and hospitals in 11 Sub-Saharan African countries, approximately a third of the health facilities reported had no access to electricity (Adair-Rohani et al. 2013).

Only 28 per cent of the health-care facilities with access to power reported reliable electricity access (defined as 'power available during service hours with no outages exceeding 2 hours on a given day a week prior to data collection') among eight of the countries reporting these data (Adair-Rohani et al. 2013); see Table 5.2.

Table 5.2. Energy access in health-care facilities in Sub-Saharan Africa

Energy access	Facility type		
	All facilities	Hospitals only	Other facilities besides hospitals
Access to electricity, %(11 countries)	74	99	72
Source of electricity, %(9 countries)			
Generator only	7	6	8
Central, solar, or other	68	93	65
Reliable electricity, % of facilities (8 countries)	28	34	26

Note: Data for access to electricity are averages among 11 countries (Ethiopia, Gambia, Ghana, Kenya, Namibia, Nigeria, Rwanda, Sierra Leone, Tanzania, Uganda, and Zambia); for source of electricity, among 9 countries (excludes Ghana and Nigeria); and for reliable electricity, among 8 countries (excludes Ethiopia, Ghana, and Nigeria).
Source: Adair-Rohani, Zukor, et al. 2013

Nigel Bruce and Chen Ding

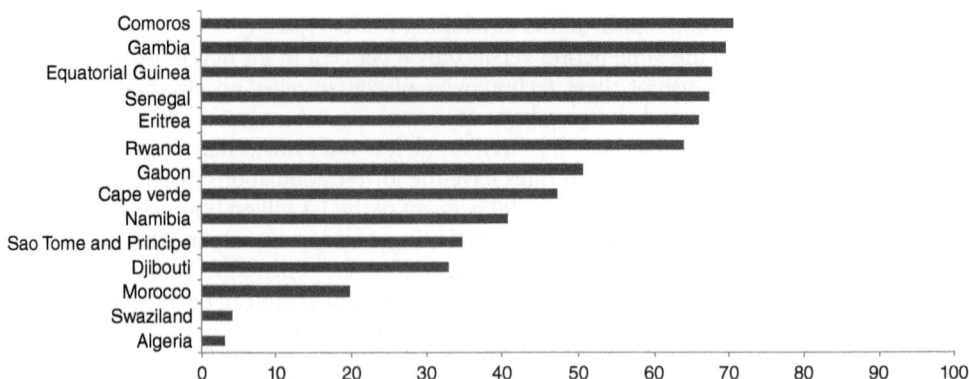

Figure 5.4. Percentage of primary schools without electricity in selected African countries

Source: UNESCO Institute for Statistics (UIS), <http//www.uis.unesco.org/datacenter>

5.2.4. Schools and public institutions

Many schools and other public institutions in LMICs also have no or inadequate electrical power. UNESCO, which regularly tracks national data related to basic services such as electricity access in schools, indicates that Sub-Saharan Africa has the lowest rate of electrification in primary schools at 29 per cent, while South Asia and Latin America had higher rates at 48 per cent and 93 per cent respectively (UN 2014; Practical Action 2013) (Figure 5.4). Access also varies markedly between urban and rural areas. For example, 27 per cent of village schools have electricity in four states in India compared to 76 per cent of schools in towns and cities in 2010 (Practical Action 2013).

Energy access is also important for the effective functioning, at the community level, of public institutions other than schools and health facilities. However, due to the wide range of existing public, private, NGO, or faith-based institutions across LMICs, little consistent data exist and the level of energy access is largely unknown (Practical Action 2013).

5.2.5. International efforts to address energy access inequalities

Growing recognition in recent years of the vast global inequalities in access to modern energy at the household and community level, and the consequences for health, economic development, and the environment, has led to the development of a number of important global and regional initiatives. Two of the most important examples are briefly described below, with links for obtaining further up-to-date information.

Sustainable Energy for All (SE4All):
<http://www.sustainableenergyforall.org/>

This special initiative of the UN Secretary-General was developed in response to recognition of the need to address energy access in order to meet the Millennium Development Goals. The overarching goal is universal access to modern household energy by 2030, with the focus on three aspects, namely to (i) ensure universal access to modern energy services (including electricity and clean, modern cooking solutions), (ii) double the global rate of improvement in energy efficiency, and (iii) double the share of renewable energy in the global energy mix. A strong emphasis on monitoring progress is reflected in the recently published report describing plans for a detailed tracking framework (UN 2013).

UNF Global Alliance for Clean Cookstoves: <http://www.cleancookstoves.org/>

Established in 2010, this public–private alliance hosted by the UN Foundation has the mission to 'save lives, improve livelihoods, empower women, and protect the environment by creating a thriving global market for clean and efficient household cooking solutions'. The goal is to foster the adoption of clean cookstoves and fuels in 100 million households by 2020. Among the key activities are the development of international standards for stoves and fuels, regional testing centres, support for the development of national action plans, and a robust evaluation strategy.

5.3. NATURE AND EXTENT OF ENERGY IMPACTS ON HEALTH

In Section 5.1, a socio-environmental model of health determinants was referred to in describing the many ways in which energy access, and the quality or lack of energy, can impact health. Section 5.2 has shown that there are billions of people, all in the world's developing countries, who lack energy for basic household needs, for everyday activities that can help feed their families and reduce poverty, and for their health care, education, and other local services. In this section, we consider in more detail how this energy access situation affects health, and we provide quantification of the impacts where this is available.

5.3.1. Basic household needs: cooking, heating, and lighting

Here we focus on the impacts of energy access in respect of cooking, heating (and cooling), and lighting. In most homes these needs will be met by a mix of devices and fuels, although in some of the poorest an open fire may be the main or sole source for all three. These uses of household energy are probably the most thoroughly studied of all the linkages between energy and health.

Cooking

The most important direct health impacts from energy used for cooking, heating, and lighting arise from air pollution and from burns, scalds, and poisoning. Kerosene is a source of HAP exposure, as well as being an important cause of burns and poisoning; more detail on the health risks of air pollution from this fuel as used in LMICs is provided in the subsequent discussion of lighting.

Household air pollution

The most recent assessment of risks from solid fuel HAP exposure were reported by Lim et al. as part of the Global Burden of Disease (2010) study of comparative risk assessment (CRA) (Lim et al. 2012). Many diseases are now linked to HAP exposure, including most of those which are known also to be caused by active tobacco smoking; this is perhaps to be expected since tobacco is a form of biomass, and while the dose may be higher in most smokers, HAP exposure starts much earlier in life (before conception and throughout pregnancy). The diseases included in the CRA (the criteria for inclusion being sufficiently strong evidence and being one of the eligible GBD project health outcomes) were child acute lower respiratory infection (ALRI), chronic obstructive pulmonary disease, ischaemic heart disease (IHD), stroke, and cataracts. Among the other conditions for which evidence is mounting of a link with solid fuel HAP, but not yet included in the CRA, are low birth-weight, pre-term birth, stillbirth, other cancers (oro-pharyngeal; uterine cervix), asthma, tuberculosis (TB) and child cognitive development. The CRA found that HAP exposure was responsible for around 3.5 million premature deaths (uncertainty interval 2.7 to 4.5) and 111 million (uncertainty interval 87 to 138) disability-adjusted life years (DALYs)[5]; furthermore 0.5 million deaths due to outdoor air pollution could be attributed to contributions from household fuel combustion.

As the number of solid fuel users, some 2.8 billion, has remained stable for the last 30 years (see Section 5.2.1), this public health burden will not diminish without radical action to assist households in switching to less polluting technologies and fuels (IEA 2012). In that regard, one of the striking findings from new research conducted on exposure–response relationships for the CRA was that for diseases such as child ALRI, and probably for IHD and stroke, HAP will need to be reduced to low levels in order to secure most of the available health benefits (Burnett et al. 2014).

Burns and poisoning

All household energy devices present a risk of burns (including electric shocks); cooking likewise will always pose a risk of scalds, and chemicals (e.g. liquid fuels) a risk of poisoning. The technologies and fuels commonly used in developing countries, however, appear to present a particularly high risk of such injuries. A systematic review of these risks, recently conducted for new WHO guidelines on household fuel combustion (Bruce et al. 2013), provide an assessment of this situation.

The recent GBD project assessment estimated that in 2010 there were almost 340,000 (UI: 235,000 to 434,000) deaths globally from fire, heat, and hot substances (Lozano et al. 2012). Furthermore, for every death, there are many more severe non-fatal injuries resulting in lasting physical and mental consequences. As a result, the tally for combined loss of life and non-fatal injury (expressed in

---

[5] One DALY represents one lost year of 'healthy' life.

DALYs[6]) was 19 million (UI: 13.3 to 24.1 million) (Murray et al. 2012). While population-based data are incomplete, it is known that more than 90 per cent of these deaths occur in LMICs. While there are common risk factors for burn injuries in both highly industrialized countries and LMICs (such as age, gender, poverty), the presence and use of cookstoves in LMICs pose unique risks, and there is considerable consistency from available studies about the critical role of cookstoves and related risk factors as a key source of mortality, morbidity, and disability from burn injuries, with a substantial proportion of these burns being sustained in the kitchen or cooking area and associated with the use of cookstoves and lamps. Women and children in particular are at increased risk for these types of injuries.

The principal cause of household-energy-related poisoning in LMICs is kerosene, used by some homes for cooking but more commonly for lighting. While it is clear that child poisoning from kerosene ingestion remains an important public health problem (the fuel is often sold and/or stored in soft-drink bottles), the incidence rates of events and adverse consequences are not well defined due to the lack of population-based studies.

Space heating and cooling

The health risk from stoves and fuels used for heating are essentially the same as those for cooking, but have not thus far been quantified in the manner reported in the 2010 CRA due to data on heating practices being very incomplete (see Section 5.2.1). These risks may be similar so long as traditional stoves and open fires provide for both cooking and heating. With the introduction of cleaner and more efficient cooking solutions (e.g. enclosed/insulated and vented solid fuel stoves and modern fuels such as LPG), however, which are either ineffective for heating or too costly to use for this purpose, there is a risk that families may revert to using the more open solid-fuel stove for warmth, drying clothes in the wet season, and so on. This would result in a more complex picture of health risks, and requires both careful monitoring of multiple devices/fuels used for multiple household needs (see for example the UN tracking framework, UN 2013) and policy responses to avoid continuing high levels of HAP exposure in regions that need heating.

As noted in Section 5.2.1, households may have few options for dealing with high temperatures as, apart from house design and materials, active control methods (i.e. fans, air conditioning) depend primarily on sufficient supplies of electricity (Haines et al. 2012). In terms of health impacts relating to temperature extremes, there is ample evidence of the consequences, with excess deaths during both winters (Healy 2003) and heat-waves (Le Tertre et al. 2006) being well-recognized in developed countries; these same risks can be expected to play out in low-income settings, although little documentation is available.

---

[6] DALYs for a health condition are calculated as the sum of the years of life lost (YLL) due to premature mortality in the population and the years lost due to disability (YLD) for people living with the health condition or its consequences.

Lighting

The main direct health risks are from household air pollution, and from burns and poisoning; the most important source of these risks is kerosene lamps, and for HAP the largest source are the simple wick types that are so widely used in low-income settings. A recent systematic review has documented the use, emissions, air pollution levels, and epidemiological evidence on health risks for kerosene use in LMICs (Lam et al. 2012). This review found that although fuel grade and contaminants (e.g. sulphur), combustion source and type (e.g. lamp or stove), and operator conditions all impact emissions, there is ample evidence that use of household kerosene devices can lead to particulate matter (PM) levels that exceed WHO guidelines (WHO 2006), substantially so in developing country homes. Levels of CO, PAH, and SO_2 may also exceed WHO guideline levels (WHO 2010). The epidemiological evidence, however, although covering a range of outcomes, does not yet allow strong conclusions nor reliably quantified risk estimates. There is some suggestion of increased risks of cancer, respiratory symptoms, eye disease (cataracts), and respiratory infections (including child ALRI and TB), but interpretation is made difficult by inconsistent results, varying outcome definitions, and sometimes uncertain exposure comparisons (Lam et al. 2012).

Overall, however, the combination of widespread use for lighting, high levels of exposure to PM and other health-damaging pollutants, and tentative epidemiological evidence suggests that HAP from kerosene combustion in the home should be considered as carrying significant health risk. This is in addition to the risks of burns, fires, and poisoning related to cooking fuels.

The indirect impacts of lighting on health, and in particular a lack of adequate light in the home, are likely to be significant but are problematic to quantify. Insufficient lighting will restrict a wide range of activities, including children doing homework, and income-generating activities conducted in the home where good light is needed and where people wish to carry this work out in the evening or early mornings. Adequate lighting can contribute to safety, for example for young children to avoid the stove or hot cook pots in otherwise dark homes, and for personal safety around and near the house.

5.3.2. Other household and community activities and services

Income-generating activities

Although the relationship is complex and not well documented, broad trends indicate that increased energy consumption over time improves human health by encouraging socio-economic development, consistent with the socio-environmental health determinant model described in Section 5.1. According to a study by Wilkinson et al. using global data from UNDP and WRI to compare infant mortality and life expectancy with per capita energy consumption (per kg of oil), it can be seen that since 1800, increasing use of fossil fuels has been associated with roughly a doubling of global life expectancy and an increase in the average human body size by 50 per cent (Wilkinson et al. 2007). A study reported in the *Poor People's Energy Outlook 2010* using global per capita energy consumption data and the UN Development Programme's human development index (HDI) shows a clear correlation between energy access and health and development, as measured by this index (Practical Action 2010). HDIs combine indicators from three dimensions: life expectancy at birth, mean years of schooling, and GNI per capita (Anand and Sen 1994). Countries with the lowest energy consumption were also those with low values of the HDI and are mostly located in Sub-Saharan Africa (Practical Action 2010).

Energy access provides the ability for households to create new earning opportunities, for example opening food stalls at night time and starting other service or manufacturing-based enterprises. Furthermore, energy access improves the efficiency and output of existing commercial activities and services and also reduces opportunity costs by freeing up time previously spent on tasks such as gathering fuel (Practical Action 2012). For example, the *Poor People's Energy Outlook* 2012 cites a UNDP project conducted in Mali in which energy access in the agricultural sector benefited women by reducing time spent physically processing the crops (Bates et al. 2009). The studies found that these women saved on average 2 to 6 hours per day, and 4 out of the 12 studies included in the project indicated that the time saved was spent on new commercial activities (Bates et al. 2009).

Water

Provision of adequate amounts of water to households, including clean water for human consumption, is at least in part dependent on access to energy. From water pumping, transportation, and distribution to water treatment including boiling and wastewater treatment, energy access is an important part of the water usage chain.

In the absence of comparable data for LMICs, and to provide a benchmark, Table 5.3 shows the average European energy costs to produce clean water. Overall, the European average electricity consumption for water corresponds to approximately 5000 kWh/year/person (Olsson 2011). In the developing world, countries which lack reliable access to electricity have difficulties in obtaining and recycling water for household use. According to the World Bank, per capita power consumption in the least developed countries is 178 kWh per year (World Bank 2010). Clearly, without sufficient power supply, boiling water with electricity for point-of-use disinfection is impossible. For water boiling using solid fuels, Mintz et al. found that one kilogram of firewood was needed to boil one litre of water for a minute (Mintz et al. 1995). Since an average person requires a minimum of two litres of drinking water per day, preparing clean water for consumption for a family of five will place significant additional demands on the need for fuel and will add to emissions of HAP into homes where traditional solid fuel stoves are used (WHO/WEDC 2011).

Table 5.3. The energy footprint of water supply and treatment

Type	Energy footprint (kWh/m^3)
Surface water	0.5–4
Recycled Water	1–6
Desalination	4–8
Bottled water	1,000–4,000

Source: WssTP 2011 as cited in Olsson 2012

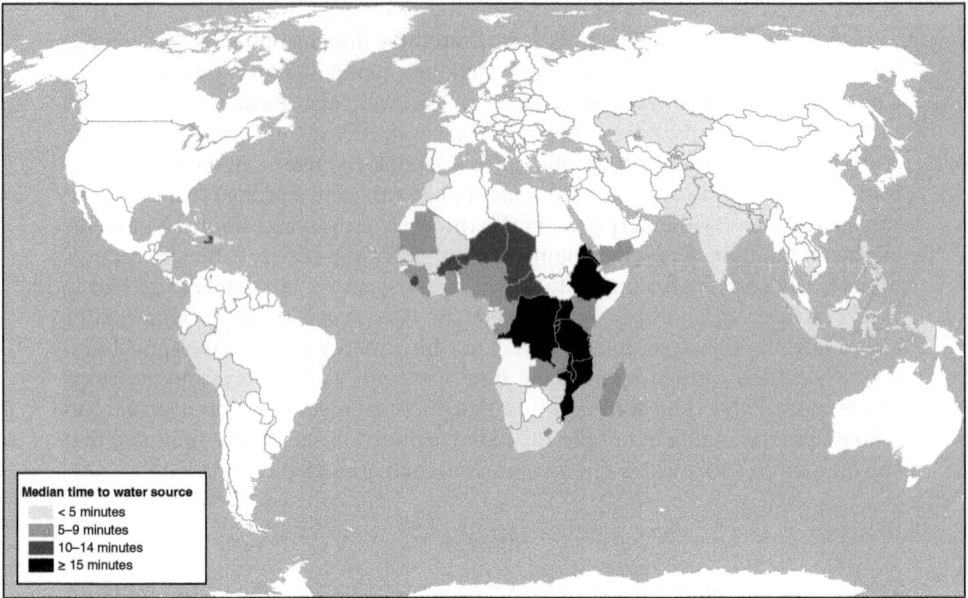

Figure 5.5. Distance to water source: median time

Thus energy access is one of the key determinants for access to clean water, which in turn is critical for the prevention of diarrhoeal disease, skin and eye infections, and typhoid (Howard and Bartram 2003; Fewtrell et al. 2005). Howard and Bartram also describe the influence of distance and time to collect water on hygiene-related health, with their study showing that households most at risk are those without immediate access to taps or piped connections who must travel more than 100 metres (more than 5 minutes) to collect water (Howard and Bartram 2003). According to USAID data using the Statcompiler tool, in 2012, many Sub-Saharan African countries required times of more than 9.5 minutes to a water source (USAID 2012) (see Figure 5.5).

Moreover, even when water is available, there is risk of disease due to improper storage before consumption as well as the possibility of contamination if the water is not properly separated from human waste. A systematic review of water, sanitation, and hygiene interventions to reduce diarrhoea in less developed countries by Fewtrell et al. concluded that diarrhoeal episodes are reduced by 22 per cent through improving water supply, 22 per cent by improving sanitation, 43 per cent through hand washing, and 17 per cent by improving water quality (Fewtrell et al. 2005).

The IEA predicts that the water sector's dependency on energy supply will only increase in the future due to the following factors: greater stress on water supply from population growth and increased standards of living; increasing transport of and pumping/drilling for water as freshwater supplies are used up; stricter national water quality standards; and a shift in irrigation practices to more energy-intensive methods (IEA 2012).

Energy use in the agricultural production chain

Energy: electricity, mechanical power, household fuels/thermal

Production	Processing/storage	Distribution
Land preparation	Drying	Infrastructure and transport
Irrigation	Milling	ICTs
Fertilizing	Pressing	Training
Crop protection	Packing	Selling
	Storing	

Agriculture–health system

Nutrition
Employment
Income

Agriculture → Health

Labour and resources

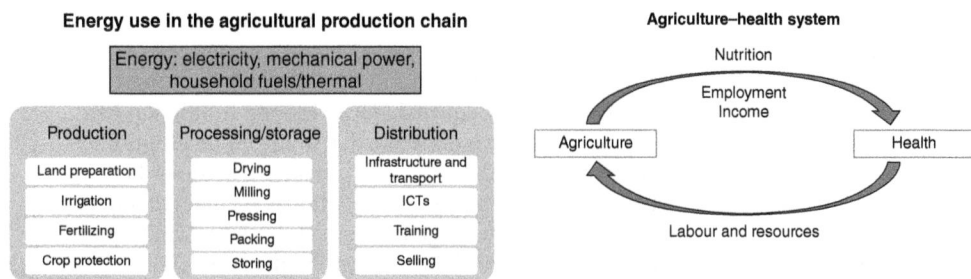

Figure 5.6. The energy–agriculture–health nexus

Source: adapted from Practical Action 2012 and Dangour et al. 2012

Agriculture

Energy sources play an integral part in agriculture–health linkages. From land cultivation and irrigation to fertilizer supply, energy inputs are an invaluable resource for agricultural production and directly impact health by increasing the quantity and nutritional composition of foods available, as well as contributing to income generation where produce is sold or bartered (Figure 5.6). An FAO study of farm power in smallholder livelihoods in Sub-Saharan Africa found that households using farm-power techniques other than the hoe gain considerable advantages in terms of area cultivated, crop diversity, yields, opportunities to redeploy family labour, and household food security (FAO 2006).

Furthermore, energy use in agricultural production can increase efficiency and reduce the health burdens on farmers. For example, in Sanchitagi, Nigeria, hoe cultivators studied in an FAO technical report complained of musculoskeletal pains and that the low rate of work possible compared to use of tractors was one reason why they remained poor (FAO 2008). Finally, agriculture also indirectly influences human health via its impact on employment and other socio-economic benefits.

In addition to the contributions that energy can make to health through increasing productivity, it is also important to recognize that poor health can also impact negatively on agriculture by restricting the ability of people with poor health to carry out effective farm work. Thus, less healthy populations have reduced labour availability and capacity, and consequently reduced output and productivity (Figure 5.6). This factor can partially explain why agricultural outputs have tended to suffer in some regions with higher prevalence of infectious diseases and low-quality health care. According to the United Nations, HIV/AIDS in Sub-Saharan Africa has been a major threat to the agricultural labour base (UN 2004).

Communications

Information and communication technologies (ICTs) can contribute to health improvement for individuals and households through the dissemination of public health information, by enabling remote consultation and calls for assistance and, at the level of service provision, by strengthening the ability to monitor the incidence of public health threats and improvements in the efficiency of health care administration (Practical Action 2010).

The mobile phone has been an especially important technology amongst the ICTs in improving health care and services in LMICs. Mobile health services[7] have gained considerable momentum in recent years due to the rapid adoption of mobile phones both by individuals and by service providers, as described in Section 5.2.2 (Mechael et al. 2010). The use of mobile phones includes standard voice and text messaging, multimedia messaging, and web browsing and email functionalities (though the latter may still be cost-prohibitive in LMICs) for treatment compliance, data collection, and disease surveillance, providing health information and point-of-care support, and providing emergency medical response (Mechael et al. 2010), Table 5.4.

Table 5.4. Mobile health applications, trends, and projects

Application areas	Trends	Project examples
Treatment compliance	• SMS remind patients to take drugs and attend appointments • LMIC studies focus more on infectious disease and drug adherence	• 'WelTel Support for Clinical Management of Patients' (Kenya) measures effectiveness of SMS in improving patient adherence to highly active retroviral therapy • 'X Out TB' (Pakistan) to promote patient treatment compliance with an incentive structure
Point of care support and diagnostics	• Mobile tools enable health-care workers to access pertinent information to increase diagnostic and treatment accuracy	• 'Pambazuko PALM' (Kenya) developed to collect patient risk assessment data and deliver counselling protocol via PDAs • 'Mobile Teledermatology Service' (Botswana, Malawi) uses mobile phones to capture and send images of patients to specialists in other African countries
Education and awareness	• Use of SMS to disseminate health information and prevention messaging • Has been critical in increasing awareness of confidential and often stigmatized issues including sexual behaviour, family planning, and sexually transmitted diseases	• 'Texting for Health' (Uganda) encouraged learning about HIV/AIDS through a short health quiz via SMS
Data collection and disease surveillance	• PDAs replace manual input of data though literature shows mixed results of reduction in errors and cost savings • Data also collected on SMS and voice and health call centres	• 'District Health Information System' (Malawi, Rwanda, South Africa) collects data on routine health-care events from clinics • 'TRACnet system' (Rwanda) aggregates data on the care of patients infected with HIV from large numbers of clinics

Source: Adapted from Mechael et al. 2010, <http://mhealthinfo.org>, and Piette et al. 2012

---

[7] Mobile health refers to 'the enhancement of health-related services using mobile phones' (see for example <http://www.mhealthinfo.org>).

While there is a growing body of literature on the subject of mobile health which suggests direct linkages between mobile telecommunications and improved patient health and behaviours including care seeking, compliance with treatment, and advice, there are still significant gaps, notably in systematic cross-country data, which prevent clear conclusions about benefits (Mechael et al. 2010).

5.3.3. Community services

Health services

Energy access, in particular reliable supplies of adequate electrical power, is critical for the delivery of high-quality health care. Table 5.5, reproduced from the Poor People's Energy Outlook (2013), shows the wide range of health facility amenities, diagnostic and treatment medical devices, and other infrastructure that depends on adequate electrical power.

Table 5.5. Energy services required for service readiness in health facilities

Purpose/service	Energy service/equipment
General amenities/infrastructure	Lighting—clinical/theatre, ward, offices/admin, public/security Mobile phone charter, VHF radio, office appliances
Basic amenities and equipment	Cooking, water heating, space heating Refrigerators, air circulation Sterilization equipment
Potable water for consumption, cleaning, and sanitation Health-care waste management	Water pump (when gravity-fed water not available) and purification Waste autoclave and grinder
Service-specific medical devices Cold chain and Expanded Programme on Immunization (EPI) refrigeration Maternity and mother/child health HIV diagnostic capacity Outpatient department (OPD) Laboratory and diagnostic equipment	Vaccine refrigerator Suction apparatus, incubator, other equipment ELISA test equipment (washer, reader, incubator) Portable X-ray, other equipment Centrifuge, haematology mixer, microscope, blood storage, blood typing equipment, blood glucose meter, X-ray, ECG, ultrasound, CT scan, peak respiratory flow meter
Surgical equipment	Equipment and facilities for: tracheostomy; tubal ligation; vasectomy; dilatation and curettage; obstetric fistula repair; episiotomy; appendectomy; neonatal surgery; skin grafting; open treatment of fracture; amputation; cataract surgery
Additional infrastructure External lighting Staff housing Emergency transportation	Security lights at front gate, main doors and around buildings, outside toilet block, walkway lights Lighting, TV, AM/FM stereo Other appliances (mobile phone charger, electric fan, etc.) Cooking and water heating Vehicle or motorbike

Source: Practical Action 2013

Thus without energy services and equipment, effective delivery of many basic health services would not be possible. For example, unreliable electricity supply compromises vaccine refrigeration, and it is estimated that half of all vaccines delivered to developing countries are ruined due to poor cold chain services (Practical Action 2013). Besides the numerous direct impacts on health, energy access in health facilities also indirectly impacts patient health by affecting the quality of the health workers that the facility attracts. As cited in *Access to Energy for Health Services*, for example, a World Bank study in 2010 found that in Bangladesh, health workers preferred to live in communities with electricity. Electricity access in health facilities also reduced the risk of absenteeism.

While the lack of electricity would seem to present an obvious barrier to the delivery of effective health care, clear evidence on the links between energy access rates of health facilities and health outcomes is lacking.

Education services

Education is a key factor in securing better health for individuals and populations, including through the opportunities it brings for employment and economic development, and through the ability to learn about and make healthier choices for self, family, and community (Feinstein et al. 2006). Energy can contribute to effective learning through the provision of adequate lighting and power for ICT equipment, although, as for health services, there is little in the way of evidence that links energy, education, and improved health outcomes.

The type and availability of energy also directly impacts children's health through, for example, food preparation, heating, and cooling in schools, which affect nutrition, concentration, and comfort. According to case studies from the *Poor People's Energy Outlook* (2013), for example, low light in schools in Bangladesh has made teaching difficult and as a consequence, teachers opened windows and shutters during the cold seasons to maximize lighting in the classroom (Practical Action 2013). Many schools prepare food on solid-fuel stoves, exposing children to additional air pollution; in a case study from rural Bolivia, it was found that students spent school time collecting firewood in order to cook their midday meal (Practical Action 2013).

Other (public institutions)

In addition to the contributions to health from well-functioning health facilities and educational establishments, a range of other public, private, and NGO-sector institutions are also important for the management and delivery of local government, services, roads, social support, and business development, all of which have the potential to impact on health (Figure 5.7).

Due to the limited availability of data on energy access in public institutions, there is little or no statistical evidence linking it to health. A few case studies reported in *Poor People's Energy Outlook* 2013, however, emphasize the potential importance of energy access in public institutions for health. In Liberia, for example, installations of small solar systems provided energy access to local police stations, which increased police presence and subsequently increased public trust in the national police and sense of safety (Practical Action 2013).

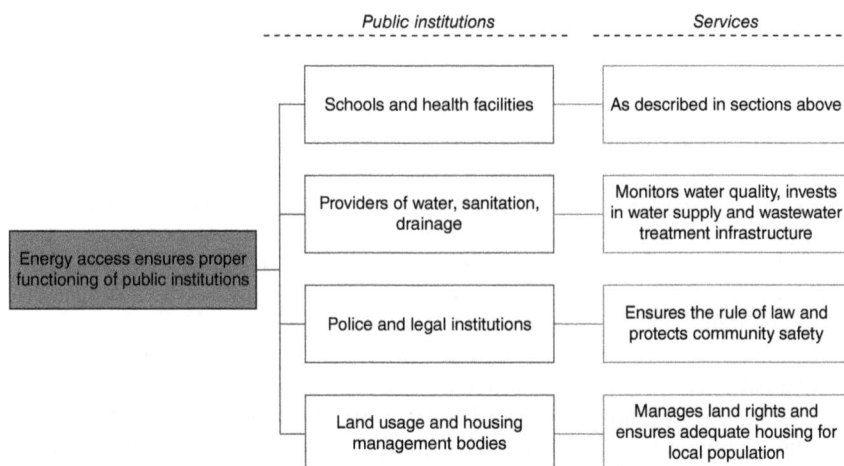

Figure 5.7. Services and contributions of public institutions
Source: Adapted from Bigg 2005

5.3.4. Health effects of energy use mediated via the environment

Local environment and forest resources

Households reliant on biomass fuel, and especially on wood, either purchase or collect this; in rural areas, the majority obtain fuelwood from forest areas and may also use crop wastes, dung, and other biomass according to location and season. These forest areas are subject to multiple demands in addition to fuelwood, including clearance for agricultural land and for building materials. The extent to which forests are managed sustainably varies greatly, but in many areas families (and it is usually women and children) have to spend many hours per week collecting fuel; where wood resources are becoming depleted, the time spent and distance covered are increasing. Women and children are vulnerable to injury and violence while collecting fuel, including sexual violence. Clearly, the type of fuel used in the home as well as the amount, which is a function of stove efficiency, will have implications for these health threats to women and children. The time saved from reducing fuel collection, or eliminating it through alternative fuels, may be used in ways which can have positive impacts on personal and family health.

Global environment

Climate change has important implications for health, particularly in low-income countries—for example, through extremes of weather, drought, and changes in the distribution of vector-borne disease. Household energy use can impact climate change in two main ways.

First, as discussed above, in some areas forest resources are not managed sustainably, so combustion of biomass in the home results in net CO_2 emissions to the atmosphere (Nabuurs et al. 2007). Globally, these contributions are not proportionally large, and the importance of this mechanism (apart from issues of forest and land management) lies in this being the basis for carbon-finance support for improved solid fuel stove programmes which reduce net CO_2 emissions through greater efficiency and hence less fuel use.

The second mechanism is through emission of substantial quantities of 'products of incomplete combustion' (PICs) from open fires and traditional solid fuel stoves, and also from kerosene lamps. These PICs include a number of important short-lived climate pollutants including black carbon, CO, methane, and non-methane volatile organic compounds. The importance of this mechanism is that not only do these PICs make a significant contribution to climate change over the short term, but they also pose an immediate risk to the health of those exposed to these pollutants (UNEP 2011).

The impacts of household energy on health through these various environmental pathways are therefore important. The direct health risks from emissions of health-damaging pollutants have been quantified as described in Section 5.3.1, but those acting via local environmental conditions and climate change are highly context-specific and thus far less amenable to quantification. One important policy implication of the linkages, however, is that action to increase the efficiency of household fuel combustion will deliver relatively quick health benefits while also contributing to mitigating climate warming (Wilkinson et al. 2009), with the latter also helping to reduce adverse health impacts of climate change over the longer term.

5.4. CONCLUSIONS

Drawing on a socio-environmental model of health determinants, this chapter has reviewed the many and varied ways in which the current energy access situation for households and communities in LMICs is damaging health and in which universal access to clean, efficient and safe energy could improve health and reduce inequalities, with all the benefits to economic development and social cohesion that this can bring.

5.4.1. Quantification of health benefits

In part due to the nature of the available evidence, but also because many of the energy and health linkages discussed are context-dependent and typically operate together in low-income settings with complex and variable interactions, quantification of these health benefits in the round is problematic. Nevertheless, the public health burden of some important components has recently been evaluated. Household air pollution (HAP) from cooking with solid fuels was estimated to account for almost 4 million premature deaths (and 111 million DALYs) in 2010, and a high proportion of the 340,000 deaths (19 million DALYs) in the same year from burns and scalds can also be attributed to household energy sources in LMICs.

Beyond these two causes, the health burden resulting from the effects of inadequate and poor-quality energy availability on many factors—water; agriculture; income generation; health, educational and other local services; and local and global environments—can be recognized as important but is currently lacking in any reliable means of quantification. While studies are available of the impacts through multiple applications of electricity access on rural development, for example in a case study of micro-grid power in rural Kenya (Kirubi et al. 2009), we are not aware of similar studies specifically focused on health improvement.

While it may therefore be useful to conduct further research on the pathways and actual health impacts of improved energy access for these additional household and community activities and services, it can also be argued that there is already a strong enough case for action. Thus, taking the example of health facilities, Adair-Rohani et al. report on the UN interagency list of essential devices for reproductive health requiring electricity (Adair-Rohani et al. 2013) and similar considerations should apply to education, water supply, agriculture, and so on.

One further angle on the overall benefits of access to clean, safe, and efficient energy can be obtained from economic cost-benefit analysis (CBA), although this has not so far attempted to address the challenge of accounting for the full range of health impacts for the reasons discussed above. CBA allows a summary of benefits to health, time-saving, the environment, and so forth, by expressing these in monetary terms and comparing this total to the costs of implementation. Published CBA studies on household energy improvements have been restricted to reducing household air pollution and have generally found that benefits exceed costs at both societal and household levels, with time-saving being a major factor (Hutton et al. 2006; Malla et al. 2011). If more complete accounting of health benefits could be included, this analytic approach could make a valuable contribution to the case for investment in rapid improvements in access to energy.

5.4.2. Policy on energy for health

Even allowing for the limitations to data and evidence, given the multiple health benefits including those for which disease burdens are already apparent, it would seem clear that the health sector has much to gain from contributing to, and advocating for, policy that accelerates access to clean, safe, and efficient energy for homes, communities, and their local services.

To date this has not been the case, with most of the effort coming from other ministries and agencies including those representing energy, environment, women and rural development, and so on. A clearer understanding of the reasons for this apparent lack of connection between the potential for health gains and public health response would help shed light on the barriers and would clarify the role that the health sector can play. These reasons should be the subject of further investigation, but would seem to include the challenges of implementing inter-sectoral collaboration and a poorly defined role for the health sector in a policy area where the main investments are perceived to be the responsibility of other sectors.

Supporting policy, including clarifying the roles of the various sectors, is part of the contribution that can be provided by the new initiatives on energy access referred to in Section 5.2.5, including the UN Sustainable Energy for All and the UNF Alliance. The Alliance, with partners such as WHO which has strong country infrastructure for supporting health ministries, is working to facilitate development of country action plans led by multi-sectoral task groups, within which the health sector can have a strong voice and a key role. Within this framework, full recognition of the benefits to health could add substantially to the priority given by governments and other stakeholders in making the necessary policy changes and in encouraging the necessary investment by the public sector, private investors, and development assistance that the IEA and others have identified (IEA 2012).

Finally, despite the enormous potential for energy access to bring benefits to health, the environment, and development, success in this field cannot be taken for granted (Hanna et al. 2012). Monitoring and evaluation should be built into budgets and planning; the plans for the new UN energy access monitoring framework are a very positive step towards this goal (UN 2013) but must also be accompanied by assessment of impacts on health (Martin et al. 2013).

REFERENCES

Adair-Rohani, H., et al. (2013). 'Limited electricity access in health facilities of Sub-Saharan Africa: A systematic review of data on electricity access, sources, and reliability', *Global Health: Science and Practice* 1(2): 249–60.

Anand, S. and Sen, A. (1994). *Human Development Index: Methodology and Measurement*, Human Development Report Office (HDRO), United Nations Development Programme (UNDP).

Bates, L., Hunt, S., Khennas, S., and Sastrawinata, N. (2009). *Expanding Energy Access in Developing Countries: The Role of Mechanical Power*. Bourton on Dunsmore, Rugby, UK: Practical Action.

Bigg, T. and Satterthwaite, D. (2005). *How to Make Poverty History: The Central Role of Local Organizations in Meeting the MDG*. London: International Institute for Environmental Development.

Bonjour, S., et al. (2013). 'Solid fuel use for household cooking: Country and regional estimates for 1980–2010', *Environmental Health Perspectives* 121(7): 784–90.

Bruce, N., et al. (2013). 'Tackling the health burden from household air pollution (HAP): Development and implementation of new WHO Guidelines', *Air Quality and Climate Change* 47(1): 32–8.

Budya, H. and Arofat, M. (2011). 'Providing cleaner energy access in Indonesia through the megaproject of kerosene conversion to LPG', *Energy Policy* 39(12): 7575–86.

Burnett, R.T. et al. (2014). 'An integrated risk function for estimating the global burden of disease attributable to ambient fine particulate matter exposure' *Environmental Health Perspectives* 122(4): 397–403.

Dahlgren, G. and Whitehead, M. (1991). *Policies and Strategies to Promote Social Equity in Health*. Stockholm: Institute for Future Studies.

FAO (Food and Agriculture Organization of the United Nations) (2003). *World Agriculture: Towards 2015/2030. An FAO Perspective*. London: FAO.

FAO (2005). *Contribution of Farm Power to Smallholder Livelihoods in Sub-Saharan Africa*. Rome: FAO.

FAO (2006). *Farm Power and Mechanization for Small Farms in Sub-Saharan Africa*. Rome: FAO.

FAO (2008). *Agricultural Mechanization in Sub-Saharan Africa: Time for a New Look*. Rome: FAO.

Feinstein, L., et al. (2006). 'What are the effects of education on health?' Organisation for Economic Co-operation and Development (OECD) [online document], <http://www.oecd.org/edu/innovation-education/37425753.pdf> (accessed 21 October 2013).

Fewtrell, L., et al. (2005). 'Water, sanitation, and hygiene interventions to reduce diarrhoea in less developed countries: a systematic review and meta-analysis', *Lancet Infectious Diseases* 5: 42–52.

Haines, A., et al. (2012). 'Promoting health and advancing development through improved housing in low-income settings', *Journal of Urban Health* 90(5), October: 810–31.

Hanna, R., Duflo, E., and Greenstone, M. (2012). 'Up in smoke: The influence of household behavior on the long-run impact of improved cooking stoves', Working Paper 12–10, Massachusetts Institute of Technology Department of Economics Working Paper Series, Social Science Research Network Paper Collection.

Healy, J. (2003). 'Excess winter mortality in Europe: A cross country analysis identifying key risk factors', *Journal of Epidemiology and Community Health* 57: 784–9.

Howard, G. and Bartram, J. (2003). *Domestic Water Quantity, Service Level, and Health.* Geneva: World Health Organization.

Hutton, G., et al. (2006). *Evaluation of the Costs and Benefits of Household Energy and Health Interventions at Global and Regional Levels.* Geneva: World Health Organisation.

IAH (International Association of Hydrogeologists) (2006). 'Groundwater and rural water supply in Africa.' Accessed September 2013.

IEA (International Energy Agency) (2012). *World Energy Outlook.* Paris: International Energy Agency.

Kirubi, C., et al. (2009). 'Community-based electric micro-grids can contribute to rural development: Evidence from Kenya.' *World Development* 37(7): 1208–21.

Lam, N., et al. (2012). 'Kerosene: A review of household uses and their hazards in low- and middle-income countries', *Journal of Toxicology and Environmental Health, Part B: Critical Reviews* 15(6): 396–432.

Le Tertre, A., et al. (2006). 'Impact of the 2003 heatwave on all-cause mortality in 9 French cities', *Epidemiology* 17: 75–9.

Lim, S., et al. (2012). 'A comparative risk assessment of burden of disease and injury attributable to 67 risk factors and risk factor clusters in 21 regions, 1990–2010: A systematic analysis for the Global Burden of Disease Study 2010.' *Lancet* 380(9859): 2224–60.

Lozano, R., et al. (2012). 'Global and regional mortality from 235 causes of death for 20 age groups in 1990 and 2010: A systematic analysis for the Global Burden of Disease Study 2010.' *Lancet* 380(9859): 2095–128.

Malla, M., et al. (2011). 'Applying global cost-benefit analysis methods to indoor air pollution mitigation interventions in Nepal, Kenya and Sudan: Insights and challenges', *Energy Policy* 39(12): 7518–29.

Martin, W., et al. (2013). 'Household air pollution in low- and middle-income countries: Health risks and research priorities', *PLOS (Public Library of Science) Medicine* 10(6).

Mechael, P., et al. (2010). *Barriers and Gaps Affecting Health in Low- and Middle-Income Countries.* Policy white paper, Columbia University Earth Institute, Center for Global Health and Economic Development (CGHED) with Health Alliance.

Mintz, E., et al. (1995). 'Safe water treatment and storage in the home: A practical new strategy to prevent waterborne disease', *Journal of the American Medical Association*, 273(12): 948–53.

Murray, C., et al. (2012). 'Disability-adjusted life years (DALYs) for 291 diseases and injuries in 21 regions, 1990–2010: A systematic analysis for the Global Burden of Disease Study 2010.' *Lancet* 380: 2197–223.

Nabuurs, G., et al. (2007). 'Forestry', in B. Metz, O.R. Davidson, P.R. Bosch, R. Dave, and L.A. Meyer, eds., *Climate Change 2007: Mitigation. Contribution of Working Group III to the Fourth Assessment Report of the Intergovernmental Panel on Climate Change,* Cambridge and New York: Intergovernmental Panel on Climate Change.

Olsson, G. (2011). 'Water and energy nexus', in R.A. Myers, ed., *Encyclopedia of Sustainability Science and Technology*, London: Springer.

Piette, J., et al. (2012). 'Impacts of e-health on the outcomes of care in low-and middle-income countries: Where do we go from here?' *Bulletin of the World Health Organization* 90(5): 365–72.

Practical Action (2010). *Poor People's Energy Outlook*. Rugby, UK: Practical Action.

Practical Action (2012). *Poor People's Energy Outlook*. Rugby, UK: Practical Action.

Practical Action (2013). *Poor People's Energy Outlook*. Rugby, UK: Practical Action.

Sanou, B. (2013). *ICT Facts and Figures*. International Telecommunications Union.

UN (United Nations) (2004). *The Impact of Aids* [online document], <http://www.un.org/esa/population/publications/AIDSimpact/AIDSWebAnnounce.htm> (accessed 23 October 2013).

UN (2013). *Sustainable Energy for All: Global Tracking Framework*. Washington, DC: United Nations.

UNEP (United Nations Environment Programme) (2011). *Near-Term Climate Protection and Clean Air Benefits: Actions for Controlling Short-Lived Climate Forcers*. Nairobi: UNEP.

UN (2014). UN data [online], <http://data.un.org/> (accessed June 2014).

UNICEF/WHO (2012). United Nations Children Fund/World Health Organisation. *Progress on Drinking Water and Sanitation: 2012 Update*. New York, UNICEF and World Health Organization.

USAID (2012). Statecompiler.

Utz, V. (2011). *Modern Energy Services for Modern Agriculture: A Review of Smallholder Farming in Developing Countries*. GIZ-HERA.

WHO (World Health Organisation) (2006). *WHO Air Quality Guidelines Global Update 2005*. Copenhagen: World Health Organisation. EUR/05/5046029.

WHO (2010). *WHO Guidelines for Indoor Air Quality: Selected Pollutants*. Bonn: World Health Organisation.

WHO (2013). 'WHO household energy database', <http://www.who.int/indoorair/health_impacts/he_database/en/index.html>.

WHO/WEDC (2011). *How Much Water is Needed in Emergencies?* Geneva, World Health Organization.

Wilkinson, P., et al. (2007). 'A global perspective on energy: Health effects and injustices', *Lancet* 370(9591): 965–78.

Wilkinson, P. et al. (2009). 'Public health benefits of strategies to reduce greenhouse-gas emissions: Household energy', *Lancet* 374(9705): 1917–29.

World Bank (2010). 'Electric power consumption (kWh per capita)' [online document], <http://data.worldbank.org/indicator/EG.USE.ELEC.KH.PC> (accessed June 2014).

6

Energy Poverty and Public Health

Assessing the Impacts from Solid Cookfuel

*Kalpana Balakrishnan, Zoë Chafe, Tord Kjellstrom,
Thomas E. McKone, and Kirk R. Smith*

6.1. INTRODUCTION

Energy use is central to human activity to prepare food, to warm and cool homes, to power transport, and to produce goods, among many other activities of society that enhance population health. Energy services provide important direct health benefits; for example, refrigeration preserves food and allows storage of vaccines. Energy for lighting enables health clinics to operate after dark and it enables school children to read and do homework, which is important for better educational results. Energy for heating and cooling helps avoid hypothermia, heat exhaustion, heat stroke, and other health impacts of extreme conditions. Access to energy improves labour productivity and the overall quality of life.

The lack of sufficient energy services, therefore, is in itself a health risk. Without modern forms of energy (access to gas or electricity), it is difficult for a community to obtain many of the energy services that can be crucial to human health, such as advanced health care and access to vaccinations. As many as half of the vaccines that are delivered to developing countries become ineffective, in part because of a lack of reliable refrigeration (Practical Action 2013: 7). Proper sterilization of surgical instruments and other medical equipment is difficult without dependable access to electricity or gas, so many of the 50 to 60 million people who suffer wounds each year are at risk of infection, as are those that must undergo surgery (Practical Action 2013: 8). It is also impossible to run medical equipment that requires electricity (incubators, ultrasounds, X-ray machines, computers and communication technologies, etc.) unless a grid connection or local generation is available. New developments in photovoltaic electricity generation and storage, as well as wind power and small-scale hydropower, will make local electricity generation more convenient and cost-effective. In cases where electricity is not available, substitutes such as paraffin (kerosene) for lighting can present health risks in themselves.

Climate change will add to the need for cooling and refrigeration to maintain health and health services in the very populations already affected by energy poverty. These are also the populations most at risk from increased heat stress in outdoor working environments due to climate change, particularly agriculture and construction, for whom economic productivity will likely fall (Kjellstrom et al. 2009).

Although energy services are required for health, dirty, dangerous, and environmentally disruptive energy supplies can lead to disease, injury, and premature death in numbers that are significant on both local and global scales. Health impacts have thus been a major consideration in the promotion of some types of energy supplies and the avoidance of others in the modern sector, but, as we argue here, need also to be considered in the traditional sector where they are even more important.

Lack of good lighting is an important aspect of energy poverty, as is the pollution from poor-quality kerosene lighting (Pokhrel et al. 2010). Both of these are being addressed by a new generation of solar-powered high-efficiency lamps. In this chapter, however, we focus on the health consequences of the energy poverty associated with use of solid fuels (such as biomass, wood, agricultural residues, and animal dung, charcoal, and coal) for cooking and other household energy needs in developing countries.

According to the Global Energy Assessment (GEA 2012), about 2.8 billion people, mostly in the least developed and developing countries, relied on solid fuels for household energy needs. Indeed, daily use of solid fuels for cooking correlates almost one-to-one with poverty globally: essentially no one in the richest third of the world's households cooks with these fuels and no one in the poorest third cooks without them.

Solid fuels in these households are often used in inefficient, poorly vented combustion devices, which results in the bulk of the fuel energy being emitted and wasted as toxic products of incomplete combustion (PICs). This has important consequences for population health, as health impacts are not a direct function of emissions or environmental concentrations but of human exposure to them. For instance, pollution over the ocean may have impacts on marine ecosystems and the climate but has no appreciable impact on human health because no one is there to breathe it. On the other hand, a relatively small amount of pollution emitted inside households may have major impacts on human health, but a small effect on ambient (outdoor) pollution levels. The exposure implications of different emissions sources has come to be termed the 'intake fraction'— that is, the fraction of the material emitted that is actually taken up by a population through inhalation, ingestion, or dermal contact (Bennett et al. 2002). Inhalation intake fractions can vary by three orders of magnitude among energy systems and thus is crucial when considering health effects (see Figure 6.1).

The use of solid fuels in traditional stoves in small and poorly ventilated kitchens, in close proximity to household members on a daily basis, thus leads to exposures that are significantly detrimental to the health of family members. Women and very young children are especially at risk, as they receive some of the highest exposures and, in the case of children, are exposed during vulnerable periods of growth and development. The global scale of the exposures, the complexity of the exposure situation (with multiple household-level determinants

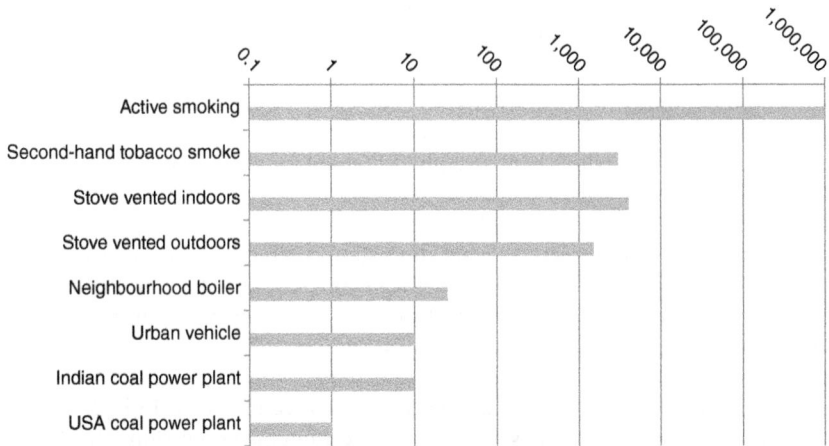

Figure 6.1. Approximate intake fractions for typical sources of air pollution, expressed as grams of pollutant inhaled for every tonne emitted; about one gram per tonne (one in a million emitted) is breathed by someone downwind of a typical power plant in the USA, while about 3,000 grams are inhaled per tonne of pollution released indoors from a stove
Source: data from Smith 1993

influencing frequency, duration, and magnitude of exposure), and the limited availability of data on exposures and health outcomes have resulted in a somewhat belated recognition of this risk factor as a major contributor to the disease burden at global, regional, and national levels. Health impacts surrounding this risk factor are thus often neglected in global energy discussions.

The Comparative Risk Assessment (CRA) of the Global Burden of Disease 2010 project (Lim 2012) found that 3.5 million people die prematurely each year as a result of direct exposure to household air pollution released by cooking with solid fuels. Cooking with solid fuels affects outdoor air quality as well, contributing approximately 16 per cent of outdoor air pollution globally and exposures in the form of 'second-hand cook smoke', with approximately another 0.5 million premature deaths as a result (Lim et al. 2012).

Multiple scientific developments have strengthened the evidence base for public health actions on the issue of household solid fuel use. With a view to consolidating information from new and ongoing efforts, this chapter describes (i) the patterns of solid fuel use in developing countries; (ii) emissions and exposures resulting from household fuel use; (iii) the range of health effects and burden of disease associated with solid fuel use; and (iv) options for interventions.

6.2. PATTERNS OF HOUSEHOLD FUEL USE IN DEVELOPING COUNTRIES

Hundreds of demographic surveys conducted in developing countries, especially over the last decade, have collected information on household fuel use. The World

Health Organization (WHO) household fuel use database now contains information for 155 countries, including 97 per cent of low- and middle-income countries (Bonjour et al. 2013). As described in this recent publication, the proportion of the world's households primarily relying on solid fuels for cooking declined from 62 per cent (95 per cent CI: 58, 66) to 41 per cent (95 per cent CI: 37, 44) between 1980 and 2010. Proportions have steadily decreased for all regions since 1980, and only in Sub-Saharan Africa (hereafter referred to as Africa, North Africa being considered part of the Eastern Mediterranean region) was the decline notably slower. Africa and South-east Asia are the regions with the highest proportion of households using solid fuels, with 77 per cent (95 per cent CI: 74, 81) and 61 per cent (95 per cent CI: 52, 70) respectively in 2010, whereas Europe and the Americas are the lowest, with less than 20 per cent. The Western Pacific and Eastern Mediterranean regions lie in the mid-range, with 46 per cent (95 per cent CI: 35, 57) and 35 per cent (95 per cent CI: 29, 40) respectively. In high-income countries, solid fuels are used by less than 5 per cent of the population. The decline has been sharpest in Asia (both Western Pacific and South-east Asia).

Despite declines in the proportions of households using solid fuels for cooking, the absolute number of people mainly using solid fuel for cooking has remained stable over the last three decades—about 2.8 billion—due to population growth. Unlike in other regions, the number of households using solid fuels almost doubled in Africa, from 333 to 646 million, and slightly increased in the Eastern Mediterranean region, from 162 to 190 million. In South-east Asia, the number has remained stable in terms of households exposed, whereas it declined in Europe, the Americas, and the Western Pacific.

Among developing countries generally, nearly 40 per cent rely on modern fuels; however, in the poorest, least developed countries, gas use is uncommon. Use of other fuels is concentrated in certain countries, for example charcoal in Sub-Saharan Africa, coal in China, dung in India, kerosene in Djibouti, and electricity in South Africa (see Table 6.1).

Socio-economic factors significantly influence household fuel choice. In most countries with per capita incomes under USD1,000, household fuel demands account for more than half of the total primary energy demand for cooking and

Table 6.1. Number of people relying on solid and modern fuels for cooking in low-, middle-, and high-income countries

Region*	Population exposed in millions (95 per cent confidence intervals)		
	1990	2000	2010
Low- and middle-income countries			
Sub-Saharan Africa	413 (395, 431)	517 (494, 539)	646 (617, 675)
Americas	119 (95, 143)	97 (69, 124)	80 (49, 111)
Eastern Mediterranean	171 (152, 190)	182 (159, 206)	190 (162, 219)
Europe	89 (66, 112)	53 (31, 75)	28 (7, 50)
South-east Asia	1,100 (979, 1,221)	1,112 (972, 1,253)	1,097 (934, 1,260)
Western Pacific	865 (711, 1,020)	809 (643, 975)	739 (563, 914)
High-income countries	15 (10, 21)	3 (0, 9)	1 (0, 7)

* Countries are grouped by WHO region and income category.

Source: Bonjour et al. 2013 (supplemental material)

heating; in contrast, such demands account for less than 2 per cent of total primary energy use in industrialized countries (UNDP 2009). As per capita incomes increase, households often switch to cleaner, more efficient energy systems for their household energy needs, that is, they move up the 'energy ladder' (Hosier and Dowd 1987; UNDP 2009). With technological progress, the income levels at which people make the transition to cleaner modern fuels has fallen.

Availability of cleaner fuels at the national level, however, does not guarantee availability of supply in rural areas (Masera et al. 2007), due to issues related to transport, reliability of supply, and socio-cultural preferences. Moreover, in every poor country the income disparities between rural and urban people are large, with most people in rural areas pursuing subsistence livelihoods. Household fuel generation, distribution, and consumption are thus closely related to the local status of energy, environment, and development. Several household factors directly influence patterns of human exposure to cooking fuel smoke, which occurs both inside and around households using poor combustion. Fuel type, kitchen location, use and maintenance of stoves, household layout and ventilation, time-activity profiles of individual household members, and behavioural practices (such as where children are located when cooking is being done) have been shown to influence pollution levels and individual exposures. Countries with low gross domestic product (GDP) also typically experience greater gender inequities in terms of income, education, access to health care, social position, and socio-cultural preferences, all of which could potentially influence exposures for vulnerable groups such as women and children.

Geographic variables can also significantly affect pollution intensity and duration. Extreme temperature differentials between seasons, rainfall, elevation, and even meteorological factors (such as wind speed, wind direction, and relative humidity) could determine whether fuels are used for both cooking and heating, as well as whether they are affecting aerosol dispersion and/or deposition. Patterns of vegetation (e.g. tropical rain forests vs. scrub) could contribute to household decisions on seeking alternative energy sources. Easy availability of wood or other biomass at little or no cost is likely to encourage continued use, especially among people living in poverty.

Although the available literature does not allow a detailed attribution of exposures to each of these variables, they can be expected to make varying contributions and should be considered when creating local or regional profiles of the exposure situation. A schematic showing the potential determinants is shown in Figure 6.2.

6.3. EMISSIONS FROM COMBUSTION OF HOUSEHOLD FUELS

A majority of households in developing countries using solid fuels for cooking burn them in poorly functioning mud or metal stoves or use open pits, usually without a chimney or other arrangement to vent the smoke from the area. Under ideal conditions, complete combustion of carbon would produce only CO_2 and water. Virtually all traditional ways of burning household biomass fuel, however,

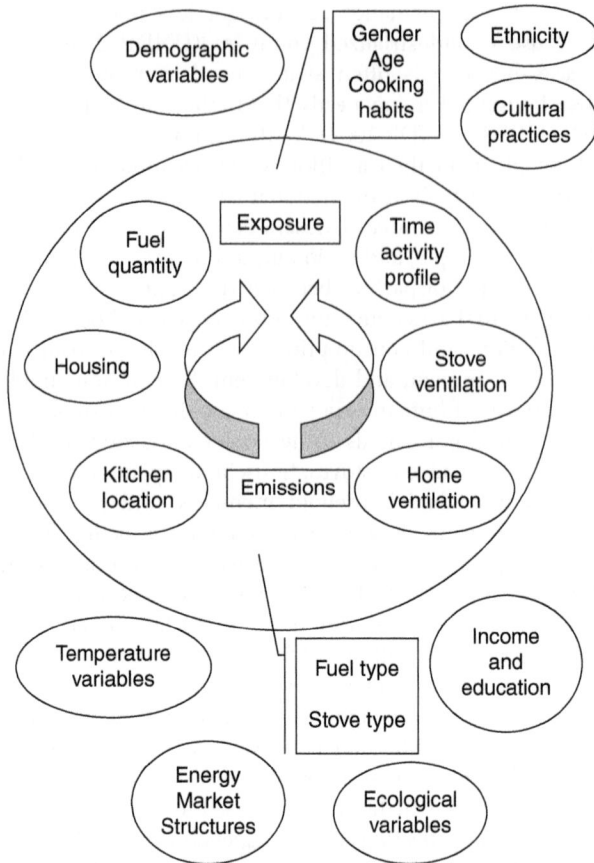

Figure 6.2. Macro- and micro-environmental determinants of exposure to solid fuel smoke

emit substantial quantities of PICs, since conditions for the efficient combustion of these fuels are difficult to achieve in typical household stoves. PICs include small respirable particles, gases such as carbon monoxide, polyaromatic hydrocarbons, phenols, quinones/semiquinones, chlorinated acids such as methylene chloride, and dioxins. Combustion of coal may release, in addition to the above pollutants, sulphur oxides, heavy metal contaminants, arsenic, and fluorine. On average, a typical solid fuel stove converts 6 to 20 per cent of the fuel into toxic substances. At least 28 pollutants present in smoke from solid fuel use have been shown to be toxic in animal studies; some 14 carcinogenic compounds and four cancer-promoting agents have been identified.

6.4. EXPOSURES TO HEALTH-DAMAGING POLLUTANTS FROM COMBUSTION OF HOUSEHOLD FUELS

Well over 100 studies over the last two decades have assessed household air pollution (HAP) levels in relation to cookfuel use. Table 6.2 provides a listing of select studies that illustrate the breadth of exposure measurements that have been performed across world regions. As may be seen in Table 6.2, over the last several

Table 6.2. Brief summary of findings from select published HAP studies reporting results on quantitative measurements of pollutant concentrations and/or exposures

Study design	Major findings	References (study locations)
Cross-sectional studies	• Provided estimates of daily average area concentrations for respirable PM and/or CO, across a wide range of household configurations using a variety of cooking fuels including biomass (such as wood, dung, crop residues), coal, kerosene, LPG, and electricity.	Smith et al. (India) Balakrishnan et al. 2002 (India) Andresen et al. (India)
	• Few also estimated exposures for women, children, and men.	Colbeck et al. (Pakistan)
	• Showed concentrations/exposures in solid fuel using homes to be consistently in excess of recommended air quality guideline levels and higher than levels in gas- or electricity-using households. Levels in gas/electricity- or kerosene-using households, however, were also often in excess of air quality guideline values. Other household variables including kitchen location were recognized to be important for exposures.	Begum et al. (Bangladesh) Kumie et al. (Nepal) Gao et al. (Tibet) Albalak et al. (Mexico) Jiang and Bell (China)
	• Many studies collected samples from multiple villages/rural habitations to generate representative exposure profiles for the population in respective countries.	Khalequzzaman et al. (Pakistan)
	• Some studies provided evidence of community-level exposures, including exposures in urban low-income communities.	
Repeat cross-sectional studies	• Provided estimates of daily average concentrations for multiple pollutants while addressing spatial and seasonal variations across multiple household configurations.	Saksena et al. (India) Jin et al. (China)
	• Provided extensive data from coal use.	He et al. (China)
	• Provided evidence on additional contributions from heating.	Lan et al. (China)
	• Intra- and inter-household variations in pollutant levels recognized to be influenced by multiple household level variables with additional differences across pollutants.	
Cross-sectional studies with modelling	• Combined household level area measurements with questionnaire-based categorical information on multiple household level variables for inclusion in models.	Balakrishnan et al. 2004 (India) Dasgupta et al. (Bangladesh)
	• Provided modelled estimates of household concentrations and/or exposures on the basis of household-level characteristics for use in long-term exposure reconstruction.	Baumgartner et al. (China)

continued

Table 6.2. Continued

Study design	Major findings	References (study locations)
	• Showed predictions from models to be modest, recognizing the need to address variability.	
Longitudinal studies	• Monitored 24- to 48-hour personal exposures and area concentrations longitudinally to address spatial and temporal variations.	Ezzati et al. (Kenya) Dionisio et al. (Gambia)
	• Provided estimates of variability in daily average/peak personal exposures to individual or multiple pollutants as well as correlations between pollutant exposures for women and infants.	Smith et al. (Guatemala) Bautista et al. (Dominican Republic)
Intervention studies using cross-sectional or paired before/after study designs	• Evaluated reductions in area concentrations after introduction of improved biomass cookstoves.	Naeher et al. (Guatemala) Clark et al. (Honduras)
	• Provided estimates for per cent reductions in household concentrations of respirable PM and/or CO with use of improved cookstoves together with reductions in fuel consumption.	Pennise et al. (Ghana, Ethiopia) Zuk et al. (Mexico)
	• Some studies also evaluated community-level or country-level cookstove programmes for emission and exposure reductions under field use conditions and provided important insights into contributions from user preference/behaviour for efficacy of interventions.	Smith et al. (HEH project –India, Mexico) Edwards et al. 2004 (China)
Intervention studies that included longitudinal monitoring of exposures and health outcomes	• Monitored area concentrations and personal exposures to CO and/or PM longitudinally to obtain individual estimate of exposure for control and intervention arms of improved biomass cookstove trials.	Smith et al. (Guatemala) Thompson (Guatemala)
	• Provided personal exposure estimates for use in generation of continuous exposure–response functions for outcomes including birth-weight, ALRI, S-T segment depression and neuro-development.	Dix-Cooper et al. (Guatemala) McCracken et al. 1999 (Guatemala) Baumgartner et al. (China)
	• Used data from longitudinal monitoring studies to make comparisons of single vs. multiple and/or group vs. multiple measures of exposures and highlighted the need for repeated long-term measures to address uncertainties in long-term exposure reconstruction.	McCracken et al. 2011 (Guatemala) Armendáriz-Arneza (Mexico)
Cross-sectional studies that included air toxics or PM size fractions	• Few studies provided estimates of area concentrations and/or exposures to PAHs, VOCs, NO$_2$ dioxins, metals, and other air toxics.	Northcross et al. (Guatemala)

• One compared particle size distributions in relation to kitchen concentrations and personal exposures among traditional and improved biomass cookstove users. Highlighted the potential for significant bias if the shift in size distribution and the change in relationship between indoor air concentrations and personal exposure concentrations are not accounted for between different stove types.	Kumie et al. (Ethiopia) Lisouza et al. (Peru) Ansari (India) Colbeck et al. (Pakistan) Fullerton et al. (Malawi) ArmendarizArnez (Guatemala)

Sources: Albalak et al. 2001; Andresen et al. 2005; Balakrishnan et al. 2002, 2004; Baumgartner et al. 2011a; Bautista et al. 2009; Begum et al. 2009; Bruce et al. 2004; Chengappa et al. 2007; Clark et al. 2011; Colbeck et al. 2009; Cynthia et al. 2008; Dasgupta et al. 2006; Dionisio et al. 2008; Dix-Cooper et al. 2012; Dutta et al. 2007; Edwards et al. 2007; Ezzati et al. 2000; Fischer and Koshland 2007; Fullerton et al. 2009; Gao et al. 2009; He et al. 2005; Jiang and Bell 2008; Jin et al. 2005; Khalequzzaman et al. 2011; Kumie et al. 2009; Lan et al. 2002; Lisouza et al. 2011; Masera et al. 2007; McCracken et al. 1999, 2007, 2011; Naeher, Leaderer and Smith 2000; Naeher et al. 2001; Northcross et al. 2010; Pennise et al. 2009; Saksena et al. 1992; Smith et al. 2000; Zhou et al. 2011; Zuk et al. 2007

years there has been a continuous evolution of methods and protocols for assessing exposure to HAP. The quantity and quality of information collected has also become considerably more detailed, with single-pollutant, cross-sectional studies measuring area concentrations being supplemented by multi-pollutant studies monitoring longitudinal exposure as part of intervention trials. A systematic review is beyond the scope of this chapter, but collectively, the evidence from these studies shows that rural women, children, and men in households using solid fuels experience extremely high levels of exposure to particulate matter and toxic gases. Often these exposures are an order of magnitude or even higher than levels generally considered safe. Further, some emissions from coal combustion (e.g. arsenic and fluorine) have additional, non-inhalational exposure routes (such as deposition on food and contamination of drinking water sources), compounding health effects (He et al. 2005).

6.5. HEALTH EFFECTS ASSOCIATED WITH HOUSEHOLD FUEL COMBUSTION

The earliest epidemiological evidence linking biomass combustion, indoor air pollution, and respiratory health came from studies carried out in Nepal and India in the mid-1980s. Since then there has been a steady stream of studies linking HAP and a range of health effects, especially in women who cook with these fuels and in young children. Recent systematic reviews describe the evidence for acute lower respiratory infections, chronic obstructive lung disease, cataracts, and lung cancer in adults (Dherani et al. 2008; Kumie et al. 2009). Based on a review in 2006, the International Agency for Research on Cancer (IARC) concluded that indoor emissions from household combustion of coal are carcinogenic to humans (group 1) for lung cancer and that indoor emissions from biomass, primarily wood, were classified as probable human carcinogens (group 2A). The Comparative Risk Assessment (CRA) organized by the World Health Organization, however, found sufficient additional evidence to include biomass emissions as a cause of lung cancer (Lim et al. 2012). Although cardiovascular disease itself has not been studied in relation to HAP, studies concerning combustion particles in relation to outdoor air pollution, environmental tobacco smoke, and active smoking strongly suggest a similar impact from household fuels as well (Baumgartner et al. 2011b; Burnett et al. 2014; Pope et al. 2009; Smith and Peel 2010; Tolunay and Chockalingam 2012). Additional evidence for other impacts is also now emerging, including low birth-weight children, cognitive function, and tuberculosis (Dix-Cooper et al. 2012; Hosgood et al. 2011; Pope et al. 2009).

The first burden of disease assessment for solid cookfuel use was part of the CRA for the year 2000 (Smith et al. 2004). Revised estimates for 2010 have been performed by a global consortium of investigators lead by the Institute for Health Metrics and Evaluation, USA (Smith et al. 2014). The latter estimation, although similar, involved some important additions.

The WHO-led CRA estimated 1.6 million excess deaths to be attributable to HAP accounting for 4 per cent of global DALYs for 2000 (Ezzati et al. 2004). In 2012, the HAP burden was estimated to be substantially higher (Lim et al. 2012), as shown in Figure 6.3. This came from being able to account for additional health outcomes (the earlier effort included only ALRI in children under 5, COPD in women, and lung cancer from coal) and choosing a lower counterfactual level for comparison (equivalent to that of levels achieved in gas-cooking households of developed countries). Thus in 2010, HAP accounted for 3.5 million (2.7 million to 4.4 million) deaths and 4.5 per cent (3.4–5.3) of global DALYs. In poorer regions of the world such as South Asia and much of Sub-Saharan Africa, HAP ranked as a leading risk factor exceeding disease burdens attributable to not only tobacco smoking and blood pressure but also child under-nutrition, water and sanitation, and ambient (outdoor) air pollution in addition to being the most important risk factor for women. The underlying epidemiologic transition from communicable to non-communicable diseases and the overall improvements in childhood mortality rates have decreased the burden on children from HAP, although it continues to be among the leading risk factors for children in Sub-Saharan Africa.

6.6. INTERVENTION EFFECTIVENESS

Household energy interventions to date in biomass-using populations have largely centred on improving fuel efficiency, either by using better fuels and stoves or using improved stoves with the same fuels (Barnes et al. 1994). Considerable evidence is available to indicate that households using gaseous (liquefied petroleum gas or biogas) or liquid fuels experience substantially lower pollution levels compared to homes using solid fuels (Albalak et al. 2001; Balakrishnan et al. 2002, 2004). A limited number of studies have also shown significant reductions with the use of electricity (Rollin et al. 2004). 'Improved' biomass stoves that

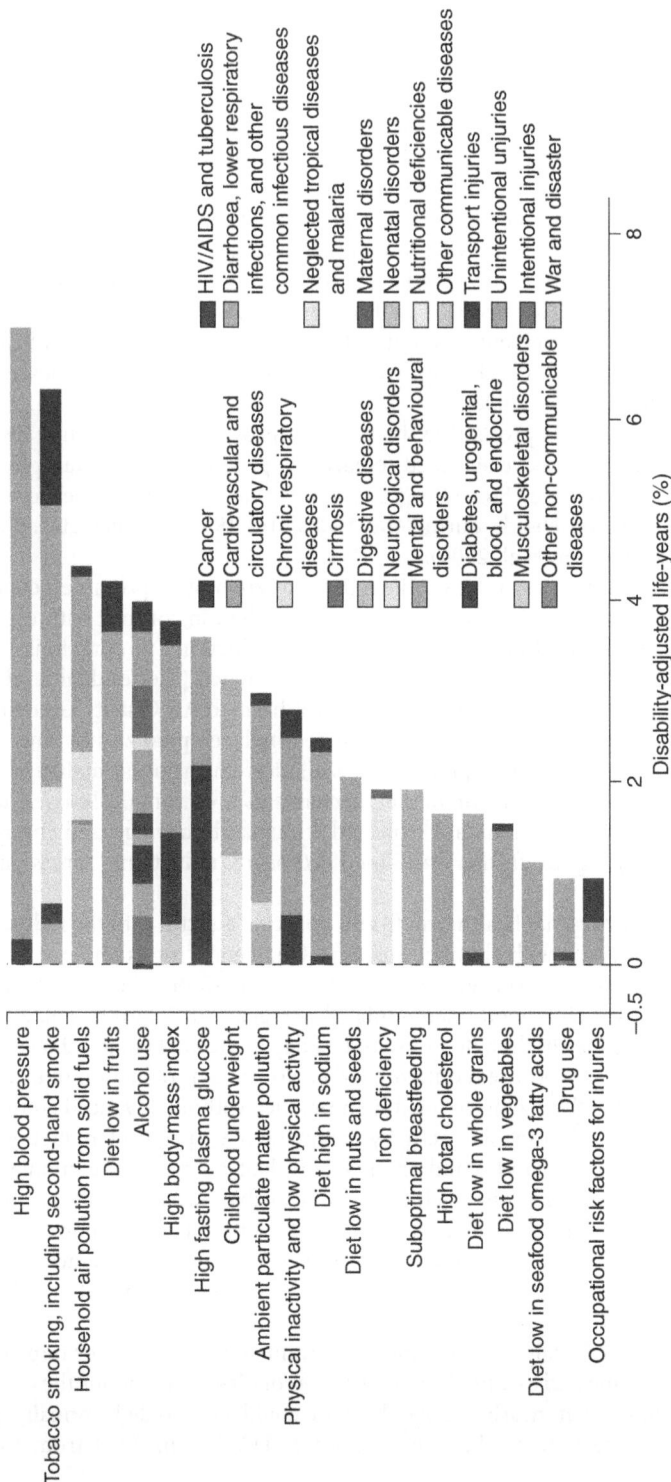

Figure 6.3. Burden of disease attributable to 20 leading risk factors in 2010 for both sexes (per cent of disability adjusted life years globally)

Source: Reprinted from *The Lancet*, 380, Stephen Lim et al., 'A Comparative Risk Assessment of Burden of Disease and Injury Attributable to 67 Risk Factors and Risk Factor Clusters in 21 Regions, 1990–2010': 2224–60 © (2013), with permission from Elsevier

use less fuel and vent emissions outside the home have been an intervention option since the early 1980s. Such programmes have been implemented in many countries, most notably in India and China. Although designed to conserve fuel, these programmes had some impacts on reducing household levels of health-damaging pollutants and CAPs (US EPA 2000; Edwards et al. 2004).

Somewhat lower indoor concentrations using stove models equipped with chimneys have also been documented in many regions, including China (Sinton et al. 2004), India (Smith et al. 1983; Ramakrishna et al. 1989), Nepal (Reid et al. 1986; Pandey et al. 1990), Latin America (Brauer and Bartlett 1996; Albalak et al. 2001; Bruce et al. 2004), Mexico (Riojas-Rodríguez et al. 2001), and Sub-Saharan Africa (UNDP 2009). Interventions that reduce exposures either through behavioural interventions or through improved ventilation have also been described (Barnes et al. 2004; UNDP 2009).

Although programmes promoting the use of improved stoves have not always proved successful, some have achieved remarkable penetration. For example, the Chinese National Improved Stove Programme was able to provide stoves to some 180 million rural households during the 1980s and 1990s (Barnes et al. 1993; Smith et al. 1993; Sinton et al. 2004).

Several programmes are underway in India that attempt to promote penetration of improved stoves using market-based approaches, in contrast with earlier, government-subsidized efforts. While substantial reductions in emissions have been achieved with many of these improved stove models, the residual levels of pollutants are still high compared to WHO's health-based Air Quality Guidelines (WHO 2006). More recent programmes increasingly emphasize not only the provision of stoves, but also support for installation and routine maintenance, training and education, and use of market mechanisms to continuously assess user preferences. These innovations are expected to expand coverage and improve performance, leading to further sustained exposure reductions across large populations.

Evidence is growing that health benefits do not scale directly with the reduction in exposure. In the trial with chimney stoves in Guatemala, for example, a 50 per cent reduction in smoke exposures resulted in just an 18 per cent reduction in physician-diagnosed childhood pneumonia (Bruce et al. 2007; Smith et al. 2011). Moreover, because the children do not spend all day in the kitchen, to achieve a 50 per cent reduction in exposure required nearly a 90 per cent reduction in kitchen levels, which is difficult to achieve reliably in practice without advanced combustion stoves. In those households actually achieving a 90 per cent reduction in exposure, on the other hand, children had only half as much pneumonia, an improvement greater than that achieved by available vaccines and nutrition supplements, which are the other major interventions for this major killer of children. To obtain this much reduction, however, will require extremely clean-burning stoves used regularly over long periods and near elimination of the existing polluting stoves.

In a striking set of analyses, simulations to estimate per cent use of improved stoves under baseline and assumed enhanced rates of dissemination for the case of India have shown that nearly 12,500 DALYs could be avoided annually per million people. The national burden of disease (DALYs) in 2020 from these three major diseases is estimated to be about a sixth lower than it would have been without the stove programme, which is equivalent to the elimination of nearly half the entire cancer burden in India in 2020 (Wilkinson et al. 2009). What is not well understood yet, however, is how well such advanced stoves actually perform in field conditions and how well they are accepted by large populations.

More recently, advanced combustion stoves—using traditional wood fuel but burning much more cleanly—have become available. One promising approach involves so-called 'gasifier' stoves that achieve very high combustion efficiency through designs that facilitate two-stage combustion. The most reliable of these use small electric blowers to stabilize combustion. Where there is no reliable electricity available, inexpensive thermal-electric generators are now being incorporated into stoves to generate the needed power from the stove heat itself. These promise to achieve emission levels that could allow biomass fuels to be used under healthy conditions in households.

6.7. CONCLUSION

The available evidence on health implications of solid fuel use provides a strong basis to advance actions that reduce the health burden attributable to solid fuel use through concerted national and global policy efforts. Advanced fuels such as electricity, liquefied petroleum gas (LPG), ethanol, and biogas are, by comparison to solid fuels, essentially pollution-free as well as more efficient. Three-fifths of the world (spread across every possible cuisine and culture) already cooks with such cleaner fuels, and thus gas or electricity probably represent a 'near universal aspirational standard' for households. However, accessibility and affordability need to be considerably increased for advanced fuels to be a viable solution for many current solid fuel users. The UNDP reported that although many developing countries have clear targets for providing access to electricity, only a handful have set targets to reduce the impacts of household solid fuel use—17 for access to modern cooking fuels and 11 for provision of improved stoves (UNDP 2009). Internationally, the Global Alliance for Clean Cookstoves has set a target of 100 million (United States Department of State 2010) clean cookstoves to be distributed by 2020, while the Partnership of the World LP Gas Association has set a target of providing 1 billion or more people with LPG by 2030 (World LP Gas Association 2013). In both cases, however, the dissemination mechanisms and funding arrangements have not been established as of this writing. Nor have the criteria been specified for what is clean and what would be the baseline for measuring or evaluating targets. Nevertheless, these are promising developments that indicate a much greater international appreciation of the extent of the problem, partly driven by the international health impact assessments described in this chapter.

There is growing understanding that considerable efforts will need to be devoted to devising robust and attractive clean-burning biomass stoves to meet local needs, a surprisingly difficult engineering task in small inexpensive devices. Evidence of health benefits from incremental improvements could be expected to provide relevant cost-effectiveness information to policymakers, greatly facilitating the acceleration of intervention efforts. In the past, unfortunately, almost any new stove could be claimed to be 'improved' when many actually did little for smoke exposure reduction, although many programmes have required some evidence of improved fuel performance for the stoves they disseminated. Many national programmes even today are not specific about the criteria of 'improved' in terms of pollution, with some notable exceptions such as the Indian National Biomass Cookstove Initiative, which declares an 'aim to achieve the quality of energy services from cookstoves comparable to that from other clean energy sources such as LPG' (Government of India 2009). In spite of this excellent framing, to date, the programme has yet to be implemented at scale or to specify the criteria by which equivalence in energy services to LPG is to be defined or determined.

Another factor that stove programmes need to address directly is adoption/usage—it does no good to disseminate a high-performance stove that no one uses, something that has happened in many past programmes. New monitoring technologies, however, are making it possible to keep track inexpensively of actual household usage (Ruiz-Mercado et al. 2011, 2012). This will allow programme developers to better understand the stove designs, training, and incentives that enhance usage.

To effectively implement stove technologies that truly lower exposures, there is thus an imminent need to define more quantitatively what is meant by various levels of 'improvement', probably separately in terms of expected emissions/exposure and fuel use per meal. In this respect, there is progress in defining standards for what constitutes a clean stove using the existing guideline/standards procedures of ISO and ANSI (ISO 2012; ANSI 2012). In addition, the World Health Organization Indoor Air Quality Guidelines (2013) provide a comprehensive review of the emissions, exposure, and health evidence for its recommendations, while specifying health-based benchmarks that could be incorporated in multiple aspects of national programmes and policies concerning household energy (WHO 2013).

Only in combination with greater shifts to truly clean fuels and by insisting on evidence of superior performance and acceptance of clean biomass stoves will it be possible to protect the world's poorest populations now burdened so greatly from the high exposures and health impacts due to burning simple solid fuels in their households.

ACKNOWLEDGEMENTS

This chapter has been adapted from Smith, K.R., Balakrishnan, K., Butler, C.D., Chafe, Z., Fairlie, I., Kinney, P., Kjellstrom, T., Mauzerall, D.L., McKone, T.E., McMichael, A.J., Schneider, M., and Wilkinson, P. 2012. Chapter 4, 'Energy and Health'. In *Global Energy Assessment—Toward a Sustainable Future*, Cambridge and New York: Cambridge University Press; Laxenburg: International Institute for Applied Systems Analysis, pp. 255–324.

REFERENCES

Albalak, R., et al. (2001). 'Indoor Respirable Particulate Matter Concentrations from an Open Fire, Improved Cookstove, and LPG/Open Fire Combination in a Rural Guatemalan Community', *Environmental Science and Technology* 35: 2650–55.

Andresen, P., et al. (2005). 'Women's Personal and Indoor Exposures to PM2.5 in Mysore, India: Impact of Domestic Fuel Usage', *Atmospheric Environment* 39: 5500–508.

Ansari, F.A., Khan, A.H., Patel, D.K., Siddiqui, H., Sharma, S., Ashquin, M., and Ahmad, I., 'Indoor Exposure to Respirable Particulate Matter and Particulate-phase PAHs in Rural Homes in North India', *Environmental Monitoring and Assessment* 170(1–4): 491–7.

ANSI (American National Standards Institute). (2012). 'First ISO International Workshop Agreement on Clean Cookstoves Unanimously Approved', press release, <http://www.ansi.org>, New York: American National Standards Institute.

Armendáriz-Arneza, C., et al. (2010). 'Indoor Particle Size Distributions in Homes with Open Fires and Improved Patsari Cookstoves', *Atmospheric Environment* 44(24): 2881–6.

Balakrishnan, K., et al. (2002). 'Daily Average Exposures to Respirable Particulate Matter from Combustion of Biomass Fuels in Rural Households of Southern India', *Environmental Health Perspectives* 110: 1069–75.

Balakrishnan, K., et al. (2004). 'Exposure Assessment for Respirable Particulates Associated with Household Fuel Use in Rural Districts of Andhra Pradesh, India', *Journal of Exposure Science and Environmental Epidemiology* 14: S14–25.

Barnes, B.R., et al. (2004). 'Testing Selected Behaviors to Reduce Indoor Air Pollution Exposure in Young Children', *Health Education Research* 19: 543–50.

Barnes, D., et al. (1993). 'The Design and Diffusion of Improved Cooking Stoves', *The World Bank Research Observer* 8: 119–42.

Barnes, D., et al. (1994). *What Makes People Cook with Improved Biomass Stoves? A Comparative International Review of Stove Programs*. Washington DC: World Bank et al.

Baumgartner, J., et al. (2011a). 'Patterns and Predictors of Personal Exposure to Indoor Air Pollution from Biomass Combustion Among Women and Children in Rural China', *Indoor Air* 21: 479–88.

Baumgartner, J., et al. (2011b). 'Indoor Air Pollution and Blood Pressure in Adult Women Living in Rural China', *Environmental Health Perspectives* 119: 1390–95.

Bautista, L., et al. (2009). 'Indoor Charcoal Smoke and Acute Respiratory Infections In Young Children in the Dominican Republic', *American Journal of Epidemiology* 169: 572–80.

Begum, B., et al. (2009). 'Indoor Air Pollution from Particulate Matter Emissions in Different Households in Rural Areas of Bangladesh', *Building and Environment* 44: 898–903.

Bennett, D., et al. (2002). 'Peer Reviewed: Defining Intake Fraction', *Environmental Science and Technology* 36: 206A–11A.

Bonjour, S., et al. (2013). 'Solid Fuel Use for Household Cooking: Country and Regional Estimates for 1980–2010', *Environmental Health Perspectives* 121: 784–90.

Brauer, M. and Bartlett, K. (1996). 'Assessment of Particulate Concentrations from Domestic Biomass Combustion in Rural Mexico'. *Environmental Science and Technology* 30: 104–10.

Bruce, N., et al. (2004). 'Impact of Improved Stoves, House Construction and Child Location on Levels of Indoor Air Pollution Exposure in Young Guatemalan Children', *Journal of Exposure Science and Environmental Epidemiology* 14 Suppl 1: S26–33.

Bruce, N., et al. (2007). 'Pneumonia Case-Finding in the RESPIRE Guatemala Indoor Air Pollution Trial: Standardizing Methods for Resource-Poor Settings', *Bulletin of the World Health Organization* 85: 535–44.

Burnett, R.T. et al. (2014). 'An Integrated risk function for estimating the Global Burden of Disease attributable to ambient fine particulate matter exposure', *Environmental Health Perspectives* <http://dx.doi.org/10.1289/ehp.1307049>. Advance Publication: 11 February 2014.

Chengappa, C., et al. (2007). 'Impact of Improved Cookstoves on Indoor Air Quality in the Bundelkhand Region in India', *Energy for Sustainable Development* 11.

Clark, M., et al. (2011). 'A Baseline Evaluation of Traditional Cookstove Smoke Exposures and Indicators of Cardiovascular and Respiratory Health among Nicaraguan Women', *International Journal of Occupational and Environmental Health* 17: 113–21.

Colbeck, I., Nasir, Z., and Ali, Z. (2009). 'Characteristics of Indoor/Outdoor Particulate Pollution in Urban and Rural Residential Environment of Pakistan', *Indoor Air* 20: 40–51.

Cynthia, A., et al. (2008). 'Reduction in Personal Exposures to Particulate Matter and Carbon Monoxide as a Result of the Installation of a Patsari Improved Cookstove in Michoacan Mexico', *Indoor Air* 18: 93–105.

Dasgupta, S., et al. (2006). 'Indoor Air Quality for Poor Families: New Evidence from Bangladesh', *Indoor Air* 16: 426–44.

Dherani, M., et al. (2008). 'Indoor Air Pollution from Unprocessed Solid Fuel Use and Pneumonia Risk in Children aged Under Five Years: A Systematic Review and Meta-analysis', *Bulletin of the World Health Organization* 86(5): 390–401.

Dionisio, K., et al. (2008). 'Measuring the Exposure of Infants and Children to Indoor Air Pollution from Biomass Fuels in the Gambia', *Indoor Air* 18: 317–27.

Dix-Cooper, L., et al. (2012). 'Neurodevelopmental Performance among School Age Children in Rural Guatemala is Associated with Prenatal and Postnatal Exposure to Carbon Monoxide, a Marker for Exposure to Woodsmoke', *Neurotoxicology* 33: 246–54.

Dutta, K., et al. (2007). 'Impact of Improved Biomass Cookstoves on Indoor Air Quality Near Pune, India', *Energy for Sustainable Development* 11.

Edwards, R., et al. (2004). 'Implications of Changes in Household Stoves and Fuel Use in China', *Energy Policy* 32: 395–411.

Edwards, R., et al. (2007). 'Household CO And PM Measured as Part of a Review of China's National Improved Stove Program', *Indoor Air* 17: 189–203.

Ezzati, M., Saleh, H., and Kammen, D. (2000). 'The Contributions of Emissions and Spatial Microenvironments to Exposure to Indoor Air Pollution from Biomass Combustion in Kenya', *Environmental Health Perspectives* 108: 833–9.

Ezzati, M., et al., eds. (2004). 'Comparative Quantification of Health Risks: Global and Regional Burden of Disease due to Selected Major Risk Factors'. Geneva: World Health Organization.

Fischer, S. and Koshland, C. (2007). 'Daily and Peak 1 h Indoor Air Pollution and Driving Factors in a Rural Chinese Village', *Environmental Science and Technology* 41: 3121–6.

Fullerton, D., et al. (2009). 'Biomass Fuel Use and Indoor Air Pollution in Homes in Malawi', *Occupational and Environmental Medicine* 66: 777–83.

Gao, X., et al. (2009). 'Indoor Air Pollution from Solid Biomass Fuels Combustion in Rural Agricultural Area of Tibet, China', *Indoor Air* 19: 198–205.

GEA (2012). *Global Energy Assessment—Toward a Sustainable Future*. Cambridge and New York: Cambridge University Press, and Laxenburg, Austria: International Institute for Applied Systems Analysis.

Government of India (2009). 'Launching of the National Biomass Cookstoves Initiative', press release, New Delhi: Ministry of New and Renewable Energy.

He, G., et al. (2005). 'Patterns of Household Concentrations of Multiple Indoor Air Pollutants in China', *Environmental Science and Technology* 39: 991–8.

Hosgood, H., et al. (2011). 'Household Coal Use and Lung Cancer: Systematic Review and Meta-Analysis of Case-Control Studies, with an Emphasis on Geographic Variation', *International Journal of Epidemiology* 40: 719–28.

Hosier, R. and Dowd, J. (1987). 'Household Fuel Choice in Zimbabwe: An Empirical Test of the Energy Ladder Hypothesis', *Resources and Energy* 9: 347–61.

ISO (International Organization for Standardization)(2012). *International Workshop Agreement 11: 2012, Guidelines for Evaluating Cookstove Performance*, Geneva: IOS.

Jiang, R. and Bell, M. (2008). 'A Comparison Of Particulate Matter From Biomass-Burning Rural And Non-Biomass-Burning Urban Households In Northeastern China', *Environmental Health Perspectives* 116: 907–14.

Jin, Y., et al. (2005). 'Geographical, Spatial, and Temporal Distributions of Multiple Indoor Air Pollutants in Four Chinese Provinces', *Environmental Science and Technology* 39: 9431–9.

Khalequzzaman, M., et al. (2011). 'Indoor Air Pollution and Health of Children in Biomass Fuel-Using Households of Bangladesh: Comparison Between Urban and Rural Areas', *Environmental Health and Preventive Medicine* 16: 375–83.

Kjellstrom, T., Holmer, I., and Lemke, B. (2009). 'Workplace Heat Stress, Health and Productivity – An Increasing Challenge for Low and Middle Income Countries during Climate Change', Global Health Action [website], <http://www.ncbi.nlm.nih.gov/pmc/articles/PMC2799237/> (accessed 15 October 2013).

Kumie, A., et al. (2009). 'Sources of Variation for Indoor Nitrogen Dioxide in Rural Residences of Ethiopia', *Environmental Health* 8: 51.

Lan, Q., et al. (2002). 'Household Stove Improvement and Risk of Lung Cancer In Xuanwei, China', *Journal of the National Cancer Institute*, 94.

Lim, S., et al. (2012). 'A Comparative Risk Assessment of Burden of Disease and Injury attributable to 67 Risk Factors in 21 Regions, 1990–2010: A Systematic Analysis for the Global Burden of Disease Study 2010', *The Lancet*, 380: 2224–60.

Lisouza, F., Owuor, O., and Lalah, J. (2011). 'Variation in Indoor Levels of Polycyclic Aromatic Hydrocarbons from Burning Various Biomass Types in the Traditional Grass-Roofed Households in Western Kenya', *Environmental Pollution* 159: 1810–15.

Masera, O., et al. (2007). 'Impact of Patsari Improved Cookstoves on Indoor Air Quality in Michoacán, Mexico', *Energy for Sustainable Development* 11: 45–56.

McCracken, J., et al. (1999). 'Improved Stove or Inter-Fuel Substitution for Decreasing Indoor Air Pollution From Cooking With Biomass Fuels in Highland Guatemala', *Indoor Air* 3: 118–23.

McCracken, J., et al. (2007). 'Chimney Stove Intervention to Reduce Long-Term Wood Smoke Exposure Lowers Blood Pressure among Guatemalan Women', *Environmental Health Perspectives* 115: 996–1001.

McCracken, J., et al. (2011). 'Intervention to Lower Household Wood Smoke Exposure in Guatemala Reduces St-Segment Depression on Electrocardiograms', *Environmental Health Perspectives* 119: 1562–8.

Naeher, L., Leaderer, B., and Smith, K. (2000). 'Particulate Matter and Carbon Monoxide in Highland Guatemala: Indoor and Outdoor Levels from Traditional and Improved Wood Stoves and Gas Stoves', *Indoor Air* 10: 200–205.

Naeher, L., et al. (2001). 'Carbon Monoxide as a Tracer for Assessing Exposures to Particulate Matter in Wood and Gas Cookstove Households of Highland Guatemala', *Environmental Science and Technology* 35: 575–81.

Northcross, A., et al. (2010). 'Estimating Personal PM2.5 Exposures Using CO Measurements in Guatemalan Households Cooking with Wood Fuel', *Journal of Environmental Monitoring* 12: 873–8.

Pandey, M., et al. (1990). 'The Effectiveness of Smokeless Stoves in Reducing Indoor Air Pollution in a Rural Hill Region of Nepal', *Mountain Research and Development* 10: 313–20.

Pennise, D., et al. (2009). 'Indoor Air Quality Impacts Of An Improved Wood Stove In Ghana and an Ethanol Stove In Ethiopia', *Energy for Sustainable Development* 13: 71–6.

Pokhrel, A., et al. (2010). 'Tuberculosis and Indoor Biomass and Kerosene Use in Nepal: A Case-Control Study', *Environmental Health Perspectives* 118(4): 558–64.

Pope, C., et al. (2009). 'Fine-Particulate Air Pollution and Life Expectancy in the United States', *New England Journal of Medicine* 360: 376–86.

Practical Action. (2013). *Poor People's Energy Outlook 2013: Energy for Community Services.* Rugby, UK: Practical Action Publishing.

Ramakrishna, J., et al. (1989). 'Cooking in India: The Impact of Improved Stoves on Indoor Air Quality', *Environment International* 15: 341–52.

Reid, H., et al. (1986). 'Indoor Smoke Exposures from Traditional and Improved Cook-stoves Comparisons among Rural Nepali Women', *Mountain Research and Development* 6: 293–304.

Riojas-Rodríguez, H., et al. (2001). 'Household Firewood Use and the Health of Children and Women of Indian Communities in Chiapas, Mexico', *International Journal of Occupational and Environmental Health* 7: 44–53.

Rollin, H., et al. (2004). 'Comparison of Indoor Air Quality in Electrified and Un-electrified Dwellings in Rural South African Villages', *Indoor Air* 14: 208–16.

Ruiz-Mercado, I., et al. (2011). 'Adoption and Sustained Use of Improved Cookstoves', *Energy Policy* 39: 7557–66.

Ruiz-Mercado, I., et al. (2012). 'Temperature Dataloggers as Stove Use Monitors (SUMs): Field Methods and Signal Analysis', *Biomass and Bioenergy* 47: 459–68.

Saksena, S., et al. (1992). 'Pattern of Daily Exposure to TSP and CO in the Garhwal Himalaya', *Atmospheric Environment* 26A: 2125–34.

Sinton, J., et al. (2004). 'An Assessment of Programs to Promote Improved Household Stoves in China', *Energy for Sustainable Development* 8: 33–52.

Smith, K. (1993). 'Fuel Combustion, Air Pollution Exposure, and Health: The Situation in Developing Countries', *Annual Review of Energy and Environment* 18: 529–66.

Smith, K., et al. (1983). 'Air Pollution and Rural Biomass Fuels in Developing Countries: A Pilot Village Study in India and Implications for Research and Policy', *Atmospheric Environment* 17: 2343–62.

Smith, K., et al. (1993). 'One Hundred Million Improved Cookstoves in China: How was it Done?' *World Development* 21: 941–61.

Smith, K., et al. (2000). 'Indoor Air Pollution in Developing Countries and Acute Lower Respiratory Infections in Children', *Thorax* 55: 518–32.

Smith, K., et al. (2004). 'Indoor Smoke from Household Solid Fuels', in M. Ezzati, A.D. Lopez, and C.L. Murray (eds), *Comparative Quantification of Health Risks: Global and Regional Burden of Disease due to Selected Major Risk Factors.* Geneva: World Health Organization, 1435–93.

Smith, K., et al. (2011). 'Effect of Reduction in Household Air Pollution on Childhood Pneumonia in Guatemala (RESPIRE): A Randomised Controlled Trial', *The Lancet* 378: 1717–26.

Smith, K. et al. (2014). 'Millions dead: how do we know and what does it mean? Methods used in the Comparative Risk Assessment of Household Air Pollution', *Annual Review of Public Health* 35: 185–206.

Smith, K. and Peel, J. (2010). 'Mind the Gap', *Environmental Health Perspectives* 118: 1643–5.

Thompson, L.M., et al. (2004). 'Impact of Reduced Maternal Exposures to Wood Smoke from an Introduced Chimney Stove on Newborn Birth Weight in Rural Guatemala, *Environmental Health Perspectives* 119(10): 1489–94.

Tolunay, E. and Chockalingam, A. (eds) (2012). 'Indoor and Outdoor Air Pollution, and Cardiovascular Health', *Global Heart* (Special Issue) 7: 197–274.

UNDP (United Nations Development Programme). (2009). *The Energy Access Situation in Developing Countries: A Review Focused on Least Developed Countries and Sub-Saharan Africa,* Nairobi: United Nations.

United States Department of State. (2010). 'The United States and the Global Alliance for Clean Cookstoves', fact sheet, Washington, DC: Office of the Spokesman.

USEPA (United States Environmental Protection Agency). (2000). *Report on Greenhouse Gases from Small-Scale Combustion Devices in Developing Countries: Household Stoves in India*, EPA/600/R-00/052, Research Triangle, NC: National Risk Management Research Laboratory.

WHO (World Health Organization). (2006). *Air Quality Guidelines: Global Update for 2005*, Copenhagen: World Health Organization Regional Office for Europe.

WHO. (2013). Indoor Air Pollution [website] <http://www.who.int/indoorair/en/index.html> (accessed 13 July 2013).

Wilkinson, P., et al. (2009). 'Public Health Benefits of Strategies to Reduce Greenhouse-Gas Emissions: Household Energy', *The Lancet* 374: 1917–29.

World LP Gas Association. (2013). *Cooking for Life: Campaign Overview*, unpublished.

Zhou, Z., et al. (2011). 'Household and Community Poverty, Biomass Use, and Air Pollution in Accra, Ghana', *Proceedings of the National Academy of the Sciences* 108: 11028–33.

Zuk, M., et al. (2007). 'The Impact of Improved Wood-Burning Stoves on Fine Particulate Matter Concentrations in Rural Mexican Homes', *Journal of Exposure Science and Environmental Epidemiology* 17: 224–32.

7

Energy and Gender

Barbara C. Farhar, Beth Osnes, and Elizabeth A. Lowry

This chapter concerns the significance of gender in energy and poverty. In particular, despite troubling impacts of the lack of women's access to modern energy, the chapter focuses on reasons for hope and recommendations for measurement of progress.

It now appears that a crucial characteristic setting human beings apart from the higher primates is the act of cooking. Unlike other species, our bodies have evolved for digesting cooked food, thus freeing our vital primal energy and time for other pursuits.

The gathering of the family around the fire has the compelling symbolism and physical satisfaction of warmth, nourishment, and nurturance. Indeed, in many countries, a baby's first experiences are on its mother's body as she cooks over the fire. In these countries, babies are born into a world of smoke. Precisely because cooking is so deeply embedded in human evolution and culture, it has been difficult to change cooking technologies. Today, 43 per cent of the world's population—3 billion people—cook their food with biomass, coal, or dung over open three-stone fires.

Yet, with the new nearly smokeless stoves—such as those produced by Envirofit—along with solar cooking, women can begin to expand their technological sophistication by reaching for the new tools that fit their own needs and help them meet the demands on them while reducing their and their children's exposure to harmful indoor emissions and the perils of wood- and dung-gathering.

Over the past decades, we've heard a good deal about the problems of women and poverty in the developing world. In September 2000, all UN member states agreed that eight international goals—called the Millennium Development Goals (MDGs)—should be achieved by the year 2015. When we review the MDGs, though, we do not find the word 'energy', despite the fact that energy issues are central to all the goals, such as 'ending poverty and hunger' and 'universal education'. Universal education, for example, depends on girls being free to go to school instead of gathering wood for their families or selling that wood for subsistence income. Even the metrics on progress toward the MDGs do not include 'energy' variables. So if we do not measure energy variables, will we accomplish changes in them?

The notion of 'energy justice' is important in this context. Energy justice means not only making energy technologies available in developing countries but also disaggregating and understanding energy problems and policies by gender in cultural and political contexts. It means supporting the potential to leapfrog ahead (in real time as developed countries learn what to do themselves) so that developing countries avoid investing in dinosaur energy infrastructures and business models, outdated when they are built. Although fossil fuel and nuclear industries want to capture developing country energy markets, distributed energy and efficiency technologies can power and energize women nimbly and precisely to meet their specific needs. Because of their versatility and portability, these technologies can also add resilience in the face of community needs to adapt to climate change impacts.

With adequate financing mechanisms and business training, men and women in the developing world are beginning to earn livelihoods by marketing and servicing energy-efficient and clean-technology cooking devices and solar energy systems. So there is hope for the future—these things *can* be done. The question for now and just ahead of us is—*will* they be done on a large enough scale to bend the curve? And, are paradigm shifts underway that will increase access to modern energy services to the 1.3 billion people currently living without electricity?

A central thesis of this chapter is that definition of key goals and measurement of progress toward them is critical in energy progress. The chapter begins by discussing global issues as they relate to gender and energy. Then, measurement of progress toward the MDGs is documented. Next, a sampling of distributed sustainable energy projects and programmes is described, focusing on those addressing cooking and distributed electrification and lighting. Finally, measurement issues and new paradigms are discussed.

7.1. KEY ISSUES IN GENDER AND ENERGY

Seven key issues are explored briefly. Women's empowerment is posited as crucial in eradicating poverty. The role of national governments in improving gender equity is measured. Women's empowerment is also enhanced through the educational advancement of women and girls. Energy access is closely linked with the availability of health care. Gender and cooking are inextricably linked, and cooking with traditional biofuels has profound consequences for the health of women and children. Women perform much of the world's domestic, agricultural, and mechanical labour.

7.1.1. National governments, energy justice, and women's empowerment

What is the role of energy poverty and energy justice in the world's climate and energy crisis? The constitutions of national governments give them responsibility for the welfare of their citizens—not just able-bodied men, but also children, the

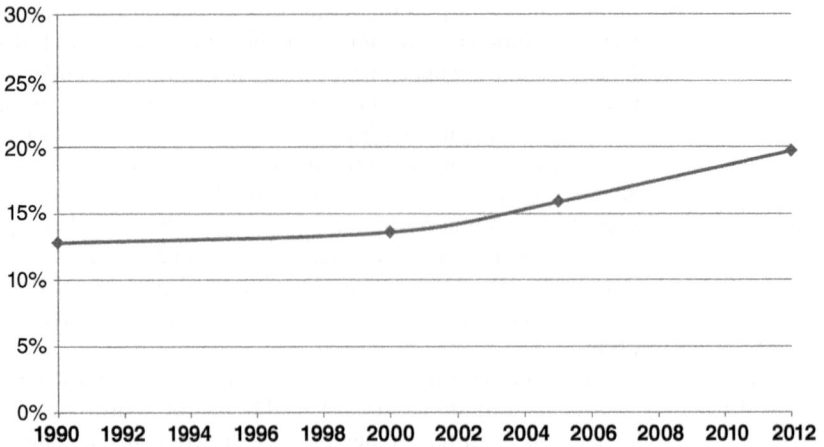

Figure 7.1. Proportion of parliamentary seats held by women worldwide, 1990–2012
Source: United Nations 2012b

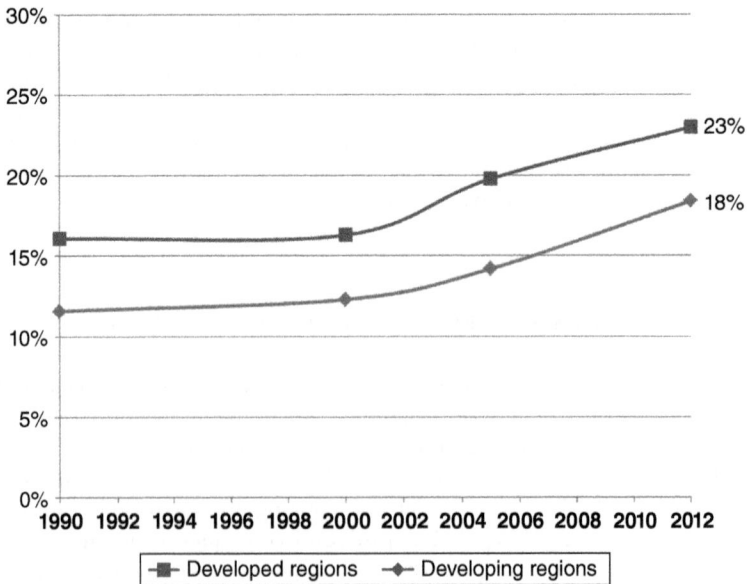

Figure 7.2. Proportion of parliamentary seats held by women, developed and developing regions, 1990–2012
Source: United Nations 2012b

elderly, the disabled, and, importantly, women. How are governments doing in empowering women in their societies?

According to data collected by the Inter-Parliamentary Union (2013), the global proportion of parliamentary seats held by women reached a high of nearly 20 per cent in 2012 (see Figures 7.1 and 7.2). The top countries were Rwanda

(56 per cent), Andorra (50 per cent), Cuba (45 per cent), Sweden (45 per cent), and the Seychelles (44 per cent). As of December 2011, women comprised 8 of 152 elected heads of state worldwide (Inter-Parliamentary Union 2011). They also comprised more than 16 per cent of ministerial positions, mostly in Europe and Africa. Underpinning these political results is the will of leaders and of the people to promote women's access to parliaments.

The increased participation of women in government may result in greater government and legal attention to gender, energy, and poverty issues. Research has shown that with relatively larger numbers of politically active women, greater importance is placed on societal issues such as health care, the environment, and economic development. This trend of women's political involvement is likely to continue over the next 15–20 years and could lead to an increase in social programmes over military ones.

7.1.2. Education and energy access

Gender equity is a double dividend for developing countries (Hanes 2007). Girls' education and the promotion of gender equality in education are vital to community development (Roudi-Fahimi and Moghadam 2003; World Bank 2013). They reduce fertility rates; lower infant, child, and maternal mortality; increase women's labour force participation; and improve household incomes. Mothers' education is a significant predictor of children's educational attainment and future opportunities.

The availability of improved lighting expands educational opportunities to women and children. Better lighting enables people to work and study during more hours of the day. It has been demonstrated that better lighting at home and school increases reading time for both children and adults, keeps children in school longer, and attracts teachers to rural areas (Independent Evaluation Group 2008; Bazilian 2013a).

According to the National Intelligence Council's report, *Global Trends 2025: A Transformed World* (National Intelligence Council 2008), the explosion of global productivity in recent years has, in part, been driven by improvement in health, education, and employment opportunities for women and girls. Higher levels of female literacy are linked with more robust GDP growth. Where mechanical energy is available to draw water, till, and transport crops, girls' school attendance and performance increases by the equivalent of one or two grades (Lallement 2008; World Bank 2009).

7.1.3. Health care and energy

Improved lighting can have positive impacts in the area of health care, especially for women. Complications during pregnancy and childbirth are a leading cause of death and disability among women of reproductive age in developing countries (World Health Organization 2013). An estimated 287,000 maternal deaths occurred worldwide in 2010; 99 per cent of them took place in developing countries, where many health clinics lack reliable electricity (World Health

Organization 2012b). The most common life-threatening complications of pregnancy (hemorrhage, obstructed labour, eclampsia, and infection) are, in fact,
treatable with proper medical attention. However, safe, effective medical procedures are challenging and, in some cases, impossible to conduct without adequate
lighting and reliable communication systems (WE CARE Solar 2011).

The World Health Organization (2012a) estimates that, in some developing
regions, more than half of health-care facilities either lack reliable electricity or do
not have access to electricity at all. Even urban hospitals in these areas may not
have electricity for hours each day. Without electricity, doctors and midwives
struggle to provide patients with adequate care, often working by the light of
candles or kerosene lanterns (Energy Map 2013). In these circumstances, patients
may not receive timely care, and procedures can be delayed until daylight hours or
conducted in rudimentary conditions, often with tragic results (Energy Map
2013). Access to reliable electricity can have a significant impact on the quality
of health care that medical professionals are able to provide to expectant mothers
in developing countries. The incidence of pregnancy-related deaths can be
reduced when health-care facilities have access to reliable lighting, mobile communication, and electricity to power medical devices.

7.1.4. Gender and cooking

Cooking is universally women's business; in 97.8 per cent of societies, women do
the domestic cooking. Wrangham (2009) posits that *Homo erectus* began controlling fire and cooking approximately 1.9 million years ago. Fossil evidence supports
such a hypothesis; at that time, *Homo erectus* began to evolve smaller teeth,
stomachs, jaws, and rib cages in concert with larger brains.

Long ago, Wrangham imagines, sparks were made or captured from natural
fires. It is easy to maintain fires during the day, and fires at night provided
protection from predatory animals as well as light and warmth. Once food was
accidentally dropped on the fire and tasted, cooking inevitably began, enjoyed by
habilines, *Homo erectus*, Neanderthals, and, some 200,000 years ago, *Homo
sapiens*.

'Cooked food is better than raw food because life is mostly concerned with
energy' (Wrangham 2009: 81). According to Wrangham, cooked food is more
nutritious than raw food, and evidence is accumulating that cooked food has
profound consequences for the energy made available to species that consume it
(Carmody, Weintraub, and Wrangham 2011; Boback et al. 2007; Jha 2012; Garner
2009).

Two major clues buttress the cooking hypothesis: (1) the fossil record over the
past 2 million years on the anatomy of human precursors and humans, and
(2) the fact that anatomy has been shaped by cooked food. 'Cooking was a great
discovery [because] it helped make our brains uniquely large, providing a dull
human body with a brilliant human mind' (Wrangham 2009: 127). The great
apes chew their food for half their time, whereas cooked food can be eaten quickly.
The great apes have large mouths and teeth whereas humans have small mouths,
jaws, and teeth. Women in subsistence societies tend to spend the active part of

their days collecting and preparing food. Men, liberated from chewing, do much as they wish.

Wrangham argues that cooking has made possible one of the most distinctive features of human society: the sexual division of labour. Women and men seek different kinds of foods, and the foods they obtain are eaten by both sexes. Women everywhere tend to provide the staples—roots, seeds, shellfish. Men search for foods especially appreciated but harder to obtain, such as meat and honey.

Wrangham hypothesizes that cooking led to marriage and, ultimately, to patriarchy. The human family, which adds a male to the female-and-young structure, is based on reciprocity of food sharing. Humans are the only species in which the sex-relation is an economic relation. Cooking in the bush would have been a conspicuous and lengthy process, making females particularly vulnerable to poachers. Males/husbands protected their females/wives from being robbed, using their bonds with other men with whom they hunted. Females prepared their husbands' meals. Men's hunted food belonged to and was distributed by the community of males, whereas women's gathered food belonged to each individual woman. Therefore, if husbands came back empty-handed, wives could feed them. Males depended on the cooked food that females provided them because chewing raw food for five to six hours, as the higher primates do, would have been too high an energy cost for them. Therefore, Wrangham argues, males were the greater beneficiaries of women's cooking.

Cooking created and perpetuated a novel system of male cultural superiority, states Wrangham. Cooking is universally viewed as 'women's work' and considered by most to be a lower-status activity. Although men appear to enjoy cooking meat, they tend to avoid other food gathering and preparation activities, and, according to Wrangham, are known to abuse women if they are displeased with their meals. Conversely, he supports the saying, 'The way to a man's heart is through his stomach'.

7.1.5. Health impacts of cooking

Given this biological basis for human cultural traditions, it is not surprising that cooking has been one of the most intractable energy problems we face. Globally, 2.8 billion people (40 per cent of the world) relied on solid fuels for their main cooking fuel in 2010, a number that has remained stable for 30 years (Smith 2013). Terrible health impacts ensue from cooking over three-stone fires in enclosed spaces. A recent systematic analysis of all major global health risks has found that household air pollution (HAP) prematurely kills 4 million people annually worldwide (Lim et al. 2012), double the number of previous estimates. Although frequently referred to in the literature as indoor air pollution, or IAP, the most recent analyses have included both the indoor and outdoor air pollution resulting from the use of solid fuels for cooking, largely biomass or coal. Approximately 500,000 of the annual HAP deaths are from outdoor air pollution or 'second-hand cookfire smoke'. Another 500,000 are childhood pneumonia deaths. Three-and-a-half-million deaths annually occur among adults from lung cancer, cardiovascular disease, and chronic obstructive pulmonary disease. Although cataracts are included, they cause very few deaths (Smith 2012).

In terms of lost healthy life years (the metric used), HAP is the second most important risk factor for women and girls globally and the fifth for men and boys (Smith 2012). It is the first risk factor for both sexes in South Asia and for women and girls in most of Sub-Saharan Africa. Nevertheless, because men have higher background rates of major diseases, in terms of absolute impacts, they suffer more than women. In relative terms, however, women are more affected by HAP (Smith 2012).

With the advent of cooking programmes and clean-cooking technologies, effective approaches to addressing HAP may be on the way. As Mortimer et al. (2012) state, historically HAP 'has not received sufficient attention from the scientific, medical, public health, development, and policy-making communities. The tide has clearly changed with the broad-based support and launch of the Global Alliance for Clean Cookstoves in 2010. There is now considerable reason for optimism that this substantial cause of cardiorespiratory morbidity and mortality will be addressed comprehensively and definitively.'

7.1.6. Associated impacts of traditional biomass cooking

The health risks associated with exposure to household air pollution are not the only burdens borne by women who rely on traditional biomass cooking technology. In many parts of the developing world, the collection of fuel (wood, dung, and other biomass material) for cooking and heating in the home is the primary responsibility of women and girls (Misana and Karlsson 2001; Solar Electric Light Fund 2003; United Nations Development Programme 2011). Time and energy dedicated to the task of fuel collection varies depending on factors such as local fuel scarcity, household size, and season (Sovacool 2012). In rural households, women typically devote at least 25 per cent of total domestic labour to wood collecting. The average woman in Africa carries 44 pounds of fuelwood three miles per day (Sagar 2005). In India, the typical woman devotes 40 hours per month to fuel collection, taking 15 separate trips, often by foot, for distances greater than three miles (Sangeeta 2008). Time spent gathering fuel is time lost to other activities, including schooling and income-generating work. These are high opportunity costs for people living in areas with limited economic opportunities to begin with.

Beyond the opportunity costs associated with this demanding work, physical injuries are also a common occurrence. Fuelwood and water collection has been linked to spinal damage in women, complications during pregnancy, and maternal mortality (Kramarae and Spender 2000). Research examining women fuelwood carriers in Addis Ababa, Ethiopia, show that they suffer from frequent falls, backaches, bone fractures, rheumatism, anemia, headaches, and other physical injuries and illnesses as a result of the arduous manual labour (Sovacool 2012). One major benefit of efficient cooking technologies is their ability to substantially reduce a household's demand for biomass fuel, thus easing the burden of fuel collection on women and girls. The associated risk of injury is lessened, and time previously spent on this task can instead be devoted to education, income-generating work, or other types of activities.

7.1.7. Domestic labour and mechanical energy

In addition to collecting and managing a household's fuel supply, women in rural areas are often responsible for other types of domestic manual labour, including fetching water, processing and hanging crops, and farming (Flavin and Aeck 2005; Misana and Karlsson 2001). The majority of farmers in Sub-Saharan Africa—in some areas up to 80 per cent—are women. The typical female farmer in the region is responsible not only for growing food but also for collecting water and firewood—putting in a 16-hour workday, on average (Worldwatch Institute 2010). Access to electricity and motorized equipment, such as pumps, can relieve women of some of the burdens of manual labour and reduce the amount of time required to accomplish domestic chores, freeing up time for schooling and other pursuits.

Although the problems appear at times intransigent and the amount of work to be done enormous, there is still reason for hope, for at least three reasons.

(1) **Measurement of MDG progress.** Upward trends document progress toward the MDGs. Although it is true that not all goals will be met by 2015, progress toward the goals *is* being measured.

(2) **Sustainable energy projects and programmes.** There are hundreds of sustainable energy projects around the world, with problems of energy and gender gaining in prominence, and a few new significant ones are being scaled up. Two important areas of energy access pertinent to women in particular are cooking and distributed electrification.[1] This chapter will present some examples of sustainable energy programmes in cooking and electrification and will discuss the issue of measurement relative to sustainable energy, gender, and poverty.

(3) **Measurement of energy access.** New global efforts have emerged such as the UN Year of Sustainable Energy for All (SE4ALL) and the development of tools for monitoring energy deficits on the one hand and monitoring progress toward total energy access on the other.

7.2. MEASUREMENT OF MDG PROGRESS

Progress toward the MDGs is being measured using some 60 indicators, and measurable progress is being made. The thesis of this chapter is that by setting goals and identifying indicators to monitor progress toward them, much more will be done to increase women's access to sustainable energy sources. The United

---

[1] Although mechanical energy is also of importance to women, it is beyond the scope of this chapter.

Nations (UN) has developed an annual report assessing progress toward achieving the MDGs (United Nations 2012a).

- Worldwide, women have more income-producing opportunities than ever
  before. Overall, women now occupy almost 40 per cent of all paid jobs
  outside agriculture compared to 35 per cent in 1990 (see Figure 7.3).
  Women now hold nearly 34 per cent of non-agricultural jobs in the developing world, compared to 29 per cent in 1990.

- Worldwide, increasing access to routine vaccinations has slashed deaths
  from measles. Today, 84 per cent of children in the developing world will
  receive at least one dose of measles vaccine before they turn 2 (see
  Figure 7.4).

- Globally, adolescent fertility has been declining since 1990; pre-natal care has
  been increasing; and access to quality reproductive health services has been
  improving (see Figures 7.5, 7.6, and 7.7).

- The demand for information and communications technology continues to
  grow worldwide. In the developing world, Internet use has increased from
  less than 1 per cent of the population in 1995 to 24 per cent in 2011 (see
  Figure 7.8). The number of mobile phone subscriptions worldwide reached 6
  billion at the end of 2011, or 87 per cent of the global population.

- Enrollment in primary education has continued to rise, reaching 89 per cent
  in the developing world (see Figure 7.9). The developing regions as a whole
  are approaching gender parity in educational enrollment. Girls' primary
  enrollment increased more than boys' in all developing regions between
  2000 and 2006.

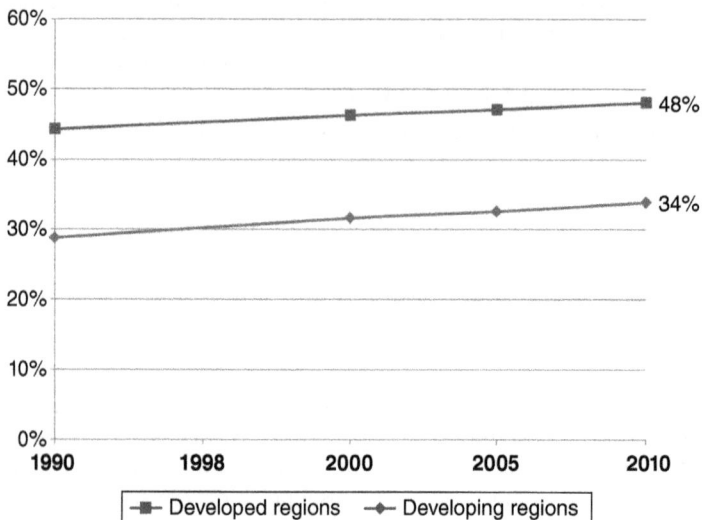

Figure 7.3. Percentages of women employees in the non-agricultural sector, developed and
developing regions, 1990–2010

Source: United Nations 2012b

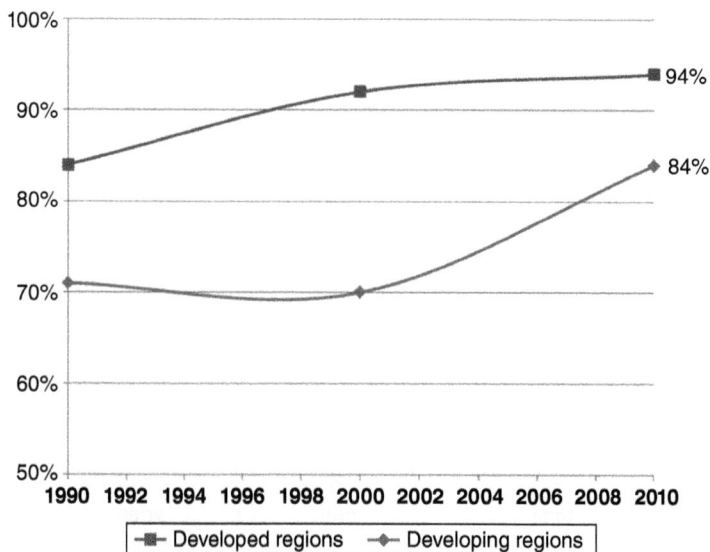

Figure 7.4. Proportion of 1-year-old children immunized against measles, developed and developing regions, 1990–2010

Source: United Nations 2012b

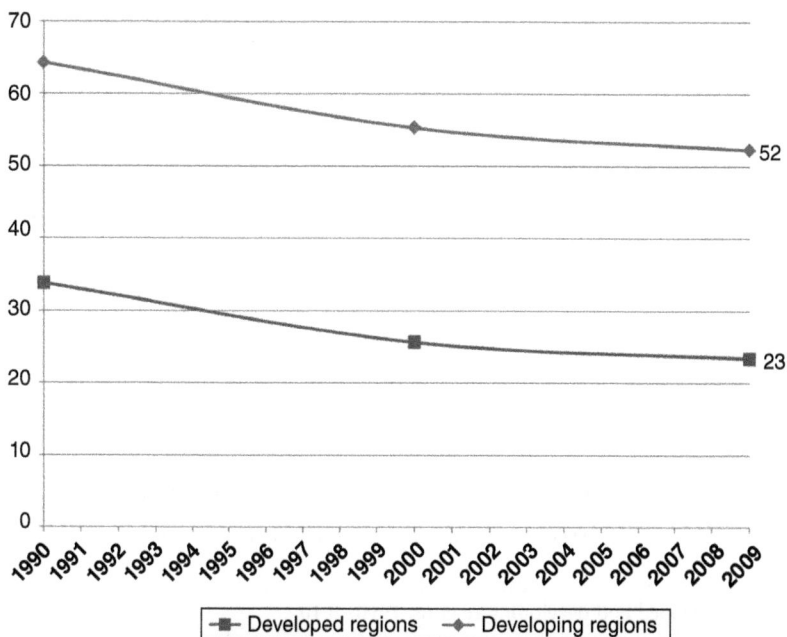

Figure 7.5. Births to women aged 15–19 per 1,000 women, developing and developed regions, 1990–2009

Source: United Nations 2012b

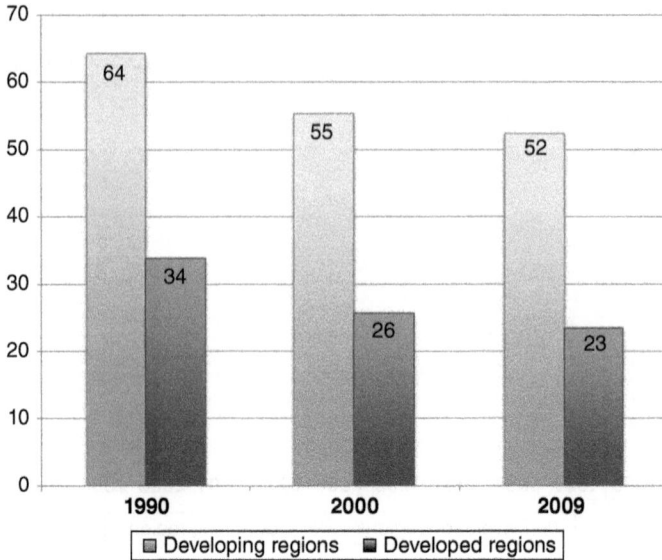

Figure 7.6. Births to women aged 15–19 per 1,000 women, developing and developed regions, 1990, 2000, and 2009

Source: United Nations 2012b

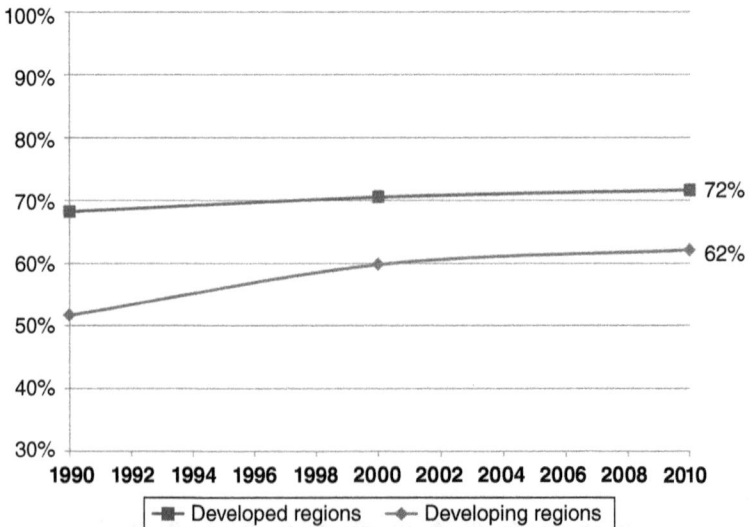

Figure 7.7. Contraceptive prevalence rate among women aged 15–49 who are married or in union, developed and developing regions, 1990–2010

Source: United Nations 2012b

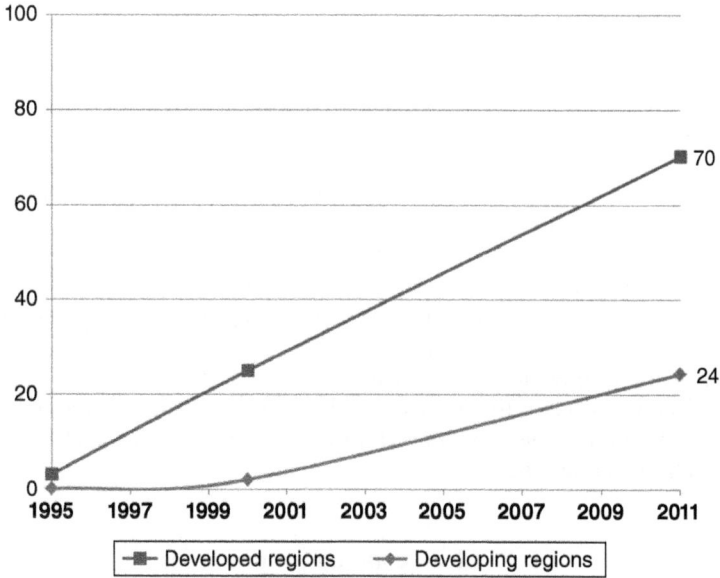

Figure 7.8. Internet users per 100 people, developed and developing regions, 1995–2011
Source: United Nations 2012b

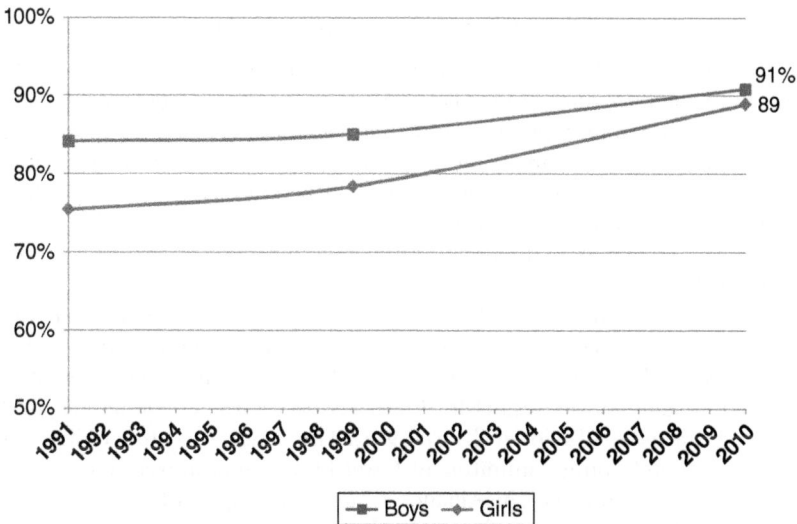

Figure 7.9. Proportion of children enrolled in primary and secondary school by gender, developing regions, 1991–2010
Source: United Nations 2012b

7.3. SUSTAINABLE ENERGY PROJECTS
AND PROGRAMMES

Hundreds, if not thousands, of sustainable energy projects are being conducted around the world. There is no way to represent in one chapter the brilliance and scope of the achievements of those engaged in these projects who are profoundly improving the human condition in education, equality, environment, health, and economic development through the use of sustainable energy technology. Nor can this chapter do justice to the thousands of private charitable, government, foundation, and non-governmental organization (NGO) programmes (not to mention individual efforts) in financing, analyzing, evaluating, and delivering sustainable energy technologies in the developing world.

Instead, the chapter focuses on a few of these projects as exemplars of the ingenuity and dedication that have marked efforts addressing gender and energy poverty. Specifically, the programmes focus on clean cooking and distributed electrification and lighting.

7.3.1. Clean cooking programmes

To exemplify efforts in clean cooking, three stove programmes are briefly described: (1) the Global Alliance for Clean Cookstoves, (2) Envirofit, and (3) the Appropriate Rural Technology Institute (India).

Global Alliance for Clean Cookstoves

On 21 September 2010, Hillary Rodham Clinton, the US Secretary of State, announced the formation of the Global Alliance for Clean Cookstoves (GACC), a more-than-$60 million public–private partnership to save lives, improve livelihoods, empower women, and combat climate change by creating a thriving global market for clean and efficient household cooking solutions (Anthony 2010). Its ambitious but achievable goal is to foster the adoption of clean cookstoves and fuels in 100 million households by 2020, thus impacting more than 3 billion people worldwide.

The effort is aimed toward addressing the fact that 'more than a century after the introduction of electric power transmission, almost 3 billion people still rely on biomass fuels to meet their energy needs. Use of this renewable fuel in unvented cooking stoves results in disastrous consequences for human health and global warming' (Taylor et al. 2011: 918). The GACC seeks to mobilize high-level national and donor commitments toward the goal of universal adoption of clean cookstoves and fuels. The Alliance is mobilizing support from a wide range of private, public, and non-profit stakeholders. More than 300 public and private partners and 35 countries have joined the Alliance (US Department of State 2012).

GACC's strategy for this transformation is to enhance demand, strengthen supply, and foster an enabling environment. GACC recognizes that it is critical to raise consumer awareness and education regarding the health, environmental, gender, and economic benefits of clean cookstoves. Acknowledging that one size does not fit all, the Alliance is working to support and encourage the development of a wide variety of cookstove designs as well as more efficient fuels to meet individual and cultural needs. Fostering an enabling environment for the dissemination of cookstoves includes such issues as addressing social barriers and devising financing models that work for those at or near the bottom of the pyramid. GACC's partners in this challenge include governments, foundations, universities, carbon project developers, multilateral organizations, NGOs, and the private sector.

GACC seeks to increase the production, deployment, and use of clean cookstoves in the developing world by addressing global needs to reach large-scale adoption, targeting countries that emphasize market-based solutions, and coordinating a global plan across all key sectors. The issue of indoor air pollution and clean cookstoves is gaining in popular awareness in part through the many celebrity ambassadors for the GACC, including Ghanaian international music star and creative activist Rocky Dawuni, Academy Award-winning actress Julia Roberts, and chef José Andrés.

Envirofit

Envirofit International, a charitable organization founded in 2003, has received a $25 million grant from the Shell Foundation to develop energy-efficient cookstoves for the developing world based on technology developed at Colorado State University's Engines and Energy Conversion Laboratory.

Envirofit applies the same product development methodology and protocols used in industry to develop products for 'bottom of the pyramid' markets. Its enterprise-based model relies on market mechanisms to guide product development and drive consumer demand. Envirofit stoves use 60 per cent less fuel than three-stone fires, reduce cooking time by half, and reduce emissions by as much as 80 per cent.

More than 450,000 Envirofit cookstoves have been sold in India, Nigeria, Kenya, Liberia, and Ghana since October 2007 through local distributors (Envirofit website). Each stove increases household income by $30 to $75 per year and reduces CO_2 emissions by 1 to 1.5 tons per year. Reduced death rates from indoor air pollution problems are expected.

Appropriate Rural Technology Institute

The Appropriate Rural Technology Institute (ARTI) is an NGO based in Maharashtra, India, founded by a group of scientists and social workers in 1996. The mission of the organization is to serve as an instrument of sustainable rural development through the application of scientific and technological knowledge. Under the leadership of Dr. A.D. Karve, the institute began with nine research projects. ARTI has since completed more than 50 projects sponsored by national and international funding agencies and has emerged as an internationally acclaimed research and development institution working in the field of rural development through innovative appropriate technologies.

Initially working in Maharashtra state, ARTI has expanded its activities to several other states in India. Its primary work has been research on and promotion of smokeless cookstoves and biodigesters (Appropriate Rural Technology Institute website). ARTI strives to make new businesses profitable for rural inhabitants. With its high-impact grassroots-level work, including the use of hands-on training and audiovisual media to engage rural artisans and village women in the adoption of innovative technologies, ARTI has become internationally recognized for its work on appropriate-scale technologies. It has also provided technical consultancy to clients in the USA, Europe, and Africa.

7.3.2. Distributed electrification and lighting

This section touches briefly on five programmes: (1) the Grameen Shakti, (2) the Solar Electric Light Fund, (3) the WE CARE Solar Suitcase Programme, (4) Elephant Energy and Eagle Energy, and (5) Grupo Fenix.

Grameen Shakti

In 1996, as a spinoff of the Grameen Bank founded by Muhammad Yunus in 1983, Dipal Barua founded the Grameen Shakti, a non-profit organization to promote renewable energy in rural areas. The Shakti went door-to-door to convince people to replace kerosene with solar power. With loans from the World Bank and the Global Environment Facility, more than 400,000 solar home systems have been installed, bringing electricity to more than 4 million rural people (Bright Green Energy Foundation website).

The Shakti brings lighting and related advantages such as mobile phones, computers, and Internet connection to rural areas, resulting in significant improvements in standards of living by increasing income-producing and educational opportunities. Solar electricity lights homes, shops, and fishing boats; charges mobile phones; runs electronics; and powers water desalination. The Shakti programme fosters solar power entrepreneurs, trains women in solar installations, and provides user-friendly financing models at low interest rates or even with no service charges at all. Barua has said that 'the solar home system plays a very effective role in bringing "green" electricity to rural households to facilitate children's education and help women to cook and participate in income-generating activities'.

Key to the Shakti's success was the deliberate inclusion of women in the uptake, installation, and servicing of solar energy systems. Barua said: 'We believe that women should be transformed from passive victims into active forces of good to bring changes in their lives and the communities in which they live' (Arthur 2010). Having won a $1.5 million Abu Dhabi Zayed Future Energy Prize, Barua has initiated the Bright Green Energy Foundation with the goal of training 100,000 women to establish their own renewable energy businesses and to showcase climate change mitigation and adaptation measures.

Solar Electric Light Fund

The Solar Electric Light Fund (SELF) was founded by Neville Williams in 1990. Within seven years, SELF had established 11 sustaining solar energy projects in Asian, African, and South American countries. Using donated funds and loans from development agencies, SELF bought home-size PV systems in bulk, usually enough for one small village at a time. SELF then sold the systems at slim mark-ups to villages through partnerships with in-country non-profit agencies.

Each participating household made a 20 per cent down payment on a system and paid off the balance, usually about $350, over several years. Their payments were pooled in a local revolving loan fund from which their neighbours could borrow to buy their own solar power. SELF used a portion of the mark-ups to establish a local dealership and train residents as solar installers and technicians. These systems provide lighting to replace kerosene lamps, which cause more than 20,000 injuries and house fires annually, emit CO_2, and expose families to hazardous waste (SELF website).

SELF has developed projects in 20 developing countries and, through its for-profit affiliate—the Solar Electric Light Company—has sold 90,000 solar home systems in India, just a start on that market.

SELF extended its vision to powering entire villages, and in 2000, the organization installed a 1.5-kW solar array in a town in South Africa, which generated enough electricity to power 20 computers and a small satellite dish that delivered Internet access to Myeka High School. This was the first solar-powered computer lab in South Africa, effecting a 'wonderful transformation' at the high school, where the pass rate jumped from 30 per cent to 70 per cent almost immediately. With an annual budget of close to $2 million, SELF continues to experiment with novel approaches to rural electrification in the developing world.

WE CARE Solar

Using solar electricity, WE CARE Solar supports safe motherhood and reduces maternal mortality in developing nations by providing health workers with reliable lighting, mobile communications, and blood bank refrigeration (WE CARE Solar website). As Mills (2012) wrote: 'Inadequate lighting in clinics poses barriers to the delivery of quality health care, discourages patients from seeking care, and compounds the risks of adverse outcomes such as maternal and infant mortality as well as infections due to the difficulty of maintaining sanitation in low-light conditions.'

WE CARE Solar co-founder Laura Stachel is an obstetrician-gynecologist who travelled to northern Nigeria in 2008 to study ways to lower maternal mortality in state hospitals. She found deplorable conditions including sporadic electricity, which impaired maternity and surgical care. Night-time deliveries were attended in near darkness, caesarean sections were cancelled or conducted by flashlight, and critically ill patients waited hours or days for life-saving procedures.

Stachel's husband, Hal Aronson, a solar energy educator in Berkeley, California, designed a suitcase-size prototype of a hospital solar electric system with solar panels, LED lights, headlamps, and walkie-talkies. Once the prototype had been field tested, WE CARE Solar distributed suitcases to midwives in outlying health clinics. Eventually, the World Health Organization, the MacArthur Foundation, the Blum Center for Developing Economies, and the Bixby Center for Population, Health, and Sustainability supported the solar suitcase programme in Nigeria and Liberia. The suitcases integrate the entire system, are safe, require little mainten-ance, are durable, and are simple to operate and to expand up to 200 watts of solar panels and a 140 amp-hour sealed battery.

In 2009, the organization developed a medical clinic photovoltaic system to provide lighting, blood bank refrigeration, and communication equipment for medical personnel. Field tested in Nigeria, the system provides lighting in operating room and delivery theatres, facilitates telecommunications between hospital staff and physicians, permits refrigeration of blood, and charges batteries for LED headlamps for night-duty workers.

Elephant Energy and Eagle Energy

Elephant Energy and its companion programme, Eagle Energy, are developing for-profit models to distribute solar lighting and mobile phone chargers in Africa and to the Navajo Nation in the Four Corners Region of the United States (Elephant Energy 2012). After seeing first-hand the tremendous appreciation for solar-powered lights in the Caprivi region of Namibia 2008, newly minted lawyer Doug Vilsack formed a non-profit organization—Elephant Energy—to bring solar energy to rural Namibia and to foster self-sustaining local shops and women entrepreneurs.

Elephant Energy has found that the main problem in providing energy in rural Namibia is not a lack of money, but rather lack of access to quality small-scale renewable energy technologies. As a result, the organization established shops to provide clean-energy products to the nearly 1 million people in Namibia that live without access to electricity, with the goal that these shops ultimately become self-supporting and profitable. Another project is developing a 'rural entrepreneur model' through which small-business owners located in areas far away from market centres are offered credit to purchase small volumes of energy products for their shops.

Elephant Energy runs a Women's Energy Action project that involves women's groups to define the most appropriate energy technologies to meet women's needs. The project operates in Namibia and Zambia with the objective of 'inspiring rural women to be leaders in sustainable development through the use and marketing of appropriate sustainable energy technologies like solar-powered lights and mobile phone chargers'.[2]

The organization is committed to viewing women not just as customers but as drivers of the energy revolution in Africa. Seeking designs of profitable business ventures for thousands of people, Elephant Energy devised the 'Rent-to-Own' and the 'Rent-A-Light' programmes to test the distribution of solar lights and mobile phone chargers using sales-agent business models. Area coordinators were paid a monthly salary, and saleswomen reported to the coordinators. The saleswomen collected monthly payments for solar lights and received the final payments as compensation, which generated significant income for them. However, the main

---

[2] Women's Energy Project and Rent-to-Own Final Report, supported by IRDNC, WWF, USAID, Open Meadows Foundation, The International Foundation, and the Robert J. & Cleetta L. Steininger Family Trust, Elephant Energy, Denver, Colorado, 2012. Elephant Energy conducted baseline surveys in the area, and found that women used wood, candles, or small batteries for lighting; 87 per cent had no indoor lighting at all. Although more than half had mobile phones, few had a reliable way to charge them. Also, women walked four to six hours to collect wood three times a week.

problem was that the saleswomen had to make repeated visits to customers to collect the monthly payments.

Elephant Energy defined the benefits of solar lighting/mobile phone charging units as:

(1) Saving money and time: women no longer had to spend large portions of their limited incomes on candles, batteries, and mobile phone charging.

(2) Protection from wildlife threats: one-third of the women studied said they successfully protected themselves and their stock from wildlife (e.g. elephant, hippo, bush pig, spring hare, duiker, hyena, leopard, wild dog, genet, and venomous snakes and scorpions) using solar lights.

(3) Improved reliability of mobile phone communications.

(4) Improved self-confidence in mastering new technologies through women's meetings.

(5) Supporting children with light for study and homework.

(6) Providing light for childbirth.

Customers were very satisfied with the installment plan programmes. One said: 'I am very satisfied. The Sun King Pro is helping me a lot. Buying candles used to finish my money each and every month. It is a long time since I have bought a candle, since last year. With the money I used to use to buy candles, I now buy food.'

Based on its experience with the rent-to-own programmes and customer surveys, Elephant Energy devised a new business model employing a technology called the 'Divi Light'. These pay-as-you-go solar lights and mobile phone chargers allow customers to make deposits on lights and then pay 30 US cents per Divi credit, which equals 24 hours of light. The lamps shut off until more credits are purchased. The lights come with 7 credits; once customers pay for 120 more credits, the lamps unlock and the light is free. The Divi Light allows lamp-to-lamp wireless connection by physically shaking and pouring credits from one lamp to another. Customers can come to buy credits in the Elephant Energy shops just as they would buy candles in shops. Elephant Energy hopes that this product, and the business model through which it is delivered, could ultimately provide lighting access to millions of people in the years to come.

The organization has also initiated a new programme, called Eagle Energy, in the vast Navajo Nation, which occupies parts of Arizona, Utah, and New Mexico. Eagle Energy works with local NGOs, businesses, and schools to provide solar lighting and mobile phone charging to remote areas lacking grid electricity. After a year of preliminary research, Eagle Energy determined that small-scale, $20–40 solar-powered lights provide safe, clean, and affordable alternatives to the kerosene that is used for lighting by some 18,000 rural Navajo families. Since 2010, Eagle Energy has distributed more than 500 solar-powered lights in the Eastern Agency of the Navajo Nation through a network of shops and schools, while seeking to devise a model for the Navajo people themselves to profit from solar light distribution.

Public health workers, who have long recognized the adverse health implications of kerosene use, are now distributing 400 solar lights and mobile phone chargers to elders in the Bennett Freeze area working through local chapter houses.

Eagle Energy is also developing market-based distribution channels for solar lighting and mobile phone charging through local Navajo shops and through the development of a Navajo Women's Energy Project. Using applied theatre techniques, the project invigorates women to imagine alternative energy futures. These techniques enhance women's participation in rehearsing various plans for action to move their nation from old energy stories to new ones.

Grupo Fenix

Grupo Fenix exemplifies another key example of grassroots sustainable energy (Grupo Fenix website). The municipality of Totogalpa is the second-poorest in Nicaragua. Since this area has some of the lowest precipitation levels in the country, wood for burning in stoves is extremely scarce. One of the greatest problems in the area is deforestation, which is exacerbated by the high use of wood-burning stoves in the area. Depletion of firewood from forests in turn causes scarcity of water sources, especially for rural areas where people have to walk long distances in search for water during the dry season. The Solar Women of Totogalpa have helped transform the lives of the women in this area by providing opportunities for women to work their way out of poverty and by helping to restore the ecological health of their area (Grupo Fenix 2013a).

Guided by their professor, Susan Kinne, students at the National Engineering University in Managua created Grupo Fénix in 1996 to explore the promise of renewable energy in Nicaragua. From these initial efforts, the programme known as PFAE (Programa Fuentes Alternas de Energía) evolved with primary emphasis on the development of a model solar community in the rural north of Nicaragua, in the community of Sabana Grande, Totogalpa, Madriz. PFAE, together with the Solar Women of Totogalpa, have created a Solar Center and have begun to offer a range of products for sale—such as solar-roasted coffee beans—based around renewable energy. In January 2012, the Solar Women of Totogalpa with the support of PFAE opened the first clean energy restaurant in Nicaragua, La Casita Solar, on the grounds of the Solar Center. They serve delicious food prepared in solar cookers, mixed with a bicycle blender and fuelled by methane from a biodigester. Completely off-grid and located on the busy Carretera Panamericana (Pan-American Highway), this centre has been extremely successful with eco-tourism by attracting customers as well as volunteers and students who want to study and work at the centre (Grupo Fenix 2013b).

7.4. MEASURING ENERGY POVERTY AND ENERGY ACCESS

As mentioned earlier, the thesis of this chapter is that measurement itself will help ensure progress toward energy goals. Recent efforts have focused on measuring energy poverty and energy access. The Sovacool (2012) review of energy poverty issues found that 40 per cent of the global population is living on less than $2 per day. The UNDP's analyses that Sovacool cites define energy poverty using two indicators: (1) the inability to cook with modern cooking fuels and (2) the lack of a bare minimum of electric lighting to read or for other household and productive activities. The poor spend 7 per cent to 15 per cent of their incomes on energy services, whereas the wealthiest spend 9 per cent.

7.4.1. Measuring energy poverty

The Oxford Poverty and Human Development Initiative (OPHI) has developed a new tool for measuring energy poverty called the Multidimensional Energy Poverty Index (MEPI) (Nussbaumer et al. 2011). The MEPI focuses on the deprivation of access to modern energy services, capturing both the incidence and the intensity of energy poverty on a per capita basis. MEPI is a composite index that generates an energy poverty score between 0 and 1 at a national level based on deprivation of access to cooking, lighting, refrigeration, entertainment/ education, and communications. The index permits comparisons among countries and regions, but is less relevant at community or project levels.

As Table 7.1 shows, the index is composed of five 'dimensions' and six indicators of those dimensions. A person is defined as 'energy poor' if a combination of deprivations exceeds a pre-defined threshold. The measure is 'the product of a headcount ratio (share of people identified as energy poor) and the average intensity of deprivation of the energy poor' (Nussbaumer et al. 2011: 9).

Table 7.1. MEPI dimensions, indicators (weights), variables, and deprivation cut-offs

Dimension	Indicator (weight)	Variable	Deprivation cut-off (poor if...)
Cooking	Modern cooking fuel (0.2)	Type of cooking fuel	Use any fuel besides electricity, LPG, kerosene, natural gas, or biogas
	Indoor pollution (0.2)	Food cooked on stove or open fire (no hood/ chimney) if using any fuel besides electricity, LPG, natural gas, or biogas	True
Lighting	Electricity access (0.2)	Has access to electricity	False
Services provided by means of household appliances	Household appliance ownership (0.13)	Has a refrigerator	False
Entertainment/ education	Entertainment/education appliance ownership (0.13)	Has a radio OR a television	False
Communication	Telecommunication means (0.13)	Has a phone land line OR a mobile phone	False

Source: Nussbaumer et al. 2011

The lack of reliable, comprehensive datasets is a key impediment to comprehensive cross-national analyses. Where available, the authors relied on existing household survey data from the International Energy Agency, MEASURE DHS (Demographic and Health Surveys), funded by the US Agency for International Development (USAID). UNICEF Childinfo reports on relevant indicators. An advantage of using survey data is that it allows comparisons between urban and rural populations, for example. Yet no comprehensive dataset exists on household lighting, among other indicators. The paucity of data will remain an issue for years to come, no matter what approach is used to measure energy poverty or progress toward improved energy access.

7.4.2. Measuring energy access

Approaching measurement from a community and project-level perspective, Practical Action (2012) has proposed total energy access (TEA) minimum standards (Table 7.2) to define the energy services people need to escape poverty and has developed the Energy Supply Index (ESI) (Table 7.3) to measure the quality of the energy supplies that households are using. These indicators are particularly useful at the community and project levels.

A crucial component of the Practical Action approach is that poor people's perspectives, experiences, and preferences will be included in measuring energy access. This is being accomplished through a Total Energy Wiki, being developed by Energypedia and Practical Action[3] to allow anyone with access to the Internet anywhere in the world to upload energy access data that has been collected using a simple standard questionnaire (Practical Action 2012: 89).

Table 7.2. Total energy access (TEA) minimum standards for 2012

Energy service		Minimum standard
Lighting	1.1	300 lumens for a minimum of 4 hours per night at household level
Cooking and water heating	2.1	1 kg wood fuel or 0.3 kg charcoal or 0.04 kg LPG or 0.2 litres of kerosene or biofuel per person per day, taking less than 30 minutes per household per day to obtain
	2.2	Minimum efficiency of improved solid fuel stoves to be 40% greater than a three-stone fire in terms of fuel use
	2.3	Annual mean concentrations of particulate matter (PM2.5) < 10 μg/m^3 in households, with interim goals of 35 μg/m^3, 25 μg/m^3, and 15 μg/m^3
Space heating	3.1	Minimum daytime indoor air temperature of 18°C
Cooling	4.1	Households can extend life of perishable products by a minimum of 50% over that allowed by ambient storage
	4.2	Maximum apparent indoor air temperature of 30°C
Information and communications	5.1	People can communicate electronic information from their households
	5.2	People can access electronic media relevant to their lives and livelihoods in their households

Source: Practical Action 2012, p. 42

---

[3] <http://www.energypedia.info/totalenergywiki>.

Table 7.3. Energy Supply Index 2012

Energy supply	Level	Quality of supply
Household fuels	0	Using nonstandard solid fuels such as plastics
	1	Using solid fuel in an open/three-stone fire
	2	Using solid fuel in an improved stove
	3	Using solid fuel in an improved stove with smoke extraction/chimney
	4	Mainly using a liquified or gas fuel or electricity, and associated stove
	5	Using only a liquid or gas fuel or electricity, and associated stove
Electricity	0	No access to electricity at all
	1	Access to third-party battery charging only
	2	Access to stand-alone electrical appliance (e.g. solar lantern, solar phone charger)
	3	Own limited power access for multiple home applications (e.g. solar home systems or power limited off-grid)
	4	Poor-quality and/or intermittent AC connection
	5	Reliable AC connection available for all uses
Mechanical power	0	No household access to tools or mechanical advantages
	1	Hand tools available for household tasks
	2	Mechanical advantage devices available to magnify human/animal effort for most household tasks
	3	Powered mechanical devices available for some household tasks
	4	Powered mechanical devices available for most household tasks
	5	Mainly purchasing mechanically processed goods and services

Source: Practical Action 2012, p. 45

'Substantial improvements in supply quality are also possible in all dimensions using decentralized technologies including solar home systems and improved biomass cooking facilities' (Practical Action 2012: 45). These can be tracked using appropriate ESI indicators (see Table 7.3) by acquiring detailed household-level data through the Wiki.

7.5. ENERGY POVERTY IMPACTS AND INDICATORS

Empowering women and improving women's access to modern energy, and especially to sustainable energy, depends, in part, on goal-setting, selection of indicators, and consistent monitoring of these indicators over time. For greatest effectiveness, this monitoring should be done by the institutions already responsible for measuring progress in human health and development, as well as the mainstreaming of gender considerations in human progress. To understand how this could most effectively be done, we must consider gender impacts.

Table 7.4 presents suggestions for a way to conceptualize the measurement of progress toward sustainable energy development goals (SEDGs). It focuses on the activities of women in energy-poor households and communities.

Table 7.4. Suggested examples of intermediate and final indicators of progress toward sustainable energy development goals (SEDGs)

Women's activity/ circumstance	Intermediate indicators	Final indicators
Fuel gathering (wood, agricultural waste, dung, charcoal)	Number of clean cookstoves diffused Number of villages reached with programmes	**Reductions in**—hours of manual labour by women; backaches; scorpion stings; snakebites; rape; attacks; bone fractures; falls; school absenteeism; headaches; rheumatism; miscarriages
Cooking (cooking over fires)	Number of clean cookstoves diffused Number of villages reached with programmes Particulate emissions within homes Air quality measures in communities/ villages	**Reductions in**—burns; scalding; cataracts; COPD; lung cancer; asthma; tuberculosis; acute/chronic respiratory infections; low birth-weights; diseases of the eye; adverse pregnancy outcomes; anemia
Lighting (kerosene, candles, the cooking fire)	Number of off-grid solar lights distributed or sold for household use Number of village solar/wind installations Number of village micro-grids Number of renewable energy utility-scale powerplants	**Reductions in**—fires; burns; particulate emissions within homes **Improvements in**—school attendance; school graduations; healthy childbirths; household income; women and girls' rates of education; access to health care; availability of data and statistics; availability of gender audits

Work on tracking progress toward SEDGs is being carried out by various steering committees of the GACC, The World We Want (SE4ALL) and others. These committees are developing indicators and methods for gauging progress, and they deserve strong support (United Nations 2013). Finally, the training and engagement of women and social scientists in developing countries must be fostered to develop the capacity to produce accurate and reliable energy data on relevant indicators for the poor and especially for women, who produce and use most of the energy in developing countries. Only by having access to robust datasets can analysts, policymakers, and national governments effectively allocate assets for energy development to achieve SEDGs.

7.6. LOOKING FORWARD—THE POST-2015 DEVELOPMENT FRAMEWORK

Energy stands at the centre of global efforts to induce a paradigm shift toward poverty eradication and sustainable development. A consultation document 'The Future We Want' (Bazilian 2013a) focuses on defining post-2015 goals and the global energy consultation for defining them. The World We Want web platform, a joint initiative of the United Nations and Civil Society,[4] will gather inputs for a

---

[4] The Civil Society coalitions include the Global Call to Action Against Poverty, World Alliance for Citizen Participation (CIVICUS), and the Beyond 2015 Campaign.

global movement to secure a post-2015 development agenda. This process will ensure that the agenda reflects the perspectives of people living in poverty. Energy is one of eleven global thematic consultations.

Bazilian (2013a) and others advocate an ambitious attack on energy poverty, arguing that 'nonlinear' growth of 19 per cent per annum in provision of grid electricity is the meaningful path forward. The World Bank, the UN, the National Renewable Energy Laboratory (NREL), and other organizations are identifying and developing strategies for accomplishing this vision. Because of the UN Year of Sustainable Energy for All (SE4ALL), political momentum has gathered and the stage has been set for countries to focus attention on energy during the decade from 2014 to 2024. 'Political aspirational goals' include:

(1) Ensuring universal access to modern energy services.

(2) Doubling the global rate of improvements in energy efficiency.

(3) Doubling the share of renewable energy in the global energy mix.

(4) Promoting integrated energy solutions to with multiple dividends in poverty reduction, gender, health, water, food security, jobs, and climate change.

Bazilian suggests that one effective approach might be multilateral investment in 'project preparation' that would pave the way for large-scale electricity generation and distribution (such as dams, CPV systems, and power lines), but leaving the responsibility for their construction to national governments.

Bazilian (2013b) believes that the mobile phone 'leapfrog' analogy is not necessarily relevant for electrification. He thinks the grid should be built in a better way using smart grids and capacity development in terms of technology *and* regulation. He argues that universal access to electricity would not increase carbon dioxide emissions very much if focused on utility-scale renewable energy installations that employ concentrating solar power, concentrating PV, geothermal energy, and large wind farms.

The Clean Energy Ministerial (CEM) is another high-level global effort to promote policies and programmes that advance clean-energy technologies and to encourage the transition to a global clean-energy economy. Its three focus areas are (1) improve energy efficiency worldwide; (2) enhance clean-energy supply; and (3) expand clean-energy access.[5] The CEM is currently the only regular meeting of energy ministers at which they exclusively discuss clean energy. The CEM also engages other ministries that play an important role in clean energy in some governments, such as ministries of science and technology or economics. The CEM builds on and informs existing multilateral technical and policy work in cooperation with institutions such as the International Energy Agency, the International Partnership for Energy Efficiency Cooperation, and the International Renewable Energy Agency. The CEM's unique focus on clean energy and the broad set of participating ministers make it an especially promising forum for collaboration.

---

[5] <http://www.cleanenergyministerial.org/>, accessed 3 March 2013.

In July 2012, 24 of the world's governments, including developed and developing countries, convened in Washington, DC, to establish the CEM. Since then, meetings have been held in Abu Dhabi (CEM2), London (CEM3), New Delhi (CEM4), and Seoul (CEM5 in 2014). One of CEM's projects is Global Lighting and Energy Access Partnership (LEAP), which works within the UN's SE4ALL campaign. LEAP focuses on the 1.6 billion people worldwide who lack access to grid electricity. Its purpose is to replace dirty, fossil-fuel-based light sources such as kerosene lanterns with solar-powered, light-emitting diode (LED) lights, facilitating access to improved lighting services for 10 million people within five years.

Another CEM effort is the Clean Energy Education and Empowerment (C3E) initiative to inspire and connect women around clean-energy issues, attract more young women into related careers, and support their advancement to leadership positions. C3E has held forums on the role of women in the clean-energy revolution.

7.7. A CLOSING WORD

At the beginning of this chapter, we asked if change will be accomplished in access to sustainable energy. Part of the answer lies in whether we embrace the concept of lack of energy access as a root cause of poverty. Another part of the answer is whether we *identify indicators* of progress toward sustainable energy development and *measure improvements* in access to resilient sustainable energy practices and products. As the post-2015 development efforts take hold, there is every reason to believe that women will be leaders and participants in developing access to new clean energy economies around the world.

A global energy transition is inevitable. The only question is one of timing. A transition from fossil fuels to sustainable energy technologies and practices, including distributed efficiency and renewable energy, will have major repercussions economically, geopolitically, and culturally. Laws promoting female participation and continuing efforts by women themselves could be a stabilizing factor to economies during the coming energy transformation. In this—and in adaptation to climate change—the role and contributions of women will be critical.

Although the situation remains deplorable and the global economic crisis has increased poverty worldwide, persistent efforts—large and small—are beginning to pay off. We do have, and should have, hope!

APPENDIX—ENERGIA

Three women founded ENERGIA in 1996 with funding from the Dutch government. ENERGIA's goal is to contribute to the empowerment of rural and urban women in the South through a specific focus on energy.

ENERGIA has developed the gender audit, a tool to examine national energy policies, government practices, and institutions to identify groups in energy and gender approaches. As a result of ENERGIA's audits, Cambodia and Uganda have created gender focal points in all of their technical ministries. These governments are collecting sex-disaggregated data to monitor progress on gender-equitable energy policies.

ENERGIA runs a network of regional and national partners or focal points in 13 countries in Africa and 9 countries in Asia. These focal points work on projects, programmes, and policies to explicitly address gender and energy issues in order to improve the sustainability of energy service and human development opportunities available to women and men. Today, ENERGIA is involved in activities that range from training to promote gender mainstreaming in the energy sector to influencing policy and network-building.

ENERGIA publishes *ENERGIA News,* a newsletter on concepts and best practice as a forum for those working on gender and energy issues.

REFERENCES

Anthony, J. (2010). 'Secretary Clinton Announces Global Alliance for Clean Cookstoves', Global Alliance for Clean Cookstoves [website], <http://www.cleancookstoves.org/media-and-events/press/secretary-clinton-announces.html> (accessed 14 January 2013).

Appropriate Rural Technology Institute [website], <http://www.arti-india.org/> (accessed 16 January 2013).

Arthur, C. (2010). 'Women Entrepreneurs Transforming Bangladesh', *Making It: Industry for Development* [online magazine], <http://www.makingitmagazine.net/?p=459> (accessed 12 July 2013).

Bazilian, M. (2013a). 'The Future We Want: Background Paper for Global Energy Consultation', Global Thematic Consultation on Energy [website], <http://www.worldwewant2015.org/node/301099> (accessed 11 July 2013).

Bazilian, M. (2013b). 'Towards Universal Energy Access by 2030', webinar presented at the Center for Science and Technology Research, University of Colorado, Boulder, <http://cirescolorado.adobeconnect.com/p1bwaq6v02l/> (accessed 11 July 2013).

Boback, S., et al. (2007). 'Cooking and Grinding Reduces the Cost of Meat Digestion', *Comparative Biochemistry and Physiology* 148A: 651–6.

Bright Green Energy Foundation [website], <http://www.greenenergybd.com/> (accessed 13 January 2013).

Carmody, R., Weintraub, G., and Wrangham, R. (2011). 'Energy Consequences of Thermal and Nonthermal Food Processing', *Proceedings of the National Academy of Sciences* 108 (48): 19101–2.

Elephant Energy (2012). 'Women's Energy Project and Rent-to-Own Final Report', Elephant Energy [website], <http://elephantenergy.org/images/documents/WE_Project_Final_Report_(Elephant_Energy_2012).pdf> (accessed 11 July 2013).

Energy Map (2013). 'WE CARE Solar', Energy Map [website], <http://energymap-scu.org/wecare-solar/> (accessed 23 February 2013).

Envirofit [website], <http://www.envirofit.org/> (accessed 16 January 2013).

Flavin, C. and Aeck, M. (2005). 'Energy for Development: The Potential Role of Renewable Energy in Meeting the Millennium Development Goals', Worldwatch Institute [website], <http://www.worldwatch.org/system/files/ren21-1.pdf> (accessed 15 February 2013).

Garner, D. (2009). 'Why Are Humans Different from All Other Apes? It's the Cooking, Stupid', *The New York Times*, 26 May [online], <http://www.nytimes.com/2009/05/27/books/27garn.html?pagewanted=all> (accessed 9 December 2012).

Grupo Fenix (2013a). 'Sabana Grande', Grupo Fenix [website], <http://grupofenix.org/about-us/where-we-work/totogalpa/> (accessed 8 January 2013).

Grupo Fenix (2013b). 'PFAE', Grupo Fenix [website], <http://grupofenix.org/pfae/> (accessed 8 January 2013).

Grupo Fenix [website], <http://www.grupofenix.org/> (accessed 12 July 2013).

Hanes, S. (2007). 'Oprah's Academy: Why Educating Girls Pays Off More', *The Christian Science Monitor*, 5 January [online], <http://www.csmonitor.com/2007/0105/p01s03-woaf.html> (accessed 12 July 2013).

Independent Evaluation Group (2008). *The Welfare Impact of Rural Electrification: A Reassessment of the Costs and Benefits*. Washington, DC: The World Bank.

Inter-Parliamentary Union (2011). 'Women in Politics: 2012 (POSTER)', Inter-Parliamentary Union [website], <http://www.ipu.org/english/surveys.htm#MAP2012> (accessed 3 March 2013).

Inter-Parliamentary Union (2013). 'Proportion of Seats Held by Women in National Parliaments (%)', The World Bank [website], <http://data.worldbank.org/indicator/SG.GEN.PARL.ZS?order=wbapi_data_value_2012+wbapi_data_value+wbapi_data_value-last&sort=desc> (accessed 3 March 2013).

Jha, A. (2012). 'Scientists Find Clue to Human Evolution's Burning Question', *The Guardian*, 2 April [online], <http://www.guardian.co.uk/science/2012/apr/02/scientists-clue-human-evolution-question> (accessed 9 December 2012).

Kramarae, C. and Spender, D., eds. (2000). *Routledge International Encyclopedia of Women: Global Women's Issues and Knowledge*, Volume 2. New York: Routledge.

Lallement, D. (2008). 'Gender Equality and Energy: Opportunities for Accelerated Sustainable Development', presentation at the World Renewable Energy Congress 2008, Glasgow, 20 July.

Lim, S., et al. (2012). 'A Comparative Risk Assessment of Burden of Disease and Injury Attributable to 67 Risk Factors and Risk Factor Clusters in 21 Regions, 1990–2010: A Systematic Analysis for the Global Burden of Disease Study 2010', *Lancet* 380(9859): 2224–60.

Mills, E. (2012). 'Technical Report #10: Health Impacts of Fuel-based Lighting', The Lumina Project [website], <http://light.lbl.gov/pubs/tr/lumina-tr10-summary.html> (accessed 12 July 2013).

Misana, S. and Karlsson, G., eds. (2001). *Generating Opportunities: Case Studies on Energy and Women*. New York: United Nations Development Programme.

Mortimer, K., et al. (2012). 'Household Air Pollution is a Major Avoidable Risk Factor for Cardiorespiratory Disease', *CHEST Journal*, 142(5): 1308–15.

National Intelligence Council (2008). *Global Trends 2025: A Transformed World*. Washington, DC: US Government Printing Office.

Nussbaumer, P., Bazilian, M., Modi, V., and Yumkella, K. (2011). 'Measuring Energy Poverty: Focusing on What Matters', Oxford Poverty and Human Development Initiative [website], <http://www.ophi.org.uk/wp-content/uploads/OPHI_WP_42_Measuring_Energy_Poverty1.pdf> (accessed 11 July 2013).

Practical Action (2012). *Poor People's Energy Outlook 2012: Energy for Earning a Living*. Rugby, UK: Practical Action Publishing.

Roudi-Fahimi, F. and Moghadam, V. (2003). 'Empowering Women, Developing Society: Female Education in the Middle East and North Africa', Population Reference Bureau [website], <http://www.prb.org/Publications/PolicyBriefs/EmpoweringWomenDevelopingSocietyFemaleEducationintheMiddleEastandNorthAfrica.aspx> (accessed 12 July 2013).

Sagar, A. (2005). 'Alleviating Energy Poverty for the World's Poor', *Energy Policy* 33: 1367–72.

Sangeeta, K. (2008). 'Energy Access and its Implication for Women: A Case Study of Himachal Pradesh, India', presentation to the 31st IAEE International Conference Pre-Conference Workshop on Clean Cooking Fuels, Istanbul, Swiss Association for Energy Economics [website], <http://www.saee.ethz.ch/events/cleancooking/Sangeeta_Istanbul_presentation.ppt> (accessed 12 July 2013).

Smith, K. (2012). Summary of the Burden of Disease Study (cf. Lim 2012, above); EPA Cookstoves and Indoor Air listserv, 21 December.

Smith, K. (2013). 'Clean Cooking Forum 2013: Kirk Smith Opening Comments (Full)', YouTube [website], <http://www.youtube.com/watch?v=Xyz9obLrmTI> (accessed 13 July 2013).

Solar Electric Light Fund (2003). 'Renewable Energy, Empowered Women', Solar Electric Light Fund [website], <http://www.self.org/SELF%20White%20Paper%20-%20Renewable%20Energy%20Empowered%20Women.pdf> (accessed 5 February 2013).

Sovacool, B. (2012) 'The Political Economy of Energy Poverty: A Review of Key Challenges', *Energy for Sustainable Development* 16: 272–82.

Taylor, M. et al. (2011). 'Burning for Sustainability: Biomass Energy, International Migration, and the Move to Cleaner Fuels and Cookstoves in Guatemala', *Annals of the Association of American Geographers* 101(4): 918–28.

US Department of State (2012). 'China Joins the Global Alliance for Clean Cookstoves', US Department of State [website], <http://www.state.gov/r/pa/prs/ps/2012/05/189275.htm> (accessed 25 January 2013).

United Nations (2012a). *Millennium Development Goals Report 2012*. New York: United Nations.

United Nations (2012b). 'Statistical Annex: Millennium Development Goals, Targets and Indicators, 2012', United Nations Statics Division [website], <http://mdgs.un.org/unsd/mdg/Host.aspx?Content=Data/Trends.htm> (accessed 10 January 2013).

United Nations (2013). 'Tracking Progress', Sustainable Energy for All [website], <http://www.sustainableenergyforall.org/tracking-progress> (accessed 12 July 2013).

United Nations Development Programme (2011). *Human Development Report 2011*. New York: United Nations Development Programme.

WE CARE Solar (2011). 'Maternal Mortality: Frequently Asked Questions', WE CARE Solar [website], <http://wecaresolar.org/wp-content/uploads/2011/09/We_Care_Solar_FAQ-_Oct-1-2011.pdf> (accessed 15 February 2013).

WE CARE Solar [website], <http://wecaresolar.org/about-us/our-story/> (accessed 8 January 2013).

World Bank (2009). *Gender in Agriculture Sourcebook*. Washington, DC: The World Bank.

World Bank (2013). 'Girls' Education'. The World Bank [website], <http://web.worldbank.org/WBSITE/EXTERNAL/TOPICS/EXTEDUCATION/0,,contentMDK:20298916~menuPK:617572~pagePK:148956~piPK:216618~theSitePK:282386,00.html#why> (accessed 12 July 2013).

World Health Organization (2012a). 'Health Indicators of Sustainable Energy in the Context of the Rio+20 UN Conference on Sustainable Development', World Health Organization [website], <http://www.who.int/hia/green_economy/indicators_energy1.pdf> (accessed 19 February 2013).

World Health Organization (2012b). *Trends in Maternal Mortality: 1990 to 2010*. Geneva: WHO Press.

World Health Organization (2013). 'Maternal Mortality Ratio (per 100,000 Live Births)', World Health Organization [website], <http://www.who.int/healthinfo/statistics/indmaternalmortality/en/index.html> (accessed 23 February 2013).

Worldwatch Institute (2010). 'Innovation of the Week: Reducing the Things They Carry', Nourishing the Planet [blog], <http://blogs.worldwatch.org/nourishingtheplanet/innovation-of-the-week-lightening-the-things-they-carry-farmers-africa-women-drought-climate-change-kenya-practical-action-stove-fuel-biogas-livestock-international-development-enterprises-ide-ethiopi/> (accessed 29 February 2013).

Wrangham, R. (2009). *Catching Fire: How Cooking Made Us Human*. New York: Basic Books.

8

Beyond Basic Access

The Scale of Investment Required for Universal Energy Access

Morgan Bazilian, Ryan Economy, Patrick Nussbaumer,
Erik Haites, Kandeh K. Yumkella, Mark Howells,
Minoru Takada, Dale S. Rothman, and Michael Levi

8.1. INTRODUCTION

A significant share of the world's population lacks access to the services provided by modern energy. Approximately 1.3 billion people do not have access to electricity (IEA 2012), and about 2.6 billion people rely on solid fuels for cooking (IEA 2012). Reliability (quality) of existing supply is also often problematic. Focusing on Sub-Saharan Africa's infrastructure, Foster and Briceno-Garmendia (2010) show that the continent's chronic power problems (i.e. inadequate generation, limited electrification, unreliable services, and high costs) significantly affect economic growth and productivity.

The international community has recognized the importance of the matter for development, and in 2011 UN Secretary General Ban Ki-moon launched the Sustainable Energy for All initiative, proposing a global target of universal access to energy services by 2030. The UN resolution designating 2012 as the International Year of Sustainable Energy for All also noted that access to modern and affordable energy services was fundamental to achieving international development goals, including the Millennium Development Goals (UN 2010). This goal was reaffirmed with the UN General Assembly's declaration of 2014–2024 as the United Nations Decade of Sustainable Energy for All (UN 2012).

Yet, despite current efforts, progress in delivering energy access is inadequate. In addition to a global political commitment (Bazilian et al. 2010a), investment and appropriate financial tools for energy access are needed to address the issue. In this chapter, we consider only the initial step in this complex agenda, namely the total cost of providing universal energy access. A number of estimates of the cost of providing universal energy access in developing countries have been produced. The methodologies and assumptions vary greatly and are generally highly abstracted. Most studies do not provide a holistic picture, but rather focus on specific aspects such as the capital cost of energy supply. This chapter reviews the literature and attempts to 'untangle' the numbers to provide a transparent basis for comparison. It then provides revised estimates for the total cost of universal energy access using the same methodology previously presented by the authors (Bazilian et al. 2010b).

Calculating large global investment requirements is a difficult task and relies on 'heroic' assumptions; still, it can often provide a useful benchmark for policy-making and international diplomacy. It is also useful for providing context for the design of financial responses. Useful precedents, at least at a conceptual level, can be drawn from similar exercises carried out in the climate change space. As an example, the United Nations Framework Convention on Climate Change (UNFCCC) commissioned a report on the scale of future financing options (UNFCCC 2007). That document included a literature review of estimated annual investment needs for both climate change mitigation and adaptation. Those estimates were then fed in to support various governmental decision-making processes.

We recognize that a solid enabling environment (including capacity building and institutional strengthening), and appropriate investment climates linked to adequate policies and regulations are crucial to delivering adequate financing for energy access. However, such considerations are beyond the scope of this chapter (see e.g. Morris et al. 2007; Morris and Kirubi 2009). It is also clear that the availability of capital is a necessary, but not sufficient condition to deliver energy access; the treatment of financial tools is also beyond the scope of this chapter, as are the sources of funding.

We begin our chapter by reviewing cost estimates related to providing access to modern energy services published in both the 'grey' and academic literature.[1] Section 8.2 provides an overview of existing estimates, including an examination of their scope and an attempt to disaggregate them for the purposes of comparison. Section 8.3 presents our own estimates. In Section 8.4, we discuss further areas of work and possible next steps for increasing the rigour of these estimates.

8.2. LITERATURE REVIEW

8.2.1. Existing estimates

Several estimates have been made of the cost of universal access to modern energy services over the past decade at the global, regional, and project levels. They focus on various aspects of energy poverty (e.g. electrification, clean cooking). A non-comprehensive review of these is provided in Table 8.1. It shows a wide variation in estimates, providing the impetus to consider further the underlying algorithms, assumptions, and parameters.

In general, the estimates that we reviewed focus on the provision of electricity to households for one or two basic uses such as lighting plus telecommunications or refrigeration. A small number consider clean cooking, but other uses such as energy for earning a living (mechanical and productive uses), energy for

---

[1] This chapter was initially drafted in early 2013, so some of its figures and data may be slightly out of date.

Table 8.1. Cumulative (unless otherwise stated) investments to facilitate access to modern energy service, in billion USD

Geographical focus	Goal	Cost estimates (billion USD)			Period	Source
		Electricity	Cooking	Others		
Global	Universal energy access (incremental)[i]	65–86 per year			2010–2030	Pachauri et al. 2013
	Universal energy access	890	89		2011–2030	SE4All 2013
	Universal energy access	890	89		2011–2030	IEA 2012
	Universal energy access	36–41 per year			2010–2030	Riahi et al. 2012
	Universal energy access	48 per year			2010–2030	Dobbs 2011
	Universal energy access	915	95		2010–2030	IEA 2011
	Universal energy access	700[ii]	56		2010–2030	IEA et al. 2010
	Improved access to reach MDG 1	223	21[iii]		2010–2015	IEA et al. 2010
	Universal energy access	35–40 per year[iv]	39–64[v]		2010–2030	AGECC 2010
	Universal electricity access	~55 per year			2008–2030	Saghir 2010
	Universal electricity access	35 per year			to 2030	IEA 2009b
	Improved access to clean cooking[vi]		1.8 per year			Birol 2007
	Universal electricity access[vii]	858			2005–2030	The World Bank Group 2006
	Improved electricity access to reach the MDGs	200			2003–2015	IEA 2004
	Universal electricity access	665			30 years	IEA 2003
Regional/local						
Africa						
Sub-Saharan Africa	Improved electricity access[viii]	17 per year[ix]			to 2030	AfDB 2008
	Universal electricity access[x]	870 NPV[xi]			to 2030	Bazilian et al 2012
	Improved energy access	6–15 per year				Brew-Hammond 2010
West Africa Power Pool	Increase household electricity access to 35%	4 per year			to 2015	UN-Energy/Africa
Central Africa Power Pool	Universal electricity access	21 per year			to 2030	Bazilian et al 2012
Southern Africa Power Pool	Universal electricity access[xii]	3 per year			to 2030	Bazilian et al 2012
East Africa Power Pool	Universal electricity access	12 per year			to 2030	Bazilian et al 2012
	Universal electricity access	12.5 per year			to 2030	Bazilian et al 2012
East African Community	Improved energy access[xiii]	1.5	0.262	0.919	to 2015	EAC 2006[xiv]
	50% electrification	1.45			10 years	CEMAC 2006

Economic Community of Central African States					
Economic Community of West African States	60% electrification; 100% improved cooking fuels; access to mechanical power in 100% of villages	2.1	0.27[xv]	10 years	ECOWAS 2005
Senegal	Increased electrification rate from 47% to 66%	0.86	2.8	10 years	ASER 2007
Bangladesh, Cambodia, Ghana, Tanzania, and Uganda	Improved energy access in line with the MDG targets	13–18 USD per capita and year[xvi]		2005–2015	Sachs et al. 2004
South Asia	Universal access to LPG	449		2010–2030	IIASA[xvii]
Brazil	Promoting LPG access to underprivileged households	0.5[xviii]		2003	Jannuzzi et al. 2004

[i] Pachauri et al (2013) calculate the incremental cost above current trends to achieve total rural electrification and universal access to clean cooking by 2030

[ii] Including both rural and urban; grid connection: generation and transmission and distribution; mini-grid: generation and distribution; off-grid: generation

[iii] Including advanced biomass stoves, LPG stoves and biogas systems

[iv] Based on IEA (2009b)

[v] Improved cookstoves: 11–31; Biogas: 30–40; LPG: 7–17; Includes capacity development costs

[vi] LPG cylinders and stoves to all the people who currently still use traditional biomass

[vii] Includes breakdown by major regions

[viii] Reliable electric power to 90 per cent of Sub-Saharan rural population, 100 per cent of the Sub-Saharan urban population, and 100 per cent of the both the rural and urban populations in the Northern African middle income countries

[ix] Considering only new generating capacity, including generation as well as transmission and distribution

[x] Includes investment (generation capacity plus T&D) cost of USD 740 NPV and operating and maintenance costs of USD 130 NPV; excluding South Africa

[xi] 5 per cent discount rate

[xii] Excluding South Africa

[xiii] Reliable electricity for all urban and peri-urban poor; Modern cooking practices for 50 per cent of population currently using traditional cooking fuels; Modern energy services for all schools, clinics, hospitals, and community centres; and mechanical power for heating and productive uses for all communities

[xiv] Including capital expenditure, programmes, and loan guarantees

[xv] For mechanical power

[xvi] Including costs of: end-use devices, fuel consumption, electrical connections, and power plants

[xvii] Updated analysis based on the methodology described in Ekholm et al. (2010)

[xviii] Subsidies for LPG access to underprivileged households in 2003

community services, and/or transport receive very limited attention. Considering the importance of these other end-uses, Practical Action (2010, 2012, 2013) has presented the concept of Total Energy Access. Their research programme presents a compelling argument for considering a more comprehensive framework for universal energy access, but it does not seek to estimate the cost of doing so. Global electricity and clean cooking estimates have been produced primarily by international organizations, while academia, regional institutions, and utilities have commonly focused on the regional and national scales. It should also be noted that two of the more recent estimates (SE4All 2013; Dobbs et al. 2011) essentially cite the 2012 and 2011 editions of the *World Energy Outlook* respectively, rather than arriving at their own estimates (IEA 2012, 2011). The figures have been produced using methodologies that range from analytical modelling to extrapolation of empirical results (i.e. multiplication).

8.2.2. Comparing the estimates

Quantitatively comparing the estimates is challenging for a number of reasons. First, the underlying methodologies and assumptions vary greatly. Second, the information required to make the estimates comparable is often not available. And third, the different studies vary widely in terms of their ambition and scope. To contend with these obstacles in part, we transform the estimates for electricity and clean cooking access into a single consistent metric—annual average USD per capita.[2,3]

Electrification costs

Figure 8.1 compares estimates of average annual cost per capita for electrification, split between capital costs, operation and maintenance (O&M), fuel and others (e.g. capacity building), when explicitly available. (A number of other studies provide insights with regard to electricity connection costs, another possible comparator; see Appendix D.[4]) In order to transform the existing varying results and 'harmonize' them to an extent, we normalized the results to $/capita/annum.

---

[2] It should be noted that in order to more accurately compare these estimates, one would want to convert these values to a common year dollar, but this was not possible with the information available.

[3] Although considering the household as a unit is more common in the literature, using per capita figures allows us to avoid further assumptions with regard to the typical size of household, which varies greatly between countries (e.g. Cameroon, 2.9, and Guinea-Bissau, 7.6; data from Banerjee et al. 2008), and allows us to consider the extension of modern energy services to areas such as productive and community uses.

[4] While a number of the connection-cost estimates are based on modelling work (e.g. Parshall et al. 2009), others rely on data from project experience (e.g. Eskom 2009). The latter provide a reliable source of information as they are not biased by normative assumptions and methodological shortcomings, but they are arguably unsuitable for extrapolation due to their context dependency. The modelling estimates are informative in that they can allow a better understanding of the cost drivers associated with electrification. Without knowing the number of people per household in these types of studies it is not possible to transform these estimates into cost per person per year used here. Therefore, we have included these estimates in Appendix D.

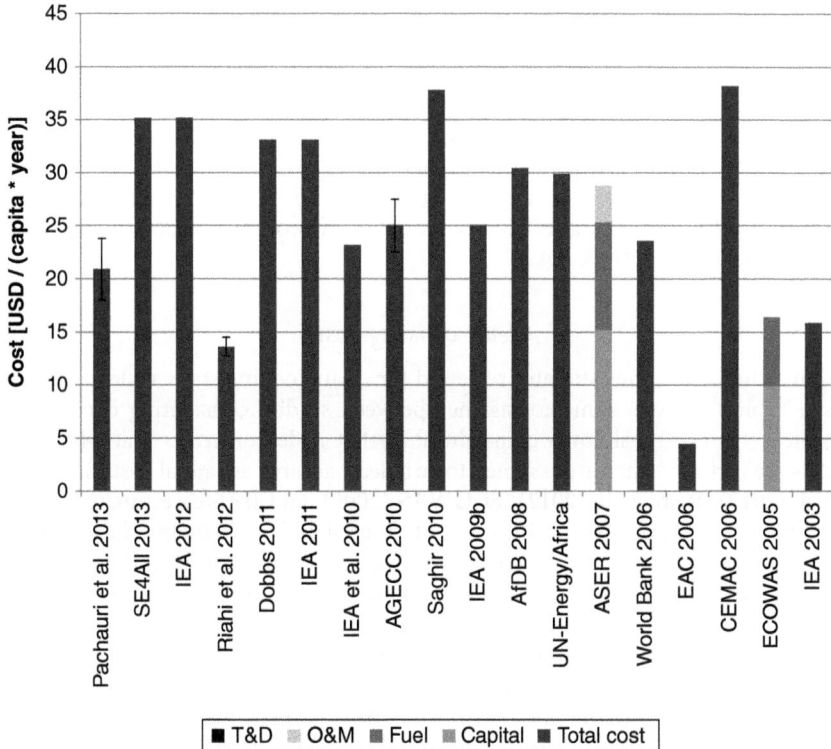

Figure 8.1. Comparison of cost estimates for electrification

Note: A number of studies do not explicitly state estimates of the component costs. In these cases, we have presented the estimated total cost without decomposing it into individual components. Riahi et al. (2012) offer a range for the annual investment required for universal energy access to both electricity and clean cooking. Here we have presented the upper and lower bounds of this range. Furthermore, the electrification estimate for the lower bound should be interpreted carefully because the proportion of the overall costs attributable to electrification is not explicitly stated in the publication. We assumed electrification constituted 60% of the costs in the low bound based on the 50/50 split in stated in the higher bound which is why the electrification estimate in the low bound is higher. Including clean cooking, the high bound is actually higher

See Appendix C to view these estimates in terms of average cumulative cost per person.

The estimates of annual costs for electrification range from 5 to almost 40 USD per capita, reflecting the large uncertainties associated with such evaluations (and their sensitivity to certain assumptions). One important insight is that the majority of studies focus solely on capital cost and do not explicitly consider recurrent (or ongoing) costs (e.g. fuel, O&M), with the notable exceptions of Pachauri et al. (2013),[5] ECOWAS (2005), and ASER (2007).

The IEA estimates are the most often cited in the literature (either directly or in a circular fashion) and have been relatively consistent over time. They are also one

---

[5] Although Figure 8.1 does not display the decomposition of the total cost between capital, O&M, and fuel, the authors stated in private communications that these costs were included in their model.

of the few sources that provide investment estimates both at the global level and disaggregated by regions. Their most recent estimates (IEA 2012; SE4All 2013) are that a cumulative USD 979 billion would be required to achieve universal electricity access by 2030, provided that appropriate policies are in place. This implies an investment of some USD 49 billion per year on average over the 2011–2030 time period. IEA (2009b) also provides insights with regard to the breakdown between generation and transmission and distribution (T&D) (IEA 2009b: 2.11; IEA 2003: 7.43). The IEA 2012 energy access scenario includes commensurate investment in T&D infrastructure (IEA 2012).[6]

Clean cooking costs

As in other areas, the estimates reviewed for clean cooking show wide variations (see Table 8.2), with some consistency between studies considering capital cost exclusively. The breakdown is insightful in that it demonstrates that when fuel costs are included, they are assumed to be at least as large as capital costs. Pachauri et al. (2013), Riahi et al. (2012), ECOWAS (2005), and IIASA's estimates, which include fuel costs or fuel subsidy costs, are thus significantly higher than those of other studies. The context dependency of the estimates must be underlined, and the fact that some studies focus on specific regions and/or types of fuel explains some of the variance.

The IEA (2012) cooking estimates include investment in advanced biomass cookstoves, in liquefied petroleum gas (LPG) stoves and canisters, and in biogas digesters, but exclude investment in infrastructure, distribution, and fuel costs. Country/regional breakdown of the investment is derived from assumptions

Table 8.2. Comparison of cost estimates for clean cooking

Cost estimates [USD/(capita*year)]		
Capital	*Fuel*	Source
23.6–31.25		**Pachauri et al.** 2013
1.8		IEA 2012
4.5–6.4		**Riahi 2012**
1.8		IEA 2011
1.0		IEA et al. 2010
0.9		AGECC 2010
0.6		EAC 2006
1.3	8.3	ECOWAS 2005
5.0	18.1	IIASA

Note: Riahi et al. (2012) include fuel subsidy costs implicitly in their estimates, and this additional cost is included in the estimates presented. The IIASA figure is based on data obtained in personal communication with Shonali Pachauri (IIASA) which stem from updated analysis relying on the methodology described in Ekholm et al. (2010)

---

[6] The methodological note for the IEA's (2012) energy access scenario can be found at <http://www.worldenergyoutlook.org/weomodel/documentation/>.

regarding the most likely technology solution in each region, given resource availability and government policies and measures.

Comparing the estimates across various energy services, the per capita cost for clean cooking is significantly lower than that for electrification. With regard to mechanical power, the paucity of data precluded our analysis.

8.2.3. Gaps in the estimates

We have identified several gaps in existing cost estimates. Many studies published prior to 2011 consider only one of the energy delivery 'vectors'. Also, as noted, most estimates both before and after 2011 pay less attention to recurrent costs. IEA (2010) estimates the cost of fuel to represent a significant share (one-quarter for coal and up to two-thirds for gas) of the total cost in the case of fossil-fuel-based power generation, while O&M represents between 5 and 10 per cent of the total cost for fossil-fuel-fired plants and up to 20 per cent for renewables. Those fractions can be much higher for rural generation. Regardless of the exact figure, the total cost of providing electricity to the poor will be significantly higher than estimates based on capital expenditure alone.

8.3. ESTIMATING THE TOTAL COSTS OF ENERGY ACCESS

This section presents revised estimates of the cost of meeting universal electrification over the 2010–2030 period using a simple algorithm previously developed in our 2010 paper (Bazilian et al. 2010a). We focus primarily on electricity because of the data challenges described previously. Our time horizon includes the years 2010–2030 in order to be consistent with the current literature, which generally considers this time horizon in light of the Sustainable Energy for All Initiative. The methodology is highly stylized, but it aims to be useful in terms of transparency and comparability and as a basis for more sophisticated estimates in the future. We then consider cooking costs and present a total figure for both services.

8.3.1. Methodology for estimating the cost of universal electricity access

We base our calculation on the full levelized costs[7] of generation as a means of capturing, besides capital costs, costs related to O&M and fuel. Additionally, we

---

[7] Levelized cost is a notion that is used for comparing the unit costs of different technologies under a number of assumptions (e.g. absence of specific market or technology risks, specific discount rate, load factors). It includes capital costs of generation, O&M costs, and fuel costs. For the purpose of this analysis, a discount rate of 10 per cent is used.

Table 8.3. Assumptions on rural and urban average consumption in each scenario

Assumption	Scenario			Unit
	Low	Medium	High	
Average urban electricity consumption	100	465	750	kWh/(capita*year)
Average rural electricity consumption	50	152	750	kWh/(capita*year)

Data source: IEA 2012; Banerjee et al. 2008; IEA 2008; Parshall et al. 2009

have revised the previous estimates to include transmission and distribution costs in the total levelized cost of each generation technology. We recognize the shortcomings of levelized cost estimates (see Bazilian and Roques 2008) for energy planning, but for this purpose they provide a sufficient and transparent calculation tool. To account for some of the uncertainty associated with such estimates, we present three scenarios (low, medium, and high) and use a static linear model to calculate the total cost of electrification.

The primary assumptions include: electricity consumption levels, the types of systems used for electrification (i.e. grid connected, off-grid, etc.), and the levelized costs of generation and power delivery. For the latter, we use updates to ESMAP's (2007) estimates derived from their Model for Electricity Technology Assessments (META) as the primary source of information.[8] Average levels of electricity consumption are assumed to be different for rural and urban population categories in the low and medium scenarios (see Table 8.3). The low scenario assumes electricity consumption levels to fulfil basic needs and nothing more by 2030, as per the initial electrification period in the IEA's universal electricity access scenario (IEA 2012).

The medium scenario depicts a case where some electricity is available for other purposes, including basic productive (economic) activities. Consistent with the IEA's (2012) universal access scenario, consumption in the high scenario reaches 750 kWh per person per year in all regions.[9] This compares to the average annual per capita consumption level in Morocco in 2010 of 781 kWh/per capita. For comparison, the average annual consumption levels in the US, Germany, South Africa, and the average among the countries in North Africa were: 13,394; 7,215; 4,803; and 1,807 kWh/per capita, respectively (World Bank 2013).

We contend that the level of per capita consumption presented in the low scenario implicitly assumes that the world's poor generally continue in poverty— we include it only as a comparator. Although this level of basic access would be better than none at all, it is important for the international community to consider the need for significant additional growth in the power sector and diversity for

---

[8] The Model of Electricity Technology Assessments is an MS Excel-based tool used to update ESMAP's (2007) previous levelized cost of generation estimates. At the time of writing, the model was available upon request and full public roll-out was anticipated for summer 2013. For the purposes of this analysis, we use the default power plant configurations for India as a proxy for other developing countries. We modify these configurations by adding T&D and removing levelized environmental costs.

[9] For some off-grid systems this level of service is high, but we assume the deployment of modern, larger-scale mini-grid and off-grid systems and technologies.

productive uses and community services (Johnson 2013; Practical Action 2010, 2012, 2013). Including higher per-person consumption levels helps us factor in a small portion of the transformation necessary in the power sector in supporting a move to a more vibrant and equitable global economy by 2030.

We use data for the year 2010 from IEA for the share of urban and rural population with access to electricity and from the UN for urban and rural population. These data are used to estimate the number of people requiring electricity access. In combination with the assumptions on average consumption, we can then estimate the total amount of electricity consumed by these new users.

In terms of connection type, urban consumers are assumed to access electricity through a grid connection. For rural communities, in the low scenario, we assume shares of grid extension, mini-grids, and off-grid systems for the low scenario based on the IEA (2009b) (see Appendix A). In the medium and high scenario we assume the shares of grid extension, mini-grid, and off-grid systems from the IEA's (2012) Energy for All scenario, which, as one can see, changed significantly from the 2009 shares. Based on these and the prior assumptions, we calculate the electricity needs per country in both urban and rural contexts by connection type, that is grid, mini-grid, or off-grid.

The levelized costs by connection type are derived from the literature (ESMAP 2013) and adjusted using the META tool. These are broken down into four categories—overnight capital,[10] O&M, fuel, and T&D. The costs for each category differ depending upon the energy source, that is coal, oil, gas, nuclear, hydro, biomass, or other renewables for grid; diesel or PV–wind hybrid for mini-grid; and gasoline, PV–wind hybrid, or PV for off-grid. The levelized cost assumptions by connection type and generation type do not differ across the scenarios. We calculate the LCOE estimates based on hypothetical power generation facilities in India using the META tool. Real-life projects may encounter significant variation in price across countries and even within countries (Rong and Victor 2012). The T&D configurations are consistent with power delivery systems available in India, and we account for power delivery losses by adopting the default values for India included in the levelized T&D cost estimates from the META tool.[11] An overview of the assumptions regarding these levelized costs is in Appendix B.

The country-level weighted average levelized cost of electricity depends upon the mix of connection types (discussed earlier) and energy sources for electricity generation. For the low scenario, we use the 2010 mix of energy sources for grid electricity from the IEA (2012); this mix varies across four major regions, namely Africa, Developing Asia, Latin America, and the Middle East. The assumptions for the medium and high variants are taken from the IEA's New Policies scenario; these also vary across the four regions (IEA 2012). We then calculate the respective weighted average costs for capital, O&M, fuel, and T&D for the four world regions. We make additional assumptions with regard to the energy mix for the generation of electricity for mini-grid and off-grid contexts, and we calculate the respective costs in a similar fashion as for grid electricity. An overview of

---

[10] The overnight cost of capital excludes financing charges and/or interest incurred during construction of the power plant ESMAP (2013).

[11] Power deliver losses can vary greatly in developing countries from 10 to 25 per cent or more (ESMAP 2007).

the assumptions regarding the levelized costs and the fuel mix for electricity generation is provided in Appendix B.

Combining all of the above, we calculate the total annual global cost of electrification as:

$$C_{tot} = \sum (P_{urb} \cdot E_{urb} \cdot c_{grid} + P_{rur} \cdot E_{rur} \cdot \sum s_{sys} \cdot c_{sys})_{region} \tag{1}$$

With:

C_{tot}: total annual global cost of electrification;
P_{urb}: urban population without electricity in given region;
E_{urb}: average urban electricity consumption;
c_{grid}: weighted average levelized cost of electrification in given region;
P_{rur}: rural population without electricity in given region;
E_{rur}: average rural electricity consumption;
s_{sys}: share of generation type (grid, mini-grid, off-grid) for rural electrification;
c_{sys}: weighted average levelized cost in given region based on the respective generation portfolio.

This methodology has a number of limitations. Notably, because of the static nature of the model, population growth is not taken into account adequately. Similarly, the changes in the access rate over time in a case where no additional policy and incentive for energy access are put in place are also not considered. Finally, technology development and learning effects are not accounted for in this analysis.

8.3.2. Results

Our estimates for the total cumulative cost of universal electrification from 2010 to 2030 range from 119 to 2,793 billion USD (see Table 8.4). Table 8.5 presents these in terms of average annual cost incurred once all people who do not currently have access to energy are connected (annual costs will be lower in the interim as people gain access).

Table 8.4. Estimates of total universal electrification cost in USD billion

	Scenario	
Low	Medium	High
119	635	2,793

Table 8.5. Estimates of annual cost to achieve universal electrification in USD billion

	Scenario	
Low	Medium	High
12	63	279

Table 8.6. Estimates of annual cost to achieve universal electrification per person in USD

Scenario		
Low	Medium	High
5.15	27.55	121.26

On a per-person-per-year basis, the investment required to achieve universal electricity access by 2030 ranges from USD5.15 to USD121.26 (see Table 8.6).

The breakdown of the costs per connection-type in different scenarios indicates that over half of the required aggregated cost is for systems based on decentralized generation, regardless of the scenario, with rural mini-grids constituting the major share in the low scenario and off-grid systems in the medium and high scenario. In the high scenario, decentralized generation accounts for over three-quarters of the required cost.

The higher costs in the medium and high scenarios are due primarily to increased electricity consumption. Specifically, just increasing the per person consumption explains 52 per cent of the difference between the low and medium scenarios; just shifting to more mini and off-grid generation explains 14 per cent, and just changing to a less carbon-intensive generation mix explains 2 per cent.[12] Similarly, increased consumption accounts for 100 per cent of the difference between the medium and high scenarios, as it is the only variable that changes between the two.

Figure 8.2 shows the share of capital, O&M, fuel, and T&D cost in the total cost of electricity for the various systems in the different scenarios. The proportion of costs allocated to each component is identical in the mini- and off-grid contexts because we have assumed the same generation technology mixes across scenarios.

In comparison to our previous estimates and a number of the studies reviewed the effect of including the levelized costs of O&M, fuel, and T&D expansion are potentially quite significant for grid- and mini-grid-connected generation. Our annual electricity estimates in the low, medium, and high scenarios are 83 per cent, 59 per cent, and 58 per cent higher, respectively, when O&M and fuel costs are included. The IEA previously estimated that T&D costs for grid-connected electricity are roughly equal to the capital costs for generation (IEA 2009b). However, our results based on ESMAP's (2013) estimates indicate that levelized T&D investments may be significantly understated in the case of grid-connected electricity and roughly equivalent to the levelized cost of generation (inclusive of O&M and fuel costs). If we then consider O&M, fuel, *and* T&D together, our annual estimates rise to being 138 per cent, 90 per cent, and 80 per cent higher in the respective scenarios than they would be compared to a capital-cost-only model.

---

[12] The additional 32 per cent is due to interactions between the consumption, connection type, and generating mix and cannot be attributed to any single factor in our analysis.

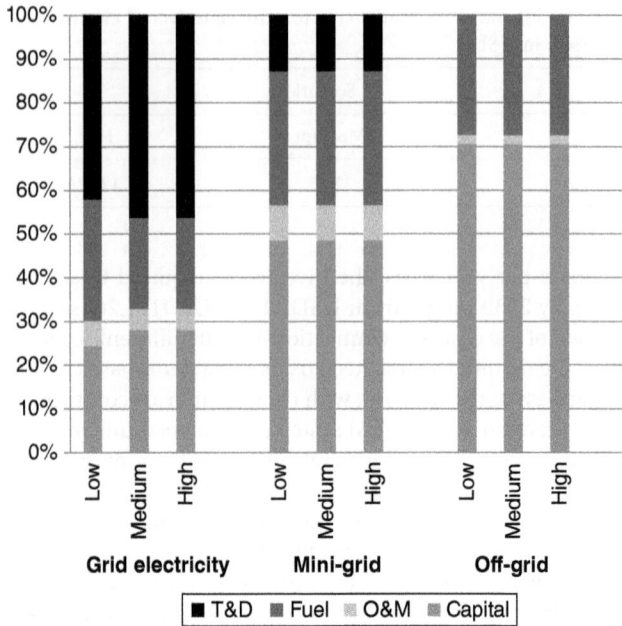

Figure 8.2. Share of capital, O&M, and fuel in the total cost in grid, mini-grid, and off-grid electricity in the low, medium, and high scenarios; all scenarios have a mix of these technologies

In addition, the results for electricity access in the revised high scenario are substantially higher than our previous estimate and the other studies reviewed. This is primarily due to the fact that this case looks beyond typical assumptions, especially with regard to electricity consumption per person. In this way, we can get a sense of the impacts of a more equitable case, and one that considers access beyond the household level to the wider economy. Furthermore, the interaction of increased rural consumption with the change in grid type adds substantially to the result. Together, the changes in these two variables plus their interaction account for almost 97 per cent of the difference between the medium and high scenarios.

8.3.3. Towards the total cost of universal energy access

We aggregate the results for both electrification and clean cooking to obtain a high-level total cost figure. With regard to clean cooking, we use global estimates published in the literature. Specifically, our estimate for the low scenario is the estimate presented in *World Energy Outlook 2012*, multiplied by a factor of two to take into account fuel costs (IEA 2012). The lower and upper bounds of IIASA's estimate provide our medium and high scenarios, respectively, and are not doubled because fuel subsidy costs are already implicitly included (Riahi et al. 2012).[13]

---

[13] Riahi et al. (2012) estimate that the total cost to provide universal energy access ranges from USD 36 to USD 41 billion. At the higher bound, 50 per cent of the cost is due to clean cooking investments. The share of the total cost allocated to clean cooking at the lower bound was not specified; therefore, we assumed that 40 per cent is due to clean cooking investments.

Table 8.7. Range of global cumulative cost of universal access to electricity and clean cooking

Scenario	Cumulative cost estimates (billion USD)		
	Electricity	Cooking	Total
Low	119	178	**297**
Medium	635	288	**923**
High	2,793	410	**3,203**

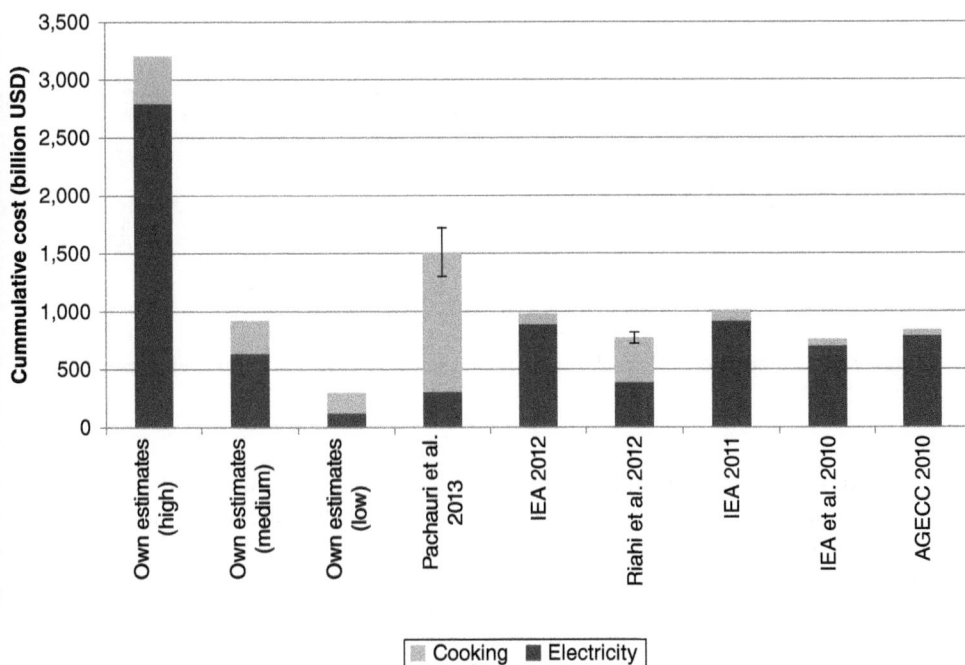

Figure 8.3. Comparison of global cumulative cost estimates for universal electricity and clean cooking access by 2030

Note: It is important to bear in mind the different scope and objective of those studies while comparing their outcomes. For example, Pachauri et al. (2013) estimates the incremental cost of achieving total rural electrification and universal access to clean cooking, while most other studies present an estimate of the total investment required

Table 8.7 provides an overview of our global cumulative cost estimates for the three scenarios. We assume that the increase in access to electricity is linear until universal access is reached in 2030.

Figure 8.3 compares our own estimates with other recently published global estimates. While our aggregated low estimate is lower than all of the other

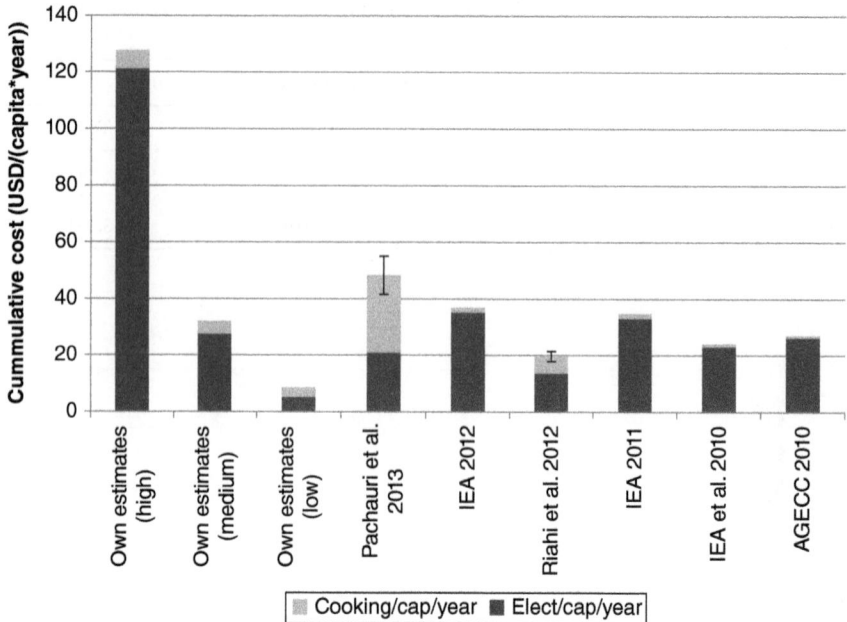

Figure 8.4. Comparison of global annual per-person cost estimates for universal energy access by 2030

estimates, our high estimate is significantly higher than the other studies. Our medium estimate is roughly equivalent to estimates in other literature, including the *World Energy Outlook 2012*. It should be noted that Pachauri et al.'s (2013) estimate only includes the incremental investment over current trends required to achieve universal access in 2030. Their estimate is therefore not directly comparable to the others, including our own, which generally present the *total* investment required to achieve universal access. Again, the large disparity between our high scenario and the available literature is mostly due to the significantly higher assumption on per capita electricity use, although our clean cooking estimate, borrowed from Riahi et al. (2012), is also higher than most of the other previously published estimates. Again, that 'high' per capita electricity use assumption is still roughly half of the average of today's North African demand—so still nearly an order of magnitude lower than an OECD average.

On an annual per-person basis, our medium scenario is also roughly equivalent to the other estimates found in the literature as well (see Figure 8.4).

8.3.4. Sensitivity analysis on electricity estimates

We undertook some sensitivity analyses of the major assumptions to help identify the key variables. We considered a variation of the average urban and rural consumption; the share of different generation systems (grid, mini-grid, and off-grid); the weighted average levelized cost for grid, mini-grid, and off-grid generation; and power generation fuel costs[14]. The medium scenario is used for this analysis, and the results are presented in Appendix E.

They indicate, as one would expect, that the average rural consumption has a strong influence on the overall result. The impact of the assumptions regarding the share of different generation systems (grid, mini-grid, and off-grid)—using the shares from IEA (2009a) as a reference—produced a result 45 per cent lower in the case of full grid extension assumed in some of the literature (Pachauri et al. 2013; Riahi et al. 2012), and 47 per cent higher in the case of the IEA's 2012 shares (our medium and high scenario) When included in the medium scenario, a 20 per cent fuel premium applied to all fossil and nuclear generation technologies in the high scenario produces a result 5 per cent higher.

8.4. FURTHER WORK

A number of aspects related to the costing of universal energy access are yet to be thoroughly assessed. This section discusses some elements that deserve further scrutiny and is presented as fodder for further work.

- The operating and fuel costs of traditional devices such as kerosene lamps are often higher than those of modern devices (e.g. solar cells and electric fluorescent lights) (Johansson and Goldemberg 2002). Therefore, transitioning to modern energy services will often actually lessen the financial burden related to energy services of households. In a similar vein, switching to more efficient energy appliances (e.g. compact fluorescent and LED lights) also allows countries to reduce requirements for investment into additional generating capacity (Goldemberg 1998). In many instances, the cost to improve end-use technologies is more than offset by capital savings due to reduced energy demand (Goldemberg et al. 1985). However, the marginal benefit of reducing national generating capacity investments is not always realized by the household investing in more efficient appliances. Moreover, efficient appliances may require more upfront capital expenditure, often made difficult by sporadic income and poor access to credit facilities.[15] Related to this, most of the estimates focus on the supply side of the energy value chain. However, investments will be required in efficient devices such as modern lighting systems or electrical drives (e.g. pumps) to allow customers to benefit from the energy service. Crude estimates, though, suggest that this additional cost, while significant, does not fundamentally change our overall conclusions, at least at the low end.

[14] The impact on the overall cost is assessed based only on the change in the fuel cost as an element of the levelized cost. Changes in fuel prices would also impact the electricity mix, which is not taken into account in this sensitivity analysis.

[15] More detailed considerations on affordability, however, go beyond the scope of this chapter and we refer to the literature on the subject (e.g. Banerjee et al. 2008). It must nevertheless be noted that the estimates provided in this study are not to be interpreted as the amount of funds that will need to be raised internationally. Indeed, the consumers will pay part, possibly even all, of the cost.

- Households typically use basic electrification to operate electric lighting. At higher levels of consumption, many adopt radios and televisions and, to a much lesser extent, electrical cooking appliances (The World Bank Group 2008). In our low-cost scenario, rural individuals are each expected to consume 50 kWh annually. That corresponds to using two 14 W compact fluorescent light bulbs for five hours each day. At typical cost of $1 per bulb, and bulb lifetimes of 10,000 hours, this adds a cost of about USD 0.40 per rural customer per year, which adds 4 per cent to our total cost estimate. An urban consumer in this scenario, meanwhile, uses 100 kWh each year, which might involve adding a small shared television. An inexpensive TV costing USD 10 and lasting five years, shared by five people, would add another USD 0.40 per person per year. The total cost of the low scenario would therefore increase by 5 per cent over our estimates.

- With increasing economic activity and standards of living associated with reduced indoor air pollution, better lighting, improved nutrition, and access to more productive means, the demand for energy is likely to rise. Such an increase would, in theory, affect the price of energy by driving it up. This effect has not been taken into account in this analysis. Along those lines, the IEA (2012) argues that the incremental increase in global energy demand in 2030 would be only 1 per cent due to universal energy access for basic needs. In the longer term, more significant increase in the energy demand—not accounted for here—can be expected due to structural changes in the economy of developing countries.

- While aggregated figures are useful to generate discussion and support policy development, investments are made based on detailed analysis. Ultimately, detailed planning and costing at utility or project level is required.[16] Such studies include a number of considerations that are commonly excluded from macro analyses such as those reviewed in this article. For instance, detailed analysis may be carried out with regard to the expected profile of the demand at different time horizons.[17] In that regard, our calculations do not take into account the possible impact on the load profile of adding a large number of a certain type of consumers (household). This would affect the load factor of power plants and therefore the price of electricity generation (depending on an enormous number of variables including market structure, regulation, etc.). Importantly, the issues of power quality, system stability, and ancillary services are not treated in our estimates or in most of the macro-literature we reviewed. These issues, central to the security of modern power systems, will probably add significantly to the cost of providing electricity. With regard to power planning, developing countries face a number of specific issues, notably uncertainties related to future demographics and technical, economic, and environmental constraints (Al-Shaalan 2009; Urban et al. 2007).

---

[16] A number of tools have been developed and applied to assessing some of the related needs of developing countries including, notably, HOMER, RETScreen, and System Advisor Model (Howells et al. 2002).

[17] Howells et al. (2006) suggest a methodology to deal with such effects.

- What are sometimes referred to as 'soft costs' are commonly not evaluated. Those include the costs associated with developing the capacity of regional, national, and sub-national institutions, a dimension that is crucial to scaling up energy access programmes (UNDP 2010). It also consists of the support to private entities related to the operation of efficient energy systems, for example. It is difficult to quantify those costs at global level,[18] but various expenditures will be required to promote energy access beside those needed to purchase and run the energy systems.

- The existing infrastructure in numerous developing countries is crumbling. Foster and Briceno-Garmendia (2010) estimate a financing gap of USD 48 billion a year for Africa alone, three-quarters of which is a shortfall in capital expenditure and the rest a shortfall of operation and maintenance spending.[19] According to AGECC (2010), around USD 15 billion of grants would be needed to overcome infrastructure backlogs and deficiencies while meeting the suppressed demand[20] in least developed countries' productive sectors. This has not been considered in most global estimates.

- Because our estimates emphasize electrification on a per-person level (and many of those surveyed only addressed needs at the household level), crucial dimensions of the economy are often left out. For instance, as societies develop, the need for heat and mechanical power is likely to increase, particularly to support the industrialization process. Our medium and high scenarios attempt to address this issue with regard to electricity needs, but still at a level below the current average of North African countries. The same applies for energy requirements for transport, which represents a significant share of the total energy consumption in industrialized countries. For context, IEA (2012) reports that, globally, final energy consumption in transport and industry each is of at least the same order of magnitude as that of the residential sector. This is important to note because without increasing access to energy for other end-uses, populations gaining access to basic lighting, refrigeration, and/or communications technology at home may still be trapped in energy poverty if the energy system cannot meet the demands of a growing economy and other community services.

8.5. CONCLUSION

We have critically reviewed estimates of the costs related to promoting energy access and have provided a basis to compare the figures. Considering some of the gaps identified, we provided three order-of-magnitude electricity access estimates based on full levelized costs, accounting for T&D investments and distribution

---

[18] UNDP (2010) notes that capacity building costs represent half or more of the total cost of the some of their energy access programmes.

[19] The authors estimate that the gap can be narrowed to USD 31 billion if certain efficiencies can be achieved. They estimate that USD 23 billion relates to the power sector.

[20] Suppressed energy demand is a situation whereby the desirable level of service cannot be reached. It is commonly due to a budget constraint or lack of adequate infrastructure.

losses, as a means of capturing dimensions absent from other analyses. It should be noted that we did not arrive at these estimates using power system planning techniques. While recognizing the coarse nature of our analysis, we find that the global cost of achieving universal access to electricity by 2030 ranges from USD 119–2,793 billion for electrification and USD 178–410 billion for clean cooking. The totals (electricity plus clean cooking) reach USD 297, 923, and 3,203 billion for the low, medium, and high scenarios, respectively. We did not undertake analysis to determine which scenario is most likely—they are 'what if' scenarios, each with its own legitimacy under a given set of assumptions. Still, if levels of service normally associated with the conditions for wealth creation is a goal, then the high scenario is closest to that aspiration. Providing further transparency to methodological approaches can help support decision-making at an international level and underpin political aspirations.

The total cost of reaching universal access to modern energy services may be significantly higher than indicated by the published studies if assumptions of per capita electricity consumption move beyond basic access. Our medium scenario is similar to other global estimates such as the IEA's energy access scenario in the *World Energy Outlook 2012* and the *Global Energy Assessment 2012*. However, a level of investment somewhere between our medium and high estimates is perhaps more realistic—due to the inclusion of T&D costs and cooking fuel, and assuming a higher level of per-person electricity consumption.

It is important to highlight that our low scenario assumes very basic access to modern energy services—one that might be better characterized as 'poverty maintenance' than providing a foundation for wealth creation. This is a recurrent implicit assumption in the literature. On the other hand, our revised high scenario presents a case that considers a more equitable future, and one that diverges from the previous literature. However, we present it for its contribution to the policy dialogue—namely the need to move to a level of energy service provision that will help spur growth in the wider economy and contribute to the provision of robust community services. As a result, the total cost for full access to energy at a level between our medium and high scenarios—at roughly 1–3 trillion in total to 2030—is significantly larger than the figures most often cited. Still, it remains a fraction of total estimated investment costs in energy-supply infrastructure, which amounts to USD 37 trillion for the period 2012–2035 (IEA 2012: 73)—and it is still a very low level of per capita consumption compared to today's OECD numbers. Notably, by explicitly considering fuel costs, we can better understand the full costs to the economy of provision of energy services.

While much remains to be done, and a significant share of the world's population still lacks access to modern energy, history has shown that progress in this regard can be both swift and wide-ranging. Providing that adequate incentives and conditions are in place, the investments required to promote energy access can unfold and bring about a number of associated developmental benefits. Past successes should serve as encouraging lessons to address the issue of energy access in the regions of the world where it remains a significant barrier to development.

ACKNOWLEDGEMENTS

A special thanks for the insightful comments by Katherine Sierra (Brookings). Also we acknowledge the very helpful support of Fatih Birol (IEA), Raffaella Centurelli (OECD), Jose Goldemberg (University of São Paulo), Daniel M. Kammen (UCB), Pedzi Makumbe (ESMAP), Shonali Pachauri (IIASA), Judy Siegel (Energy and Security Group), Bipulendu Singh (ESMAP), and Marina Ploutakhina and Alois P. Mhlanga (UNIDO).

DEDICATION

This chapter is dedicated to our dear friend Professor Abeeku Brew-Hammond. He taught us all how to better understand this issue, and he will be deeply missed.

DISCLAIMER

The views expressed herein are those of the authors and do not necessarily reflect the views of any organization. Designations such as 'developed', 'industrialized', and 'developing' are intended for statistical convenience and do not necessarily express a judgement about the stage reached by a particular country or area in the development process.

APPENDIX A—ASSUMPTIONS
ON THE SHARE OF
GENERATION SYSTEMS

We assume that electricity for urban electrification is provided for through the grid. In rural contexts, we assume the following share in the respective scenarios:

Table 8A.1. Assumptions on the share of electricity generation systems

		Scenario		
Assumption		Low	Medium	High
Share of rural electrification through	grid	0.61	0.30	**0.30**
	mini-grid	0.34	0.46	**0.46**
	off-grid	0.05	0.25	**0.25**

Data source: based on IEA 2009b and 2012 and own estimates

APPENDIX B—ASSUMPTIONS ON THE LEVELIZED
COST OF ELECTRICITY

Table 8B.1. Assumptions on levelized costs of grid electricity per region for the low scenario

Energy source	Levelized cost (USc/kWh)					Share in electricity mix			
	Capital	O&M	Fuel	T&D	Total	Africa	Developing Asia	Latin America	Middle East
Coal[xix]	2.44	0.63	1.07	4.06	8.20	0.45	0.77	0.04	0.00
Oil	1.93	0.98	7.33	4.05	14.29	0.14	0.03	0.20	0.41
Gas[xx]	1.33	0.52	7.06	4.08	12.99	0.31	0.09	0.24	0.58
Nuclear	3.60	0.90	0.65	3.97	9.12	0.02	0.02	0.04	0.00
Hydro	5.04	0.36	0.00	4.14	9.54	0.06	0.06	0.39	0.01
Biomass and waste	2.11	0.39	5.56	5.84	13.90	0.00	0.01	0.07	0.00
Others[xxi]	21.22	0.59	0.00	34.50	56.31	0.01	0.02	0.02	0.00
	Weighted average of total levelized cost (USc/kWh)					11.1	*9.9*	*12.6*	*13.5*

Data source: own estimates and calculations based on IEA 2012, ESMAP 2013

[xix] Based on a 500 MW supercritical plant in India with 40 per cent efficiency and an 80 per cent capacity factor
[xx] Based on a 450 MW combined cycle gas turbine plant in India with 50 per cent efficiency and and 80 per cent capacity factor
[xxi] Based on a 5 MW utility-scale solar PV plant in India

Table 8B.2. Assumptions on levelized costs of grid electricity per region for the medium and high scenarios

Energy source	Levelized cost (USc/kWh)					Share in electricity mix			
	Capital	O&M	Fuel	T&D	Total	Africa	Developing Asia	Latin America	Middle East
Coal	2.44	0.63	1.07	4.06	8.20	0.35	0.62	0.05	0.00
Oil	1.93	0.98	7.33	4.05	14.29	0.06	0.01	0.07	0.23
Gas	1.33	0.52	7.06	4.08	12.99	0.32	0.11	0.21	0.67
Nuclear	3.60	0.90	0.65	3.97	9.12	0.06	0.10	0.08	0.04
Hydro	5.04	0.36	0.00	4.14	9.54	0.09	0.06	0.41	0.01
Biomass and waste	2.11	0.39	5.56	5.84	13.90	0.05	0.04	0.11	0.01
Others	21.22	0.59	0.00	34.50	56.31	0.07	0.06	0.08	0.03
	Weighted average of total levelized cost (USc/kWh)					*14.0*	*12.1*	*14.5*	*14.3*

Data source: own estimates and calculations based on IEA 2012, ESMAP 2013

Table 8B.3. Assumptions on levelized costs for mini-grid for the low, medium, and high scenarios

Energy source	Levelized cost (USc/kWh)					Share in electricity mix
	Capital	O&M	Fuel	T&D	Total	
Diesel generator	1.21	3.37	15.61	3.91	24.10	0.6
PV–wind hybrid	35.28	1.10	0.00	3.91	40.29	0.4
	Weighted average of total levelized cost (USc/kWh)					30.58

Data source: own estimates and calculations based on ESMAP 2013

Table 8B.4. Assumptions on levelized costs for off-grid for the low scenario

Energy source	Levelized cost (USc/kWh)					Share in electricity mix
	Capital	O&M	Fuel	T&D	Total	
Diesel generator	0.78	4.30	99.21	0.00	104.29	0.2
PV–wind hybrid	75.98	0.86	0.00	0.00	76.84	0.5
PV based on 300 W PV plant	42.87	0.70	0.00	0.00	43.57	0.3
	Weighted average of total levelized cost (USc/kWh)					72.35

Data source: own estimates and calculations based on ESMAP 2013

APPENDIX C—ASSUMPTIONS
ON THE SHARE
OF GENERATION SYSTEMS

Figure 8C.1 presents the electrification estimates found in the literature in terms of average cost per person, as compared to the average cost per person per year estimates.

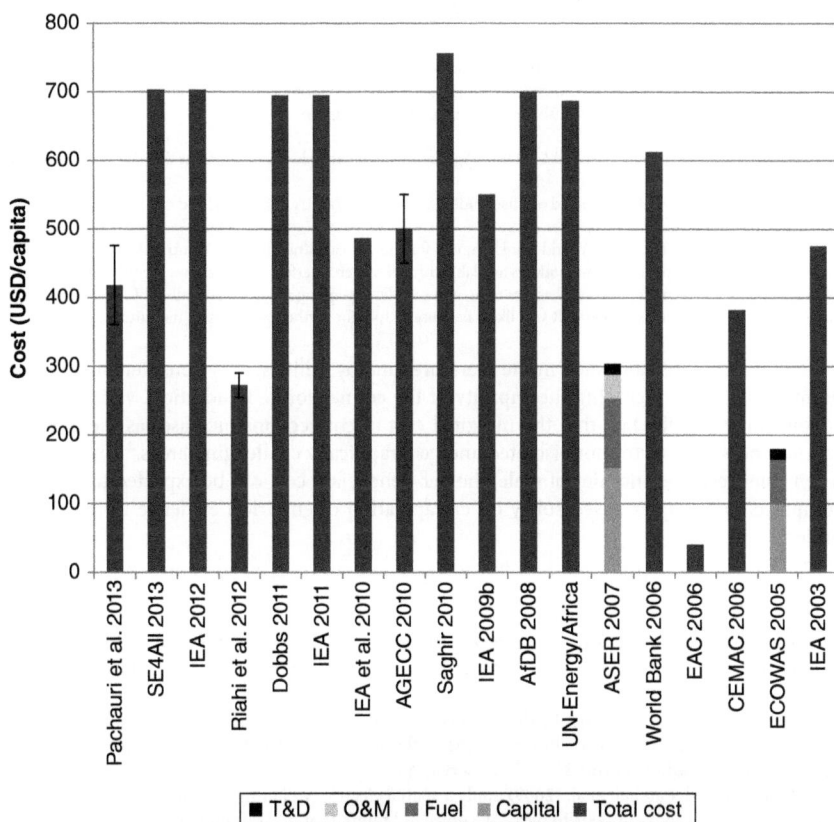

Figure 8C.1. Comparison of cost estimates for electrification in terms of USD per capita

APPENDIX D—CONNECTION COST ESTIMATES

Table 8D.1 provides an overview of the surveyed estimates for electricity access presented in terms of cost per connection.

Figure 8D.1 provides an overview of the electricity connection cost in the studies reviewed. Our non-comprehensive literature review identified no new connection cost estimates published since 2010; however, it is possible that individual utilities, municipalities, or universities have published such estimates.

Table 8D.1. Estimated cost of electricity access per connection in USD

| Geographical focus | Cost per Connection (USD) | | | |
	Goal	Electricity	Period	Source
Regional/local				
Africa	Electrification	806 USD	1998–2005	**AICD 2008**
South Africa	Electrification	1,000 USD[xxii]	Not specified	**Eskom 2009; IEA 2009a**
Kenya	Electrification	1,900 USD[xxiii]	Not specified	**Parshall et al. 2009**
Botswana	Electrification	1,100 USD[xxiv]	Not specified	**Krishnaswamy et al. 2007**
Mali	Rural electrification	776 USD[xxv]	Not specified	**AMADER quoted in Foster et al. 2010, p. 199**
(unspecified)	**Electrification**	**above 1,200 USD[xxvi]**	**Not specified**	**Practical Action**

[xxii] The average is expected to increase as the electrification process moves to communities in more remote rural areas
[xxiii] Average cost per household in a so-called realistic penetration scenario, with USD 1,500 and 2,615 for infilling and grid extension, respectively; based on modelling of grid extension
[xxiv] Based on project experience
[xxv] Based on project experience from AMADER (Agence malienne pour le développement de l'énergie domestique et l'électrification rurale)
[xxvi] New connection to electricity; based on case studies; varies from country to country, and can be as much as 6,000 USD in some cases
Not included in Table 8D.1 is the World Bank's model for the Africa Infrastructure Country Diagnostic Initiative (The World Bank Group 2010). The model is available through a web interface which allows the user to alter major assumptions, including urban and rural access rates, and calculate spending needs for a number of African countries and compare them against a baseline. It will likely be a useful tool for further estimates in the future

Connection costs published in the literature (and by utilities) vary considerably due to a number of factors, including the capacity of the connection.[21] In addition, various publications underline the fact that the marginal cost of connections increases as the electrification process moves to more isolated and geographically challenging areas.[22] In contrast, when considering economies of scale, the per-beneficiary cost can be expected to decrease compared to estimates based solely on extrapolation of empirical evidence from isolated projects.

[21] For comparison, the Durban municipality charged a fee of USD 897 for a single phase connection, and between USD 1,636 and 98,645 for a three phase connection as of 1 July 2012. Data source: <http://www.durban.gov.za/Resource_Centre/Services_Tariffs/Electricity%20Tariffs/2012%202013%20 20Tariff%20Book.pdf>. Similarly, Hydro Québec charged from USD 2,043 for a basic connection to a detached house. The price rises based on the scale of demand. Data source: <http://www.hydro quebec.com/publications/en/rates/pdf/frais_service.pdf>.
[22] In that regard, Parshall et al. (2009) apply a spatial electricity planning model to estimate the cost of connection in various contexts in parts of Kenya and found that it varies greatly between settlements around major cities and more isolated rural areas.

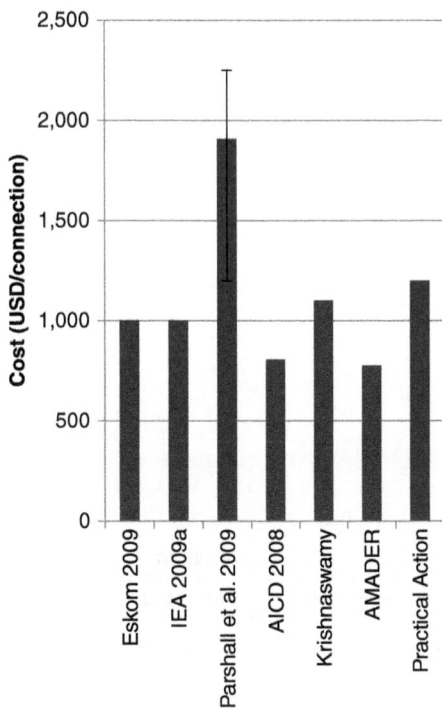

Figure 8D.1. Comparison of connection cost estimates

APPENDIX E—RESULTS OF SENSITIVITY: THE ANALYSIS

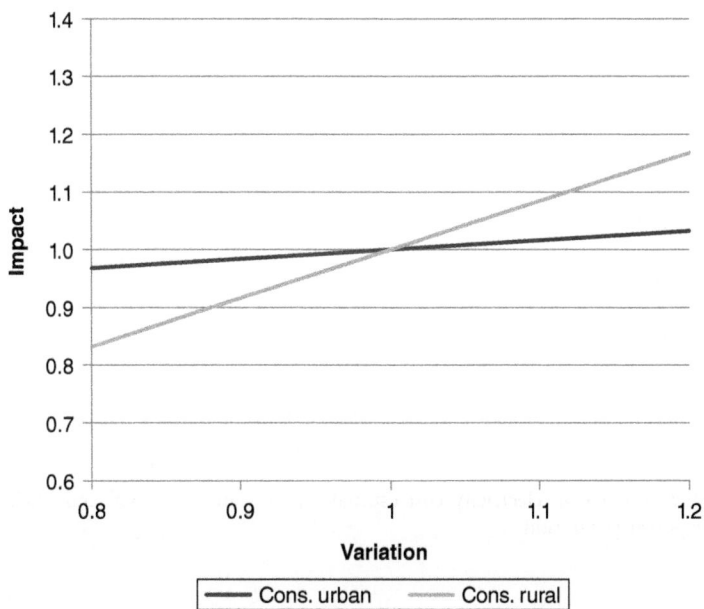

Figure 8E.1. Impact on cost estimates of variation of rural and urban consumption

Morgan Bazilian et al.

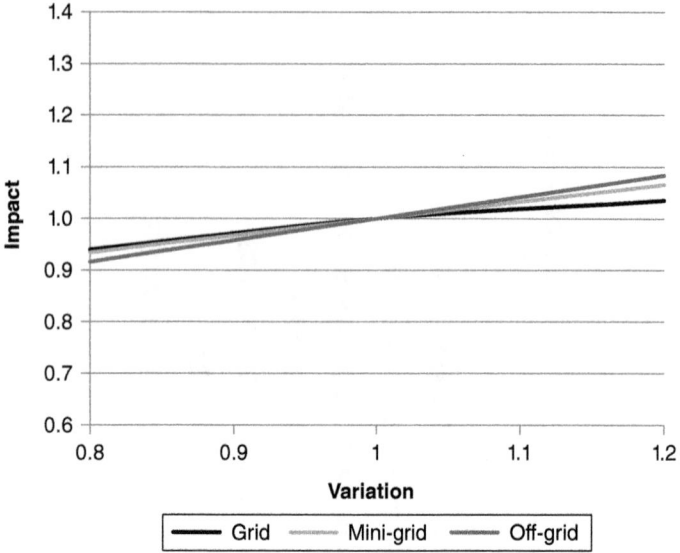

Figure 8E.2. Impact on cost estimates of variation of levelized cost for grid, mini-grid, and off-grid systems

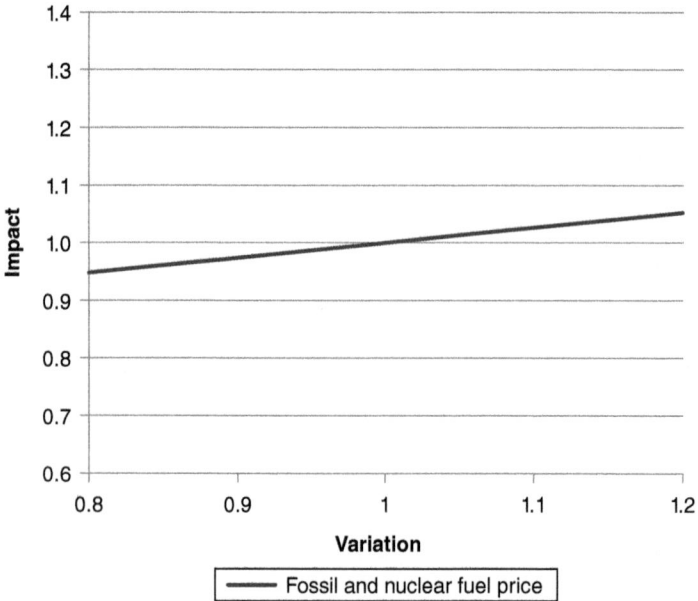

Figure 8E.3. Impact on electricity cost estimates of variation in fossil and nuclear fuel prices for power generation

Table 8E.1. Impact of cost estimates of the three different electricity connection-type configurations

	Share of grid, mini-grid, off-grid (by scenario)		
	Full grid expansion	IEA 2009 (low)	IEA 2012 (medium and high)
Annual cost when used in medium scenario	29.7	43.1	63.5
Per cent change	−45%	0%	47%

REFERENCES

AfDB. (2008). *Clean Energy Investment Framework for Africa—The Role of the African Development Bank Group*. Tunis: African Development Bank. Available at: <http://www. afdb.org/fileadmin/uploads/afdb/Documents/Policy-Documents/10000025-EN-PRO POSALS-FOR-A-CLEAN-ENERGY-INVESTMENT-FRAMEWORK-FOR-AFRICA.PDF>.

AGECC. (2010). *Energy for a Sustainable Future*. New York: The Secretary-General's Advisory Group on Energy and Climate Change. Available at: <http://www.unido.org/ fileadmin/user_media/Services/Energy_and_Climate_Change/EPP/Publications/AGECC_ Report.pdf>.

AICD. (2008). *Unit Costs of Infrastructure Projects in Sub-Saharan Africa*. Background Paper 11: Africa Infrastructure Country Diagnostic. Washington, DC: The World Bank Group.

Al-Shaalan, A.M. (2009). 'Problem Associated with Power System planning in Developing Countries'. *Journal on Electronics and Electrical Engineering* 1(1): 43–8.

ASER. (2007). *Costing for National Electricity Interventions to Increase Access to Energy, Health Services, and Education—Senegal Final Report*. Agence Sénégalaise d'Electrification Rurale; Energy Group, Columbia Earth Institute. Available at: <http://modi.mech. columbia.edu/wp-content/uploads/2010/01/Senegal_WorldBank_Report_8-07.pdf>.

Banerjee, S., et al. (2008). *Access, Affordability, and Alternatives: Modern Infrastructure Services in Africa*. Washington, DC: The World Bank. Available at: <http://www.in frastructureafrica.org/aicd/library/doc/367/access-affordability-and-alternatives-modern­infrastructure-services-africa>.

Bazilian, M. and Roques, F. (2008). *Analytical Methods for Energy Diversity and Security*. Oxford: Elsevier.

Bazilian, M., Nussbaumer, P., Haites, E., Levi, M., Howells, M., and Yumkella, K. (2010b). Understanding the Scale of Investment for Universal Energy Access. *Geopolitics of Energy* 32(10–11): 19–40.

Bazilian, M., Sagar, A., Detchon, R., and Yumkella, K. 2010a. 'More Heat and Light'. *Energy Policy* 38(10): 5409–12.

Bazilian, M. et al. (2012). 'Energy Access Scenarios to 2030 for the Power Sector in Sub-Saharan Africa', *Utilities Policy* 20: 1–16.

Birol, F. (2007). 'Energy Economics: A Place for Energy Poverty in the Agenda?' *The Energy Journal* 28(3). Available at: <http://www.iea.org/papers/2007/Birol_Energy_Journal. pdf>.

Brew-Hammond, A. (2010). 'Energy Access in Africa: Challenges Ahead'. *Energy Policy* 38 (5): 2291–301.

CEMAC. (2006). *Plan d'action pour la promotion de l'accès à l'energie dans la région CEMAC.* Economic Community of Central African States.

Dobbs, R., Oppenheimer, J., Thompson, F., Brinkman, M., and Zornes, M. (2011). *Resource Revolution: Meeting the World's Energy, Materials, Food, and Water Needs.* New York: McKinsey Global Institute.

EAC. (2006). *Strategy on Scaling Up Access to Modern Energy Services in Order to Achieve the Millennium Development Goals.* East African Community. Available at: <http://www.eac.int/energy/index.php?option=com_docman&task=doc_download&gid=15&Itemid=70>.

ECOWAS. (2005). *White Paper for a Regional Policy—Geared towards Increasing Access to Energy Services for Rural and Periurban Populations in order to Achieve the Millennium Development Goals.* Economic Community of West African States. Available at: <http://www.gm.undp.org/Reports/ECOWAS%20energy%20white%20paper.pdf>.

Ekholm, T., Krey, V., Pachauri, S., and Riahi, K. (2010). 'Determinants of Household Energy Consumption in India'. *Energy Policy* 38(10): 5696–707.

Eskom. (2009). *Annual Report.* Available at: <http://www.eskom.co.za/annreport09/ar_2009/index_annual_report.htm>.

ESMAP (Energy Sector Management Assistance Program). (2007). *Technical and Economic Assessment of Off-grid, Mini-grid and Grid Electrification Technologies.* Washington, DC: Energy Sector Management Assistance Program, The World Bank Group. Available at: <http://siteresources.worldbank.org/EXTENERGY/Resources/336805-1157034157861/ElectrificationAssessmentRptAnnexesFINAL17May07.pdf>.

ESMAP. (2013). *Model for Electricity Technology Assessments (META) and Guide.* Washington, DC: The World Bank Group. Available upon request at: <https://www.esmap.org/node/3051>.

Foster, V. and Briceno-Garmendia, C. (2010). *Africa's Infrastructure—A Time for Transformation.* Washington, DC: The International Bank for Reconstruction and Development/The World Bank. Available at: <http://www.infrastructureafrica.org/aicd/library/doc/552/africa%E2%80%99s-infrastructure-time-transformation>.

Goldemberg, J. (1998). Leapfrog Energy Technologies. *Energy Policy* 26(10): 729–41.

Goldemberg, J., Johansson, T.B., Reddy, A.K., and Williams, R.H. (1985). Basic Needs and Much More with One Kilowatt per Capita. *Ambio* 14(4): 190–200.

Howells, M.I., Alfstad, T., Cross, N., Jeftha, L., and Goldstein, G. (2002). *Rural Energy Modeling.* Program on Energy and Sustainable Development, Standford University. Available at: <http://pesd.stanford.edu/publications/rural_energy_modeling/>.

Howells, M., Victor, D.G., Gaunt, T., Elias, R.J., and Alfstad, T. (2006). Beyond Free Electricity: The Costs of Electric Cooking in Poor Households and a Market-friendly Alternative. *Energy Policy* 34(17): 3351–8.

IEA. (2003). *World Energy Investment Outlook 2003.* Paris: International Energy Agency.

IEA. (2004). *World Energy Outlook 2004.* Paris: International Energy Agency.

IEA. (2008). *Energy Statistics of Non-OECD Countries.* International Energy Agency.

IEA. (2009a). *Comparative Study on Rural Electrification Policies in Emerging Countries.* Paris: International Energy Agency. Available at: <http://www.iea.org/papers/2010/rural_elect.pdf>.

IEA. (2009b). *World Energy Outlook 2009.* Paris: International Energy Agency.

IEA. (2010). *Projected Costs of Generating Electricity.* Paris: International Energy Agency, Nuclear Energy Agency, Organisation for Economic Co-operation and Development.

IEA. (2011). *World Energy Outlook 2011.* Paris: International Energy Agency.

IEA. (2012). *World Energy Outlook 2012.* Paris: International Energy Agency.

IEA, UNDP, and UNIDO. (2010). *Energy Poverty—How to Make Modern Energy Access Universal?* Paris: International Energy Agency. Available at: <http://www.worldenergyoutlook.org/media/weowebsite/2010/weo2010_poverty.pdf>.

Jannuzzi, G.M. and Sanga, G.A. (2004). 'LPG Subsidies in Brazil: An Estimate.' *Energy for Sustainable Development* 8(3): 127–9.

Johansson, T.B. and Goldemberg, J. (2002). *Energy for Sustainable Development—A Policy Agenda*. United Nations Development Programme.

Johnson, Oliver. (2013). 'Universal Energy Access: Moving from Technological Fix to Poverty Reduction.' Briefing Paper 6/2013. Bonn: German Development Institute.

Krishnaswamy, V. and Stuggins, G. (2007). *Closing the Electricity Supply-Demand Gap*. Washington, DC: The World Bank. Available at: <http://siteresources.worldbank.org/EXTENERGY/Resources/336805-1156971270190/EnergyandMiningSectorBoardPaper No20.pdf>.

Morris, E. and Kirubi, G. (2009). *Bringing Small-Scale Finance to the Poor for Modern Energy Services: What is the Role of Government?* United Nations Development Programme. Available at: <http://content.undp.org/go/newsroom/publications/environ ment-energy/www-ee-library/sustainable-energy/bringing-small-scale-finance-to-the-poor-for-modern-energy-services.en>.

Morris, E., Winiecki, J., Chowdhary, S., and Cortiglia, K. (2007). *Using Microfinance to Expand Access to Energy Services*. The SEEP Network. Available at: <http://www.arcfinance.org/pdfs/pubs/Energy_Summary_FINAL.pdf>.

Pachauri, S., van Ruijven, B., Nagai, Y., Riahi, K., van Vuuren, D., Brew-Hammond, A., and Nakicenovic, N. (2013). 'Pathways to Achieve Universal Household Access to Modern Energy by 2030.' *Environmental Research Letters* 8(2). Available at: <http://iopscience.iop.org/1748-9326/8/2/024015/pdf/1748-9326_8_2_024015.pdf>.

Parshall, L., Pillai, D., Mohan, S., Sanoh, A., and Modi, V. (2009). National Electricity Planning in Settings with Low Pre-existing Grid Coverage: Development of a Spatial Model and Case Study of Kenya. *Energy Policy* 37(6): 2395–410.

Practical Action. (2010). *Poor People's Energy Outlook 2010*. Total Energy Access Project. Available at: <http://practicalaction.org/totalenergyaccess>.

Practical Action. (2012). *Poor People's Energy Outlook 2012*. Total Energy Access Project. Available at: <http://practicalaction.org/totalenergyaccess>.

Practical Action. (2013). *Poor People's Energy Outlook 2013*. Total Energy Access Project. Available at: <http://practicalaction.org/totalenergyaccess>.

Riahi, K., et al. (2012). 'Energy Pathways for Sustainable Development'. In *Global Energy Assessment: Toward a Sustainable Future*, Cambridge and New York: Cambridge University Press; Laxenburg: International Institute for Applied Systems Analysis, pp. 1203–1306.

Rong, F. and Victor, D. (2012). *What Does It Cost to Build a Power Plant?* ILAR Working Paper 17. San Diego: Laboratory on International Law and Regulation.

Sachs, J., et al. (2004). *Millennium Development Goals Needs Assessments—Country Case Studies of Bangladesh, Cambodia, Ghana, Tanzania and Uganda*. Millennium Project. Available at: <http://www.unmillenniumproject.org/documents/mp_ccspaper_jan1704.pdf>.

Saghir, J. (2010). Energy and Development: Lessons Learned. Paris: International Energy Agency.

SE4All (Sustainable Energy for All Initiative). (2013). *Global Tracking Framework*. Chapter 2: Universal Access. New York: United Nations. Available at: <http://www.sustainableenergyforall.org/images/Global_Tracking/7-gtf_ch2.pdf>.

The World Bank Group. (2006). *An Investment Framework for Clean Energy and Development: A Progress Report*. Washington, DC. Available at: <http://siteresources.worldbank.org/DEVCOMMINT/Documentation/21046509/DC2006-0012(E)-CleanEnergy.pdf>.

The World Bank Group. (2008). *The Welfare Impact of Rural Electrification: A Reassessment of the Costs and Benefits*. Washington, DC. Available at: <http://

lnweb90.worldbank.org/oed/oeddoclib.nsf/DocUNIDViewForJavaSearch/EDCCC33082
FF8BEE852574EF006E5539/$file/rural_elec_full_eval.pdf>.

The World Bank Group. (2010). AICD Spending Needs Model. Available at: <http://info.
worldbank.org/etools/aicd/electricity.asp>.

The World Bank Group. (2013). *World Development Indicators*. Available at: <http://data.
worldbank.org/indicator>.

UN. (2010). *International Year of Sustainable Energy for All*. Resolution 65/101. New York:
United Nations General Assembly, sixty-fifth session.

UN. (2012). *Promotion of New and Renewable Sources of Energy*. Resolution 67/215. New
York: United Nations General Assembly, sixty-seventh session.

UNDP. (2010). *Capacity Development for Scaling Up Decentralized Energy Access Programmes*. New York: United Nations Development Programme. Available at: <http://
content.undp.org/go/cms-service/stream/asset/?asset_id=2625476>.

UNDP and WHO (World Health Organization). (2009). *The Energy Access Situation in
Developing Countries—A Review on the Least Developed Countries and Sub-Saharan
Africa*. New York. Available at: <http://content.undp.org/go/cms-service/stream/asset/?
asset_id=2205620>.

UN-Energy/Africa. *Energy for Sustainable Development: Policy Options for Africa*. Available
at: <http://www.uneca.org/eca_resources/publications/unea-publication-tocsd15.pdf>.

UNFCCC. (2007). *Investment and Financial Flows to Address Climate Change*. United
Nations Framework Convention on Climate Change. Available at: <http://unfccc.int/
files/cooperation_and_support/financial_mechanism/application/pdf/background_paper.
pdf>.

Urban, F., Benders, R.M.J., and Moll, H.C. (2007). 'Modelling Energy Systems for Developing Countries.' *Energy Policy* 35(6): 3473–82.

9

Energy and Water

A Critical Linkage

Allan Hoffman

'Energy poverty refers to the situation of large numbers of people in developing countries whose well-being is negatively affected by very low consumption of energy, use of dirty or polluting fuels, and excessive time spent collecting fuel to meet basic needs . . . it is distinct from fuel poverty, which focuses solely on the issue of affordability' (Wikipedia). What is missing from this definition and from most energy poverty discussions to date is the inextricable linkage between energy and water, commonly referred to as the water–energy nexus. It is the recognition that energy requires water and that water requires energy, and that energy and water must be considered jointly if we are to optimize their societal use. Access to both is critical to poverty alleviation and economic development.

Until fairly recently those focused on energy thought about water in limited ways, and those focused on water rarely thought about energy. While this is understandable, and while it worked for most of the twentieth century, it will not work in the twenty-first. As we make progress on alleviating energy poverty we may be creating another problem: water poverty. This arises from the fact that many forms of energy production and use depend on the availability of water, and it is significant that the InterAction Council, composed of 20 former heads of state, has recommended 'placing water at the forefront of the global political agenda' (InterAction Council 2011). The resulting conundrum that must be addressed is: As we progress into the twenty-first century and as population, per capita incomes, and global demands for energy and potable water increase, can we provide the quantities that are needed in a timely, cost-effective and environmentally safe manner, given that energy and water goals often conflict? The answer is yes, if we plan carefully and take the nexus fully into account. It is also true that energy shortages and water shortages often do not coincide geographically—some areas are poor in water but rich in energy and some are the reverse. On a global basis we will not run out of either water or energy but will undoubtedly have to pay more for both. The purpose of this chapter is to provide the context for these conclusions and their implications for economic development.

We must also look at the role of energy and water in security and economic development (Hoffman 2009). Personal security refers to an individual's employment, health, or ability to be shielded from violence. It is a concept that also applies to nations. Historically, national security had a military connotation—that is, whether a country could protect itself against internal disruptions and foreign invaders. Today it is widely accepted that a nation's security also depends on the state of its economy, of which energy and water are critical parts, and the quality of its governance. Recognition of the critical relationship among economics, governance, and security was a major outcome of the Bretton Woods conference of 1944 that led to the creation of the International Monetary Fund and the International Bank for Reconstruction and Development. More recently it has been recognized that states with little security, often referred to as fragile states, can undermine their neighbours and regions economically as well as politically.

A theme often expressed, especially since the attacks of 9/11, is that poverty leads to terrorism and that addressing poverty will diminish instability and the terrorism threat. This is certainly true to some extent, as most people need to have hope of a better tomorrow if they are not to be receptive to extreme measures. However, several recent studies have concluded that poverty alone does not automatically lead to terrorism and instability, as evidenced by the relative affluence of many of the 9/11 terrorists and many others in the ranks of Al-Qaeda and other terrorist organizations. The missing factor appears to be governance and its link to economic development.

In a statement to the United States Institute of Peace in 2009, Robert Zoellick, President of the World Bank, defined the problem as follows:

> Fragile states are a witches' brew of ineffective government, poverty, and conflict.... Weak governance, corruption, and insecurity combine in a downward cycle. Fragility does not just mean low growth, but a failure in the normal growth process, such that grinding, hopeless poverty becomes a persistent condition.... Too often, the development community has treated states affected by fragility and conflict simply as harder cases of development.... Yet these situations require looking beyond the analytics of both security studies and development—to a different framework of building security, legitimacy, governance and economy. This is not security as usual or development as usual.... This is about 'securing development'—bringing security and development together first to smooth the transition from conflict to peace and then to embed stability so that development can take hold over a decade and beyond. Only by securing development can we put down roots deep enough to break the cycle of fragility and violence.
>
> (Zoellick 2009)

To secure development and sustainably create the jobs that reduce poverty, governments must first establish their legitimacy by providing basic services (energy, water) in an environment safe for investment and other economic activity. This must go hand-in-hand with establishing the rule of law, including respect for property rights.

9.1. THE ENERGY CONTEXT

Any discussion of energy and energy poverty must start with the recognition that people value not energy itself but rather the services that energy makes possible— heating; cooling; cooking; lighting; transportation of people, water, and goods;

communication; and a broad range of commercial activities. In addition, it is often said that energy is the lifeblood of modern societies. What is more true is that energy in various forms has been critical to human activities over the centuries, but that modern societies provide a high level of energy-dependent services that go well beyond human and animal power.

It follows that governments will want to provide these services with the least amount of energy feasible, to minimize energy costs and environmental and national security impacts. Global energy today is provided largely by fossil fuels (coal, oil, natural gas) and this will be true for several decades into the future, given large reserves and devoted infrastructure. Nevertheless, fossil fuel resources are finite and non-renewable and their combustion releases carbon dioxide, a greenhouse gas, into the atmosphere. Unless captured and sequestered, these emissions will eventually have to be restricted. Cost increases and volatility, as well as global warming concerns, are likely to limit fossil fuel use before resource restrictions become dominant, and increasing concentration of supplies in just a few countries raises, for many countries, serious national security concerns. In addition, the world's current energy delivery infrastructure is highly vulnerable to natural disasters, terrorist attacks, and other breakdowns, and energy imports constitute a major drain on financial resources.

It is also important to recognize that on a global basis energy is not in short supply. The sun, a modest star sitting 93 million miles from earth, pours 6 million quads of radiation annually into the earth's atmosphere, and this is only 4 parts in 10 billion of the sun's total output. While some of this insolation bounces off the earth's clouds back into space, most of it (approximately 70 per cent) enters the atmosphere and becomes part of the earth's energy balance with the sun. When too much solar energy is captured by the earth rather than reradiating into space as infrared radiation, we get global warming and associated climate change. There is also considerable geothermal energy under our feet in the form of hot water and hot rock. Thus, the reality is that energy is not in short supply on the earth. What is in short supply is inexpensive energy that people can afford to buy (Hoffman 2006).

9.2. THE WATER PROBLEM

Water supply issues associated with energy arise from the fact that many forms of energy production and use depend on the availability of water. These include hydropower, the use of falling or moving water to generate electricity (which diverts some water from other uses such as food-growing and fish ladders, although associated dams can provide controlled quantities of water for recreational and agricultural uses); water requirements of fossil fuel extraction and processing (including the increasing use of water in deep shale fracking for new supplies of natural gas and oil); the use of water to cool the hot exhausts of thermal power plants (the largest use of water in the US); the irrigation of crops to be used for conversion to ethanol and other biofuels; water requirements of carbon capture and sequestration; and eventually, in a hydrogen economy, the production of hydrogen from electrolysis of water. New energy-related requirements will put additional strain on currently used aquifers, rivers, and lakes and will divert water use from additional food-growing and urban consumption, a rapidly growing source of increased global demand.

Other, indirect, linkages exist as well between energy and water. Energy production and use can lead to contamination of underground and surface water supplies. For example, the use of water in fracking is still a small part of the overall demand, but it is growing and does raise serious concerns about contamination of large sources of fresh water due to toxic and radioactive materials contained in fracking water returned to the surface for disposal. If competing water uses limit use of waterways for transport of goods (as is happening now on the Mississippi River), rail and truck will require more energy to move those goods. Competition for water resources is already limiting licensing and operation of thermal power plants—fossil, nuclear, and concentrating solar. Another critical linkage is that energy production and use are major contributors to greenhouse gas emissions, which have the potential to disrupt the hydrological cycle which delivers water to its many uses. Early signs of such disruption are already appearing (NRDC 2012).

9.3. THE WATER CONTEXT

'Water is life' is a truism. Without water, life as we know it would not exist. The hard reality is that there are no substitutes for water (as there are for energy sources), and if you do not have water you die. It is also true that we live on a water-rich planet—more than 300 million cubic miles of water, with each cubic mile containing more than one trillion gallons—and the amount of water on the planet hasn't changed over millions and perhaps even billions of years. To quote *National Geographic*: 'The water the dinosaurs drank millions of years ago is the same water that falls as rain today' (Kingsolver 2010). So is there a problem with water supply?

The biggest problem is that 96 per cent of all the water on earth is found in the oceans, with an average salt concentration of 35,000 parts per million. That is quite salty, and human beings, animals, and many plants are harmed when drinking or ingesting water that has more than about 500–1,500 parts per million of dissolved solids. Until the human organism and other organisms can ingest salt water safely, something that would require genetic changes that take a while to take hold, we have to deal with available supplies of fresh water or pay the price, energy- and cost-wise, for converting saline to fresh water. This identifies part of the conundrum.

With regard to fresh water, which constitutes the remaining 4 per cent of water on the planet, most is not easily available for our use. Some is tied up in icecaps, glaciers, and permanent snow cover in mountainous regions (although global climate change seems to be addressing that part of the problem), a not insignificant amount is tied up as water vapour in the atmosphere, and the rest is in groundwater, lakes, and rivers. The result is that 99.7 per cent of all the water on earth is not available for human and animal consumption. And of the remaining 0.3 per cent, much is inaccessible due to unreachable locations and depths. Thus the good news and the bad news is that we make productive use of much less than 1 per cent of our global water resources.

Another problem is that fresh water is not distributed uniformly around the globe. Some areas have plenty, some have little to none. The struggle to control water resources has shaped human economic and political history, and water has been a source of tension (not always related to water poverty) wherever water resources are shared by neighbouring peoples (Wolf 1998). Globally there are 215 international rivers and 300 groundwater basins and aquifers that are shared by two or more countries, and in the volatile Middle East water is a source of conflict not only between the Israelis and Palestinians, but also between Egypt and Sudan, and among Turkey, Syria, and Iraq. Even in the United States there is tension between and among states not usually associated with water poverty, but the issue is of increasing concern. Many other examples can be cited.

In addition, as mentioned above, precipitation patterns that bring much of the water will change as the climate does. How those patterns will change is something many scientists and planners are working hard to understand, and climate change has the potential to disrupt the hydrological cycle and impact water resources long before other climate change impacts are felt. By altering the timing of winter snows, snowmelt, and spring rains, climate change could overload reservoirs early in a year, forcing unplanned releases of water and leaving water-dependent areas high and dry in summer. Coastal areas and island nations also face a serious threat. Rising water levels, before they destroy property and flood low-lying areas, will cause saltwater intrusion of freshwater supplies, putting the drinking water of millions at risk. One observer has labelled the anticipated changes as a 'hydroclimatic time bomb that is already ticking' (Sandford 2011), indicating more of a looming threat than an immediate crisis compared to energy poverty.

We must also ask how the demand for fresh water is changing and what the implications are of not having enough. Current global annual demand for fresh water is estimated to be about 1,000 cubic miles, approximately 30 per cent of the world's total accessible fresh water supply, and has more than tripled in the past 50 years. In addition, the earth's total human population continues to rise. It has already passed the 7 billion mark, with another 1 billion anticipated by 2030 and another 1 to 2 billion by 2050. An additional problem mentioned in the Economist Intelligence Unit's report 'Water for all' is that 'the world's middle class is expected to grow from less than 2bn today to nearly 4.9bn' by 2030, with even more growth by 2050. 'As this more affluent population increases demand for water will surge, not least due to a greater appetite for meat and other goods that are more water-intensive to produce. In developing countries, where the vast majority of both population growth and rising incomes can be found, a 50 per cent increase in water withdrawals is expected by 2025, while developed counties will increase by 18 per cent. As a result, as UN-Water highlights, water use continues to expand at more than twice the rate of population growth' (Economist Intelligence Unit 2012).

A significant part of this expansion is for food production (the third leg of the energy–water–food nexus)—agriculture accounts for nearly three-quarters of global water use on average—and it is important to note that over-pumping of groundwater by the world's farmers, from aquifers in India, China, the US and elsewhere, already exceeds natural replenishment by more than 4 per cent of withdrawals and is growing. China differs from India in that energy poverty is receding rapidly while water poverty is increasing, whereas India is struggling to make discernible progress on both.

The implications of too little fresh water are significant, as are the implications of energy poverty. The World Health Organization (WHO) estimates that, globally, more than 1 billion people lack access to clean water supplies and more than 2 billion lack access to basic sanitation today. An additional challenge is increasing urbanization: the WHO estimates that six in ten people will live in cities in 2030, up from 50 per cent in 2010 (WHO 2006). The world's megacities are already experiencing severe water stress and this situation will only worsen. The amount of water deemed necessary to satisfy basic human needs is 1,000 cubic metres per capita annually. In 1995, 166 million people in 18 countries lived below that level. By 2050, experts project that the availability of potable water will fall below that level for 1.7 billion people in 39 countries (WHO 2006). Water shortages currently plague almost every country in North Africa and the Middle East.

These shortages have major health effects. Water-borne diseases account for roughly 80 per cent of infections in the developing world. Nearly 4 billion cases of diarrhea occur each year, with diarrheal diseases killing millions of children. Another 60 million children are stunted in their development as a result of recurrent diarrheal episodes. In addition, 200 million people in 74 countries are infected with the parasitic disease schistosomiasis, intestinal worms infect about 10 per cent of the population in the developing world, and an estimated 6 million people are blind from trachoma, with an at-risk population estimated to be 500 million (WHO 2006).

A number of voices have sought to sound the alarm on energy and water issues for many decades, with some noticeable progress on energy poverty. Only recently, though, has the world begun to focus significant attention on water. World Water Forums have been held every three years starting in 1997. The UN Millennium Summit in New York in 2000 identified water and energy availability as critical global issues, as did the 2002 World Summit on Sustainable Development in Johannesburg. The UN declared 2003 the International Year of Freshwater, and designated the period 2005–2015 the UN Decade of Water. At its 2000 summit, the United Nations adopted two Millennium Development Goals related to water and sanitation: to reduce by half, by 2015, the proportion of people without access to (a) safe drinking water and (b) basic sanitation. Assuming a world population in 2015 of 7.2 billion, to meet these goals 1.6 billion more people will need to be supplied with access to safe drinking water and an additional 2.2 billion with access to basic sanitation. Even if both 2015 goals are reached, which is questionable, 600 million people in 2015 will still lack access to clean water and 1.5 billion to adequate sanitation.

More recently the high level global InterAction Council (IAC), meeting in Canada in 2011, warned of an impending 'water crisis' and agreed to establish a panel to address a worldwide leadership gap on the issue. This was followed in 2012 by the release of a report entitled 'The Global Water Crisis: Addressing an Urgent Security Issue' (InterAction Council 2012). In the foreword to this report, Gro Brundtland, former Prime Minister of Norway, underlined the danger in many regions, particularly Sub-Saharan Africa, West Asia, and North Africa, where critical shortages already exist: 'As some of these nations are already politically unstable, such crises may have regional repercussions that extend well beyond their political boundaries. But even in politically stable regions, the status quo may very well be disturbed first and most dramatically by the loss of stability in hydrological patterns.' IAC Co-Chair Jean Chretien, former Canadian Prime Minister, further stated when the report was released that 'The future political impact of water scarcity may be devastating. Using water the way we have in the past simply will not sustain humanity in the future. The IAC is calling on the United Nations Security Council to recognize water as one of the top security concerns facing the global community. Starting to manage water resources more effectively and efficiently now will enable humanity to better respond to today's problems and to the surprises and troubles we can expect in a warming world.' Nevertheless, despite this growing awareness, and growing appreciation of the energy–water linkage, international support for water projects in recent years has been marginal and declining.

9.4. GENDER CONSIDERATIONS

Another issue critical to development is gender equity (Hoffman 2004). In the context of this chapter gender is a social and not a biological construct, and refers to a set of relations, including power relations, which define social function on the basis of sex. Thus gender relations can be changed, and while gender relations are not inherently oppressive, all too often they are oppressive of women. Where gender equity is missing, meaning that women and men do not have equal conditions for realizing their full human rights and potential to contribute to national, political, economic, social, and cultural development and to benefit from the results, there are serious negative consequences for development and for addressing the issues of energy and water poverty.

Women head one-third of the world's families (in parts of Latin America families headed by women are the majority) and frequently are the financial mainstays of and principal energy- and water-providers for their families. They are responsible for half of the world's food production, and they produce between 60 and 80 per cent of the food in most developing countries. To produce adequate sanitation, food, and energy for cooking, women and girls must first 'produce' water, firewood, charcoal, and dung, and in developing countries they spend many hours each day doing so. This reduces significantly the time they might otherwise use for education, community involvement, and cottage industries. If safe and reliable water sources do not exist nearby, people are forced to pay exorbitant prices to street vendors or rely on unsafe local water resources. This has major implications for hygiene and the spread of diseases among poor women and their families. They are also harmed by inhaling the particulates associated with cooking in confined spaces. Finally, poor women's access to energy and water is less than that of poor men because decisions are most likely made by men and the needs of women are often ignored or undervalued. This has led to a situation where women are among the poorest of the poor in most parts of the world, leading to a 'feminization of poverty'.

9.5. ADDRESSING THE CONUNDRUM

From the discussion above, it is clear that water and energy are in abundant but not necessarily inexpensive supply. We will not run out of water or energy, but we will undoubtedly have to pay more for both. The dilemma that makes it hard to address the conundrum is that government officials are too often reluctant to tell people the hard truths if it involves higher costs and adverse political reaction, and people generally want more energy and clean water but are reluctant to pay more for them. The issue is not technological but economic and political. The good news is that we have options, but they are not cost-free, and many parts of the developing world lack the means to make the necessary investments. Developed countries often face similar constraints.

The responses to these challenges clearly imply an inevitable transition to a global energy system that, over time, will rely less and less on traditional fossil fuels and more and more on renewable energy resources which offer important advantages: stabilization of long-term energy costs, reduced market uncertainties, reduced international competition for energy resources, reduced water consumption and greenhouse gas emissions, enhanced job creation, and the ability to keep an increasing share of payments for energy supplies at home where they can be used for domestic investment. A thoughtful discussion of these issues was first put forward in 2004 by Dr. Donald Aitken in his ISES white paper (Aitken 2004). Nuclear power may also play a major role in our future energy system if safety, cost, waste storage, and nonproliferation issues can be adequately addressed.

More specifically, what can we do in the near to medium term to address the conflicts inherent in energy–water issues? Reducing leakage from water supply systems is an obvious target. There are cities around the world, especially but not exclusively in underdeveloped and developing countries, that on average lose more than 40 per cent of domestic water supply due to leaky networks. In some countries this leakage exceeds 50 per cent. Reducing these losses will be critical to meeting future fresh water demands and reducing associated energy consumption.

Fortunately, there are many other options to pursue as well, and the costs are not beyond reach. As former US President Bill Clinton pointed out at the 2011 IAC meeting: 'It would not take a lot of money in relative terms for the world to show solidarity in addressing the world's water supply and sanitation shortfall.' In fact, one recommendation of the follow-up IAC report released in 2012 is to 'increase annual investment in water supply and sanitation-related efforts by approximately US \$11 billion', a relatively modest amount. Another major recommendation is to endorse the human right to water.

There are many research opportunities as well. We can reduce steam power plant cooling requirements; reduce the energy requirements of desalination; develop less energy-intensive technology for water decontamination, treatment, and reuse; reduce water use in agriculture; understand the water requirements of emerging energy technologies (biofuels, carbon capture and sequestration, oil and gas shales, tar sands, hydrogen economy); and better understand the impact of global climate change on spatial and temporal variability of water resources.

A technology that is already important and will be of increasing importance in the future is that of desalination, the removal of salts from saline water to produce fresh water suitable for human consumption or irrigation. It is a rainfall-independent water source. Extensive use of desalination, of both brackish and sea water, will be required to meet the needs of a growing world population, both in developing and developed countries. The two most widely used desalination technologies are reverse osmosis (RO) and multi-stage flash distillation (MSF). According to the International Desalination Association Yearbook 2011–2012 there are 16,000 desalination plants operated worldwide, producing 66.4 million cubic metres per day, 40 million by RO and 23 million by MSF (IDA 2012); 40 per cent of these plants each produce more than 50,000 cubic metres per day, and a 65 per cent increase in installed capacity is expected by 2015.

Finally, we can build on lessons learned in recent years by the development community. For energy and water services to be sustainable, community members, both male and female, must participate in decisions about the design, management, and maintenance of the services. Demand for water and energy must drive our strategies, rather than supply, and solutions must be tailored to local conditions. Energy and water must also be recognized as scarce resources, with costs attached to their provision.

In his *Science* article of 28 November 2003, 'Global Freshwater Resources: Soft-Path Solutions for the Twenty-first Century', Peter Gleick wrote that 'The most cited estimate of the cost of meeting future infrastructure needs for water is $180 billion per year to 2025 for water supply, sanitation, wastewater treatment, agriculture, and environmental protection' (Gleick 2003). He proposes a different, 'soft path', approach to global water security, analogous to the soft-path approach for energy first proposed by Amory Lovins in the 1970s (Lovins 1976). The soft path assumes, correctly, that people's fundamental interest is in satisfying demands for energy-dependent or water-dependent services, and that society's focus should be not on energy or water use per se but on maximizing the services and benefits provided per unit of energy and water used. Gleick estimates that if a soft-path approach is used, the cost to improve global water security could be in the range of $10 billion to $25 billion per year for the next two decades, a much more achievable level of investment and consistent with the more recent IAC recommendation.

9.6. ENERGY FOR WATER

Energy used in delivering water is approximately 6 per cent of global energy consumption and is used in the following ways:

- Lifting groundwater. Power needed = (water flow rate) × (water density) × (head). For example, lifting water from a depth of 100 feet (30.5 metres) at a flow rate of 20 gallons (75.7 litres) per minute, and assuming an overall pump efficiency of 50 per cent, requires 1 hp (0.75 kW).

- Pumping water through pipes. Power needed = (water flow rate) × (water density) × (H+HL), where H is the lift of water from pump to outflow and HL is the effective head loss from water flow in the pipe. For example, moving water uphill 100 feet at 3 feet (30.5 metres at 0.9 metres) per second through a pipeline that is 1 mile long (1.6 kilometres) and 2 inches (5 centimetres) in diameter requires 4.8 hp (3.6 kW).

- Treating water. Average energy use for water treatment, according to southern California studies, is 652 kWh per acre-foot (AF), where 1 AF = 325,853 gallons (1,233.5 kilolitres). In many remote parts of the world, treatment must be more basic and is therefore less expensive than in developed regions.

- Desalination. Energy costs are the principal barrier to greater use of desalination. The energy required to produce a cubic metre of potable water, exclusive of energy required for pretreatment, brine disposal, and water transport, is approximately 5 kWh via RO and 25 kWh via MSF. Costs of desalinated water from large-scale plants are typically in the range 60–90 US cents per cubic metre.

9.6.1. Where is the energy to come from?

Historically, the answer has been the grid in developed regions, and either human power (treadle pumps), hydraulic ram pumps (that require no electricity), or diesel generators in remote regions. The use of diesel generators is neither inexpensive nor environmentally benign. Even in developed regions that draw power from the grid, reliability can be an issue as frequent grid outages demonstrate.

Renewable energy can play a key role in meeting this challenge, both in developing and developed countries. Solar-powered water pumping can raise clean water from depth and transport it to where it is needed, and already does so in many locations. In addition, in many parts of the developing world, two plentiful resources are brackish water and solar energy. Use of solar energy to power local desalination of brackish water can provide significant new sources of water in areas that have few if any other potable water supply options. Jordan, Israel, the Palestinian Authority, and the United States, in a joint effort, have undertaken such activity in the Middle East. These results are easily replicated in many other parts of the world (Hoffman 2007).

Similarly, solar energy can be used in remote locations to power disinfection systems for contaminated water. WaterHealth International's UV Waterworks technology, for example, uses ultraviolet radiation in a 60-watt system to kill bacteria and viruses via DNA disruption (DoE 2010). A single solar panel (or wind energy or low-head hydro) can power this system, which is capable of disinfecting 4 gallons per minute. Even less energy will be needed in the future for UV disinfection of contaminated water as UV LEDs become available, where the emitted radiation is narrowly focused on biologically active UV wavelengths.

In developed countries, WorldWater & Power Corporation has demonstrated the value of solar energy in providing water services (WorldWater 2013). The electricity crisis in California in the 1990s, which led to limited power availability for the state's billion-dollar agriculture, winery, and water utilities markets, presented an opportunity for WorldWater to provide solar-powered water-pumping systems (up to 600 hp) to these markets. Such systems keep the water flowing even when the grid is down; they also reduce costs by providing solar electricity during peak demand periods when electricity is more expensive, thus reducing peak demand charges.

Other promising renewable technologies include concentrating solar thermal power systems that heat water to drive steam turbines. Such systems can be operated jointly as providers of electricity and potable water. An ocean energy technology, the open cycle version of OTEC (Ocean Thermal Energy Conversion), which taps the temperature differences at selected locations in the ocean to generate electricity, can also produce large quantities of distilled water. In remote island locations, the water produced is likely to prove more valuable than the electricity. Other ocean technologies (wave energy, tidal, ocean current) can also be used to provide electricity and/or water.

9.7. ENERGY SUPPLY AND DEMAND

Looking at trends in energy supply and demand in more detail, we note that in the twentieth century population growth (1.8 billion in 1900 to more than 6 billion in 2000), increased average income, and increasing urbanization (13 per cent in 1900 to 48 per cent in 2000) led to a rapid rise in electrification and dramatically increased global energy demand (44 quads in 1900 to 406 quads in 2000; a quad is one quadrillion BTUs or 25.2 Mtoe). Transportation proved to be the fastest-growing consumer of energy supplies, with well over 90 per cent of transportation energy needs provided by petroleum. This pattern is continuing in the twenty-first century.

Projections by the International Energy Agency, the European Commission, the World Energy Council, the US Energy Information Administration, and others all point to the same general conclusions: there will be increased consumption of all primary energy sources over the next several decades; fossil fuels will remain dominant, accounting for most of the increase in energy use; natural gas demand will grow fastest, but oil will still be the largest individual fuel source; nuclear power will grow, but slowly; global emissions of carbon dioxide will grow more rapidly than primary energy supply; and use of renewable energy will grow rapidly but will not displace fossil fuels as the principal energy source.

Specifically, the US Department of Energy's Energy Information Administration, in its International Energy Outlook 2010 (EIA 2010), projects that, under business-as-usual, total world energy demand will rise from 510 quads in 2010 to 740 quads in 2035. Coal will provide 28 per cent of the total, petroleum 30 per cent, natural gas 22 per cent, nuclear 6.6 per cent, and renewables (solar, wind, biomass, hydropower, etc.) 13.4 per cent. Most of this growth will take place in the developing world.

These projections mask several critical questions. When will conventional world oil production peak, with attendant impacts on oil price and competition for resources? What do tar sands and fracking developments mean for future oil and natural gas supplies? How urgent is it to reduce growth in global energy demand and related emissions of carbon dioxide and other greenhouse gases? How vulnerable to disruption is our energy infrastructure, on which we depend so heavily? How quickly can renewable and advanced nuclear energy technologies be brought on line to replace fossil fuels? Conventional oil production (90 million barrels per day at the end of 2011) is currently holding its own due to major new discoveries at great depths beneath the ocean floor, but this is a finite, non-renewable resource and most analysts expect this production to peak out within the first half of the twenty-first century. It is important to note that 'peaking out' still leaves lots of oil in the ground to be pumped, and petroleum will be continue to be a major global energy source for the rest of the century and perhaps beyond. Demand for petroleum today is driven largely by its use as a transportation fuel, which is growing as automobile, truck, and aviation use grows in many countries, but it will decrease as more fuel-efficient cars, alternative fuels, and electric drive propulsion systems enter the transportation market.

Tar sands (also known as oil sands or, more technically, bituminous sands) are a type of unconventional petroleum deposit that has only recently been considered to be part of the world's oil reserves as higher oil prices and new technology enable profitable extraction and processing. They consist of loose sand or partially consolidated sandstone saturated with a dense and extremely viscous form of petroleum technically referred to as bitumen. Natural bitumen deposits are reported in many countries, with large reserves in Canada, Russia, and Kazakhstan. Total natural bitumen reserves are estimated at 250 billion barrels globally, of which 177 billion (71 per cent) are in Canada. Current Canadian production is approaching 2 million barrels per day.

Fracking, the term commonly used for induced hydraulic fracturing, is a technique using pressurized fluids to release natural gas, petroleum, and other substances for extraction from subsurface rock layers. First used in 1947, modern horizontal slickwater fracking has made the extraction of shale gas economical. It was first used commercially in 1998 in the Barnett Shale in Texas. Today it is being widely used in several shale regions in the US and its use is being explored actively in many other countries. It is also a large fossil fuel resource, and according to the International Energy Agency technically recoverable resources are estimated to be 7.3 quadrillion cubic feet for shale gas, 2.7 quadrillion cubic feet for tight gas, and 1.7 quadrillion cubic feet for coalbed methane. Current annual global consumption of natural gas is 116 trillion cubic feet.

Oil shale deposits are found in all world oil provinces, although most of them are too deep to be exploited commercially. According to the International Energy Agency's 2010 World Energy Outlook, world oil shale resources may be more than 5 trillion barrels of oil, of which more than 1 trillion barrels may be technically recoverable (IEA 2010). For comparison, the world's proven conventional oil reserves are estimated to be 1.3 trillion barrels. There are approximately 600 known oil shale deposits.

Renewable energy comes in many different forms and has significant potential for replacing our current fossil-fuel-based energy system. Most analysts consider this transition inevitable, but it will take time, as have all earlier energy system transitions. Renewable energy technologies derive mostly from the direct and indirect impacts of radiation from the sun: solar energy in the direct form of photovoltaics and concentrated solar power; wind energy, an indirect form of solar energy arising from uneven heating of the earth's surface; biomass energy derived from growth of organic materials that capture and utilize solar radiation; ocean energy systems that utilize the energy in waves, tides, and deep ocean temperature differences; and hydropower, driven by the earth's hydrological cycles. Another energy form considered renewable is geothermal energy, derived from radioactive decay heat released in the earth's core. It is potentially a very large energy resource and currently is the principal energy source in Iceland.

A major concern about widespread use of renewable energy has been the issue raised by some proponents of traditional energy sources: can renewables supply enough energy to meet most or all of our needs? This question has recently received a definitive answer with the June 2012 publication of a transparent and well-documented study by the US National Renewable Energy Laboratory entitled 'The Renewable Electricity Futures Study' (NREL 2012), which concludes that 'Renewable energy sources, accessed with commercially available technologies, could adequately supply 80 per cent of total US electricity generation in 2050 while balancing supply and demand at the hourly level.'

Nuclear power, a non-CO_2-emitting generating technology, currently supplies about 6 per cent of the world's energy and 14 per cent of the world's electricity. There are more than 400 nuclear power reactors currently in operation of 1 GWe size or greater. With the potential to supply much of the world's electricity without contributing to global climate change, nuclear power has its strong proponents. It also has critics, who point to the potential of radiation releases during nuclear accidents (e.g. Chernobyl, Three Mile Island, Fukushima), uncertainties about the safe long-term storage of nuclear wastes, the high cost of building nuclear power plants, and the potential for diversion of nuclear materials for use in nuclear weapons and so-called 'dirty bombs'. These are serious issues that are stimulating much discussion and leading to proposals for the development of smaller, modular nuclear reactors (200–400 Me in size) that may mitigate some of these problems. The long-term hope for nuclear power is fusion, the process that powers the sun, which is being researched actively in many countries. The world's oceans contain enough deuterium to supply endless amounts of fusion energy, but reproducing the sun's energy process on earth is proving to be mankind's most difficult technological challenge.

REFERENCES

Aitken, D. (2004). 'Transitioning to a Renewable Energy Future'. ISES White Paper [online], <http://www.ises.org/shortcut.nsf/to/wp> (accessed 7 August 2013).

DoE (United States Department of Energy). (2010). 'Technology Transfer and Intellectual Property Management: Success Stories—Waterhealth International Inc'. [online], <http://www.lbl.gov/tt/success_stories/articles/WHI_more_no_Mex.html> (accessed 7 August 2013).

Economist-Intelligence Unit. (2012). 'Water For All?' [online], <http://digitalresearch.eiu.com/water-for-all/> (accessed 7 August 2013).

EIA (United States Energy Information Agency). (2010). 'International Energy Outlook 2010' [online], <http://www.eia.gov/forecasts/archive/ieo10/index.html> (accessed 7 August 2013).

Gleick, P. (2003). 'Global Freshwater Resources: Soft-path Solutions for the twenty-first Century'. *Science* 302, November.

Hoffman, A. (2004). 'Water and Related Energy Issues: Gender Implications'. Workshop on Gender and Poverty Reduction: Issues and Roles of the North and South. World Renewable Energy Congress, Denver.

Hoffman, A. (2006). 'Energy Security Considerations'. Presentation to the Industrial College of the [US] Armed Forces, March.

Hoffman, A. (2007). 'Solar Powered Desalination in the Middle East and Northern Africa'. Speech delivered to Environment 2007: International Conference on Integrated Sustainable Energy Resources in Arid Regions, Abu Dhabi, January.

Hoffman, A. (2009). 'Energy Poverty and Security'. *Journal of Energy Security*, Institute for Analysis of Global Security, April.

IDA. (2012). International Desalination Association. Yearbook 2011–2012 [online], <http://www.idadesal.org/publications/ida-desalination-yearbook/> (accessed 7 August 2013).

IEA. (2010). International Energy Agency. 'World Energy Outlook 2010' [online], <http://www.iea.org/publications/freepublications/publication/name,27324,en.html> (accessed 7 August 2013).

InterAction Council. (2011). 'Former Heads of Government: World Needs Water Leadership' [online], <http://interactioncouncil.org/former-heads-government-world-needs-water-leadership> (accessed 8 August 2013).

InterAction Council. (2012). 'The Global Water Crisis: Addressing an Urgent Security Issue' [online], <http://www.interactioncouncil.org/global-water-crisis-addressing-urgent-security-issue-1> (accessed 8 August 2013).

Kingsolver, B. (2010). 'Water is Life'. *National Geographic*. April.

Lovins, A. (1976). 'Energy Strategy: The Road Not Taken?' *Foreign Affairs*, October.

NRDC (Natural Resources Defence Council). (2012). 'Global Warming' [online], <http://www.nrdc.org/globalwarming/> (accessed 7 August 2013).

NREL (National Renewable Energy Laboratory). (2012). 'Renewable Electricity Futures Study' [online], <http://www.nrel.gov/analysis/re_futures> (accessed 7 August 2013).

Sandford, R. (2011). Speech delivered to 29th Annual International Interaction Council Forum, Toronto: March 20–23. Available at: <http://opencanada.org/foreign-exchange/r-w-sandford-speech/> (accessed 7 August 2013).

WHO (2006). World Health Organization. 'Meeting the MDG Drinking Water and Sanitation Target: The Urban and Rural Challenge of the Decade' [online], <http://www.who.int/water_sanitation_health/monitoring/jmp2006/en/> (accessed 7 August 2013).

Wolf, A. (1998). 'Conflict and Cooperation Along International Waterways'. *Water Policy* 1(2), Oregon State University.

WorldWater. (2013). WorldWater and Solar Technologies, Inc. [website], <http://www.worldwater.com> (accessed 7 August 2013).

Zoellick, R. (2009). 'Securing Development'. Passing the Baton Conference, US Institute of Peace, 8 January.

Part II

Lessons Learned

10

Striving Towards Development

China's Efforts to Alleviate Energy Poverty

Wenke Han, Luo Zhihong, and Lijuan Fan

A major hallmark of energy-poor countries and regions in the world is their dependency on traditional biomass fuels. Lack of access to modern forms of energy hinders the ability of nations and regions to develop industrial activities, resulting in slow employment growth and long-term perpetuation of general poverty. As long as energy access is lacking, countries' economic and social development are constrained, and negative impacts on human health and the environment continue; energy poverty thus remains a key impediment to development throughout the world.

China is the largest developing country in the world by population, and its per capita utilization of modern energy resources is far below the world average. Before China's 'Reform and Opening Up' under Deng Xiaoping in the late 1970s, economic and social development occurred at a slow pace, and the country was troubled by energy poverty.

As Chinese social and economic development began to prosper under Deng, the central government commenced focusing its efforts on addressing energy poverty. In this period, the Chinese government introduced many policies and measures to expand and optimize the energy system to match the rapid pace of its economic development. Over time, this has resulted in a gradual decline in energy poverty among the general population and an accompanying improvement in environmental protection and economic and social welfare.

10.1. BACKGROUND OF RURAL ENERGY ACCESS IN CHINA

China has long been tortured by issues of energy poverty. At the time that the People's Republic was established in 1949, total power generation capacity and electricity generated were low at 1.85 million kW and 4.3 billion kWh, respectively. In this context, a large proportion of the population in the countryside and small urban centres were without access to electricity. Their daily energy supply was dependent on traditional biomass utilization, such as direct combustion of crop stalk and wood.

Over the following decades, thanks to government-led investments and careful planning of the power, coal, and oil industries, China's modern energy supply has greatly improved, providing rural areas and towns greater access to electricity, coal, oil, and other modern fuel sources. By 1992, there were only 28 counties, 1,453 towns, and 63,120 villages—with a total population of 120 million—left without electricity access. As these areas are mostly in the western and central remote mountainous areas, energy access is difficult due to their great distance from China's power grid.

10.1.1. Historical overview of rural electrification (1949–present)

The process of rural electrification in China can be divided into four stages.

Early beginnings (1949–1957)

Shortly after the founding of modern China, the government's work focused on urbanization and relieving food shortages; rural electrification was of secondary importance. There was no specific national institution to manage or invest in rural electrification at this time. The percentage of rural electricity consumption out of total national electricity consumption was thus low at 0.5 to 0.8 per cent of China's total power generation.

Gradual electrification under strict central planning (1958–1978)

In the 1960s, the government initiated a centralized programme of agricultural production and began to invest in the construction of new power facilities to support irrigation, water conservation, and drainage. In this period, rural electrification developed modestly, and the rural electricity consumption rate grew from 0.8 to 10 per cent. By the 1970s, however, severe power shortages in urban areas were diverting government attention away from rural electrification; with little support from the central government, local governments were left on their own to develop small local power stations (mostly small hydro).

New Great Leap Forward (1979–1998)

The late 1970s signified a turning point in the Chinese economic development model and rural electrification. Under the new leadership of Deng Xiaoping and the new policies of 'Reform and Opening Up', the economic system was transformed from a central planning system to a more market-oriented model. As a result, local governments at the county level were given greater autonomy and funding for construction projects, which included rural grid construction and small hydropower development. Furthermore, with the creation of the Township and Village Enterprises as the new economic model in the countryside, rural demand for electricity also took off in this period as economic growth spurred a

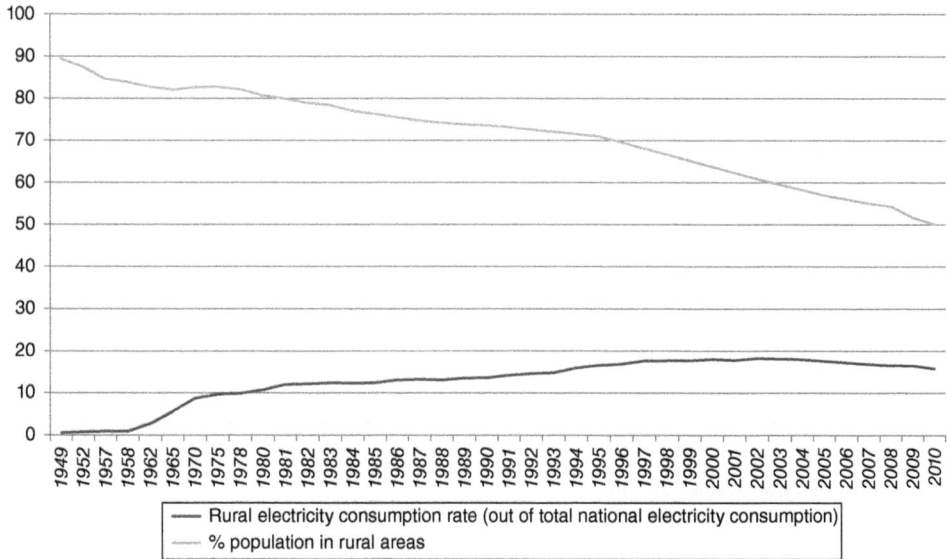

Figure 10.1. Chinese rural electricity consumption rate vs. percentage rural population

rise in income and in electricity consumption. The rural electricity consumption rate grew from 10 per cent in 1979 to 17.6 per cent in 1998 (see Figure 10.1).

Reform and upgrading (1999–present)

With the total rural household electricity access rate finally reaching over 97 per cent in 1999, the central government began to focus on tackling the major problems associated with such a rapid but haphazard development of the rural grid network. In this period, the central government implemented its rural grid renovation programme, focusing on reforming the grid management structure and tariff system, improving supply quality and grid efficiency, and upgrading the physical grid infrastructure. This grid network upgrading is still in place and functioning today.

The number of people in China without access to electricity has been reduced significantly in the twenty-first century. By the end of 2009, only 5.3 million or 0.4 per cent of the total population in China were without power. Electricity is now available in all county-level areas, 99.9 per cent at the township level, and 99.8 per cent at the village level. 99.7 per cent of all rural households now have power access, a higher percentage than in other developing countries.

10.1.2. Rural household energy consumption

Growth in household rural energy consumption has also been rapid. As shown in Table 10.1, household energy consumption in rural areas grew by 3 per cent per year on average from 1995 to 2009, to 523.58 million tons coal equivalent (tce). On

Table 10.1. Comparison of household energy consumption in rural China: 1995 and 2009

Items	1995			2009		
	Total (million tce)	Percentage (%)	Per capita (tce)	Total (million tce)	Percentage (%)	Per capita (tce)
Total consumption	382.83	100.00	0.45	523.58	100.00	0.73
Commercial energy resources[1]	130.59	34.15	0.15	256.39	48.97	0.36
Non-commercial energy resources[2]	251.05	65.57	0.29	250.63	47.87	0.35
Renewable energy[3]	0.69	0.28	N/A	16.60	3.16	0.02

[1] Coal, gas, oil products, and electric power generation. [2] Firewood and crop stalks. [3] Solar energy, small wind power, and biogas.
Source: National Statistics Bureau 2010

a per capita basis, average household energy consumption grew even more rapidly, by 4.5 per cent annually to 0.73 tce, a total increase of 64 per cent over this period.

As shown above, most of the growth in household energy consumption in rural China has been met by commercial energy resources, including coal, gas, and oil products. Total commercial energy resource consumption reached 256.39 million tce, 1.96 times that of 1995, with an average annual growth rate of 6.8 per cent, 3.8 percentage points higher than the growth rate of overall household energy consumption. Commercial energy resources as a percentage of total rural energy consumption grew from 34 per cent in 1985 to 49 per cent in 2009. Conversely, the share of non-commercial energy resources fell to 47.8 per cent from 65.5 per cent. Thus the energy resource structure in rural areas is increasingly moving towards commercial energy resources and away from traditional wood fuel and crop stalk.

Renewable energy resources are also becoming an increasingly significant part of the overall daily residential energy structure in rural areas. In 2009, total renewable energy deployment, including methanol, biogas, solar energy, and small wind power, was 16.58 million tce, 24 times 1995 levels. By then, 31.6 per cent of total rural household energy production was from renewable resources.

10.2. CHINA'S MAIN ACTIONS TO ADDRESS ENERGY POVERTY

10.2.1. Accelerating rural electrification through infrastructure investment and government subsidies

Rural grid infrastructure upgrade and expansion

The central government considers improvement of the rural grid infrastructure a major part of solving China's energy poverty issue in rural areas. As previously mentioned, since 1998 the government has carried out rural power grid network construction and upgrading, which has included overhauling the rural power management system to promote unified electricity tariffs between rural and urban areas.

Between 1998 and 2002, local and central government investments in rural power grid reconstruction totalled 288.5 billion RMB. After 2003, government projects focused on upgrading county-level grids and those in China's midwestern regions. In mid-2008, rural power grid reconstruction and power infrastructure construction received another boost in public investment: 56.24 billion RMB in total was spent, of which 13.2 billion RMB was from the central government. By 2010, Chinese investments in power grid reconstruction and rural power infrastructure upgrading totalled 462.2 billion RMB.

Overall, in the last decade, these improvements have provided energy access to about 350 million people who had previously no access to electricity. According to the National Energy Administration, the voltage qualification rate in rural areas has increased from 78 per cent to over 95 per cent, and the power supply reliability rate has improved from 87 per cent to 99 per cent—all based on efforts made since 1998.

Rural grid infrastructure upgrading and power tariff adjustments have in turn spurred rural economic development by expanding rural demand and farmers' consumption levels and by promoting employment and production. This is evidenced by the gradual growth of rural power consumption as a percentage of total national power consumption. Additionally, rural families increased consumption of modern household appliances including colour televisions, refrigerators, washing machines, and cooking appliances, which previously could not be used without reliable power access.

While these are all achievements for which the nation should take credit, this rapid pace of development brought along with it new challenges, such as ensuring that grid construction and reconstruction can keep up with the population's rapidly growing energy consumption. In light of this challenge, the Chinese central government has launched a new round of rural power grid reconstruction to be undertaken between 2011 and 2015. New engineering standards will be applied to all facilities, and the overall infrastructure will be upgraded for expanded supply capacity and power quality. Funding for projects in central and west China will come from the central government, while funding along the more prosperous eastern coast will be supplied by project developers. In addition to ensuring increased energy access, this new round of power grid reconstruction is also intended to bridge the rural–urban development gap.

'Appliances to the Countryside' subsidy programme

To complement the government's efforts to upgrade the rural power grid and expand electricity access to the countryside, the central government also introduced a subsidy programme to subsidize purchases of household appliances. In 2008, the central government launched the 'Appliances to the Countryside' programme to promote the purchase and use of appliances such as televisions, refrigerators, and washing machines among rural households. As of 15 December

2009, there had been a total of 34.3 million sets of electric appliances sold in rural areas, with a total of 6.29 billion RMB in subsidies allocated.

Prior to the programme, usage of household appliances in rural areas was low due to limited electricity access and low household incomes. In 2007, there were as many television sets per capita in rural areas as there had been in urban areas in 1996, and as many washing machines per capita as could be found in urban centres in 1985. Through direct subsidies to rural households, the programme expanded rural electricity consumption while at the same time improving living standards and raising household energy and resource efficiency. Most of the electric appliances in the programme are energy efficient and tailored to rural power and water features. The programme has not only contributed to energy poverty alleviation but has done so in a responsible and efficient manner.

10.2.2. Government projects to improve energy supply to poor and remote areas

There are many key projects led by the Chinese government to improve energy supply to the poor and remote areas without electricity access, of which four are discussed in this section.

Brightness Programme and Township Electrification Programme

The Brightness Programme was China's first large-scale government initiative to promote off-grid renewable power technology in rural areas. The programme provides off-grid solar, wind, and small hydro household and village systems to remote areas in China without access to the grid, including Gansu, Qinghai, Inner Mongolia, Tibet, and Xinjiang provinces. From 2001 to 2005, the programme invested approximately 10 billion RMB in installing 1.78 million household systems, 2,000 village systems for village electricity use, and 200 power station systems. The programme resolved power use issues for 2,000 villages with a total population of 8 million. It also provided power to 100 isolated military posts and 100 microwave communication stations.

In continuation of the efforts of the Brightness Programme to provide off-grid renewable energy access to remote areas, the central government launched the Township Electrification Programme in 2002. Totaling 4.7 billion RMB, the programme became one of the largest renewable energy programmes in the world. The government subsidized the capital cost of equipment, which led to the installation of 268 small hydropower stations and 721 solar and wind hybrid power stations in Tibet, Xinjiang, Qinghai, Gansu, Inner Mongolia, Shaanxi, and Sichuan provinces, which now provide electricity to more than 1,000 towns, 300,000 households, and a population of 1.3 million.

County-level Integrated Rural Energy Construction Programme

Another rural energy programme introduced by the central government to address energy shortages, the programme sought to promote the development and utilization of renewable energy technologies while gradually developing local energy systems and markets at the county level. Between 1991 and 1995 during the 'Eighth Five Year Plan' period, the central government invested a total of 5.85 billion RMB in 139 counties across 7,900 different projects.

Upon the projects' completion, China's newly added power supply capacity increased by 10.8 million tce, and its total power supply capacity reached 1.63 million kW. The programme also led to 11.37 million tce in energy savings while traditional biomass energy installations decreased by 13 per cent. Between 1996 and 2000 under the 'Ninth Five Year Plan', the government invested in rural energy construction projects in another 207 counties.

County hydropower construction of national rural electrification

From 1983 to 2000, the central government established in three phases 653 'primary electrification counties' which had access to electricity primarily through small hydropower installations (less than 50 MW). By 2000, China's small hydropower installation capacity had reached 220 million kW, and annual electricity supply capacity reached 70 billion kWh, 6.4 per cent of the total national electricity supply. In 2006, all 653 of these rural counties saw large GDP growth, fiscal revenue increases, and greater electricity use per capita. From 1984 to 2000, the first phase of 109 counties saw their GDP rise from 7.9 billion RMB to 148.9 billion RMB and their fiscal revenue grow from 800 million RMB to 9.6 billion RMB. From 1995 to 2000, the third phase of selected counties saw GDP increase from 234.5 billion RMB to 477.8 billion RMB and fiscal revenue rise from 15 billion RMB to 25.5 billion RMB. All 653 counties saw significant increases in their industrial production value (by 10 per cent) in only five years. Growth in local industry also brought along with it new employment for 30 million workers in rural areas. Participating counties also saw increases in appliance usage, with 90 per cent of households purchasing televisions and other appliances. Another benefit was improved access to energy for schools in mountainous, remote areas; this helped propel primary and middle school attendance levels to beyond 95 per cent.

On top of the success of the 653 rural electrification counties, the government installed another 400 hydropower systems in 400 counties between 2001 and 2005. By the end of 2005, total investment in these systems reached 115.1 billion RMB with 10.6 million kW in small hydropower capacity. Between 2006 and 2010, the government approved another 500 hydropower electrification counties, which relied primarily on small hydropower. One important secondary benefit was the project's impact on deforestation. As more than 20 million households with access to the hydropower systems replaced their traditional biomass sources such as fuelwood, the programme led to a decrease in overall deforestation by over 120 million cubic metres.

The Green Energy County Programme

Under its Medium and Long Term Renewable Energy Development Plan (2005–2020), the central government launched the Green Energy County Programme in 2010 to further expand the utilization of renewable energies across China. The programme is intended to increase the proportion of renewable energies as a percentage of the total energy mix in counties with abundant renewable resources. Also, the programme aims to promote greater awareness of energy sustainability and low-carbon living among the rural population. As of today, the first 108 'green counties' have been designated by the central government, and their budgets are in place. The criteria for participating counties is that more than 50 per cent of household energy must come from renewable energy and that various biomass residues and wastes are treated and utilized in reasonable ways. Overall, the government aims to increase the number of green counties to 500 by 2020.

Besides the aforementioned programmes, the central government is also co-operating with the European Union, World Bank, and other international organizations to address energy poverty, specifically working in the area of biomass energy utilization efficiency improvements.

10.2.3. Promoting greater utilization of renewable energies

In 2001, China launched its 'Go West' strategy, and in 2006 it implemented the plan for 'New Rural Construction'. The Go West strategy aimed to promote further economic and social development in China's underdeveloped western regions, while the New Rural Construction Plan aimed to strengthen the rural economy. Using energy as the primary basis to ensure greater social and economic development, the two programmes targeted the further construction of a modern energy system using local renewable energy resources to enhance local energy supply.

China has a rich endowment of renewable energy resources (see Table 10.2), most of which are distributed throughout rural areas.

In the implementation of the Go West strategy and the New Rural Construction plan, small hydropower, biogas, wind power, and distributed solar PV systems were widely deployed in rural areas.

China has rich small-hydropower resources, with technical availability of 128 million kW, approximately 23 per cent of the country's total hydropower resources. The central government has been attentive to small hydropower deployment, which it regards as an important source to increase rural energy supply. Efforts have been made to accelerate the progress. By the end of 2009, China had

Table 10.2. Renewable energy deployment potential

	Resources	Technical availability	100 million tce
Solar energy	2300 billion tce	Unlimited	23,000
Wind energy	3.2 billion kW	1 billion kW	8
Small hydropower	180 million kW	128 million kW	1.4
Biomass energy			
Crop stalk	700 million t	350 million t	1.7
Fuelwood		220 million t	1.3
Animal dung	1.8 billion t	20 billion m3	0.17
Trees and aquatic plants	11.2 billion t	1.5 billion t	6

Source: China Ministry of Agriculture 2011

built about 45,000 hydropower stations, with more than 55 million kW in installations, and annual electricity generation of more than 150 billion kWh, or one third of total power generation. This made China the top-ranked electricity generating nation in the world. Small hydropower deployment is present in all remote and mountainous regions and addresses the electricity needs of over 300 million people.

China's north-west and coastal regions and islands enjoy abundant wind resources, making them suitable for wind power development. Small wind turbines have proven popular and feature simple structure, straightforward operations and maintenance, and low capital costs. By 2009, approximately 250,000 small wind turbines had been erected with a cumulative capacity of 150,000 kW to generate about 270 million kWh of electricity. The basic power supply issue has been solved for remote areas and activities without grid access: watch-houses, inland lakes, isolated islands and inshore aquaculture, TV and meteorological stations, microwave communication stations, and agricultural irrigation.

Small solar photovoltaic (PV) power systems do not consume conventional energy, have no spinning components, and feature long life, not to mention easy maintenance and application. These characteristics are appropriate for power supply in mountains, grasslands, deserts, frontier areas, and islands. Since the 1980s, many small PV power systems in a range of capacities have been installed in Tibet, Qinghai, Inner Mongolia, Xinjiang, Ningxia, and Gansu provinces. By 2009 over 80,000 kW of solar capacity had been installed, serving approximately 1 million households in such areas.

Modern biomass energy applications are also an important part of China's rural energy development. These applications mainly include biogasification, briquette, and biomethanol technologies. In the last few years, China has encouraged efficient development of biomass energy applications. By 2009, China had 888 crop stalk gasification stations and 259 briquette stations. Biogas users have reached 35 million households, with annual biogas production of 12.4 billion cubic metres, and 56,000 biogas-engineering projects, which are generating 917 million cubic metres of biogas annually. In total, biomass energy amounts to 9.76 million tce.

Renewable energy resource deployment works to optimize rural energy consumption and is another means that the central government has developed to lessen energy poverty.

10.2.4. Addressing energy poverty through employment creation and coal subsidies

In addition to the previously mentioned policies and measures to address rural energy poverty, the Chinese central government also introduced two novel programmes—the 'Coal Relief' and 'Employment Relief' programmes—which indirectly aid rural living conditions through subsidized coal and employment. The Employment Relief Programme was a government initiative in which public funds were invested into infrastructure projects, which provided local employment opportunities for the poor. The Coal Relief Programme, created under the

Ministry of Finance, subsidized farmers' purchases from the coalmines while the government directly paid the coal companies. By providing subsidized coal, the government hoped to stem rampant deforestation in rural areas by replacing traditional biomass with coal, though admittedly the programme did not solve emissions issues. Affordable coal is being provided from coal-rich provinces such as Shanxi, Shaanxi, Gansu, Inner Mongolia, Ningxia, Guizhou, Xinjiang, and Yunnan.

10.3. SUMMARY OF CHINA'S EXPERIENCES IN REDUCING ENERGY POVERTY

There are five key learning points from China's efforts to eradicate energy poverty within its borders.

10.3.1. Government leadership and orientation

To address energy poverty, China has benefited from coordinated, centralized policy and implementation. The central government led a series of national key programmes and projects focusing on rural power grid reconstruction, small hydropower development, and county-level electrification in the twentieth century; in the twenty-first century, advances have been made through the 'Go West' and 'New Rural Construction' programmes. All these programmes and projects are closely related to economic and social development, environmental protection, and rural development in China. Throughout the process, the central government has set very clear targets, arranged institutional settings, publicized scientific planning, and issued well-organized policies to ensure smooth implementation and obtain close-to-expected outcomes.

10.3.2. Fiscal and tariff support

To address energy poverty issues, the central government provides significant fiscal and tariff support to related programmes and projects. In total, the central government has allocated approximately 600 billion RMB in energy-related programmes, from engineering projects and technical training to information services. Furthermore, the central government has issued preferential fiscal and tariff policies such as a favourable 6 per cent value-added tariff for small hydropower development, a much lower rate than the standard value-added tax of 17 per cent.

10.3.3. Promotion of advanced and practical technologies tailored to local situations

A range of renewable energy technologies have been adopted, such as small hydropower, biomass energy, solar energy, and small wind power. Such technologies were tailored to local and rural resources, environments, and customs in order to foster public acceptance and support from local residents. For instance, small hydropower development technologies were implemented according to local needs—dams were not built when run-of-river technology was deemed most appropriate, and vice versa. Renewable energy deployment was well integrated into existing agricultural production and rural economic development. New production models were formulated, promoting local economic development and environmental protection, and at the same time residents benefited financially.

10.3.4. Support for technological innovation and improved rural energy standards

The central government integrated available resources to organize research institutes and other enterprises engaged in rural energy research and development. Through this systemized, centralized approach, advances were made in areas such as bio-briquettes, crop stalk gasification, biogas generation, medium and large sized biogas engineering infrastructure and equipment, PV power generation, and wind energy. The central government also led the R&D transition into practical applications. By 2009, there were more than 200 renewable energy national and industrial standards under development and pending, including biomass, solar, wind, micro-hydropower, new liquid biofuels, and rural energy conservation. As the quality of rural energy technologies has risen, rural energy standardization has become an integral part of China's energy standards system.

10.3.5. Urbanization and future challenges

For most rural areas, urbanization is a key part of addressing energy poverty over the long run. Low urbanization rates mean more of the rural population is left behind in China's quest for economic growth.

Biomass consumption is a case in point. Due to the high availability and customary usage of traditional biomass as well as the lack of public acceptance of the often more expensive and less familiar new biomass fuels, it is difficult for people to change the way they use this source of energy. With the acceleration of urbanization, old attitudes concerning lifestyle and energy consumption will change. As people consume more advanced products, which require high and stable power, their choice of energy consumption will also modernize. The central government has made urbanization a priority, especially among smaller cities and rural areas. By 2009, 622 million people were considered to be urbanized. This translates to 46.6 per cent of the entire population, close to the average level in middle-income countries.

While urbanization solves the problem of energy poverty in most areas, the urbanization process and the rapid pace of economic development, which have gone hand-in-hand in China, have brought other challenges, including rapid growth of energy demand. While working to ensure that there is sufficient energy supply, the government must at the same time continue to improve China's energy consumption per capita and improve rural grid infrastructure in remote areas. Finally, efforts must be made to alleviate environmental pressures from fossil fuel consumption. These issues must all be resolved gradually over time in order to fully eliminate energy poverty. Even though unknown challenges still lie ahead, the accomplishments of the past should fuel optimism for the future.

REFERENCES AND FURTHER READING

Chen, X., Xiao, W.Z., and Liu, G. (2010). 'China Rural Energy Industry Development Review 2009', Beijing: New Energy Industry.

China Ministry of Agriculture. (2009). *Agriculture in Modern China: 60 Years*. Beijing: China Agricultural Press.

China Ministry of Agriculture (2011). *Agricultural Year Book: 2010*. Ministry of Agriculture, PRC, available at: <http://english.agri.gov.cn/service/ayb>.

China Poverty Alleviation Development and Service Center. (2011). 'Features and Poverty Criteria of China's Poor Areas'. Document no longer available online (accessed 22 May 2011).

Li, H. (2002). 'Coal Relief Program: Exploration of China's Western Ecological Development', *Development* 3.

Ming, Y. (2003). 'China's Rural Electrification and Poverty Reduction', *Energy Policy* 31(3): 283–95, available at: <http://dx.doi.org/10.1016/S0301-4215(02)00041-1> (accessed 8 August 2013).

National Statistics Bureau. (2009). *Modern China 60 Years*. Beijing: National Statistics Press.

National Statistics Bureau. (2010). *China Rural Statistical Yearbook 2010*. Beijing: China Statistics Press.

National Statistics Bureau. (2010). *China Statistical Yearbook 2010*. Beijing: China Statistics Press.

National Statistics Bureau. (2011). *China Statistical Yearbook 2011*. Beijing: China Statistics Press.

Peng, W. and Pan, J. (2006). 'Rural Electrification in China: History and Institution', *China and World Economy* 14: 71–84.

Xiao, Q. (2010). 'Launching the Rural Power Grid Reconstruction and Upgrading', *China Energy News*, 23 November.

Zhou, X., et al. (2007). *China Electricity Plan*. Beijing: China Hydropower Press.

11

Indian Approaches to Energy Access

Debajit Palit, Subhes C. Bhattacharyya, and Akanksha Chaurey

11.1. INTRODUCTION

India is predominantly a rural country—home to the largest rural population in the world—with approximately 70 per cent of the total population living in the countryside. India's economic and social development is thus inherently linked to growth in the rural economy. In order to contribute to India's overall development, the rural economy must have access to modern energy and cleaner fuel sources. Balachandra (2011) observes that India faces rural energy challenges on three fronts: the presence of large-scale rural energy poverty where the population lacks access to modern energy carriers; the need for expanding the energy system to bridge the access gap as well as to meet the requirements of a fast-growing economy; and the desire to be in league with global economies for mitigating the threat of climate change. The best possible outcome would be to achieve all the three objectives without compromising on any one.

Statistics from Census 2011 indicate that around 77 million households in India were living without electricity in 2011 (Figure 11.1). Despite the government's efforts for more than a half century to improve electricity services, household electrification levels and electricity availability is still far below the world average. Further, 836 million people do not have access to modern cooking fuel in India and depend on traditional biomass (Rehman et al. 2012). Energy poverty is exacerbated by the lack of an integrated policy framework, division of the energy sector across multiple agencies, overemphasis on serving urban customers, misdirected subsidy regimes, ineffective implementation, poor governance of the sector, resource constraints, and other structural factors (Balachandra 2011; Krishnaswamy 2010; Chaurey et al. 2004; Kemmler 2007).

This chapter examines the trends of rural energy programmes in India and attempts to capture comprehensively the development of rural electrification and energy for cooking. It starts by highlighting the rural energy situation in India and thereafter summarizes some of the key rural energy programmes, covering grid and off-grid electrification as well as energy for cooking programmes. The chapter then shares the experience of the Rural Energy Programme in India and highlights some specific challenges in enhancing access to draw lessons for cross-learning potential. Finally, the conclusion section summarizes the study from the

Households by main source of lighting (in millions)

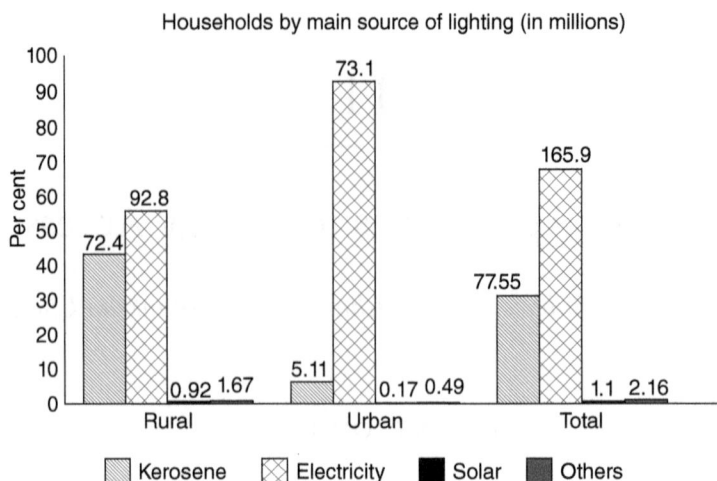

Figure 11.1. Electricity access in India: households by main source of lighting, in millions
Source: Census of India 2011

wide range of experiences and suggests means of improving the rural energy access level.

11.2. THE RURAL ENERGY SITUATION IN INDIA

Households in rural India generally need energy for three purposes: cooking, lighting, and productive uses. Despite conscious efforts of the central and provincial governments since the start of the planning process in 1951, the pace of modern energy provision in rural India has been somewhat sporadic, and past efforts in terms of both policies and programmes have achieved only marginal success (Modi 2005; Bhattacharyya 2006). Though most villages—around 94 per cent of a total of 593,732 villages—have electricity access (CEA 2013), the household electrification rate is much lower—about 67 per cent (Census of India 2011). Only nine out of 28 states have achieved more than 90 per cent household electrification, with larger states including Assam, Bihar, Madhya Pradesh, Rajasthan, Uttar Pradesh, and West Bengal lagging behind in terms of their rural electrification efforts. Specifically, more than 50 per cent of the total non-electrified households are in Uttar Pradesh (with 20 million households), Bihar (15 million) and West Bengal (9 million households). Some researchers note that structural factors may explain disparities in the share of electrified villages between regions and states (Chaurey et al. 2004; Kemmler 2007). Further, even where electricity access is available, the quality of supply remains poor and power is often not available during the evening hours when people need it the most (Palit and Chaurey 2011).

The low household electrification level in many states reflects the fact that historically the level of electrification has been measured as a percentage of electrified villages (and not as a percentage of electrified households) with grid

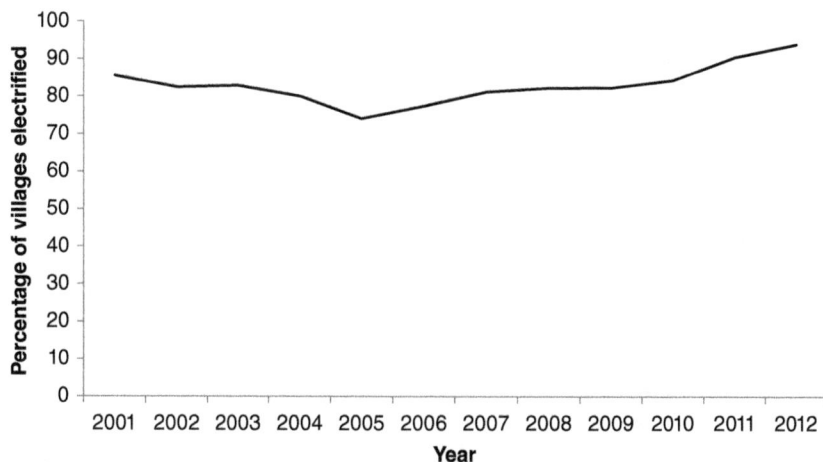

Figure 11.2. Trend of electrification in India

Source: Prayas 2011

extension to any point within the revenue boundary of a village, irrespective of whether any household is getting connected or not. In fact, some researchers argue that electrification as a part of the Green Revolution in agriculture was the main driver for rural electrification (Bhattacharyya 2006; Krishnaswamy 2010). The official 'definition' of village electrification has evolved over the years and is a significant factor in understanding electrification efforts in India. However, since 2004, the Government of India has adopted a more comprehensive definition of 'village electrification'.[1] As a result many villages that were previously considered electrified were reclassified as un-electrified, with a drop of almost 10 per cent in the village electrification rate (Figure 11.2).

However, rural energy problems are not limited to electricity access only. Lack of clean fuels for cooking seems to be the more critical issue: a large majority of the rural population uses biomass fuels in traditional cookstoves. Almost 76 per cent of the rural households still use solid fuel in inefficient cookstoves to meet their cooking requirements (Figure 11.3). Specifically, five states (Andhra Pradesh, Bihar, Madhya Pradesh, Uttar Pradesh, and West Bengal) account for nearly 50 per cent of all households using solid fuel in India (Census of India 2011). The sad part of it is that there has been almost no decrease in the consumption of solid fuels between the census of 2001 and the census of 2011. As per the latest (66th) round of the National Sample Survey Organization (NSSO), about 29.50 kg of firewood and chips (per capita per month) are consumed in rural households (NSSO 2011). Of the modern cooking fuels, liquefied petroleum gas (LPG) is used for cooking purposes in around 26 million urban households and only 7.5 million rural households, whereas kerosene is used in 12.5 million households, out of

---

[1] A village will be deemed to be electrified if: basic infrastructure such as distribution transformers and distribution lines are provided within the inhabited locality; electricity is provided to public places such as schools, panchayat offices, health centres, dispensaries, community centres and so on; and the number of households electrified is at least 10 per cent of the total number of households in the village.

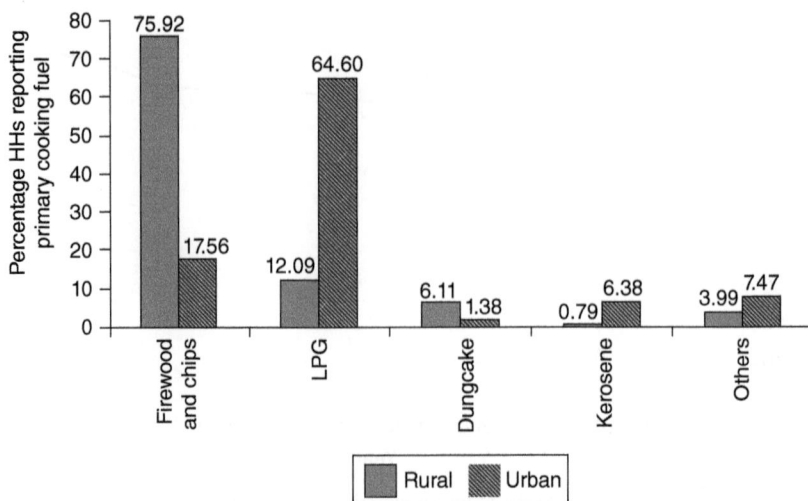

Figure 11.3. Cooking energy use by Indian households (HHs)
Source: National Sample Survey 66th Round, 2009–10

which around 10 million are urban households. A Dalberg analysis (2012) observes that Indians spend around 7 to 10 per cent of their total expenditure on cooking fuel and light, which is significantly higher than households in other developing countries. Some rural households have expanded their access to LPG stoves, especially after the introduction of the National Rural LPG Distribution Scheme in 2009, but merely having an LPG connection does not guarantee usage of or dependence on it. For instance, in a classic example of fuel 'stacking', many Indian households with LPG connections still use firewood as their primary cooking fuel, generally due to affordability issues; LPG is then used sparingly for quick cooking (Joon et al. 2009).

There is also a geographic and income-based divide in terms of energy access. Urban areas or upper-income households consume more electricity than do rural areas or lower-income households (Figure 11.4) (Pachauri 2007; Ramji et al. 2012). Even among urban and rural households with comparable incomes, the former consume more electricity. Generally, however, electricity consumption per capita increases with higher levels of income (Ramji et al. 2012). As Figure 11.5 indicates, there is a clear link between the economic development of a state and its electrification rate. States with low per capita income have performed poorly compared to those with higher income. Further, low-income groups appear to use electricity mostly for lighting, whereas elevated electricity consumption among upper-income groups can be attributed to appliances and productive use. Also, the rural–urban inequity is significantly greater for modern cooking than for electricity. For example, while the rural–urban electricity access gap is 28.22 percentage points, the rural–urban divide for modern cooking access, such as LPG, is much higher at 52.5 percentage points (NSSO 2011).

Ramji et al. (2012) also observe that the energy access situation in India, while showing clear signs of improvement in urban areas, is still a challenge in rural

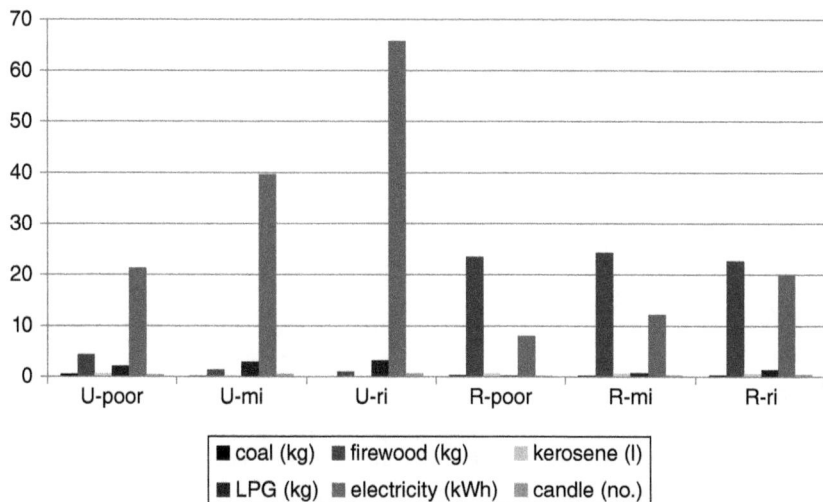

Figure 11.4. Fuel consumption pattern by income class in India

Note: U-poor = urban poor, U-mi = urban middle class, U-ri = urban rich; R-poor = rural poor, R-mi = rural middle class, R-ri = rural rich.
Source: NSSO 2012

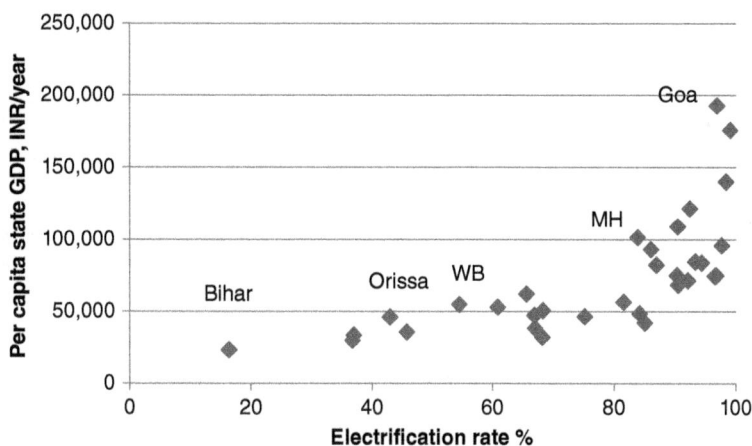

Figure 11.5. Electrification improves with per capita income
Source: Census 2011, Economic Survey 2012–13

India. In the case of kerosene and electricity, they note that in the NSSO 55th round (1999–2000) the intersection point between the graph of kerosene and electricity occurred around the 8th Monthly Per Capita Expenditure (MPCE) class; in the 61st round (2004–2005) the intersection point occurred around the 6th MPCE class; and in the 66th round (2009–2010) the intersection point occurred around the 3rd MPCE class. This means that the switch (denoted by the intersection point of kerosene and electricity in the graph) to modern lighting fuels is

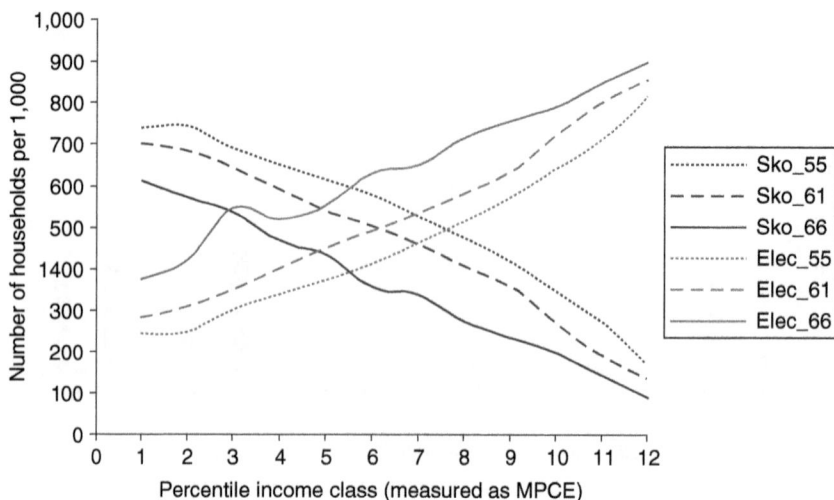

Figure 11.6. Penetration of primary lighting fuel in rural India

Note: Sko = kerosene; Elec = electricity.
Source: Ramji et al. 2012; National Sample Survey 66th Round, Government of India

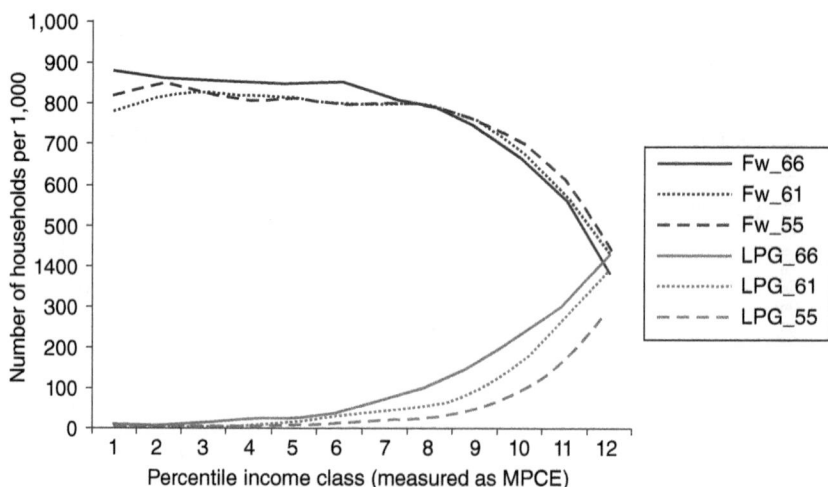

Figure 11.7. Penetration of primary cooking fuel in rural India

Note: Fw = firewood; LPG = liquefied petroleum gas.
Source: Ramji et al. 2012; National Sample Survey 66th Round, Government of India

occurring at lower income classes over time, indicating improved access to modern lighting fuels and a clear transition path (see Figure 11.6). However, the same researchers show that the intersection point between the graph for firewood and LPG occurs only at the higher MPCE classes. This indicates that the switch from firewood to LPG is occurring only among the higher income classes and thus there is no real transition in terms of access to modern cooking fuels (see Figure 11.7).

Clearly, the transition witnessed in lighting fuel usage is not replicated in the case of cooking fuels. Balachandra (2011) observes that though the Integrated Energy Policy has recommendations for expanding rural cooking energy access, these recommendations get ignored among the many so-called 'critical recommendations'. Providing access to modern cooking fuels and, more importantly, effecting a transition towards modern cooking fuels will remain a challenge in terms of energy access at the household level, as the factors governing a household's decision to use a particular fuel are very different from the case of lighting; these factors also differ from one region to another, making the goal of energy access more challenging.

11.3. RURAL ENERGY PROGRAMMES IN INDIA

In the rural electrification sector, over the years, a number of government programmes[2] (such as Kutir Jyoti, Minimum Needs Programme, Accelerated Rural Electrification Programme, etc.) have attempted to enhance access either as part of overall rural development or by specifically targeting rural electrification. However, Bhattacharyya (2006) argues that the multiplicity of programmes made funding for each of them inadequate and that programme implementation was not properly coordinated or managed. During the last decade, though, rural electrification has become a political priority, driven by the realization of its neglect over the years, with the central government creating the necessary enabling environment through the REST (Rural Electricity Supply Technology) Mission[3] in 2001, Electricity Act 2003[4], National Electrification Policy 2005[5], and Rural Electrification Policy 2006. In 2001, the government declared the objective of 'power for all' by 2012 under the REST Mission and continued it

---

[2] The Minimum Needs Programme started in the Fifth Five-year Plan period (1974–1979), which had rural electrification as one of the components. The Kutir Jyoti Programme was initiated in 1989 to provide single-point light connection to all Below Poverty Line (BPL) households. This programme provided a 100 per cent grant for the one-time cost of internal wiring and service connection charges. The Accelerated Rural Electrification Programme (2003), which was initiated to offer interest subsidy to states for rural electrification, was combined with the Kutir Jyoti Programme in February 2004 to create the Accelerated Rural Electrification of 100,000 villages and 10 million households.

[3] The REST Mission was launched for the electrification of 100,000 villages and 10 million households. REST was designed to ensure a holistic and integrated approach to providing electricity for all by 2012, by identifying and adopting technological solutions, changing the legal and institutional framework, and promoting, financing, and facilitating alternative approaches. Under the programme, electrification projects based on grid extension as well as stand-alone electrification based on distributed generation was eligible for capital subsidy.

[4] The Electricity Act 2003 made the government (both state and central) obligated to supply electricity to all rural areas. The Act further mandates the formulation of national policy on rural electrification, focusing on management of local distribution networks through local institutions. The Act also opened the door to off-grid generation to a much greater extent than before and specifies distributed generation through stand-alone energy systems as a model for rural electrification in addition to grid extension. The Act also allows for the provision of electricity to 'notified' rural areas—from generation through to distribution—with no prior need for a license.

[5] The National Electricity Policy 2005 aims at achieving a minimum consumption of one kWh per household per day by 2012.

with the Ministry of Power's launch of a large-scale electrification effort, the Rajiv
Gandhi Grameen Vidyutikaran Yojana (RGGVY) scheme, in April 2005.

11.3.1. Rajiv Gandhi Grameen Vidyutikaran Yojana

The RGGVY was launched by merging all other existing schemes of rural electrification, with the goal of electrifying all un-electrified villages/hamlets, providing
access to electricity to all households in five years, and providing 23.4 million free
connections to households below the national poverty line. The scheme attempted
to address some of the common ailments of rural electrification in the country
such as poor distribution networks, lack of maintenance, low load density with
high transmission losses, rising costs of delivery, and poor quality of power
supply. The RGGVY provides grants that cover up to 90 per cent of electrification
costs, while the remaining 10 per cent is provided through loans from the Rural
Electrification Corporation, the nodal agency for implementation of the scheme.
To ensure revenue sustainability, a novel concept of rural electrification distribution franchises was also introduced and made responsible for metering, billing,
and revenue collection (MBC) for particular territories. In some cases, input based
franchises (IBF) were also installed, where the IBF procure electricity in bulk from
the distribution utility and distribute it in their operational areas. At the same
time, decentralized distributed generation (DDG) was for the first time considered
under the mainstream power sector to electrify villages that cannot technologic-
ally or cost-effectively be electrified through grid-based modalities. Both hardware
and soft costs are supported under the scheme, with 90 per cent as grant payment
and 10 per cent to be arranged by the project developer. The DDG scheme also for
the first time attempted to address the issue of the operational sustainability of off-
grid projects through an annuitized subsidy delivery mechanism and by providing
an operational subsidy—to address the gap between the high cost of generation in
remote areas and the revenue inflow in such projects. Financial viability is also
ensured through the creation of anchor load in the villages, such as providing
energy services to the telecom tower to ensure optimum load as well as financial
return.

11.3.2. Remote Village Electrification Programme

While the Ministry of Power is the nodal ministry for extension of the electricity
grid, the Ministry of New and Renewable Energy (MNRE) has also been enhancing electricity access through decentralized renewable energy technologies such
as solar home systems, solar photovoltaic power plants, small hydropower plants,
and biomass gasification, under the Remote Village Electrification Programme
(RVEP), wherever grid extension is not feasible. The RVEP was initiated in 2001
for provision of basic lighting facilities in un-electrified census villages whether or
not these villages were likely to receive grid connectivity. The scheme was
subsequently modified to cover only those un-electrified census villages that are
not likely to receive grid connectivity. As of December 2011, the RVE programme
has reportedly covered 12,369 villages and hamlets (MNRE 2012). However,

Palit (2013) observes that the vast majority, more than 90 per cent, of the villages taken up for electrification under RVE were provided with solar home system or solar power plants. Similar to RGGVY, the central financial assistance of up to 90 per cent of the project cost is provided as a grant with specific benchmarks as applicable in respect of the technologies adopted for electrification, with the balance of project costs being met by the beneficiaries and/or the state governments.

11.3.3. Village Energy Security Programme

The Village Energy Security Programme (VESP) was conceptualized by MNRE as a step forward to the RVE programme and attempted to address the total energy need for electricity, cooking, and motive power in remote villages through use of locally available biomass. Undoubtedly the programme was ambitious, having set itself a mandate of meeting a rural community's complete demand for energy services. Appropriately for such a pioneering and unprecedented programme, the initial phase of VESP was intended to test the concept and capacity of various institutions to deliver energy to remote and inaccessible communities. However, this test phase met with very limited success and most of the test projects could not be sustained. The programme was discontinued, and no new test projects have been sanctioned since 2010 (Palit 2011).

11.3.4. Jawaharlal Nehru National Solar Mission

MNRE is also implementing the Jawarlal Nehru National Solar Mission (JNNSM), one of the eight National Missions comprising India's National Action Plan on Climate Change. On the launch of the JNNSM, all solar energy programmes promoted by MNRE were integrated under the Mission. It has the twin objectives of contributing to India's long-term energy security and its environmentally sustainable growth. The Mission also aims to incentivize the installation of 22,000 MW of on- and off-grid solar power, using both solar PV and concentrating solar power technologies by 2022, along with a large number of other solar applications such as solar lighting, heating, and water pumps. The first phase (up to 2013) focused on promoting off-grid systems to serve populations without access to commercial energy as well as on adding capacity to grid-based systems, augmenting the supply with 'clean' energy.

11.3.5. Clean cookstoves programme

In terms of the dissemination of cleaner cooking sources, MNRE has been running since the early 1990s the National Programme of Biogas Development (NPBD) and the National Programme on Improved Cookstoves (NPIC) for providing clean energy to the rural population. The NPBD was broadened and rechristened as the National Biogas and Manure Management Programme (NBMMP) in 2010. The NBMMP programme caters mainly to setting up family-size biogas plants to provide fuel for cooking purposes and organic manure to rural households. A cumulative total of 4.47 million family-type biogas plants had been set up in the country as of 31 December 2011 against an estimated potential of 12 million plants (MNRE 2012). On the other hand, MNRE discontinued the NPIC in 2002, as the programme did not achieve much success, and put the responsibility of promoting improved cookstoves on the local governance institutions. At the time of discontinuation, around 33.8 million smokeless stoves (which were not necessarily improved as far as thermal efficiency is concerned), around 27 per cent of the target, have reportedly been distributed in rural India. MNRE launched the National Biomass Cookstoves Initiative (NBCI) in December 2009 to enhance the availability of clean and efficient energy for the energy deficient and poorer sections of the biomass-using population.

11.4. EXPERIENCES FROM THE RURAL ENERGY PROGRAMME IN INDIA

11.4.1. Centralized electricity grid

Electricity is in the concurrent list of the Indian constitution and hence both state as well as central governments have jurisdiction over it. With the reforms in the electricity sector, the Central Electricity Regulatory Commission (CERC) has been established to regulate central and interstate-level power-related activities, while the State Electricity Regulatory Commissions (SERCs) work on state-level licensing, state-level electricity tariffs, and so on. In the rural electrification sector, the principal actor has traditionally been the state electricity utilities. However, in accordance with the Electricity Act 2003, the role of the central government has grown larger, as both central and state governments now hold joint responsibility for rural electrification. For example, as part of the RGGVY, state governments prepared rural electrification plans which were then coordinated between state governments, state utilities, and other agencies by REC, the nodal agency for the RGGVY. State governments were also required to ensure tying up generation from various sources and ensure the supply of electricity for at least 8 hours. Additionally, in case of supply of electricity at a tariff below what is set by the SERC, provision for subsidy to the state utilities was to be ensured by the state. Further deployment of franchisees[6] was also made obligatory to ensure revenue

---

[6] Under the Electricity Act 2003, a franchisee means a person authorized by a distribution licensee to distribute electricity on its behalf in a particular area within his area of supply. Broadly there are two types of franchise operating in India: revenue franchises (RF) and input based franchises (IBF). The role of a RF is limited to billing, revenue collection, complaint redressal, facilitating release of new service connection, and keeping vigil on the status of the distribution network in the franchised area in order to provide appropriate feedback to the utility. The franchisees are paid a fixed percentage of collections on achievement of the target. On the other hand, the IBF buys electricity from the utility at a pre-determined rate and distributes the electricity within the prescribed territory. The IBF collects revenues from the consumers based on the regulated tariff and keeps the surplus after paying the utility for the energy received at a bulk supply tariff. It also handles all commercial activities related to issue of new service connections, metering, meter reading, billing, collection, realizing bad debts, disconnection, reconnections, customer complaint handing, and so on.

sustainability. At the central level, the Ministry of Power formulated the rural electrification policies, sanctioned projects under RGGVY, and provided funds through REC. Progress of work is monitored following a three-tier process by the central Ministry, REC, and the state implementing agency.

While the achievement under RGGVY is indeed impressive, with 91 per cent of the targeted villages energized and 84 per cent of the BPL households connected to the grid (as of May 2013), the percentage of total rural households (including BPL households) who were connected to the grid during the programme period is around 56 per cent (MoP 2013). There are also questions over the quality of power supply, sustainability of infrastructure, revenue generation from the connected households, and the contribution of this initiative to rural development. One of the reasons may be because a substantial section of the unserved consumers in grid-connected areas are not taking the electricity connection—due either to household financial constraints or to the perception that electricity services (quantity and quality) are inadequate. Added to this is the ambiguity in interpreting the term 'electricity access'. As part of RGGVY, the government is extending the electricity grid with the understanding that the necessary infrastructure has been created and now the onus of taking the connection rests with individual households, where many consumers are waiting to get free connection as has been done for below-poverty-line households.

Further, while the creation of franchisees for the management of local power distribution in rural settings is reported to have introduced efficient billing and revenue collection in some places, thereby ensuring stable delivery of electricity, and has been able to generate local employment and business opportunities for the village youth (TERI 2007a; 2007b; 2010), the overall growth of franchisees (just in terms of numbers, not necessarily in quality) has not been fast enough. Franchisees cover only 38 per cent of the RGGVY villages and 19 per cent of the total villages in the country.

Despite the above developments under RGGVY, the Indian experience of grid extension to rural areas cannot be considered a grand success given the large number of non-electrified households in the country. The focus on pump energization has led the country to become self-sufficient in food production, but the system is not sustainable. This is due to the practice of subsidized and unmetered supply to agriculture that has proven costly for the utilities. The recent emphasis on electricity access by creating subsidized infrastructure has brought limited benefits because of the top-down approach used with very limited local participation, local resources integration, and inadequate focus on quality of supply. Financially weak utilities have no incentive to enhance access when there is limited prospect of recovering costs through such business expansion. Lack of integration of rural development with any electrification effort and politically motivated decision-making has adversely affected India's fight against electricity access.

11.4.2. Renewable energy mini-grids

Grid connection has been the most favoured approach to rural electrification for the majority of rural households in India, but renewable off-grid technologies have also been disseminated for areas which are either inaccessible for grid connectivity or hamlets not recognized as villages as per national census records. While grid extension has primarily followed the utility model, most of the renewable energy mini-grids, a model pioneered in India, are either structured around community-based models or run by private and social enterprises (Palit and Chaurey 2011). The model has evolved over time based on the experience of previous years; now the private sector is playing a greater role in setting up localized renewable-energy-based mini-grids to serve the population.

A majority of publicly-supported projects promoted by MNRE, such as VESP and RVE, have followed the community-based model to manage the mini-grid projects in off-grid mode. Here, the VEC plays the role of power producer, distributor, and supplier of electricity. It also collects payments from end-users and resolves disputes in the case of a disruption in power supply. The PIA sets up the energy production systems and hands over the hardware to the VEC for day-to-day operation and management. As the off-grid sector is outside the purview of the current regulatory regime, often the tariff is set by the VEC in consultation with the PIA such that it takes care of the fuel and O&M costs.

On the other hand, the private operators' models are based on the 'fee for service' model. These agencies provide electricity or lighting services on a flat-rate basis (e.g. INR 150–200 per household per month) or by metered rates in rural areas, through installation of solar DC micro-grids, biomass gasifier-based systems, or diesel generators (Palit and Chaurey 2011; Palit 2013). These agencies operate mainly in villages where there are supply constraints from the grid due to inadequate generation capacity. Recent policy development has enabled such social enterprises to accelerate local power production and the distribution of supply through renewable energy, both for enhancing access in off-grid areas and for covering un-electrified households in grid-connected areas. A major part of the project cost in these cases is borne by financing from banks, venture capital investors, and equity from the company. Wherever a subsidy is available, private companies receive it subject to the norms of the government. Aside from that, the entire cost is recovered through a retail tariff which is usually relatively high. In addition, the projects are set up in areas that are 'not so remote' that consumers do not have the ability to pay.

Additionally, O&M being a critical determinant of the success, mini-grid projects have also evolved their own mechanisms for smooth and uninterrupted plant operation. Mini-grid operation is traditionally managed by local-level operators who are trained for the job, while minor maintenance is expected from the operator, and the entire management of O&M lies with the VEC. However, WBREDA (West Bengal Renewable Energy Development Agency) and CREDA (Chhattisgarh Renewable Energy Development Agency), two state-level agencies that have been successfully operating mini-grids in their respective states, have also involved qualified technicians as third parties for local power plant O&M. Either these technicians are from the original equipment suppliers or the task is contracted to local service providers who engage trained personnel for the job. CREDA has moved a step further, developing a cluster approach for maintenance to reduce transaction costs, since their power plants are located in very remote areas in the state. In general, one cluster consists of 10 to 15 villages with each such cluster employing one technician, one assistant to the master technician, and operators and VECs for each mini-grid.

While the subsidy pattern may be comparable for similar technologies, the tariff structures for consumers under mini-grids do not follow a uniform pattern. The tariff is usually based on a flat rate ranging from USD0.6 per light point per month to USD2.5 for 100 W of connected load. This fixed tariff is much easier to administer than metered tariffs in areas of very low consumption. However, a disadvantage of the system has been overloading by some households, which puts extra pressure on the entire system. This has been observed, for example, in the case of Sunderban, where there are better economic conditions and the local people have higher levels of aspiration to use various appliances.

Nevertheless, despite concerted efforts in promoting renewable options in the country, according to Census 2011, only 0.4 per cent of the households in India rely on decentralized solar energy systems for lighting purposes. These initiatives therefore remain marginal in the overall electrification context of the country, and unless a targeted approach is taken to improve electricity access through renewable energies, the pilot studies and demonstration activities will not be sufficient to bring the desired change.

11.4.3. Energy for cooking

Electricity access programmes in India have received the required attention from policy planners, but energy access for cooking has not been a priority. The national biogas and cookstove programmes were technology-focused, with dissemination being the objective and numbers deployed as the target for measuring success (Balachandra 2010). Further, Srivastava and Rehman (2006) observe that the subsidies on LPG and kerosene have mostly benefited the middle- and high-income rural households and the urban poor, and have not been targeted to fully cover the energy-poor. In the case of biogas, programmes have been more focused on technological aspects and not on cleaner energy service provision. Policymakers have thus failed to influence a shift from biomass-based cooking in rural areas, and even the limited success achieved is confined to rich rural families (Planning Commission 2002; Srivastava and Rehman 2006). While India launched the National Biomass Cookstove Initiative (NBCI) in late 2010, incorporating lessons learned from prior policy initiatives, it has not been well organized, nor has it received anywhere near the resources that were released to RGGVY to expand electricity access.

Energy access programmes in India tend to be successful as long as the individual households accrue perceived and real benefits. Poor rural households do not consider it beneficial to shift from free biomass to commercial cooking fuels such as kerosene or LPG, even if the commercial fuels are highly subsidized. The remote rural areas provide low opportunity cost of labour and there is also minimal opportunity for cash income. Thus the savings achieved in terms of time and efforts in biomass collection because of reduced biomass use in improved cookstoves do not actually translate into benefits as the local population cannot utilize the time for other income-generating activities. Also, the health benefit of clean cooking fuels is often not valued because of lack of awareness by many poor beneficiaries. Electricity, on the other hand, is regarded as an aspirational demand which can also enable people the opportunity to earn more—cleaner cooking devices, however, are considered expenditure and thus avoided.

11.5. SPECIFIC CHALLENGES IN ENHANCING ACCESS

There are many challenges—technical, financial, regulatory, and institutional—hindering energy access in India. In spite of moderate to high village electrification rates, household connections in rural India continue to be low. Further, a large population continues to use biomass fuels for meeting their cooking energy needs. Some of the specific challenges inhibiting 'sustainable' energy access in the country are discussed here.

a) Subsidy trap: The Indian electrification programme is deeply anchored in the subsidy culture. The expansion of the system to rural areas has always been supported with grant capital which utilities may access to meet their electrification objectives. The capital subsidy level reached its peak in recent times with 90 per cent support available for grid extension as well as some off-grid options. However, the subsidy syndrome is even more visible in the case of electricity tariffs for grid-based supply, where agricultural consumers as well as most of the households in many states pay a minimum charge for their consumption. The problem is compounded by the practice of un-metered supply, especially in the bigger states of Uttar Pradesh and Bihar, which makes billing and revenue collection difficult. As the grid-based supply remains the preferred choice of consumers, mainly because of the abovementioned subsidy culture, it becomes practically impossible for any off-grid system to achieve parity with grid-based supply. The issue is so politically influenced that tariff rationalization has become a major chal-lenge for the electricity regulators, and as a consequence most of the utilities have become practically financially unviable entities. The expectation that the grid might reach one day and that the subsidized supply will be available creates undue pressure on off-grid supply. While a case for supporting the poor with infrastructure creation can be made, such a system essentially has to be selective and targeted to ensure sustainability. However, this has never been ensured in the Indian case, causing a major obstacle for sustainable electrification. The same also applies to cooking energies, where LPG sub-sidies have reached the middle-income and the richer section of the popu-lation. Moreover, while the fuel is subsidized, the cookstove is not and the poor often cannot afford the stove, thereby making the subsidy an ineffect-ive tool for the poor. Nevertheless, subsidy removal faces huge political resistance.

b) Poor linkage with the rural development agenda: The Indian approach has remained top-down, where the central government designs the programme that is implemented by state agencies or others to achieve the programme targets. However, the agenda has always failed to consider the issue of electrification in a holistic manner and has imagined it as a process of taking the infrastructure or gadgets to the rural people without appreciating the fact that energy is a derived demand that requires appropriate appliances to create the demand. Consumers cannot procure such equipment unless they have sufficient buying power, and the process has not focused on facilitating rural development that could enhance incomes and buying power.

c) Lack of local resource integration: Although India has been pursuing various policies on renewable energies and enhancing access, efforts in harnessing local resources for local-level supply have not been very effective. Large programmes have often relied on conventional technologies and fuels, whereas funding and political support have been limited for programmes based on local resources. This has created failed projects and sent the wrong signals, although the potential remains high.

d) Reliable supply: While the village electrification level has grown to around 94.4 per cent since the RGGVY was initiated, there still exist structural deficiencies in implementation and in sustaining the rural electrification sector; this impacts growth in the household electrification level and the supply of electricity to rural areas. An adequate supply of power and also quality service are important to sustain the rural electricity distribution network, but field assessment of the programme indicates that measures have not been taken in a concurrent parallel fashion to improve the quality of rural supply and service[7] (TERI 2012).

e) Financing challenge: Despite continued efforts to improve electricity access, along with deployment of different technologies and institutional and financial models, there are still hurdles that hinder the scaling-up and proliferation of renewables-based electricity generation and distribution systems in developing countries (Chaurey et al. 2004; Palit and Chaurey 2011). Jaisinghani (2011) observes that most companies active in mini-grid/off-grid distribution are not able to access sufficient capital to expand. He goes on to argue that off-grid electrification is also hampered by non-uniform technical approaches and undeveloped non-technical processes (such as tariff collection and responses to system abuse) which further hinder access to finance at the early stage of projects. In case of improved cookstoves, the prices ranges from $25 to around $70, which poor households find difficult to raise in the absence of financing mechanisms at affordable interest rates.

f) Policy barriers: In spite of the policy push over the years, it is sometimes argued that the full potential of renewable energy to enhance energy access in remote areas cannot be realized under existing conditions. For example,

---

[7] As per Rural Energy Policy, State Governments should within six months prepare and notify a Rural Electrification Plan. The goal of such plans is to provide electricity access to all households by establishing transmission systems, distribution systems, and adequate power supply. While 17 states have so far announced their plans, they have not taken the required steps to fulfil commitments made under the plans. Further, to avail themselves of capital subsidies under the RGGVY scheme, the states have to ensure a minimum of 6–8 hours of daily power supply. While the states have committed to such arrangements, in reality the states are not procuring enough power to supply in the rural areas. This is because building supply in rural areas is often deemed to be a loss-making proposition; with no entity creating generation capacity, there is often chronic supply shortage

current policy frameworks and interconnection standards do not fully allow excess generation from local mini-grid systems to be fed to the conventional grid at the lower-voltage level. Also, under the current policy and regulatory regime, decisions on tariffs for an independent mini-grid fall outside of the regulatory regime. Off-grid systems are entirely free of licensing obligations and regulatory oversight, leaving retail tariffs to be determined solely by market forces or negotiation between the electricity service provider and consumers. In most cases, the village energy committee sets the tariff in consultation with project implementation agency such that it will cover fuel and O&M costs. At the same time, the remoteness of these projects increases their capital and O&M costs and hence the cost of generation and supply. The limited ability of rural consumers to pay is an additional challenge. As a result, projects sometimes fail after a few months of operation, as has been observed in the case of VESP (Palit et al. 2013). In terms of the legal framework, the benefits of cross-subsidization are limited to the grid-supplied consumers and cannot be extended to the consumers of off-grid systems. This benefit could have helped the financial viability of mini-grid systems in remote areas where user payments are insufficient.

g) Institutional barriers: Institutional and organizational shortcomings also have a major negative effect on the operation of many projects. Cust et al. (2007) argue that even economically viable projects fail simply because the importance of appropriate organizational structure and institutional arrangement of these projects are not adequately appreciated. Past experiences show that a large number of energy access projects have seen limited success because focus has been generally on technical installation without paying sufficient attention to long-term sustainability (Kumar et al. 2009; Bhattacharyya 2006; Balachandra 2011). Similarly, for grid-based electrification projects, TERI studies indicate that transformer burnout has been common in many areas where grid was extended under RGGVY; proper load estimation is not done during the planning stage, or electrons do not flow in the distribution line as augmentation of supply is not done along with implementation of grid infrastructure. In the case of improved cookstoves, while there is a very large market for cookstoves, there are a limited number of players in the market and the majority of them are small and have yet to scale up to meet the magnitude of the problem.

h) Technical design issues: In the case of cookstoves, improved models in many cases have not been appropriate to the cooking habits or customs of local people, which has restricted sales. For example, in many cases, the user wants both fast and slow cooking capability. With many biomass stoves not having proper regulators, this is not possible. Additionally, Raman et al. (2013) observe that while forced-draft stoves can provide remarkably clean burning when used with the right fuel and adequate air supply, they require smaller wood chips. Users at large find this fuel processing tedious, an issue that restricts the adoption of these cookstoves.

11.6. KEY LESSONS

First, these Indian experiences suggest that it is important to provide an ecosystem of innovation beyond 'physical access'. The rate of success is directly dependent on the government's commitment to create an enabling environment, which includes having a clear-cut policy framework and milestones, systems for defining and enforcing appropriate technical standards, financial support mechanisms, and support for capacity building. For instance, the absence of standard specifications for biomass gasification systems or policy for fuel sourcing may have resulted in the large-scale failure of biomass-based mini-grid systems as compared to the solar PV mini-grids in India. Similarly, the discontinuation of the national cookstoves programme since the early 2000s has restricted the dissemination of cleaner stoves with almost no increase in the number of households getting access to cleaner cooking fuels. On the other hand, the Electricity Act 2003, along with the National and Rural Electrification Policy, the launch of the REST mission, and later the RGGVY, assisted in sharply increasing the electrification rate. At the same time, economic linkages, access to credit with flexible designs, bundling of smaller projects, and community-centric institutional arrangements also need to be organized appropriately to facilitate successful outcomes.

Second, the renewable energy electricity generation projects in India are mainly community-centric projects or involve NGOs and thus lack an organized delivery model compared to utility-driven, conventional grid-based projects. This can be said to be a limitation of the current institutional model, as implementation metrics and operational practices differ from organization to organization and agencies are not able to benefit from a standardized set of implementation guidelines. It is, however, also important to note that buy-in and acceptance by the community and the ability to see the benefits have helped in the success of projects. The community approach has been more successful where the project has also focused on improving the productive uses of electricity. Furthermore, examples from Sunderban and HPS show that collecting revenues is comparatively more successful where villagers receive some income, either because of their existing income-generating activities or because newly created activities arise from electrification. Also, a divided ownership model, where operation and revenue collection are kept independent of each other, generally brings better focus on generation and distribution. Thus service delivery models need to be designed and structured taking into account the uniqueness of the locality or region within which the plant is to be installed. A uniform approach may prove counterproductive and may not fulfil the stated objectives. Additionally, evidence drawn from rural electrification experiences in India reveals that an appropriate support system should be a mixture of both participatory and multi-level approaches. While local issues could be better addressed through a participatory mode of governance structure, policy, regulatory, and financing matters can be dealt with at appropriate intermediary and/or higher levels. In addition, private projects can be financially sustainable in areas with adequate income generation, while remote areas with minimal possibility of cash generation require top-down support, both technical and financial, for building up local institutions to manage and sustain projects.

Third, decentralized energy access projects' size, technology, and design are influenced by various actors and factors: at the macro level, these are the prevailing policy and implementing agencies; at the middle level, service companies are influential; and at the micro level, household socio-economic characteristics are important. A sufficient and steady fuel supply, especially through involvement of the local community, is critical for the sustainability of biomass-based interventions. Also, in the case of solar mini-grids, the storage batteries are found to be the technically weakest part of the systems. This has created additional challenges for the whole operation and sustainability of solar mini-grids: they are difficult to operate, which creates a need for the development of advanced technical understanding among the operators; furthermore, to extend battery life, they require appropriate drawing of electricity by consumers. All these challenges illustrate the close interconnections that exist between technical and non-technical matters and thus the importance of focusing on these connections to obtain viable solutions. Also, some renewable energy technologies require certain economies of scale and scope in order to work properly. One cannot promote biogas or gasifier systems in the same way as solar home systems or improved cookstoves. Depending on the proximity of the habitations, the merit of setting up a local mini-grid and a central power plant of higher capacity might be beneficial over smaller-capacity systems. In addition, off-grid renewable energy technologies and affiliated infrastructures need to be compatible with future grid synchronization to ensure that communities can still rely on them when the grid reaches the village. This serves as an enduring reminder that development and energy practitioners should remain flexible about the sizes, capacities, and configurations of the technology they intend to deploy and should maintain strict quality specifications for the systems they select.

Fourth, local stakeholder capacity building has ensured better project performance for most of the decentralized energy access projects—whether for electricity distribution or enhancing clean energy for cooking—as has been observed in the Indian experience. In addition to good technical solutions, dedication and skill and the ability to make good decisions in daily O&M are crucial for the performance of a whole energy supply system so as to best benefit community members. Merely providing training to the remote communities may not address all issues, as community members may not grasp a proper understanding of the technology because of their lack of exposure and familiarity. Continuous training for operation and adaptation may be more useful for ensuring sustained use of any technology, especially the decentralized technologies.

Last, to address the challenges of access for electricity, it is essential that energy provision and development are synchronized in such a manner that the designed delivery model is attuned to the current developmental levels (and ability to pay) of the community; simultaneously, the community needs to be prepared for a bigger and more complex delivery model. Also, maximizing the load of rural energy systems becomes a crucial factor for ensuring financial viability. Thus operational sustainability goes hand in hand with high-capacity factors, efficient systems of revenue collection, or coupling energy services with additional productive uses. Wherever measures are taken to help communities increase load through income-generating activities or acquiring modern appliances, the chances of project viability increase. This finding underscores how the provision of energy services to rural areas must be linked with the creation of livelihood opportunities and adequate capacity building.

ACKNOWLEDGEMENTS

This chapter is based primarily on research carried out on rural electrification efforts in India, conducted as part of a multi-consortium research project titled 'Decentralized off-grid electricity generation in developing countries: Business models for off-grid electricity supply', supported by the Engineering and Physical Sciences Research Council (EPSRC)/ Department for International Development (DFID) research grant (EP/G063826/2) from the Research Council United Kingdom (RCUK) Energy Programme. The chapter also draws, in part, from papers by the same authors published in *Energy for Sustainable Development* 15(3) (special issue on off-grid electrification in developing countries).

REFERENCES

Balachandra, P. (2010). 'Climate Change Mitigation as a Stimulus for Expanding Rural Energy Access in India'. Report submitted to the Belfer Center for Science and International Affairs, Harvard Kennedy School, Harvard University, Cambridge, MA.

Balachandra, P. (2011). 'Modern Energy Access to All in Rural India: An Integrated Implementation Strategy', *Energy Policy* 39: 7803–14.

Bhattacharyya, S. (2006). 'Energy Access Problem of the Poor in India: Is Rural Electrification a Remedy?' *Energy Policy* 34: 3387–97.

Census of India. (2011). Government of India.

Central Electricity Authority, Government of India. (2013). Monthly Review of Power Sector, New Delhi.

Chaurey, A., Ranganathan, M., and Mohanty, P. (2004). 'Electricity Access for Geographically Disadvantaged Rural Communities—Technology and Policy Insights', *Energy Policy* 32: 1693–705.

Cust, J., Singh, A., and Neuhoff, K. (2007). 'Rural Electrification in India: Economic and Institutional Aspects of Renewables', EPRG 0730 and CWPE 0763.

Dalberg Global Development Advisors. (2013). India Cookstoves and Fuels Market Assessment; <http://www.dalberg.com/documents/Dalberg-india-cookstove-and-fuels-market-assessment.pdf> (last accessed 8 August 2014).

Jaisinghani, N. (2011). 'Islands of Light—The Experience of Micro-grid Power Solutions', *Solar Quarterly* 3(3): 10–18.

Joon, V., Chandra, A., and Bhattacharyya, M. (2009). 'Household Energy Consumption Pattern and Socio-Cultural Dimensions Associated with it: A Case Study of Rural Haryana, India', *Biomass and Bioenergy* 33: 1509–12.

Kemmler, A. (2007). 'Factors Influencing Household Access to Electricity in India', *Energy for Sustainable Development* 11(4): 13–20.

Krishnaswamy, S. (2010). 'Shifting of Goal Posts—Rural Electrification in India: A Progress Report', New Delhi: Vasudha Foundation.

Kumar, A., et al. (2009). 'Approach for Standardization of Off-Grid Electrification Projects', *Renewable Sustainable Energy Review* 13: 1946–56.

Ministry of Power, Government of India. (2013). 'Rajiv Gandhi Grameen Vidyutikaran Yojana—at a glance' [online], <http://rggvy.gov.in/rggvy/rggvyportal/index.html> (accessed 31 May 2013).

MNRE (Ministry of New and Renewable Energy). (2012). *Renewable Energy for Rural Applications: Annual Report 2011–2012*, New Delhi: MNRE.

Modi, V. (2005). 'Improving Electricity Services in Rural India' [online], <http://web.iitd.ac.in/~pmvs/rdl722/RuralEnergy_India.pdf> (accessed 4 July 2013).

NSSO (National Sample Survey Organization). (2011). *Level and Pattern of Consumer Expenditure, 2009–10* New Delhi: Ministry of Statistics and Programme Implementation, Government of India.

NSSO. (2012). NSS 67th Round (July 2010–June 2011) (Uniform and Mixed Reference Period, Unit Level Data, Household Consumer Expenditure Round). Data on CD, National Sample Survey Organization, Ministry of Statistics and Programme Implementation, Government of India.

Pachauri, S. (2007). 'Global Development and Energy Inequality Options', *Options* Winter 2007. International Institute for Applied Systems Analysis [website], <http://www.iiasa.ac.at/Admin/INF/OPT/Winter07/opt-07wint.pdf> (Accessed November 20, 2011).

Palit, D. (2011). 'Performance Assessment of Biomass Gasifier-Based Power Generation Systems Implemented under Village Energy Security Program in India'. Proceedings of the International Conference on Advances in Energy Research, Indian Institute of Bombay, Mumbai.

Palit, D. (2013). 'Solar Energy Programs for Rural Electrification: Experiences and Lessons from South Asia', *Energy for Sustainable Development* [online journal], <http://www.sciencedirect.com/science/article/pii/S0973082613000045> (accessed 4 July 2013).

Palit, D. and Chaurey, A. (2011). 'Off-Grid Rural Electrification Experiences from South Asia', in S. Bhattacharyya, ed., *Rural Electrification Through Decentralised Off-Grid Systems in Developing Countries*, London: Springer-Verlag.

Palit, D. et al. (2013). 'The Trials and Tribulations of the Village Energy Security Programme (VESP) in India', *Energy Policy* 57: 107–17.

Planning Commission, Government of India. (2002). 'Evaluation Study on National Project on Biogas Development' [online], <http://planningcommission.nic.in/reports/peoreport/peoevalu/peo_npbd.pdf> (accessed 4 July 2013).

Prayas Energy Group. (2011). 'Rajiv Ghandi Rural Electrification Programme: Urgent Need for Mid-course Correction', discussion paper, July. Prayas Energy Group.

Raman, P., et al. (2013). 'Performance Evaluation of Three Types of Forced Draft Cook Stoves Using Fuel Wood and Coconut Shells', *Biomass and Bioenergy* 49: 333–40.

Ramji, A., et al. (2012). *Rural Energy Access and Inequalities: An Analysis of NSS Data from 1999–00 to 2009–10*. New Delhi: The Energy and Resources Institute.

Rehman, I.H., et al. (2012). 'Understanding the Political Economy and Key Drivers of Energy Access in Addressing National Energy Access Priorities and Policies', *Energy Policy* 47: 27–37.

Srivastava, L. and Rehman, I. (2006). 'Energy for Sustainable Development in India: Linkages and Strategic Direction', *Energy Policy* 34: 643–54.

TERI (The Energy and Resources Institute). (2007a). *Evaluation of Franchise System in Selected Districts of Uttar Pradesh, Uttaranchal, and Karnataka*. New Delhi: The Energy and Resources Institute.

TERI. (2007b). *Evaluation of Franchise System in Selected Districts of Assam, Karnataka, and Madhya Pradesh*. New Delhi: The Energy and Resources Institute.

TERI. (2010). *Analysis of Rural Electrification Strategy with Special Focus on the Franchise System in the States of Andhra Pradesh, Karnataka, and Orissa*. New Delhi: The Energy and Resources Institute.

TERI. (2012). *Executive Summary of RGGVY Evaluation*, Project Code 2011ER01. New Delhi: The Energy and Resources Institute.

12

Modern Energy Services to Low-Income Households in Brazil

Lessons Learned and Challenges Ahead

Gilberto M. Jannuzzi and José Goldemberg

12.1. INTRODUCTION

The final document of the Rio+20 UN Conference recognized once more the role of energy in the process of poverty eradication, improving the quality of life, and satisfying basic human needs—all of these being critical towards achieving sustainable development (United Nations 2012).

Universal access to electricity and liquefied petroleum gas (LPG) in Brazil has been pursued over several decades and under different governments. The strong urbanization movement in Brazil somewhat facilitated the connection of more households to the electricity grid and access to LPG distribution services; nevertheless, even the rural areas showed remarkable improvements. In 1960 less than 40 per cent of households had access to electricity and only 18 per cent used LPG; in 2010 these percentages had been raised to 98.7 per cent and 98 per cent respectively. The country has therefore provided access to modern cooking fuels and electricity to practically every household spread across a large and diverse territory. This chapter reviews the main policies that shaped this successful development.

It is interesting to note that LPG access expanded more rapidly than electricity in Brazil, which may have been the result of the different business model adopted and the less expensive infrastructure needed. This is a different situation compared to countries that have more recently promoted increased access to modern fuels for cooking and electricity, as has been the case in several developing countries such as Angola, Sri Lanka, and Vietnam (IEA et al. 2010). Apparently governments and international financing agencies give higher priority to electrification when compared to access to modern cooking fuels (Jannuzzi and Goldemberg 2012).

The Brazilian experience shows also that access to modern energy services is not sufficient to promote the required economic and social development, as has been argued by researchers regarding the Brazilian context (Pereira, Freitas, and Silva 2011; Obermaier et al. 2012). This has also been the case reported in other countries (Sokona, Mulugetta, and Gujba 2012). The role of energy in poverty eradication is increasingly being understood as a critical requirement for productive processes and as a way to generate income—not simply as a means to satisfy basic needs (Srivastava et al. 2012; Bhattacharyya 2012).

Access to modern fuels and electricity has improved indeed, but in spite of this, poverty has persisted in many areas of the country. This fact points towards the need for new and more innovative approaches for the role of energy in socio-economic development considering the evolution of technologies available today—especially regarding electricity generation and end-use technologies. These technologies allow, for example, distributed generation to be connected to conventional grids, new ways to bill customers, and the introduction of smart appliances and smart metering, among many other applications (Welsch et al. 2013). Together with new tools and practices supported by new business models and regulatory networks, these energy technologies can potentially place the low-income consumer in a very different position. They allow for an approach to energy services that may go far beyond the usual concept of providing energy for basic needs and consumerism, permitting these households to use energy services to generate income and engage in productive activities.

Analysing past efforts, we distinguish three components in Brazilian public policy to address the energy needs of the low-income population: access to modern energy services (electricity and LPG), affordability issues focusing on tariff subsidies, and, more recently, electricity affordability based on end-use efficiency improvements. The two first components are supply-side goals that aim to connect and help maintain customers with energy services. These were in fact the main components of public policies until recently. The third component is essentially a demand-side approach, and it has been more specifically oriented towards electricity services. However, the main motivation may not have been to reduce the customer energy bill and hence to improve affordability, but rather to avoid the high commercial losses to the utilities in low-income areas due to power theft.

We discuss in the next sections the evolution of access to LPG and electricity services during the last decades, describing the main policies that helped to expand these services. We discuss the concept of affordability for both energy services and the instruments used to subsidize the low-income population in particular. As discussed by other authors (Winkler et al. 2011; Welle-Strand et al. 2012; Obermaier et al. 2012; Bambawale and Sovacool 2012), energy affordability is an essential part of providing economic development and alleviating poverty.

The rationale for investing in end-use efficiency in low-income households is a more recent component (since 2005), and although it may contribute to lowering current household energy bills, hence favouring affordability, it also open new opportunities for income generation and integration of new services without marginalizing poor households.

The final section of the chapter presents new challenges and the need to move beyond the concept of supplying energy for basic needs and consumerism in favour of integrating energy services to generate income and productive activities leading to new and renewable technologies.

12.2. COOKING ENERGY: THE CREATION OF A MARKET FOR LPG, ACCESS AND AFFORDABILITY POLICIES

In the 1960s, only 18 per cent of Brazilian households had access to LPG (or gas sold by city providers in a smaller proportion), and this was highly concentrated in the urban areas of the country. Over the next decades, the penetration of this cooking fuel reached almost every household in the country; today an estimated 98 per cent of total households use LPG (94 per cent in the rural areas), as indicated in Table 12.1.

This successful introduction of a cleaner fuel replacing fuelwood and charcoal was the result of policies conceived in the 1960s that had the following objectives:

1. Creation of a national infrastructure for production and distribution of LPG.
2. Creation of a retail market with the participation of private entrepreneurs.
3. Ensuring affordable prices to the final consumer (subsidies).

Unlike many other developing countries, there was no special effort made in Brazil to improve biomass cookstoves. The strategy at that time recognized LPG for cooking as a public good and therefore it was necessary to ensure security of supply and means to reach households at affordable prices. The state oil company PETROBRAS was mobilized to produce (and import when needed) LPG and distribute it to private companies and retailers, which were in charge of the final commercialization to end-use consumers.

The government administered and controlled LPG consumer prices, which were uniform throughout the country, and guaranteed profit margins to distri-buters and retailers by subsidizing production costs—that is controlling the ex-factory LPG production cost from PETROBRAS (see Table 12.2). In fact, during most of the period 1980–2000, real LPG prices dropped as inflation was high and LPG was an administered price (Brown 2011; Jannuzzi and Sanga 2004).

LPG pricing to final consumers provides an example of the concept of energy affordability: energy for cooking is a basic need and as a public good should be

Table 12.1. Evolution of access to modern energy services, 1960 and 2010

Total HH (millions)	1960 Number HH	average annual % growth rate (1960–2010) % per year	2010 Number HH
total	13.5	2.8	57.3
urban	6.3	4.2	49.2
rural	7.2	0.2	8.1
Electricity access	HH with access		HH with access
total	38.5%	12.7	98.7%
urban	72.5%	12.6	99.7%
rural	8.4%	13.4	92.6%
LPG (and natural gas)	HH with access		HH with access
total	18.0%	17.0	98%
urban	35.1%	16.8	100%
rural	0.3%	19.5	94%

Note: HH = household.
Sources: IBGE 2012a; ANEEL 2005; IBGE 2012b

Table 12.2. Price structure for a 13 kg bottle of LPG before and after price liberalization (2001–2002) in R$

Price components	December 2001	January 2002	
		Average	Low-income class
Production costs	9.00	6.67	6.67
Levies (federal and state taxes)	3.76	3.36	3.36
Distribution and profit margins	13.02	13.71	13.71
Subsidy	−3.47 (PPE)	0	−7.50 (gas voucher)
Final retail price	22.30	23.74	16.24
Subsidy (as % of production costs and levies)	27%	0	75%

Note: PPE was a fund from collected taxes on other fuels.
Source: Jannuzzi and Sanga 2004

affordable and equitably distributed among all households, irrespective of their income and location.

Although access has become widespread since the 1970s, it is likely that among poorer Brazilian households several other fuels have been used during this period to meet demand. The national established price was still expensive, as some local studies indicated (Jannuzzi 1988). Even today, fuelwood and charcoal are used to complement the energy needs of poorer and rural households, but the consumption of solid biomass fuels in Brazil is marginal, and most of it is collected without financial costs to the user (Uhlig 2008; Brown 2011).

Initially, the government created regional franchises for LPG distribution and allowed private entrepreneurs to participate (similar to regional concessions) with exclusivity. Later on, commercialization quotas were given, which introduced some competition. With greater competition at the retail level, services were improved and the retailers' brand names became important. Several companies sought quality certificates. A delivery system was implemented in practically all of the urban areas such that the retail company would visit its customer at regular intervals (Lucon, Coelho, and Goldemberg 2004).

Since its very beginning, the engagement of the private sector was important as it helped to shape a business model for this service and thus developed a market for LPG. The role of government was critical—regulating final prices to consumers and attracting private entrepreneurs for commercialization and distribution in such a large and diverse territory. LPG was subsidized for more than four decades in Brazil by means of a cross-subsidy scheme from funds collected from sales of various petroleum fuels,[1] and it is likely that PETROBRAS also assumed some of the production costs during the period, especially during the oil price hikes of the 1970s. LPG's final cost was the same in all parts of the country. At the same time, the government guaranteed a profit margin to the private sector.

In May 2001, end-user prices for LPG were liberalized as a part of a gradual process of deregulating the petroleum sub-sector. As LPG prices were deregulated and collective subsidies to all customers were eliminated, the federal government

---

[1] During the period 1973–2001 it was estimated that the annual per capita LPG subsidy was US $0.73 (in 2001 USD) (Jannuzzi and Sanga 2004).

started implementing a policy to assist only low-income families to purchase LPG through a voucher given on a monthly basis. Unlike under the previous subsidy system, which subsidized all LPG users, the new programme benefits were available only to families with a monthly per capita income no more than half the minimum-wage income. The share of subsidies increased to 75 per cent of the final price (see Table 12.2) but was limited only to those families registered in federal social programmes. An estimated 8.5 million households received the LPG voucher that allowed them to purchase a 13 kg LPG bottle, sufficient to meet cooking energy demand for one month.

However, this scheme resulted in an immediate increase of 17 per cent in the average retail price and, as a consequence, a decrease of 5.3 per cent in household LPG consumption in 2002 equivalent to 9.3 PJ (MME, 2013). It was only in 2004 that LPG consumption started to increase, as the government included the gas voucher in a wider social programme.[2]

12.3. ELECTRICITY ACCESS

Electricity programmes, both for rural and for low-income urban areas, have been introduced continually since the 1960s despite limited resources (Goldemberg, Rovere, and Coelho 2004; Winkler et al. 2011; Obermaier et al. 2012). The approach, as was the case with LPG, has included two components: *access* and *affordability*. In this section, we discuss the evolution of electricity access.

Access policies included waiving connection fees for low-income and rural households, either to the main grid or to stand-alone systems. Funding to promote access (connection fees) usually came from the public sector (federal or state agencies) and later from taxes included in the electricity tariff to all customers. This is still the practice, as current regulation maintains special conditions for low-income households such as free connection charges, discounted tariffs, and special provisions for disconnect/reconnect and debt negotiations (ANEEL 2010).

Incentives for rural electrification started in the 1960s using a cooperative system. Rural cooperatives were considered another type of customer, and a profitable one for utilities to connect. These cooperatives had the purpose of creating enough aggregate demand by combining the individual electricity requirements of its members. Once a cooperative was connected to the grid, they were in charge of connecting and re-selling electricity to its associates.

As a very urbanized country, Brazil also experienced the challenge to provide regular access to electricity services for low-income households within urban and peri-urban areas. Although these areas were frequently close to the grid, due to the poor housing settlements utilities had to be forced by governmental programmes and specific regulation to connect these households. Illegal connections and power theft were common for decades in these areas, and in some places these practices still represents a challenge. At the end of the 1970s some utilities and

---

[2] The 'Bolsa-Família' programme ('Bolsa Família' 2013).

state governments started to promote the electrification of these areas, particularly in São Paulo and Rio de Janeiro (Boa Nova and Goldemberg 1999).

In 1994 a large federal programme called PRODEEM was launched to promote access to electricity to schools, health centres, and other community installations, but it excluded individual rural households. Essentially donor agencies and the federal government funded this programme, which consisted of the installation of solar PV stand-alone systems free of charge to end-users. Another programme called Luz no Campo was launched some four years later to provide connection to the electricity grid for rural households mainly in the north-east region of the country. Both programmes presented several difficulties, were heavily criticized, and were revealed as too expensive. They achieved modest results (Obermaier et al. 2012; Goldemberg, Rovere, and Coelho 2004).

In 2001, a national law was approved by the congress that determined regional targets for universal access (Law 10.438/2002) for each utility. The utilities were divided into five groups, considering their electrification rates at the time together with an appraisal of their technical and economic capacity to meet full universalization in their concession areas. Utilities with already high electrification (above 96 per cent) rates were given less time to meet full universalization (two years). Those few utilities with less than 80 per cent electrification rate were given 12 years. These targets were set in 2003 by the regulator, which was in charge of the supervision of utilities' universal access plans (ANEEL 2003).

More recently, the 'Luz para Todos' (Light for All) programme was launched by the Federal Government in November 2003. It had the objective of connecting the remaining households without access to electricity—mainly in rural areas. This was estimated to be over 10 million people by 2008 (or 2 million households, 90 per cent of them located in rural areas) at that time (Fugimoto 2005). By 2009, according the governmental sources, the targeted number of households had been met at a total cost of about USD10 billion (MME 2013).

The 'Luz para Todos' also helped utilities to comply with their assigned universalization targets, established in 2003. Several approaches and technologies were used to connect remote customers, from solar PV stand-alone systems to mini-grids. New regulation was also established to guarantee similar service standards for remote consumers and those using conventional connections to the grid. For example, the final tariff paid by the customer is the same whether they are using a stand-alone system or a regular connection. Interestingly, customers with stand-alone systems were not satisfied with the smaller PV kits that could deliver up to 13 kWh/month, which was just enough for lighting and small appliances. They wanted systems that could also at least power a refrigerator (Varella, Gomes, and Jannuzzi 2009).

12.4. AFFORDABILITY: SUBSIDY SCHEMES AND THE LOW-INCOME SOCIAL TARIFF FOR ELECTRICITY

12.4.1. Subsidies and special tariff to low-income households

Affordability, as mentioned, has been another component of public policies to ensure continuing access to electricity services in Brazil.

Table 12.3. Scheme of discounts given to residential consumers

Consumption (kWh/month)	Level of discount
0–30	65%
31–100	40%
100–regional limit	10%
Above regional limit	0%

Source: Fugimoto 2005

There was an initial understanding that a lower 'social tariff' should benefit all residential consumers up to the level of their 'basic needs', as was the case with LPG. However, unlike cooking energy requirements, there was not a consensus on the amount of electricity required to meet basic needs. In 1986 a minimum limit was set to be 80 kWh/month; later this was changed to 30 kWh/month. All customers in the country benefited from these discounts until 1995, irrespective of their geographical location or income.

The general principle behind this subsidy was that tariffs were cross-subsidized by a scheme whereby consumers with higher consumption levels would pay higher tariffs, and discounts were given to lower monthly consumption levels, as is shown in Table 12.3. These discounts were given up to a certain consumption level set regionally. The idea was to protect poor households of poorer regions whose consumption levels were also low.

For some utilities, the cross-subsidy scheme has not been enough to cover the subsidies given to the social tariff and in these cases the federal government has complemented the costs to the utilities.

From 1995 onwards, the understanding was that these discounts should be given only to low-income households and efforts were made to define more precisely the category of residential households that would be entitled to the lower tariffs. Most utilities opted to continue to use consumption levels as a convenient criteria, but others included appliance ownership surveys and socio-economic indicators. Some studies suggested setting 4 per cent of household disposable income[3] as the upper limit for their electricity bill (Fugimoto, 2005). The criteria were not uniform across the several utilities and presented distortions and clear indications that subsidies were not targeting the desired population (Fugimoto 2005).

Several approaches to address affordability concerns have been implemented over time considering rural and urban households. This indicates that it is not a clear-cut concept, as also can be concluded from other studies analysing this issue in developing and industrialized countries (Pachauri et al. 2004; Bambawale and Sovacool 2012; Bouzarovski, Petrova, and Sarlamanov 2012; Winkler et al. 2011).

Since 2002, several modifications have been introduced in the way low-income families have been classified and entitled to special tariffs (social tariff), but the

---

[3] The practice in Europe is increasingly to adopt a 10 per cent affordability threshold as defining 'energy poverty' (House of Commons Energy and Climate Change Committee 2010), another example that illustrates that affordability is not an absolute concept.

Table 12.4. Amount of subsidies for low-income consumers, by region and total

Region	2004	2005	2006
South	13%	12%	12%
South-east	27%	27%	26%
North-east	43%	44%	43%
North	4%	4%	4%
Centre-West	13%	14%	15%
Total (R$ million)	1.126	1.307	1.400

Note: Estimated to December 2006 from partial results, up to June 2006 (R$708 millions).
Source: Jannuzzi 2007

important fact is that there has been a continuous public policy to provide lower tariffs to these households. At this time, a more uniform concept of affordability was introduced and a target audience was defined for all utilities. Since then, only households enrolled in the federal social programme 'Bolsa Família' are entitled to the 'social tariff', which is roughly half the regular residential tariff (Fugimoto 2005).

The Low-Income Social Tariff[4] is a benefit created by the Federal Government in 2002, which concedes energy tariff discounts to low-income families. The consumers that can be granted the social tariff should fulfil the following requirements:

- All households supplied with monophase power supply whose average monthly consumption ranges between 0 and 80 kWh based on the previous 12 months, without exceeding 220 kWh more than one time within this period.

- All households supplied with monophase power supply whose average monthly consumption ranges between 80–220 kWh based on the previous 12 months; these households must also be registered in the National Unified Register for Social Programmes of the Central Government.

There are almost 18 million consumers classified as 'low-income' in Brazil, of which 43 per cent are concentrated in the north-east region, followed by the south-east (36 per cent).

The available data shows an annual rising trend in paid subsidies (Table 12.4). The largest part of the subsidies are provided in the north-east region, followed by the south-east.

12.4.2. Low-income energy efficiency programmes

The third component of public policies that have targeted low-income households and energy affordability is related to improvements in end-use efficiency—reductions in electricity consumption and consequential reduction in the impact

---

[4] It was created by Law n. 10.438, as of 26 April 2002, and by the Resolutions n. 246 (30 April 2002) and n. 485 (29 August 2002).

of energy bills. It should be noted that there is a more recent, but increasing, perception of the role of end-use energy efficiency in energy poverty alleviation efforts: because end-use efficiency helps low-income customers to reduce their bills, governments can reduce the amount of subsidies in tariffs (Birner and Martinot 2005; Sathaye et al. 1994) and utilities can minimize their commercial losses and theft (Kelman 2010). Energy efficiency also enhances energy security (Bambawale and Sovacool 2012) and provides other benefits to the consumer, including improved health through a less-polluted indoor environment (Smith and Haigler 2008; Smith, Rogers, and Cowlin 2005; Boardman 2010).

In Brazil, efforts to secure funds and activities in energy efficiency started in a more systematic way in 1998 when the regulator ANEEL established mandatory investments in energy efficiency programmes, initially only to privatized utilities and later to all. A percentage of the net annual revenues (0.05 per cent) of each utility is dedicated to fund end-use energy-efficiency programmes, the PEE (Programa de Eficiência Energética—the national Energy Efficiency Programme). This programme is implemented by each utility and supervised by ANEEL. Since 2005, a minimum of 50 per cent of those investments for each utility must be allocated to low-income energy efficiency programmes, and in 2010 this was raised to 60 per cent. About USD80–100 million annually has been invested in energy-efficiency programmes to low-income households. Only households enrolled in the federal social programme 'Bolsa Família' are entitled to participate in these energy efficiency programmes.

These programmes have introduced modern and more efficient technologies to households living in slums, including remote digital metering and real time demand monitoring of households, more efficient transformers, new cabling systems, and other high-quality materials (Kelman 2010). Refrigerator and lamp replacement programmes, and more recently solar water heating systems, have been the preferred end-use energy efficiency programmes targeting these populations. Initially these programmes were reasonably cost-effective, especially because the existing equipment was obsolete and inefficient, but in many areas these programmes are now presenting increasing costs. Until recently, they consisted mostly of give-away programmes, free of any cost to participants; they had little impact in transforming local markets and improving final consumer behaviour. More importantly, lack of payment and illegal connections are still recurrent problems, even amongst households participating in the programmes.

12.5. BEYOND ACCESS AND AFFORDABILITY: INCOME GENERATION AND PRODUCTIVE USES OF ENERGY

Although Brazilian programmes have indeed promoted energy access, improved energy affordability, and reduced energy consumption among the targeted households (Jannuzzi 2010), they have not been sufficient to prevent electricity theft and lack of payment. Lower tariffs (social tariff) and investments to improve efficiency have not been enough to solve completely the affordability issues. It has been clear from the Brazilian experience that the approach taken so far seems to provide limited impact (Obermaier et al. 2012). In particular, end-use energy efficiency programmes need to be coupled with other social and housing programmes.

It is important to note that that are still good opportunities to enhance the role and impact of the PEE, for example. The regulator has revised periodically the set of rules under which every utility designs and implements their programmes; there is now also the possibility of combining low-income efficiency programmes with distributed generation projects. This is possible due to the introduction of net metering of regulation in early 2012, which allows consumers connected in low-tension grids to install micro-generation systems—for example photovoltaic panels—and sell excess electricity to the grid. This is an opportunity to generate income in these areas while at the same time fostering entrepreneurship via cooperatives, for example, among these households.

Welsch et al. (2013) explore in great detail the possibilities of smart-grid and renewable energy technologies as means to connect low-income households into wider electricity systems. Although they focus their discussion on options to improve access to energy services in Sub-Saharan Africa, most of the suggested alternatives are valid for more urbanized countries, such as Brazil. These technologies include advanced two-way communication, automation and control technologies, solar and wind generation, and storage systems. New billing systems can be envisaged using telecom services integrated with energy technologies; electricity consumption information can be provided via mobile phones, including payments. Remote metering can reduce the costs of conventional readings and also prevent energy theft (Kelman 2010), thus reducing overall administrative costs, which will help lower final tariffs. The introduction of smart appliances and time-of-use tariffs could help low-cost access to electricity during off-peak hours.

New technologies can create flexible tariffs and payment schemes, thus still helping affordability of energy services and integrate smart (and more efficient) appliances. Onsite generation using renewable resources can be integrated into the grid, reducing technical and commercial losses currently common in Brazilian urban and peri-urban low-income areas, and at the same time increasing the potential to generate local income.

Some Brazilian utilities are incorporating into the mandatory low-income energy efficiency programmes smart-metering technologies, new cabling and connection systems, and the replacement of inefficient appliances (Kelman 2010). Their focus has been to reduce power theft as well as metering and billing costs. One utility started an innovative programme in 2007 whereby customers can trade recyclable materials (paper, glass and metals) as part of their electricity bill and all these transactions are mediated by the customers' mobile phone, a digital card, and the utility billing system (Portal Coelce—Ecoelce 2012). These schemes also address affordability issues and at the same time can create new productive opportunities.

12.6. CONCLUSIONS

The Brazilian experience shows that public policies that have addressed energy access and affordability, albeit successful, are not, and cannot be, a substitute for

Level 3
Modern society needs

Level 2
Productive uses

Level 1
Basic human needs

Electricity for lighting,
health, education,
communication, and
community services (50–100
kWh per person per year)

**Modern fuels and
technologies for cooking
and heating** (50–100 kgoe of
modern fuel or improved
biomass cookstove)

**Elctricity, modern fuels
and other energy services**
to improve productivity
e.g.

- Agriculture: water
  pumping for irrigation,
  fertilizer, mechanized
  tilling
- Commercial: agricultural
  processing, cottage
  industry
- Transport: fuel

Modern energy services
for many more domestic
appliances, increased
requirements for cooling
and heating (space and
water), private
transportation
(electricity usage is
around 2000 kWh per
person per year)

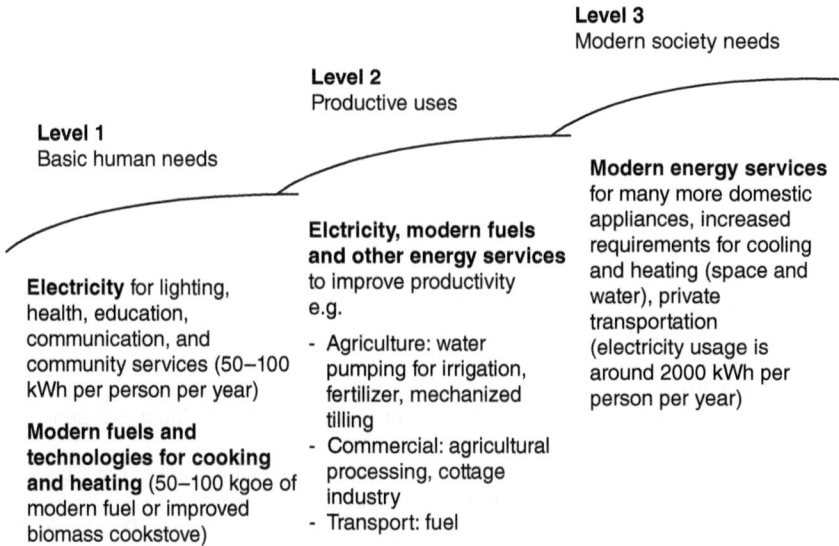

Figure 12.1. Incremental levels of access to energy services
Source: AGECC 2010.

social programmes in order to tackle poverty alleviation issues—they are but a part of the effort. Access and affordability policies should also be accompanied by efforts to create jobs, generate local income, and foster productive uses of energy. Several experts are pointing towards this direction, remarking that the 'energy access' discussion needs to take place in the context of energy transitions and envisaging the productive use of energy as an important step to change the poverty situation (Bhattacharyya 2012; Sokona, Mulugetta, and Gujba 2012).

Energy efficiency, smart technologies, and onsite micro-electricity generation might help to create innovative and new opportunities for income generation amongst the low-income population. New technologies, new business models, and appropriate regulation can help to move more low income households to levels 2 and 3 as depicted in Figure 12.1. This figure presents a schematic and useful view of the three main levels of energy requirements as conceived by the Advisory Group on Energy and Climate Change (AGECC 2010) and shows an improvement in the understanding of the 'energy ladder' concept, which has been used to model a progressive improvement in energy use as household income increases (see, for example, Masera, Saatkamp, and Kammen 2000).

Using this framework, we can evaluate that the challenge in Brazil now is to promote the access to level 2 and 3 to all. As the objective is to move more and more households to the third level of the 'energy ladder', energy efficiency, smart grids, distributed generation—and all of this supported by new business models and regulation—need to become an integral aspect of public energy policies.

Finally, although Brazil has experience in implementing programmes that have improved access and affordability, and has been doing this for some time, these efforts need to be consistent with an exit strategy out of the highly subsidized electricity tariffs and end-use energy efficiency programmes that are still required for low-income households.

REFERENCES

AGECC (Advisory Group on Energy and Climate Change). (2010). *Energy for a Sustainable Future* [online document], <http://www.un.org/wcm/webdav/site/climatechange/shared/Documents/AGECC%20summary%20report%5B1%5D.pdf>.

ANEEL (Agéncia Nacional de Energia Elétrica). (2003). 'Resolu/ww Normativa 223' [online document], <http://www.aneel.gov.br> (accessed 15 October 2013).

ANEEL. (2005). *Atlas de energia* [online document], <http://www.aneel.gov.br> (accessed 15 October 2013).

ANEEL. (2010). *Direitos e deveres do consumidor de energia elétrica* [online document], <http://www.aneel.gov.br> (accessed 15 October 2013).

Bambawale, M. and Sovacool, B. (2012). 'Energy Security: Insights from a Ten Country Comparison.' *Energy and Environment* 23(4) June: 559–86.

Bhattacharyya, S. (2012). 'Energy Access Programmes and Sustainable Development: A Critical Review and Analysis.' *Energy for Sustainable Development* 16(4) September: 260–71.

Birner, S. and Martinot, E. (2005). 'Promoting Energy-efficient Products: GEF Experience and Lessons for Market Transformation in Developing Countries.' *Energy Policy* 33(14): 1765–79.

Boa Nova, A. and Goldemberg, J. (1999). 'Electrification of Shanty Towns in São Paulo.' Presented at INTA—International Urban Development Association—23rd Annual Congress, Lyon.

Boardman, B. (2010). *Fixing Fuel Poverty: Challenges and Solutions*. Oxford and New York: Earthscan.

'Bolsa Família.' (2013). *Wikipedia, the Free Encyclopedia*, <http://en.wikipedia.org/w/index.php?title=Bolsa_Fam%C3%ADlia&oldid=527638814> (accessed 15 October 2013).

Bouzarovski, S., Petrova, S., and Sarlamanov, R. (2012). 'Energy Poverty Policies in the EU: A Critical Perspective.' *Energy Policy* 49 October: 76–82.

Brown, D. (2011). *Economia da energia no segmento residencial rural no Brasil com enfoque na lenha*. Salvador: Universidade Federal da Bahia, Faculdade de Ciências Econômicas.

Fugimoto, S. (2005). 'A Universalização Do Serviço De Energia Elétrica—Acesso e Uso Contínuo'. São Paulo: Escola Politécnica, Universidade de São Paulo [online document], <http://www.teses.usp.br/teses/disponiveis/3/3143/tde-27052009-150949/pt-br.php> (accessed 15 October 2013).

Goldemberg, J., Rovere, E., and Coelho, S. (2004). 'Expanding Access to Electricity in Brazil'. *Energy for Sustainable Development* 8(4): 86–94.

House of Commons Energy and Climate Change Committee. (2010). 'Fuel Poverty'. Fifth Report of Session 2009–10, Volume 1, 24 March, HC 424-1, <http://www.publications.parliament.uk/pa/cm200910/cmselect/cmenergy/424/424i.pdf>.

IBGE. (2012a). Instituto Brasiliero de Geografia e Estatística 'Situação Dos Domicílios—Censo 2010'. <http://www.ibge.gov.br/home/estatistica/populacao/censo2010/caracteristicas_da_populacao/caracteristicas_da_populacao_tab_pdf.shtm> (accessed 15 October 2013).

IBGE. (2012b). 'Habitação (1960–70)'. <http://www.ibge.gov.br> (accessed 15 October 2013).

IEA, UNDP, and UNIDO (International Energy Agency, United Nations Development Programme, and United Nations Industrial Development Programme). (2010). 'Energy

Poverty: How to Make Modern Energy Access Universal?' *World Energy Outlook, 2010 Edition*, Paris: International Energy Agency [online document], <http://www. worldenergyoutlook.org/publications/weo-2010/> (accessed 15 October 2013).

Jannuzzi, G. (1988). 'Uso de lenha em áreas urbanas'. *Ciência e Cultura* 40(3): 289–91.

Jannuzzi, G. (2007). 'Energy efficiency programs for low-income household consumers in Brazil: Considerations for a refrigerator-replacement program'. USAID Contract No. EPP-1-03-03-00007-00 Sub Activity N4.

Jannuzzi, G. (2010). 'Energy Poverty and Technology Leapfrogging: A Look on End-use Efficiency Programs for Low Income Households in Brazil'. *Geopolitics of Energy* 32(10–11): 51–6.

Jannuzzi, G. and Goldemberg, J. (2012). 'Has the Situation of the "Have-nots" Improved?' *Wiley Interdisciplinary Reviews: Energy and Environment* 1(1) July: 41–50.

Jannuzzi, G. and Sanga, A. (2004). 'LPG Subsidies in Brazil: An Estimate'. *Energy for Sustainable Development* 8(3): 127–9.

Kelman, J. (2010). 'Electric Energy Theft Control in Rio De Janeiro'. *Geopolitics of Energy* 32(10–11): 57–64.

Lucon, O., Coelho, S., and Goldemberg, J. (2004). 'LPG in Brazil: Lessons and Challenges'. *Energy for Sustainable Development* 8(3): 82–90.

Masera, O., Saatkamp, B., and Kammen, D. (2000). 'From Linear Fuel Switching to Multiple Cooking Strategies: A Critique and Alternative to the Energy Ladder Model'. *World Development* 28(12) December: 2083–103.

MME (Ministério de Minas e Energia). (2013). 'Luz para todos'. <http://luzparatodos.mme. gov.br/luzparatodos/Asp/o_programa.asp> (accessed 15 October 2013).

Obermaier, M., Szklo, A., La Rovere, E., and Rosa, L. (2012). 'An Assessment of Electricity and Income Distributional Trends Following Rural Electrification in Poor Northeast Brazil'. *Energy Policy* 49: 531–40.

Pachauri, S., Mueller, A., Kemmler, A., and Spreng, D. (2004). 'On Measuring Energy Poverty in Indian Households'. *World Development* 32(12) December: 2083–104.

Pereira, M., Freitas, M., and da Silva, N. (2011). 'The Challenge of Energy Poverty: Brazilian Case Study'. *Energy Policy* 39(1) January: 167–75. <http://www.coelce.com.br/ coelcesociedade/programas-e-projetos/ecoelce.aspx> (accessed 23 June 2013).

Sathaye, J., et al. (1994). 'Economic Analysis of Ilumex a Project to Promote Energy-efficient Residential Lighting in Mexico'. *Energy Policy* 22(2): 163–71.

Smith, K. and Haigler, E. (2008). 'Energy-efficient Residential Lighting in Mexico'. *Annual Review of Public Health* 29(1) April.

Smith, K., Rogers, J., and Cowlin, S. (2005). *Household Fuels and Ill-health in Developing Countries: What Improvements can be Brought by LP Gas?* Practical Action and World LP Gas Association.

Sokona, Y., Mulugetta, Y., and Gujba, H. (2012). 'Widening Energy Access in Africa: Towards Energy Transition.' *Energy Policy* 47, Supplement 1, June: 3–10.

Srivastava, L., Goswami, A., Diljun, G., and Chaudhury, S. (2012). 'Energy Access: Revelations from Energy Consumption Patterns in Rural India.' Energy Policy 47, Supplement 1, June: 11–20.

Uhlig, A. (2008). 'Lenha e carvão vegetal no Brasil: Balanço oferta-demanda e métodos para a estimação do consumo'. D.Sc., Instituto de Eletrotécnica e Energia, Universidade de São Paulo.

United Nations. (2012). *United Nations Rio+20: The Future We Want.* <http://www.un. org/en/sustainablefuture/> (accessed 15 October 2013).

Varella, F., Gomes, R., and Jannuzzi, G. (2009). *Sistemas fotovoltaicos conectados à rede elétrica no Brasil: Panorama da atual legislação.* Campinas: International Energy Initiative and Universidade de Campinas.

Welle-Strand, A., Ball, G., Hval, M., and Vlaicu, M. (2012). 'Electrifying Solutions: Can Power Sector Aid Boost Economic Growth and Development?' *Energy for Sustainable Development* 16(1) March: 26–34.

Welsch, M., et al. (2013). 'Smart and Just Grids for Sub-Saharan Africa: Exploring Options'. *Renewable and Sustainable Energy Reviews* 20: 336–52 [online document], <http://www.sciencedirect.com/science/article/pii/S1364032112006120> (accessed 15 October 2013).

Winkler, H., et al. (2011). 'Access and Affordability of Electricity in Developing Countries'. *World Development* 39(6): 1037–50.

13

Energy Poverty in the Middle East and North Africa

Laura El-Katiri

The Middle East and North Africa (MENA) region, one of the world's most energy-rich, has been understandably overlooked in much of the energy-poverty literature. Massive oil and gas reserves are widely but wrongly thought to have allowed the region's governments to provide their citizens with universal access to modern fuels and electricity.[1] A closer look reveals a different picture. Parts of MENA suffer from serious shortcomings in the infrastructure and supply of modern fuels and in the provision of access to electricity, particularly in mountainous and rural areas in parts of North Africa and the Levant.[2] Large segments of the population continue to rely on biomass, again especially in rural areas.[3] Available data suggest that energy poverty in the region is highly concentrated in the lower Arabian Peninsula (specifically Yemen), across parts of North Africa, and in parts of the Levant. Yemen stands out as the MENA's poorest member, both in terms of income and in terms of energy access.

This chapter aims to fill a gap in the literature by supplementing the scarce available data with country-based observations. One of the most important points the MENA case illustrates is that energy poverty results not from a single cause but rather from a combination of different but interrelated factors. To solve energy poverty, one must address all of these factors, including such often under-appreciated obstacles as under-education and the weight of custom in perpetuating the use of traditional fuels. Section 13.1 provides an overview of the scope and depth of energy poverty in the MENA based on the limited data available. Section 13.2 discusses the various factors accounting for the persistence

---

[1] The MENA under our definition includes Algeria, Bahrain, Djibouti, Egypt, Iran, Iraq, Israel, Jordan, Kuwait, Lebanon, Libya, Morocco, Oman, Qatar, Saudi Arabia, Syria, Tunisia, the United Arab Emirates (UAE), West Bank and Gaza, and Yemen. Historical data is available from World Bank (2013).

[2] The Levant sub-region in this chapter includes Syria, Lebanon, Israel, West Bank and Gaza, Jordan, and Egypt.

[3] In this chapter, we define energy poverty as 'lack of an efficient supply and distribution infrastructure for modern fuels; no access to reliable and affordable supply of electricity; low consumption of modern energy per capita; and high reliance on traditional biomass for cooking' (OPEC Fund for International Development 2010: 19).

of energy poverty in the region. Section 13.3 presents possible solutions based on the region's experience. Concluding remarks follow.

13.1. ENERGY POVERTY IN THE MENA: AN OVERVIEW

Although energy poverty is not one of the development problems traditionally associated with the Middle East and North Africa, modern energy systems, including access to electricity, have in fact been a relatively recent introduction in most of the region, following the rapid expansion of public investment in infrastructure in the 1950s and 1960s. That is the case in the oil-rich Arabian Peninsula, which today is fully electrified. Until the mid-1990s, lower-middle-income countries such as Morocco reported rural electrification rates of less than a fifth of all households. Rural areas, where some 38 per cent of the MENA's population live today, have long been neglected (Achy 2010; World Bank 2013).

Recent statistics, despite significant shortcomings in household-level data availability in all but a few cases, suggests a dramatic improvement in energy access.[4] The MENA region today is highly stratified across the income range. At the top of the ladder are the economically advanced, high-income countries of the Gulf, such as Qatar, the world's richest country on a per capita basis, and the other monarchies of the Gulf Cooperation Council (GCC). These are followed by upper-middle and middle-income countries such as Iran, Saudi Arabia, Libya, and Algeria. At the bottom of the scale come lower-middle-income and low-income countries such as Morocco, Jordan, Syria, Tunisia, Egypt, and Yemen (see Table 13.1 below).

The energy access situation in the MENA region is as contrasted as income ranges and rural–urban geography. In the high-income Gulf states, with urbanization rates of typically above 80 per cent, electrification rates have reached nearly 100 per cent with virtually universal access to all types of modern energy services.[5] In the MENA as a whole, around 95 per cent of the population was estimated to have access to electricity in 2009, well above the averages for South Asia and Sub-Saharan Africa (see Table 13.1). Similarly, Arab countries enjoy relatively high rankings in comparative energy-development indices such as the International Energy Agency's (IEA) Energy Poverty Index, where mostly Sub-Saharan and South Asian countries rank at or near the bottom (see Appendix A).

Yet the picture in the MENA is not uniformly rosy. In other parts of the region, lack of network coverage and low incomes leave around 20 million people without access to even basic levels of electricity. Those include 1 million Moroccans, 1.4 million Syrians, 1 million Iranians, more than 4 million Iraqis and more than

---

[4] The World Bank's 2005 survey of household energy use in Yemen is a major exception and provides detailed data covering the early 2000s.

[5] Among the Gulf monarchies, only Oman displays a share of below 80 per cent urban population. With some remaining small settlements living in remote mountainous regions, Oman is hence the only GCC economy with some remaining non-electrified parts of the country, albeit small and by no means comparable to neighbouring Yemen (World Bank 2013).

Table 13.1. Basic energy development indicators in the MENA, 2010

Country	Population (mn)	GDP per capita, PPP (constant 2005 international $)	Land area (km²)	Total electricity production (TWh)	Electricity consumption (kWh per capita)	Electrification rate (%*)
The Gulf States						
Bahrain	1.3	21,345	760	13,230	9,814	99.4
Iran	74.0	10,462*	1,628,550	232,955	2,652	98.4
Iraq	32.0	3,195	434,320	50,167	1,183	86.0
Kuwait	2.7	45,623	17,820	57,029	18,320	100.0
Oman	2.8	24,559	309,500	19,819	5,933	98.0
Qatar	1.8	69,798	11,590	28,144	14,997	98.7
Saudi Arabia	27.4	20,534	2,149,690	240,067	7,967	99.0
United Arab Emirates	7.5	42,353	83,600	97,728	11,044	100.0
Yemen	24.1	2,373	527,970	7,757	249	39.6
The Mashreq						
Egypt	81.1	5,544	995,450	146,795	1,608	99.6
Israel	7.6	25,995	21,640	58,566	6,856	99.7
Jordan	6.0	5,249	88,780	14,779	2,226	99.9
Lebanon	4.2	12,619	10,230	15,712	3,569	99.9
Syria	20.4	4,741	183,630	46,413	1,905	92.7
West Bank and Gaza	3.9	n/a	6,020	n/a	n/a	n/a
North Africa						
Algeria	35.5	7,564	2,381,740	45,560	1,026	99.3
Libya	6.4	15,361*	1,759,540	31,613	4,270	99.8
Morocco	32.0	4,227	446,300	22,308	781	97.0
Tunisia	10.5	8,508	155,360	16,096	1,350	99.5
MENA	**382.1**	**18,336**	**11,212,490**	**1,144,738**	**5,319**	**94.8**
Regional context						
East Asia & Pacific	2,201.6	9,629	24,323,100	7,110,198	3,063	90.9
Europe & Central Asia	891.0	24,448	27,380,709	5,303,520	5,527	n/a
Latin America & Caribbean	588.8	11,422	20,142,390	1,402,513	1,982	93.4
North America	343.5	45,862	18,240,980	4,963,271	13,567	n/a
South Asia	1,632.9	3,070	4,771,220	1,120,056	555	62.2
Sub-Saharan Africa	853.9	2,257	23,615,950	441,886	553	32.4

* The last percentages available are from 2009.
Source: World Bank

14 million Yemenis (see Table 13.2).[6] Yemen, the Middle East's poorest country and an oil and natural-gas exporter, illustrates the depth of the problem of energy poverty in the MENA. The country suffers from the most acute form of energy poverty: over half of the population still has no access to the electricity grid.

---

[6] Estimates based on World Bank (2013).

Table 13.2. Household energy consumption in Yemen by fuel (percentage of households reporting the use of each fuel), 2005

	Urban	Rural	Total	Lowest decile	Highest decile
Electricity	92	42	53	22	82
PEC grid	80	23	36	11	62
Non-grid, incl. self-generation	12	19	18	9	19
No access to electricity	8	58	47	78	18
LPG	93	74	78	49	93
Diesel	13	4	11	3	34
Kerosene	46	83	75	92	57
Fuelwood	36	85	74	80	66
of which purchased	24	24	24	13	36
Charcoal	12	6	8	2	18
Dung	3	23	18	12	21
Crop residue	3	26	23	24	20
of which purchased	<0.5	1	1	<0.3	2

Source: World Bank

Exceptionally detailed household data for the country show the extent of household reliance on traditional biomass fuels, including firewood, animal dung, and waste materials, for all energy needs, as well as the widespread use of kerosene, the liquid fuel of the poor. This picture closely resembles that of other energy-poor regions, including Sub-Saharan Africa and East Asia (El-Katiri and Fattouh 2011).

Traditional fuels such as firewood and animal dung, despite being perceived as of inferior quality in much of the generic economic literature, remain widely used by all Yemeni income groups, including the top income deciles, sometimes alongside higher-quality liquid fuels when those are available. Charcoal is often mixed with other forms of biomass for cooking and heating, contributing on its own between 1 and 15 per cent of the Yemeni household's energy mix. Rural households collect and use crop residues and dung with little variation across income groups. Both are typically mixed with wood, and sometimes charcoal, for heating and cooking. They are used in higher proportions during the rainy season, when wood goes up in price and must be collected over longer distances. Kerosene may be added where available as a popular starter fuel (World Bank 2005b: 120). This contradicts the traditional representation of the energy ladder according to which higher-income households climb up to higher-quality fuels at the expense of inferior sources of energy (World Bank 2005a: 66; World Bank 2005b: 117).

The use of biomass is also widespread in rural Egypt and Tunisia, and to a lesser extent in Morocco, Lebanon, Libya, and Iran, a fact concealed by most datasets of aggregate energy consumption for those countries.[7] In Morocco, a growing body

---

[7] The overall energy consumption balance by country typically estimates biomass consumption vis-à-vis such alternatives as oil, natural gas, and coal, all of which weigh much higher than estimated levels of biomass. Disaggregated data on biomass consumption volumes, as opposed to the relative share of biomass in overall energy consumption, shed more light on the region's actual biomass consumption and its dependence on traditional fuels. Needless to say, national estimates of biomass use are often imprecise—as are population estimates—in the absence of systematic censuses and state-administered data collection (Sara 2011; Fritzsche et al. 2011; El-Moudden 2004: 51; World Bank 2013).

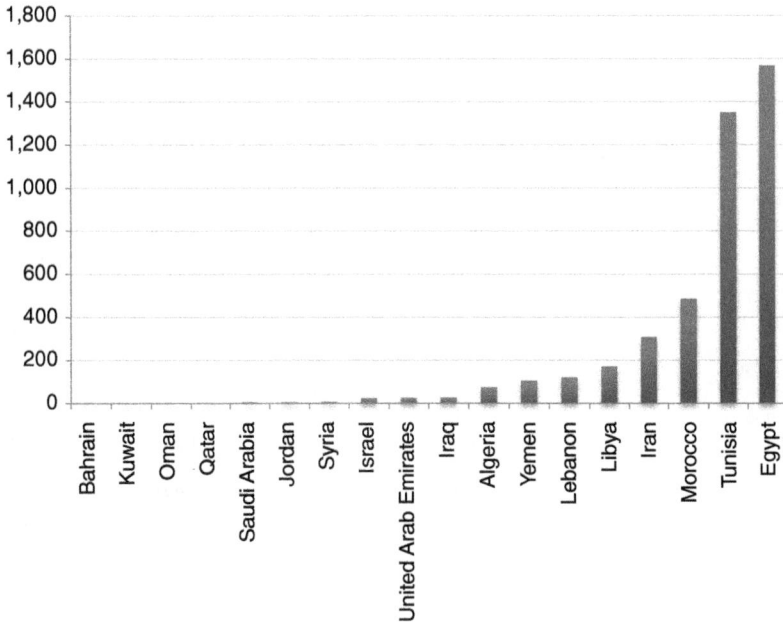

Figure 13.1. Estimated biomass use in the MENA (metric tons of oil equivalent), 2010

Note: 'Biomass' is defined as combustible renewables (dung, crop residue, fuelwood) and waste.
Source: World Bank

of literature deals with the problem of rural deforestation, which so often stems from the persistence of firewood use in rural households. The resulting environmental damage undermines living conditions and spreads social conflicts (Zejli and Bennouna 2009; Le Polain de Waroux and Lambin 2012).[8] Modern fuels such as liquefied petroleum gas (LPG), while widespread in the MENA today, are a recent development; during the 1960s and 1970s, much of the rural populations of North Africa and the Levant relied almost exclusively on firewood, biomass, and animal power. Since then, with the increasing availability of liquid fuels such as kerosene, diesel, or residual fuel oil at relatively low cost, their use as a complement to traditional biomass has rapidly spread for heating, lighting, and cooking, and as fuel for small power-generating units (see Figure 13.1) (Akash and Mohsen 1999; Al-Mohamad 2001; Karaki et al. 2005; Kabarati 2005: 9).

In most MENA countries, per capita electricity consumption lies comfortably above the low-income benchmark of 242 kWh per capita, but national averages are distorted by the high energy-intensity of local industry and hide vast

---

[8] Deforestation as a result of fuelwood collection is also reported in Yemen, where the growing radius of fuelwood collection in deforested areas gives rise to numerous social conflicts over land and forest ownership. Both Morocco and Yemen have for this reason promoted the use of LPG bottles via raised subsidies (World Bank 2005b: 59).

Table 13.3. Electricity consumption (kWh per capita) in the MENA, 1980–2010

Country	1980	1990	2000	2010
The Gulf States				
Kuwait	6,028	8,253	14,822	18,320
Qatar	9,806	9,643	14,385	14,997
United Arab Emirates	5,771	8,594	12,722	11,044
Bahrain	4,637	6,573	8,994	9,814
Saudi Arabia	1,949	4,041	5,840	7,967
Oman	609	2,121	3,207	5,933
Iran	539	967	1,553	2,652
Iraq	785	1,253	1,199	1,183
Yemen	63	123	139	249
The Levant				
Israel	3,022	4,176	6,323	3,569
Lebanon	908	475	2,610	2,226
Jordan	399	1,050	1,377	1,905
Syria	357	696	1,093	1,608
Egypt	368	669	994	n/a
West Bank and Gaza	n/a	n/a	n/a	n/a
North Africa				
Libya	1,177	1,614	2,276	4,270
Tunisia	402	638	991	1,350
Algeria	338	541	695	1,026
Morocco	242	360	490	781
MENA	**2,428**	**3,350**	**5,169**	**6,173**
High-income average	5,861	7,394	8,946	9,414
Middle-income average	374	940	1,063	1,823
Low-income average	136	224	174	242

Source: World Bank

differences between income groups. Nevertheless, steep gains in per capita electricity consumption in the MENA reflect the rapid development of electricity use and access across the region (see Table 13.3).

13.2. FEATURES OF ENERGY POVERTY IN THE MENA

Energy poverty in the MENA region has many faces, depending on geography, economic and demographic structures, and other local, cultural, and circumstantial factors. Some of these factors are short-term and conflict-related, for instance temporary income poverty and displacement, none of which can easily be measured with currently available statistical tools. Others are structural and longer-term, such as structural poverty, geographical divides, and the impact of tradition and custom.

13.2.1. Income poverty

Income poverty remains one of the main obstacles to energy access, although the interaction of income poverty and lacking access to energy is by no means one-directional. In various regional contexts, studies have shown the intricate relationship between households' disposable income levels and where they sit on the energy ladder. Disposable income determines the ability of households to move up in the fuel continuum towards more expensive, higher-quality fuels and electricity (Bruce 2005; IEA/UNDP/UNIDO 2010). Rises in income levels are frequently associated with increased energy access, although this is by no means automatic (Ramon and Toman, 2006). Access to modern fuels and electricity, on the other hand, also conditions socio-economic development and hence income growth. Energy poverty can hence become part of a vicious circle that continuously reinforces poverty, particularly in rural areas.

Energy poverty in the MENA is most prevalent, and persistent, in countries with high rates of income poverty, particularly in North Africa, parts of the Levant, and Yemen. Around 15 per cent of Moroccans and Egyptians are estimated to live below the $2-per-day poverty line, down from around twice as many during the 1990s. In Iran and Tunisia, that proportion is estimated at more than 8 per cent, and it is more than 40 per cent in Yemen and Djibouti (see Table 13.4; World Bank 2013).

Table 13.4. Poverty levels in the MENA

	Poverty headcount ratio at $2 a day (PPP) (% of population)	Population (millions)	Gini coefficient	Last year available
The Gulf States				
Yemen	46.6	9.62	37.7	2005
Iraq	21.4	6.27	30.9	2007
Iran	8.0	5.60	38.3	2005
Bahrain	n/a	n/a	n/a	n/a
Kuwait	n/a	n/a	n/a	n/a
Oman	n/a	n/a	n/a	n/a
Qatar	n/a	n/a	41.1	n/a
Saudi Arabia	n/a	n/a	n/a	n/a
United Arab Emirates	n/a	n/a	n/a	n/a
The Levant				
Syria	16.9	3.03	35.8	2004
Egypt	15.4	12.09	30.8	2008
Jordan	1.6	0.10	35.4	2010
West Bank and Gaza	0.3	0.01	35.5	2009
Lebanon	n/a	n/a	n/a	n/a
North Africa				
Djibouti	41.2	0.32	40.0	2002
Morocco	14.0	4.35	40.9	2007
Tunisia	8.1	0.81	41.4	2005
Algeria	n/a	n/a	n/a	n/a
Israel	n/a	n/a	39.2	n/a
Libya	n/a	n/a	n/a	n/a

Source: World Bank

Wealth in the MENA is unevenly distributed both across the region and, within individual countries, across society. Most of the wealth is concentrated in the oil-rich, sparsely populated Gulf countries. In many North African and Levantine economies, a small fraction of the population accounts for most of the national wealth. In Egypt, Jordan, and Morocco, the top 10 per cent earners are estimated to hold around 30 per cent of their countries' wealth, while the bottom 10 per cent account for only 2–3 per cent of the wealth (a situation captured in the Gini coefficients shown in Table 13.4; World Bank 2013; Jaber and Probert 2001; Achy 2010). An account from Jordan during the early 2000s illustrates similar income cleavages:

> The distribution of income in Jordan . . . is very unequal. The wealthiest 10% of households earn more than 50% of the total national household income, while the poorest 48% earn only about 10% of the total household income, and live below what is accepted to be the poverty level: the average monthly income per family is less than 300 USD. Lack of adequate housing and access to basic services are also indicators of poverty: at present about 30% of the population live in miserable conditions, e.g. permanent Palestinian refugee camps or marginal houses, far below basic acceptable levels. More than 40% of all households have no access to a sewage network (Jaber 2001: 312).

Demographic growth intensifies the region's poverty problem. The MENA's population continues to grow at among the highest rates in the world—around 1.8 per cent per annum in the late 2000s. For some countries, the growth is even faster. The populations of Yemen and Iraq grew by more than a third in the 2000s alone; Egypt's population expanded by more than a fifth over the decade (World Bank 2013).

Income inequality is just part of the problem. Land ownership is just as unevenly distributed, as is employment, which tends to be overwhelmingly concentrated in informal and seasonal sectors. In Morocco, data on the distribution of agricultural land indicate that the 5 per cent largest farmers own one-third of the land (Achy 2010: 13). Across much of North Africa, the concentration of land ownership results in highly unequal access to land and resources and in the traditional reliance of much of the rural population on seasonal and precarious farming jobs, with little income security through the course of the year. In Morocco, seasonal agricultural work makes up 40 per cent of employment (Achy 2010). Other sectors of the economy, such as tourism, hospitality, and construction, offer similarly seasonal and informal employment. As a result, incomes tend to be highly volatile and uncertain, posing an additional problem for many households' investment in energy.

Even in parts of the MENA where there is a local market for modern fuels and electricity, the combination of income poverty, job insecurity, and high seasonal fluctuation in earnings restricts access to modern energy services in two ways. First, many households lack the disposable income to cover the cost of modern energy, such as higher-quality liquid fuels, LPG, and electricity. More importantly, income poverty means that many households cannot afford the initial investment required to access modern energy services, such as the cost of an electricity connection, a new stove, or the equipment needed for liquid fuels and LPG supplies. In Yemen, an initial electricity connection costs from YR10,000 (rural) to YR25,000 (urban), more than many poor households can afford (World Bank 2005a: 33). In many MENA countries, the initial connection fee is added to a flat monthly fee, irrespective of actual household consumption and income levels. This escalates the per-unit cost of electricity for many poor households, whose consumption for simple devices, such as light bulbs and radios, is typically very low. Illegal connections to the national grid offer a widespread alternative across much of the Levant, Yemen, and North Africa. Another popular option is an illegal connection to a neighbour's line at a low informal fee (World Bank 2005a: 33; Kandil 2010).

Similarly, the initial cost of a full-size LPG bottle, sufficient to supply a household for a month, may exceed households' incomes at the time of purchase. In Yemen, many low-income households for this reason prefer to resort to inferior fuels such as kerosene, despite a lower calorific value per unit out of energy, because it is sold in small 2-litre bottles at a considerably lower one-off cost than the standard 11 kg LPG cylinder (World Bank 2005a: 27). These factors may frequently explain why even in the presence of comparably low nominal unit prices, many households fail to move up the energy ladder.

13.2.2. Geography

As in other parts of the world, geography forms another major driver of energy poverty in the MENA.[9] Besides regional disparities in overall access to modern fuels and electricity, Yemen and parts of North Africa and the Levant show a vivid contrast between well-supplied urban areas and the often undersupplied country-side. Even where the willingness and ability to pay exist, infrastructure constraints often cut off entire communities from access to modern fuels and electricity.

The MENA's rural–urban divide in electricity access is striking: more than 98.5 per cent of the MENA's urban areas in 2009 were estimated to have access to electricity, compared with just 72 per cent for rural areas (IEA 2011). These numbers include not only national grid access but also time-limited access through local mini-grids and self-generation, suggesting that 28 per cent of the MENA's rural communities lack access even to most basic forms of electricity. Around 38 per cent of the MENA's population live in rural areas, although this share is considerably larger in countries such as Egypt, Syria, and Yemen, where typically more than half of the population lives in the countryside (World Bank 2013). In Yemen, where electricity access is particularly problematic, only about half of the country's population had access to electricity in 2005, including 92 per cent of urban residents but just 42 per cent of the rural population. A mere 23 per cent of the population had access to Yemen's main electricity grid, the remainder relying on local mini-grids and self-generators (World Bank 2005a, b).[10]

---

[9] See Ramon and Toman (2006) for a general study and Gupta and Sudarshan (2009) in the case of India.

[10] The data found by this survey was largely collected in 2003, but the extent of households interviewed and the quality of the data mean the survey is the single most detailed, comprehensive, and reliable assessment of household energy use available for Yemen. A total of 3,540 households participated in the survey.

In many cases, location does correlate with overall poverty levels, with 7 out of 10 poor Moroccans and 8 out of 10 Yemenis living in rural areas (Achy 2010: 2; Government of Yemen, World Bank, and UNDP 2007: 25). This concentration of income poverty in the countryside reinforces the lack of access to energy. Many rural energy markets remain small and geographically dispersed. This makes it all the more financially challenging to expand the grid to those areas. Transport and logistics, particularly across scarcely inhabited, mountainous territory, raise the cost of local fuel supply. This cost must be borne either by suppliers or, more frequently, by local communities, despite national price controls (Dougherty 1994: 119).

Even in urban areas, energy access in many MENA countries may be lacking. That is the case in shanty towns or 'informal' urban housing. Populated by the urban poor, often refugees from conflict zones, informal housing is frequently associated with a lack of infrastructure such as electricity and sewage access. The size of the urban population living in informal housing in MENA countries is significant and, in all likelihood, vastly underestimated. A Jordanian study from 2001 suggests that informal housing, including the country's vast refugee camps housing several million Palestinian and Iraqi refugees, comprise more than half of the urban building stock (Jaber and Probert 2001: 124). Syrian data from the early 2000s shows a vast divergence in electricity coverage in informal housing across regions: while electrification rates in Latakia, Aleppo, and Suweyda were reported as between 65 and 78 per cent, in the province of Daria they fell to as low as 31 per cent (El-Laithy and Abu-Ismail 2005: 91). With urbanization rates as high as 6 per cent in part of the MENA, electrification rates in urban areas are likely to fall as infrastructure fails to keep up with the continuing pressure on land and affordable living space (World Bank 2013).

13.2.3. Energy supply volatility

One of the most endemic features of energy supply in the MENA, particularly in rural parts of the Levant and North Africa, is its intrinsic unreliability. Unreliable supplies may be due to a variety of reasons and mostly affect the electricity sector. Grid suppliers lack adequate generation capacity to supply all customers, and thus they ration supplies, particularly in the countryside. Underinvestment in maintenance and upgrading can increase the frequency of technical faults and outages. In many rural areas with access to mini-grids and neighbourhood grids, such as are commonly employed in Yemen and parts of rural Jordan, generation capacity is often designed to supply electricity for only a few hours per day subject to availability of fuel (the supply of which can be delayed due to transport issues, particularly in remote rural areas). The widespread practice of illegal connections of houses or entire neighbourhoods to the grid further exacerbates the shortage of electricity supplies and increases the occurrence of unplanned electricity outages due to overload.

Yemen's electricity coverage is particularly problematic. Yemen's total installed generating capacity in 2009 stood at only 1,551 MW for a population of 24 million people (Republic of Yemen, Ministry of Energy and Electricity 2009: 18). Per capita consumption of electricity, at 203 kWh, was only one-tenth of the Arab world's average of some 2,000 kWh (Republic of Yemen, Ministry of Energy and Electricity 2009: 19 and comparison numbers from IEA 2010). The country's main PEC grid connects mainly urban areas and still entirely excludes the former south Yemen. Where there is grid access, both residential and industrial users experience frequent shortages and load-shedding. Supply disruptions occur many times during the year as a result of old, inefficient generation capacity, inadequate transmission and distribution (T&D) infrastructure, and fuel shortages.[11] This situation is particularly damaging for businesses and industries, which must invest in back-up generators to avoid prolonged production disruptions.[12]

This situation is the legacy of long-term underinvestment in Yemen's utility sector, including in new capacity, the maintenance and repair of old T&D infrastructure, and the expansion of Yemen's electricity grid towards southern and particularly rural communities. Yemen's public utility PEC is severely under-funded, not least due to Yemen's government-regulated pricing system, originally intended to help poor people access electricity. With electricity prices artificially held down for years under an extensive electricity subsidy system, PEC has for many years been unable to recover its costs (World Bank 2005b: 94). In conse-quence, it has neither the financial nor the physical capacity to extend its main grid to remote provinces. In many provinces, the only alternative, self-generation or village-based mini-grids, offers electricity access for just a few hours a day. Electricity access is thus not only rare, particularly for the rural poor, but also intermittent and of poor quality (World Bank 2005b: 91).

Even in comparatively richer areas of the MENA, electricity supplies can be highly volatile. Thus Lebanon suffers from the region's highest rate of registered annual blackouts, translating into daily disruptions of 3 to 12 hours. Rural households and urban areas outside Beirut experience the longest power cuts. As in Yemen, recurring electricity failures in Lebanon and other parts of the Levant have driven up households' and businesses' reliance on back-up gener-ation, fuelled by diesel or fuel oil at substantial cost. The additional burden to households' expenditure in Lebanon as a result of blackouts was estimated by a World Bank study at more than double their expenditure for normal grid-based electricity (World Bank 2009: 7–9). Low-income households are least able to pay for backup generation, and are hence left behind. The World Bank commented on Lebanon's case:

> While the cost of electricity provided by EdL [Electricite du Liban] has remained low for consumers, the burden resulting from EdL's service decline (high and increasing frequency of supply interruptions) has risen due to reliance on back-up generation, damaged appliances resulting from power surges and opportunity costs to households (World Bank 2009: 12).

---

[11] This widespread problem was for the first time thoroughly acknowledged at an unusually outspoken press conference in January 2010 by Yemen's energy and electricity minister, Awadh Al-Suqatari (Assamiee 2010).

[12] An IFC Enterprise Survey found that in 2010 more than 50 power outages were experienced countrywide, most of which lasted several hours, causing significant commercial losses for businesses (IFC Enterprise 2011; Assamiee 2010).

Unreliable electricity access can deter households from investing in an initial electricity connection, as in the case of Morocco and Yemen. Self-generation, which entails no binding ties with state suppliers and enables electricity generation on demand, can thus offer a greater appeal for rural households, especially where electricity access is required only for certain activities such as radio-listening or lighting. While not altogether depriving low-income households from energy access, unreliable electricity supplies nevertheless raise hurdles that they must overcome to move up the energy ladder, while also increasing the cost of energy solutions.

In a vicious circle of sorts, poverty often feeds into, and exacerbates, the unreliability of electricity supply. The World Bank estimates that Yemen's public electricity provider, PEC, suffers technical and non-technical electricity losses of as much as one-third, including via illegal connections and electricity theft by households unable or unwilling to pay for their power supplies (Shaher 2011). In Lebanon, around 23 per cent of electricity generated by Lebanon's EdL is thought to be lost through illegal connections, in addition to technical losses of around 15 per cent. EdL thus collects revenues for only around 60 per cent of its production, which has to cover 100 per cent of its production costs (World Bank 2009: 5). In the case of both PEC and EdL, price controls aimed to help provide low-income households with access to electricity have added to the financial burden of non-payment, effectively preventing the companies from recovering their costs (Fattouh and El-Katiri 2012: 29–30; Assamiee 2010; Shaher 2011). The structural long-term problem behind these companies' financing frameworks prevents them from playing the role desired by policymakers: to ensure stable and equitable access to electricity including through systematic investment in the maintenance and expansion of existing generation and distribution capacity.

Yemen and Lebanon illustrate how energy poverty can become entrenched over the long term.[13] In theory, underinvestment in the energy sector should be temporary, as the resulting energy shortfalls should encourage investment in production capacity, import, and infrastructure. In the MENA, however, under-investment in the electricity sector tends to become systematic, leading to long-term supply shortfalls that adversely affect these countries' development prospects not only at household level but at the industrial level as well.

13.2.4. Custom and convenience

In addition to income levels and service reliability, custom and convenience may also determine a household's choice of energy services. Those are powerful, if often overlooked, factors in the MENA region. Elsewhere, too, in such places as India or Sub-Saharan Africa, it has been observed that income gains and fuel availability did not automatically translate into a move up the energy ladder. Many factors can dampen consumers' interest in modern fuels, including personal preferences, the unreliability (real or perceived) of fuel supply, the volatility of fuel

---

[13] Similar observations could be made for Morocco, Syria, Jordan, Egypt, and Yemen (Fritzsche et al. 2011; Fattouh and El-Katiri 2012; World Bank 2005a).

prices, and the cost of switching. In many developing countries, even wealthy households may not have access to modern fuels, because the infrastructure to supply them (regularly) is not in place (Heltberg 2005; Masera et al. 2000; Mekonnen and Köhlin 2008; Barnes and Toman 2006).

In the MENA, particularly in rural areas, the management of household members' time and the composition of household goods are still frequently subject to time-honoured traditions, for instance the ancestral division of labour assigning to women and children the task of collecting fuel such as biomass and firewood locally (El-Moudden 2004: 66; Kabarati 2005: 9). The persistence of these customary work divisions, rooted deeply in many traditional communities' heritage, may contribute to the use of preferred fuels such as fuelwood over kerosene or LPG, and a general perception that different fuels, possibly requiring a change in cooking habits and equipment, and electricity were simply 'not needed' (Dougherty 1994; Akash and Mohsen 1999; World Bank 2005b).

That households should expose themselves to the potentially devastating health and environmental effects of prolonged reliance on biomass and inferior liquid fuels such as kerosene often reflects an information deficit, especially in households with low levels of education or that lack access to radio or other media. Lack of access to radio is itself a secondary effect of lack of access to electricity. A number of studies lament the detrimental health effects of the widespread use of biomass and kerosene for cooking and heating in rural Jordan for instance, but also the lack of public awareness of the health risks associated with those fuels (Jaber and Probert 2001: 119; Kabarati 2005: 9).

Similarly, the environmental effects of the over-collection of fuelwood, and the resulting deforestation of large areas in countries such as Morocco, Jordan, Syria, and Yemen are understood by governments and scientific research, but are often entirely disregarded by local communities relying traditionally on this source of energy (El-Moudden Saloua 2004; Fritzsche et al. (2011); Le Polain de Waroux and Lambin 2012; Schilling et al. 2012). An account from Morocco's High Atlas mountains illustrates this effect:

> The effects of historical patterns of firewood exploitation are seen in today's disproportional fuel energy hardship. By continued adherence to some well-practiced conventions rooted in a more wood-abundant past, clans reinforce the unsustainable exploitation of wood from communal lands (Dougherty 1994: 133).

In view of the considerable time commitment that biomass and fuelwood collection represent for rural households, the use of commercial fuels would provide female household members with increased spare time and would foster school attendance by children. Lack of access to commercial fuels is thus poverty-reinforcing.[14] While current income levels can be an important prohibiting factor,

---

[14] Many Yemenis point out that fuelwood collection contributes to low school enrolment rates for girls. The average time spent per household and month on the collection of fuelwood, according to a 2005 World Bank household survey in Yemen, is some 100 hours, thus on average 25 hours per week—the equivalent of half a school day each day. Yemen still displays a vast gender gap in education. Only 55 per cent of primary-school-aged girls were enrolled in primary education, compared to nearly 75 per cent of boys in 2005 (World Bank 2005a: 49, 52). For a similar account from Jordan, see Kabarati (2005: 9).

it is often also custom, and non-awareness of the long-term financial benefits of women's education, that keep some rural households from accessing potentially affordable sources of energy.

Some low-quality-fuel consumption can also be the result of informed, deliberate household choice. Fuelwood, in particular, is often the cooking fuel of choice for households across all income ranges in the MENA. Among higher-income households, it tends to be valued for giving food a better taste than liquid fuel and LPG stoves. Many low-income households prefer it to kerosene and fuel oil, which they associate with toxic fumes, and which a Yemeni household described in a survey as 'causing headaches'.[15]

13.2.5. Political conflict

In parts of the MENA, the persistence of energy poverty does not stem from lack of government attention, underinvestment, geographical factors, or low incomes. Rather, it is driven by socio-political instability, whether short-term or ongoing. Little is known about the scope and duration of conflict-driven energy poverty, especially in the case of conflicts that result in years of instability and a lack of effective domestic institutions. The consequences must nevertheless be seen as severe for local populations, adding to socio-economic neglect which in turn perpetuates and feeds into social conflict. The Arab-Israeli conflict is a case in point. The conflict caused hundreds of thousands of Palestinian refugees to settle in the Gaza strip and camps in Jordan and Lebanon, where many of them ended up spending decades in provisional housing, often with little if any access to electricity and sewage. There are no available data on electricity service rates in the West Bank and Gaza, but official Israeli reports estimated operating rates at the Gaza Strip's sole power station's at 20 per cent of capacity at the end 2012, suggesting significant undersupply of Gazan households (Israel Ministry of Foreign Affairs 2012).

Political conflict in Syria, ongoing at the time of writing, and continued domestic conflict in Iraq since the 2003 war have delayed and at times halted necessary investment in electricity networks and power generation capacity, as well as in upstream oil and natural gas capacity to fuel power generation. These conflicts have also interfered with fuel transportation (Al Bawaba Business 2012; Reuters 2012). Many households and businesses in Iraq were reported in 2012 to experience up to 15 hours of power cuts per day (Al Sayegh 2012). In Moroccan-claimed Western Sahara, most of the energy infrastructure has been installed by Morocco, but electricity and fuel supply remains largely limited to urban areas, leaving vast parts of Western Sahara undersupplied, a situation rarely discussed due to the virtual absence of data on the region.[16] In Yemen, long-term conflict between the government and various tribes has turned energy supply into a

---

[15] The World Bank's report reads as follows: 'In the PRA, women expressed a strong dislike for cooking with kerosene, explaining that its bad smell affects the taste of food and causes headaches. Respondents throughout the study area also described kerosene as a safety risk' (World Bank 2005a: 43).

[16] For a rare comment, see Hagen (2008).

weapon, targeting pipelines and power stations for recurring attacks in the provinces and allegedly leading the central government to purposefully under-develop energy infrastructure in the south (Boucek 2009: 6; Phillips 2007: 3, 14).

13.3. APPROACHES TO TACKLING ENERGY POVERTY IN THE MENA

Lack of energy access has been recognized as a serious impediment to socio-economic development by many MENA governments over the past decades, leading to the Abu Dhabi Declaration on Environment and Energy, a joint Arab declaration that sets 'the alleviation of poverty and the achieving of sustainable development' as 'the ultimate priority of the Arab countries'. The signatories pledged to support the 'promotion of the supply of energy to rural and remote areas in the Arab world and diversification of the sources of such energy' (Arab Ministerial Conference 2003). Among the MENA countries most affected by energy poverty in the past, Morocco and Jordan have run extraordinarily success-ful rural electrification programmes. These success stories allow some tentative conclusions about future approaches that the region might adopt.

13.3.1. The importance of local solutions

The experience of Morocco, Jordan, and in part Yemen suggests that the com-mercial feasibility of energy access schemes often depends on a combination of local factors, such as geography, the concentration of low-income households, and energy consumption patterns. The most significant form of energy poverty in the region remains the lack of access to electricity, often due to the prohibitive cost of extending national grids to rural areas, especially in the case of highly scattered small villages.

Morocco's Programme d'Électrification Rurale Global (PERG), first launched in 1996 when estimated rural electrification rates were as low as 18 per cent, specif-ically catered to a range of village needs based on the long-term commercial viability of village-based electricity access. The programme identified villages lack-ing electricity access across the country, classifying them into different categories based on whether their connection to Morocco's main electricity grid was finan-cially viable. In areas where the connection to the grid was deemed uneconomical, the programme reviewed local conditions to assess the viability of alternative solutions such as photovoltaic generators, small hydro turbines, wind turbines, diesel generators, and hybrid systems. Over a period of 15 years, more than 35,000 villages and some 1.9 million rural households were electrified, lifting rural electri-fication rates to 97 per cent by 2009 (Agence Française de Développement 2013).

Jordan obtained similarly positive results following its 2002 launch of a rural photovoltaic (PV) electrification programme, aimed at improving both access to electricity and the life quality of rural electricity users. Departing from previous electrification programmes, which had supplied rural communities with diesel-powered self-generators, the programme focused on renewable energy resources instead. Diesel generators had been found to be highly polluting and required frequent repair and maintenance, often in geographically remote areas where technical defaults led to frequent interruptions for long periods, resulting in high costs. Many rural communities welcomed the introduction of cleaner and supposedly low-maintenance PV generators which avoided the running costs associated with diesel fuel (Al-Soud and Hrayshat 2004: 593). The photovoltaic potential of both Morocco and Jordan, as well as other parts of the MENA region, is high, owing to high sunshine rates through most of the year. A similar potential exists for wind power in many parts of North Africa (Benkhadra 2009; Al-Soud and Hrayshat 2004). A French development report summarizes the advantages of local solutions to electricity access, pointing to the many local benefits of renewables-based project in addition to their advantages from a climatic point of view:

> The use of renewable energy sources therefore addresses less the issue of climate constraints than financial considerations: rural electrification based on renewable energy enables local resources to be used to best advantage while remaining free from the price volatility of fossil energy and the extra costs associated with its transport to isolated areas. Accordingly, in many rural regions, recourse to low-carbon solutions is an economically rational way of enabling poor households to gain access to a quality electricity service (Raspaud 2012: 3).

Local energy solutions can be limited in scope and service quality compared to grid-based solutions, however. Morocco's PERG has facilitated electricity access for more than 1 million households since the 1990s, but many local generators, including photovoltaic, diesel, and wind, provide an inferior service compared to grid-based access, including shorter operating hours and less reliable access (Raspaud 2012: 3).[17] Many village-based mini-grid solutions in Yemen report similar failings, exposing households to outages on a daily basis. The most adverse effect of unreliable electricity supply in those villages relates to health, namely the inability to keep spoilable medicines refrigerated, to obtain light when needed during medical emergencies, and to operate basic medical electrical devices (World Bank 2005a: 2).

In the absence of emergencies, relatively simple information and training can often improve energy access. Information and training can also reduce some of the negative consequences of continued use of biomass such as poor indoor air quality and the long-term exposure of women and infants to polluting energy sources. Among many inexpensive, small-scale solutions is the use of biomass biodigesters as observed in rural Morocco. Simple micro-hydro installations, improved cookstoves, and information regarding the airing of houses can help local communities significantly improve the quality of their energy consumption while offering better access (Dougherty 1994: 120).

---

[17] This circumstance has also been observed by the author in rural Morocco.

13.3.2. Financing energy access

The relative price of various energy services, as well as switching costs, continue to be important determinants of households' access to energy. Key to enabling access to energy is finding ways to render energy financially affordable to low-income households, wherever they may be located. Many MENA governments have responded to this challenge since the 1960s through fuel subsidies. Most governments have favoured universal policies benefiting all social categories, but varying by fuel type and in the size of the subsidies. While undoubtedly having helped millions of low-income households in the MENA in accessing modern forms of energy, energy subsidies have become increasingly criticized for their inefficiency and, at times, ineffectiveness in tackling energy poverty.

The universal dimension of most energy subsidies in the MENA has come under particular criticism. More often than not, liquid fuels and LPG are subsidized irrespective of household income, implying that that majority of fuel subsidies leak to comparatively large consumers such as middle- and high-income households. Statistical evidence from Yemen shows that 40 per cent of LPG subsidies and 70 per cent of diesel subsidies are captured by the three highest income deciles, compared to 21 per cent and 6 per cent respectively by the lowest decile (El-Katiri and Fattouh 2011: 55). A 2005 World Bank study estimates that around 76 per cent of Egypt's electricity subsidies, and 87 per cent of fuel subsidies, leak to non-poor households (World Bank 2005c: 30).

These leakages are all the more serious in view of the significant fiscal burden fuel and electricity subsidies in many MENA countries constitute. Government funds are diverted away from other pro-poor investments such as targeted social assistance, public health programmes and education. In Yemen, for instance, around one third of governmental expenditure is allocated to fuel subsidies— funds which appear to benefit middle- and high-income households more than low income ones. In many MENA countries, the size of energy subsidies outweighs fiscal allocations to education and health, with Jordan's 2007 price reform showing evidence of the rebalancing effect of lower fuel subsidies (see Figure 13.2).

While there is a lot to say for pro-poor energy subsidies in general, the MENA region shows that targeted subsidies aimed at specific social groups, either via vouchers or by selecting special fuels used mostly by the poor, could both improve the subsidies' effectiveness and reduce their fiscal burden. Self-targeting programmes based on specific fuels, such as kerosene, have come under criticism for cementing the poor's reliance on those low-quality fuels at the bottom of the energy ladder, however. Household data from Yemen supports this view, although difficulties in accessing cleaner LPG also reflect the high one-off price of a bottle of LPG, as explained above.

On the other hand, Morocco's use of LPG subsidies demonstrates the opposite case of a relatively successful policy of broadening energy access to higher-quality fuels. Morocco has increasingly subsidized LPG bottles since the 1990s as the preferred fuel for household cooking, owing to LPG's features as a relatively efficient and clean domestic fuel. LPG subsidies in the Moroccan case are self-targeting and avoid to some extent those leakages associated with diesel and gasoline: LPG is almost exclusively used by households for cooking, allowing little

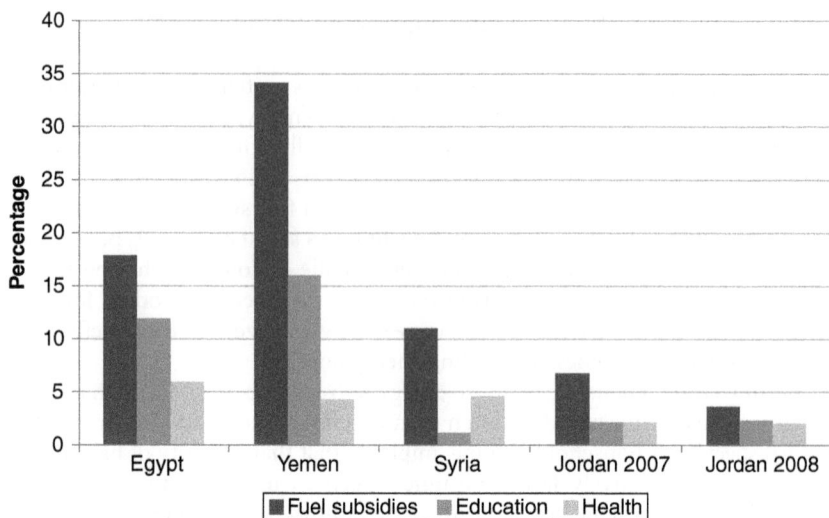

Figure 13.2. Share of governmental expenditure on fuel subsidies, education, and health in selected MENA Countries, 2008

Source: Egypt: World Bank 2013; Yemen: Breisinger et al. 2011, author's own calculations; Syria: IMF 2010a, MEES, World Bank 2013; Jordan: IMF 2010b, World Bank 2013

other use, for instance in agriculture or other industries. Household size, which correlates negatively with income levels, is a key factor in determining household use of LPG, suggesting larger (hence poorer) families benefit more from the subsidy than smaller ones. Even in the case of higher-income households, the use of LPG primarily for cooking precludes over-consumption.

However imperfect it may be, this system is a good way for Morocco to target low-income households in the absence of comprehensive national censuses and household income data. An account of the Atlas mountains' tribal communities in Morocco during the 1990s illustrates the effectivity of LPG subsidies in promoting LPG over kerosene and other liquid fuels (Kojima 2011):

> There is a primary dependence on non-commercial, organic fuels such as freely gathered firewood. Agricultural residues (walnut shells and corn cobs) and cattle dung are also used, although in comparatively minimal quantities...The major commercial fuel is liquid petroleum gas (LPG). All households own at least a small canister, refillable at the Imlil market...In earlier times, kerosene was the commercial fuel of choice but it has now been almost completely replaced by LPG. State subsidies provide a partial explanation: 36% of the LPG production cost is subsidized, compared to 3% for kerosene. The only other type of commercial fuel consumed in Imlil is charcoal...However, rarely do households opt to pay the price of 3 dirhams/kg ($0.35/kg), considered exorbitant compared to available substitutes (Dougherty 1994: 126).

Jordan, where the government raised kerosene prices during the 1990s, provides a similar experience:

> In 1993, under the energy-adjustment programme, kerosene's unit-price was increased in order to reduce the subsidies, and to reflect more accurately its true

economic cost. Hence kerosene consumption rates fell dramatically. Simultaneously, with the high growth-rate of the population, LPG emerged as a more attractive fuel, with reliable supplies for domestic, commercial and even industrial use, partly because it is a cleaner fuel compared with kerosene. In particular, LPG consumption increased sharply after the establishment of several factories, which manufacture LPG heaters for space-and-water heating' (Jaber and Probert 2001: 122).

One of the greatest impediments to the study of energy poverty in the MENA, and a major hurdle undermining the effectiveness of energy subsidy programmes intended to combat it, remains the lack of comprehensive household consumption data by country, both historic and current—a lack all the more surprising given the heavy fiscal burden of those subsidies in government budgets, particularly in countries with large social safety deficits such as Egypt, Morocco, Jordan, and Yemen. While energy subsidies per se, when used and targeted effectively, can be a blessing for households whose energy access would otherwise be restricted by the prohibitive cost of modern fuels and electricity, inadequate statistics can severely undermine their effectiveness.

Electricity access offers a whole range of additional pricing challenges. Morocco's rural electrification programme, PERG, was from the beginning based on the goal of full economic recovery of all electrification costs. This was one of the factors that allowed the programme to run successfully for over 10 years and cover the greatest possible number of households. PERG was co-financed by local and federal stakeholders: OME, which contracted most local service providers directly, contributed 55 per cent of the funds; local councils supplied 20 per cent, and local recipients 25 per cent. The financial burden of the programme on final recipients was alleviated by allowing households to spread their payments over seven years, free of interest, at a moderate rate of around 40 Dh per month and household (Agence Française de Développement 2013). Such approaches demonstrate that rural access to electricity can be both commercially viable for the government and supplying companies and financially feasible for low-income households. Other options exist to help electricity access without subsidizing electricity tariffs universally. Examples are connection subsidies, which assist poor households to overcome initial high investment costs via grants of government-sponsored development loans, or the subsidization of the first set of units of electricity consumption under a progressive tariff system based on actual household consumption.

13.3.3. Wider socio-economic solutions

Besides targeting access to energy itself, a more structural and increasingly more important goal must necessarily be the expansion and improvement of social welfare, which—if targeted well—could enable many more households to overcome income poverty as a structural hurdle to energy access. It is in this context that strengthened and more transparent governing institutions in the Middle East could significantly contribute to alleviating both poverty in general and energy poverty in particular, by enhancing their capability to register and survey households' income levels, assess their specific socio-economic needs, and respond to these needs in an accountable way. Part of this response could be significantly improved social safety systems as well as targeted investment in both urban and rural infrastructure. Similarly, political stability and peace are prerequisites for poverty alleviation and long-term investment in energy supply, alongside other sectors promoting economic growth and prosperity democratically, such as universal access to health and education. Iran's energy pricing reform of 2010–2011,[18] which raised energy prices across fuels and electricity and has since begun to redistribute the savings via direct cash transfers to households and industries, illustrates another way in which wasteful subsidy systems can be eliminated while compensating the poor for the loss of income resulting from surging energy prices.

13.4. CONCLUDING REMARKS

Over the past decades, MENA governments have invested immensely to provide affordable modern energy services to their citizens. Thanks to this investment, the region today does not rank among the emerging economies primarily associated with energy poverty. Yet the problem of energy poverty has not been eradicated from the region. Significant pockets of population, from Yemen to parts of North Africa and the Levant, continue to suffer from lack of electricity access and a high reliance on traditional biofuels. As shown in the case of Morocco, the persistence of this traditional use, at a time of demographic growth, is unsustainable, not only from a health standpoint but also in view of its long-term environmental effects, including deforestation and overall worsening living situations at the margins of society.

The many different facets of energy poverty in the MENA suggest a wide array of reasons for its persistence across the region. Those factors traditionally associated with energy poverty by the literature, primarily income poverty, play an important part in this explanation. Yet, as energy poverty also halts socio-economic development, the relationship is by no means linear or one-directional. Arguably, where incomes have been high and state expenditure on infrastructure and energy supply has been high, socio-economic progress has been high, which raises once more the question why indices such as the UN Millennium Goals do not factor in basic energy access as a condition to, rather than primarily the result of, higher development. The MENA region also uniquely combines other contributing factors, including the high concentration of energy poverty in rural areas, underlining the MENA's characteristic rural–urban divide in terms of income levels and infrastructure access; and the frequently overlooked role of local culture, often the result of a lack of education and access to the outside world in many rural areas particularly.

The MENA region's most powerful tool in combating energy poverty consists in access to finance, including via government-funded extension programmes and comparably low domestic energy prices. Electrification programmes such as those

---

[18] For a summary of the reform and its proceeds, see Tabatabai (2011) and Guillaume et al. (2011).

initiated by Morocco and Jordan, while imperfect in outreach, demonstrate the success well-conceived initiatives can have as well as the enormous social impact associated with universal electricity access. The MENA region's other policy tool aimed at facilitating energy access, energy subsidies, has been by contrast considerably more problematic. While arguably lowering income hurdles to energy access, many economic distortions associated with energy subsidies, including their relative inefficiency in targeting key consumer groups, means that in many MENA countries, subsidies constitute an ever-increasing fiscal burden. By crowding out other potential pro-poor investment in sectors such as education and health, the usefulness of this tool to combat energy poverty in the future must be questioned—making fiscal reform aimed at alleviating poverty levels across the region an ever more essential good.

APPENDIX A

Table 13A.1. The IEA's Energy Development Index (EDI), 2011

COUNTRY	EDI rank	EDI value	Commercial energy use per capita index	Electrification index	Electricity consumption index	Modern fuels for cooking index
Libya	1	0.923	1.000	1.00	0.812	0.882
Islamic Republic of Iran	2	0.889	0.932	0.98	0.644	1.000
Lebanon	3	0.850	0.478	1.00	1.000	0.924
Venezuela	4	0.844	0.740	0.99	0.687	0.959
Argentina	5	0.798	0.570	0.97	0.675	0.976
Jordan	6	0.773	0.392	1.00	0.701	0.997
Malaysia	7	0.741	0.733	0.99	0.642	0.596
Algeria	8	0.706	0.353	0.99	0.485	0.993
Syria	9	0.703	0.333	0.92	0.560	1.000
Uruguay	10	0.692	0.343	0.98	0.845	0.599
South Africa	11	0.681	0.838	0.72	0.669	0.497
Egypt	12	0.668	0.264	1.00	0.482	0.928
Costa Rica	13	0.616	0.310	0.99	0.610	0.551
Brazil	14	0.590	0.295	0.98	0.438	0.645
Cuba	15	0.581	0.285	0.97	0.475	0.595
Ecuador	16	0.563	0.246	0.91	0.284	0.808
Mongolia	17	0.550	0.363	0.63	0.268	0.941
Thailand	18	0.547	0.414	0.99	0.371	0.412
People's Republic of China	19	0.547	0.477	0.99	0.301	0.415
Morocco	20	0.532	0.136	0.97	0.187	0.839
Colombia	21	0.528	0.192	0.93	0.348	0.643
Panama	22	0.517	0.254	0.87	0.443	0.503
Dominican Republic	23	0.515	0.203	0.95	0.350	0.553
Tunisia	24	0.498	0.230	1.00	0.284	0.481
Jamaica	25	0.490	0.327	0.91	0.333	0.389
Paraguay	26	0.480	0.152	0.96	0.520	0.283
Bolivia	27	0.397	0.173	0.75	0.158	0.510
Peru	28	0.390	0.137	0.84	0.203	0.379

continued

Table 13A.1. continued

COUNTRY	EDI rank	EDI value	Commercial energy use per capita index	Electrification index	Electricity consumption index	Modern fuels for cooking index
Philippines	29	0.383	0.105	0.89	0.154	0.387
Vietnam	30	0.381	0.132	0.97	0.276	0.141
Yemen	31	0.378	0.092	0.32	0.100	1.000
El Salvador	32	0.361	0.182	0.85	0.215	0.201
Indonesia	33	0.297	0.205	0.60	0.194	0.189
India	34	0.294	0.139	0.72	0.098	0.221
Honduras	35	0.285	0.113	0.67	0.238	0.124
Guatemala	36	0.284	0.126	0.78	0.141	0.089
Botswana	37	0.280	0.246	0.39	0.337	0.153
Pakistan	38	0.270	0.102	0.58	0.164	0.235
Sri Lanka	39	0.258	0.085	0.74	0.112	0.097
Nicaragua	40	0.241	0.100	0.69	0.100	0.076
Gabon	41	0.230	0.187	0.29	0.370	0.074
Ghana	42	0.193	0.042	0.56	0.077	0.098
Zimbabwe	43	0.175	0.075	0.34	0.241	0.042
Bangladesh	44	0.168	0.031	0.34	0.051	0.253
Senegal	45	0.156	0.036	0.35	0.048	0.190
Côte d'Ivoire	46	0.135	0.030	0.41	0.043	0.061
Nigeria	47	0.134	0.043	0.44	0.046	0.004
Cameroon	48	0.130	0.031	0.42	0.029	0.035
Congo	49	0.120	0.049	0.29	0.046	0.090
Benin	50	0.110	0.044	0.15	0.022	0.221
Sudan	51	0.110	0.034	0.28	0.042	0.085
Angola	52	0.110	0.075	0.17	0.103	0.093
Nepal	53	0.102	0.006	0.37	0.023	0.012
Haiti	54	0.093	0.021	0.31	0.000	0.042
Cambodia	55	0.081	0.023	0.15	0.037	0.118
Eritrea	56	0.075	0.000	0.24	0.008	0.055
Zambia	57	0.074	0.045	0.09	0.134	0.029
Togo	58	0.053	0.011	0.10	0.042	0.058
Kenya	59	0.037	0.029	0.06	0.024	0.038
United Republic of Tanzania	60	0.022	0.022	0.03	0.021	0.014
Myanmar	61	0.018	0.020	0.02	0.021	0.008
Ethiopia	62	0.017	0.000	0.07	0.003	0.000
Mozambique	63	0.013	0.024	0.01	0.017	0.003
Democratic Republic of Congo	64	0.010	0.018	0.00	0.019	0.002

Source: IEA (2011)

REFERENCES

Achy, L. (2010). 'Morocco's Experience With Poverty Reduction: Lessons for the Arab World'. Carnegie Papers, Carnegie Endowment for International Peace, December [online], <http://carnegieendowment.org/files/morocco_poverty1.pdf> (accessed January 2013).

Agence Française de Développement (2013). *Le Programme d'électrification rurale global (PERG) au Maroc* [online], <http://www.afd.fr/home/AFD/L-AFD-s-engage/rioplus20/ projets-rio20/electrification-maroc> (accessed January 2013).

Akash, B. and Mohsen, M. (1999). 'Energy Analysis of Jordan's Rural Residential Sector', *Energy Conversion & Management* 40: 1251–8.

Al Bawaba Business (2012). 'Will the Lights go Out in Iraq?' [online], <http://www.albawaba.com/business/will-lights-go-out-iraq-445413>.

Al-Mohamad, A. (2001). 'Renewable Energy Resources in Syria', *Renewable Energy* 24: 365–71.

Al Sayegh, H. (2012). 'Iraq Pays High Price for Lack of Electricity', *The National*, 13 July 2012 [online], <http://www.thenational.ae/business/energy/iraq-pays-high-price-for-lack-of-electricity>.

Al-Soud, M. and Hrayshat, E. (2004). 'Rural Photovoltaic Electrification Program in Jordan', *Renewable and Sustainable Energy Reviews* 8: 593–8.

Arab Ministerial Conference (2003). *Abu Dhabi Declaration on Environment and Energy* [online], <https://www.ead.ae/TacSoft/FileManager/Conferences%20&%20Exhibitions/2003/44-AbuDhabiDeclaration-Eng.pdf>.

Assamiee, M. (2010). 'Yemen's Electricity Problem in Detail', *Yemen Times*, 18 January.

Barnes, D. and Toman, M. (2006). 'Energy, Equity, and Economic Development', in Lopez, R. and Toman, M.A., eds., *Economic Development and Environmental Sustainability: New Policy Options*. Oxford: Oxford University Press.

Benkhadra, A. (2009). 'Moroccan Project for Solar Energy', speech at the project launch ceremony, 2 November [online], <http://www.masen.org.ma/?Id=42&lang=en> (accessed June 2014).

Boucek, C. (2009). *Yemen: Avoiding a Downward Spiral*, Carnegie Endowment for International Peace, Report No. 102, September 2009.

Breisinger, C., Engelke, W., and Ecker, O. (2011). *Petroleum Subsidies in Yemen: Leveraging Reform for Development*, International Food Policy Research Institute, March.

Bruce, N. (2005). 'The Health Burden of Indoor Air Pollution: Overview of the Global Evidence', in World Health Organization, ed., *Indoor Air Pollution and Child Health in Pakistan*. Karachi: Aga Khan University seminar report, September.

Dougherty, W.W. (1994). 'Linkages between Energy, Environment, and Society in the High Atlas Mountains of Morocco', *Mountain Research and Development* 14(2): 119–35.

El-Katiri, L. and Fattouh, B. (2011). *Energy Poverty in the Arab World: The Case of Yemen*, MEP1, Oxford Institute for Energy Studies, August.

El-Laithy, H. and Abu-Ismail, K. (2005). *Poverty in Syria: 1996–2004*. United Nations Development Programme, June.

El-Moudden Saloua, M. (2004). *Impact du prélèvement du bois de feu sur les parcours steppiques cas d'Ighil n'Mgoun, province de Ouarzazate*, Institut Agronomique et Veterinaire Hassan II.

Fattouh, B. and El-Katiri, L. (2012). *Energy Subsidies in the Arab World*. Arab Human Development Report Research Paper Series, United Nations Development Programme [online], <http://www.undp.org/content/dam/undp/library/Environment%20and%20Energy/UNDP-EE-AHDR-Energy-Subsidies-2012-Final.pdf> (accessed January 2013).

Fritzsche, K., Zejli, D., and Taenzler, D. (2011). 'The Relevance of Global Energy Governance for Arab Countries: The Case of Morocco', *Energy Policy* 39: 4497–506.

Government of Yemen, World Bank, and United Nations Development Programme (2007). *Yemen Poverty Assessment*, Volume I, November.

Guillaume, D., Zytek, R., and Farzin, M. (2011). 'Iran—The Chronicles of the Subsidy Reform', IMF Working Paper, WP/11/167, International Monetary Fund.

Gupta, E. and Sudarshan, A. (2009). 'Energy and Poverty in India', in Noronha, L. and Sudarshan, A., eds., *India's Energy Security*, London: Routledge.

Hagen, E. (2008). 'The Role of Natural Resources in The Western Saharan Conflict, and The Interests Involved', conference paper, International Conference on Multilateralism and International Law. Pretoria: 4–5 December.

Heltberg, R. (2005). 'Factors Determining Household Fuel Choice in Guatemala', *Environment and Development Economics* 10: 337–61.

IEA (International Energy Agency) (2010). *Energy Statistics of Non-OECD Countries 2010.* OECD/IEA.

IEA (2011). *The Energy Development Index* [online], <http://www.iea.org/publications/worldenergyoutlook/resources/energydevelopment/theenergydevelopmentindex/> (accessed March 2013).

IEA, UNDP, and UNIDO (International Energy Agency, United Nations Development Programme, and United Nations Industrial Development Organization) (2010). 'Energy Poverty—How to Make Modern Energy Access Universal?' *World Energy Outlook, 2010 Edition*, Paris: International Energy Agency [online], <http://www.worldenergyoutlook.org/publications/weo-2010/>.

IFC Enterprise (2011). 'Enterprise Surveys: Infrastructure' [online], <http://www.enterprisesurveys.org/data/exploreTopics/Infrastructure> (accessed June 2014).

IMF (International Monetary Fund) (2010a). 'Syrian Arab Republic: 2009 Article IV Consultation—Staff Report; and Public Information Notice', IMF Country Report No. 10/86, March.

IMF (2010b). 'Jordan: 2010 Article IV Consultation—Staff Report and Public Information Notice', IMF Country Report No. 10/297, September.

Israel Ministry of Foreign Affairs (2012). 'The Humanitarian Situation in Gaza', 18 November [online], <http://www.mfa.gov.il/MFA/HumanitarianAid/Palestinians/Humanitarian_situation_Gaza_18-Nov-2012.htm?DisplayMode=print> (accessed January 2013).

Jaber, J. (2001). 'Prospects of Energy Savings in Residential Space Heating', *Energy and Buildings* 34: 311–19.

Jaber, S. and Probert, S. (2001). 'Energy Demand, Poverty and The Urban Environment in Jordan', *Applied Energy* 68: 119–34.

Kabarati, M. (2005). 'Identification of National Energy Policies and Energy Access in Jordan', conference paper, Energy Research Group, American University of Beirut, 25 January.

Kandil, M. (2010). 'The Subsidy System in Egypt: Alternatives for Reform', Policy Viewpoint Series, Cairo: Egyptian Centre for Economic Studies, December.

Karaki, S., et al. (2005). 'Electric Energy Access in Jordan, Lebanon and Syria', draft paper prepared for 'Energy Access II', Working Group Global Network on Energy for Sustainable Development.

Kojima, M. (2011). *The Role of Liquefied Petroleum Gas in Reducing Energy Poverty*. World Bank, Extractive Industries for Development Series No. 25.

Le Polain de Waroux, Y., and Lambin, E. (2012). 'Monitoring Degradation in Arid and Semi-Arid Forests and Woodlands: The Case of the Argan Woodlands (Morocco)', *Applied Geography* 32: 777–86.

Masera, O., Saatkamp, B., and Kammen D. (2000). 'From Linear Fuel Switching to Multiple Cooking Strategies: A Critique and Alternative to the Energy Ladder Model', *World Development* 28(12): 2083–103.

Mekonnen, A. and Köhlin, G. (2008). 'Determinants of Household Fuel Choice in Major Cities in Ethiopia', Environment for Development Discussion Paper Series. Environment for Development, and Resources for the Future, 18 August [online], <http://www.rff.org/RFF/Documents/EfD-DP-08-18.pdf>.

OPEC Fund for International Development. (2010). 'Energy for Sustainable Development: Opportunities and Challenges', paper presented at the 9th Arab Energy Conference, Doha, 9–12 May.

Phillips, S. (2007). 'What Comes Next in Yemen? Al-Qaeda, the Tribes, and State-Building', Carnegie Paper No.103, Carnegie Endowment for International Peace, March 2010 [online], <http://carnegieendowment.org/files/yemen_tribes1.pdf> (accessed 15 October 2013).

Ramon, L. and Toman, V. (2006). 'Energy, Equity and Economic Development', in Ramon, L. and Toman, V., eds., *Economic Development and Environmental Sustainability: New Policy Options*. Oxford: Oxford University Press.

Raspaud, L. (2012). 'Sustainable Energy and the Fight Against Poverty', *Field Actions Science Reports*, Special Issue 6.

Republic of Yemen, Ministry of Energy and Electricity (2009). *Al-Taqrir al-Sanawy 2009* [Annual Report 2009, in Arabic].

Reuters (2012). 'Turkey says Syria Halts Power Imports; Network Damaged', Reuters, 11 October 2012.

Sara, F. (2011). *Poverty in Syria. Towards a Serious Policy Shift in Combating Poverty*. Strategic Research and Communication Center [online], <https://www.cimicweb.org/cmo/ComplexCoverage/Documents/Syria/Poverty_in_Syria.pdf> (accessed January 2013).

Schilling, J., Freier, K., Hertig, E., and Scheffran, J. (2012). 'Climate Change, Vulnerability and Adaptation in North Africa with Focus on Morocco', *Agriculture, Ecosystems and Environment* 156: 12–26.

Shaher, M. (2011). 'Billions of Rials Lost from Electricity Sector', *Yemen Times*, 12 February.

Tabatabai, H. (2011). 'The Basic Income Road to Reforming Iran's Price Subsidies', *Basic Income Studies* 6(1).

World Bank (2005a). *Household Energy Supply and Use in Yemen*, Volume I, Main Report, No. 315/05, December.

World Bank (2005b). *Household Energy Supply and Use in Yemen*, Volume II, Annexes, Report No. 315/05, December.

World Bank (2005c). 'Egypt—Toward a More Effective Social Policy: Subsidies and Social Safety Net', No. 33550-EG. [online], <http://siteresources.worldbank.org/INTPSIA/Resources/490023-1171551075650/Egypt_PSIA_121605.pdf> (accessed January 2013).

World Bank (2009). *Lebanon: Social Impact Analysis - Electricity and Water Sectors*. Washington, DC: World Bank.

World Bank (2013). *World Development Indicators* [online], <http://data.worldbank.org/data-catalog/world-development-indicators> (accessed January 2013).

Zejli, D. and Bennouna, A. (2009). 'Wind Energy in Morocco: Which Strategy for Which Development?', in Mason, M. and Mor, A., eds., *Renewable Energy in the Middle East*. Dordrecht: Springer, 151–73.

14

Energy Poverty in Sub-Saharan Africa
Poverty Amidst Abundance

Abeeku Brew-Hammond, Gifty Serwaa Mensah,
and Owusu Amponsah

14.1. INTRODUCTION

Many have been the efforts made towards eradicating or reducing global poverty; however, poverty is still around. A myriad of factors and strategies must be put in place in order to reduce poverty. Although access to modern energy services is not the only precondition for economic development and thus poverty reduction, it is, and remains, a key prerequisite for economic development and welfare creation. There is an established relationship between access to energy services, or energy poverty reduction, and economic development (UNDP 2010; Brew-Hammond 2007; Practical Action 2010; UNDP and WHO 2009).

Much literature published recently has made reference to the 1.3 billion and 2.7 billion people who lack access to electricity and clean cooking fuels respectively. Over 95 per cent of these energy-poor people reside either in developing Asia or Sub-Saharan Africa (United Nations 2011). According to UNDP and WHO (2009), a greater percentage of the world's energy-poor live in the least developed countries (LDCs). Of the 48 countries designated as LDCs by the United Nations, about 69 per cent (i.e. 33) are in Sub-Saharan Africa (UNCTAD 2012), which implies a high level of energy poverty in the region; more precisely, 585 million of the people in Sub-Saharan Africa lack access to electricity and some 653 million people lack access to modern cooking fuels (OECD/IEA 2011). Figure 14.1 shows that apart from South Africa and Ghana, all the other Sub-Saharan African countries have at least 50 per cent of the population living without access to electricity. In fact, Africa's average energy consumption per capita of 0.66 tons of oil equivalent (TOE) is approximately one-third of the global average of 1.8 TOE in 2008 (UNIDO 2009). This low level of access to electricity and modern cooking fuels hampers the economic and social development as well as the health of the population living in such poverty.

Among the many factors that account for energy poverty is the absence of energy resources that can be exploited by a nation to make energy services available to its citizenry. Fortunately for Africa, this does not seem to be the factor responsible for the high level of energy poverty.

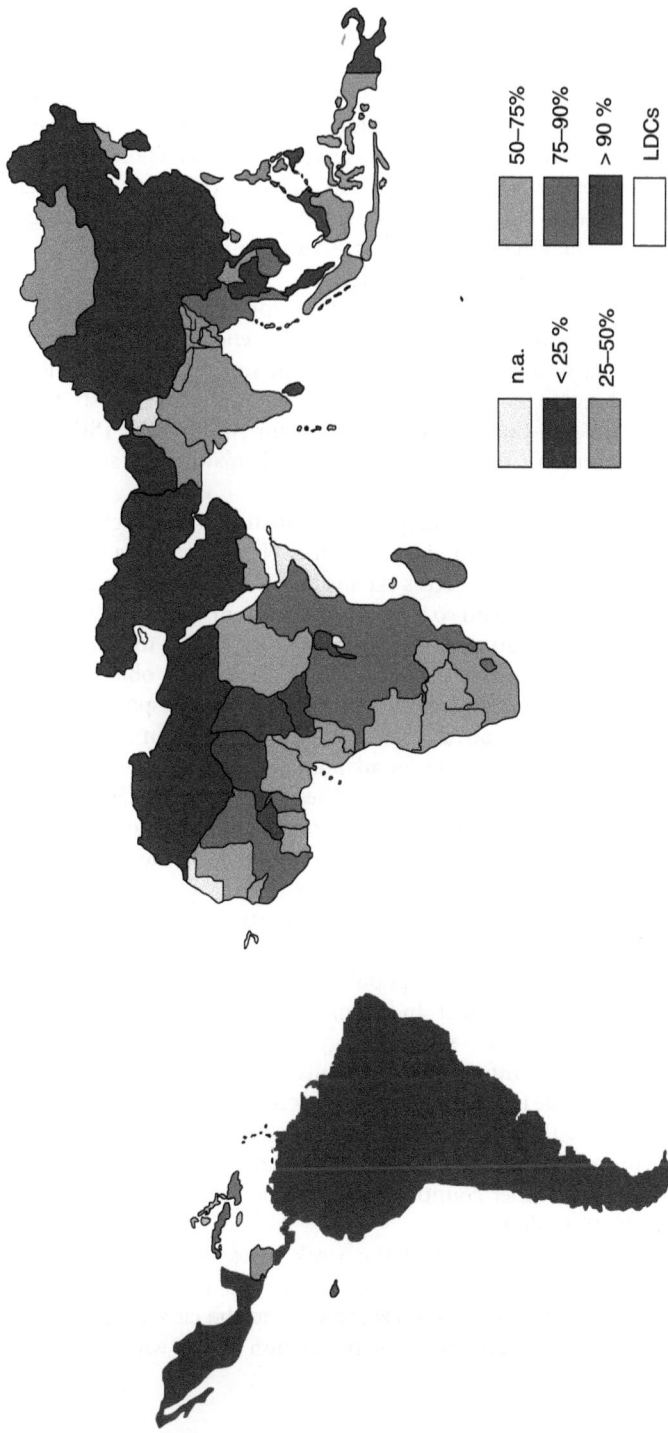

Figure 14.1. Share of people without electricity access in developing countries, 2008

Source: UNDP and WHO 2009

Legend:

n.a.

< 25 %

25–50%

50–75%

75–90%

> 90 %

LDCs

In fact, the region is endowed with tremendous quantities of renewable and non-renewable energy resources. In terms of renewable energy, sunshine is readily available throughout the year. Wind energy potential is also strong. Africa appears to have over 1,750 TWh potential of hydropower and 14,000 MW of geothermal potential (UNIDO 2009). In terms of fossil fuel reserves, there is an abundance of coal, over 117 billion barrels of proven oil reserves, and 14.6 trillion cubic metres of proven gas reserves in Southern Africa as of 2007 (UNIDO 2009). Other countries such as Ghana, Uganda, and Mozambique are now discovering new oil and gas reserves. The electricity needs of Africa can be met with either fossil energy resources or renewable energy resources.

So why does Africa, and Sub-Saharan Africa for that matter, suffer such extreme energy poverty in spite of all the abundant energy resources available on the continent? Take solar power, for example. The sun shines always and there is available technology to harness the energy from the sun; why is this not happening? In the wake of the Sustainable Energy for All initiative (SE4ALL) by the UN Secretary General, it is important to find answers to such relevant questions.

It should be emphasized that the energy access situation is not all gloomy across all of Africa. We may observe from Figure 14.1 that Northern Africa has achieved considerable success in electricity access expansion. In addition, most of Southern Africa is also well illuminated, and so is Ghana to an extent. In fact only 1 per cent and 3 per cent of the population in North Africa lack access to electricity and modern cooking fuels respectively (OECD/IEA 2011). The objective of this chapter is thus to examine the factors which have been responsible for the successful reduction of energy poverty among some countries in Africa (notably Egypt and Ghana) and to draw lessons for adaptation and replication. The chapter also examines the cases of some selected countries which experience gross energy poverty and assess the factors that have contributed to such levels of energy poverty. Lessons will then be drawn from the findings of these case studies, which will inform the chapter's recommendations.

The chapter will discuss Egypt and Ghana as success stories on one hand, and review the cases of Nigeria and Uganda as countries with widespread energy poverty on the other hand. With a population of about 81 million people[1] (World Bank World Development Indicators 2011, cited in BTI 2012) and electricity consumption rate of 7 per cent, Egypt has attained almost universal access to electricity (Energy Information Administration 2012). The government of Ghana has also been able to scale up electricity access about fivefold from 15 per cent to 73 per cent within two decades. Electricity access programmes and strategies in Egypt and Ghana clearly provide useful cases from which lessons can be drawn to enhance other countries' efforts in achieving universal access to electricity by the 2030 deadline.

Nigeria has been chosen as one of the weak cases due to the paradox it represents. With all the fossil resources the country has at its disposal to make both electricity and modern cooking fuels available to the citizenry, the country still falls in the 50–75 per cent range of population living without electricity

---

[1] Ranked third in Africa after Nigeria and Ethiopia.

(Figure 14.1). Usually, developing countries place more importance on electricity access provision at the expense of modern cooking fuels. This implies that Nigeria will have an even higher rate of poverty in terms of access to modern cooking fuels. The country therefore proves to be an interesting case study for the chapter.

Uganda, on the other hand, has been chosen basically for its very low level of access to electricity (9 per cent), which implies that access to modern cooking fuels is even lower. It also provides an intriguing case to examine what has been responsible for such very low rates of access to modern energy services.

Senegal offers very useful lessons for other African countries to reduce their overdependence on wood fuels in favour of modern and cleaner cooking fuels. For example, in Senegal, with a population of 10.2 million people, a total of 110,000 tonnes of LPG is used annually, while Nigeria, with a population of about 150 million people, consumes just 58,000 tonnes of LPG per annum. Senegal's success in reducing over-dependence on wood fuels provides important lessons to other countries in how to reduce carbon emissions caused by wood fuel extraction.

14.2. SUCCESS STORIES

Hailu (2012) maintains that there is a need to evaluate existing energy access practices and processes with a view to developing models, policies, and institutional set-ups that take into account and respond to the peculiar problems and conditions across the continent. Brew-Hammond (2007), after reviewing energy access experiences in Africa, suggests that practically all countries in Sub-Saharan Africa would need to do more or less as South Africa and Mauritius have done in order to expand energy access. What can be found in common in the two views is the need for poorly performing countries to draw lessons from the experiences of the better-performing nations towards expanding access to modern energy. To us, the first step in bringing energy access to those that need it is to review the experiences of the countries that have made significant progress in reducing energy poverty. The intention is to highlight their good practices in order for other countries within the continent to replicate universal access to modern energy services by the deadlines set by regional bodies such as New Partnership for Africa Development (NEPAD) and sub-regional bodies such as the Economic Community for West African States (ECOWAS) and the Economic Community of Central African States (CEMAC). The authors thus review the experiences of Egypt and Ghana, which have been identified as well-performing countries and have thus achieved significant levels of access to electricity (see Table 14.1). We also review Senegal's national butanization programme and its effects on Liquefied Petroleum Gas (LPG) access and use. In reviewing the countries' experiences, we place emphasis on what was done right—to assist other developing countries in their energy planning process.

Table 14.1. Energy access rates in Africa

Country	Electrification rate	Country	Electrification rate
Malawi	9.0	Gabon	36.7
Uganda	9.0	Congo	37.1
DR Congo	11.1	Zimbabwe	41.5
Mozambique	11.7	Senegal	42.0
Tanzania	13.9	Botswana	45.4
Burkina Faso	14.6	Côte d'Ivoire	47.3
Lesotho	16.0	Cameroon	48.7
Kenya	16.1	Nigeria	50.6
Ethiopia	17.0	Ghana	60.5
Zambia	18.8	Mauritius	99.4
Madagascar	19.0	Algeria	98.0
Togo	20.0	Libya	97.0
Benin	24.8	Morocco	85.0
Angola	26.2	Egypt	98.0
Eritrea	32.0	South Africa	66.0
Namibia	34.0	Tunisia	99.0
Sudan	35.9		

Source: Data taken from WEO 2011, cited in Hailu 2012; IEA 2006, cited in Brew-Hammond 2007

14.2.1. The Egyptian electrification experience

The literature indicates that the Maghrebian (North African) countries have high electricity access rates. Egypt, among other North African countries, is one of the best-performing countries in Africa in terms of electricity access rate (IEA 2006, cited in Brew-Hammond 2007). According to the International Energy Agency (cited in Energy Information Administration 2012), the Egyptian electrification rate was approximately 99.6 per cent in 2009, comprising 100 per cent urban access and 99.3 per cent rural access. To underscore the significance of target-setting in expanding access to energy services, up to the 1970s the Egyptian government's policies and strategies for the electricity sector focused on providing electricity to all end-users. This was three decades before the NEPAD's 2001 target to increase to 35 per cent or more the access to reliable and affordable commercial energy supply of Africa's population by 2020. Egypt's success may partly be attributed to its progressive investment in the electricity sector and its commitment to achieving universal access regardless of how remote some communities may be from the grid. By 2006, 98 per cent of Egyptians had access to electricity, about three times higher than the NEPAD's target of 35 per cent and about 23 per cent more than the Forum of Energy Ministers of Africa's (FEMA) 75 per cent target by 2015.

The rural electricity access rate, which has been low in many developing countries, was increased in Egypt through the rural electrification programme, which was piloted in 1971 but scaled up in the early 1990s. The dispersed nature of both houses and rural communities coupled with their low demand, however, rendered economically unfeasible the extension of grid electricity to the most remote and under-populated communities. Nevertheless, decentralized energy options (such as solar photovoltaic and the utilization of solar thermal energy

for heating and cooling purposes in tourist resorts and new villages) were introduced. Furthermore, the role that decentralized systems have played in electricity
generation in Egypt is not unique, as other successful countries such as Ghana and
Morocco have similar experiences.

The country's renewable energy (mainly from wind) installed capacity is 550
MW and is about 25 per cent of Ghana's total electricity installed capacity (Energy
Commission 2011). The target of the government of Egypt is to generate 20 per
cent of its energy needs from renewable energy sources, mainly wind, paralleling
the European countries (OECD 2010).

To facilitate poor people's access to electricity in Egypt, the government
subsidizes power. Tariffs for the first segment of domestic consumption—less
than 50 kWh per month, mainly among low-income families—have remained
unchanged at 5 Pt (approximately 0.62 €) for a number of years. The success of the
rural electrification programme is evident in the 99.3 per cent rural access rate.

Egypt encouraged the private sector to enter into the electricity market in the
1990s in the forms of Independent Power Producers (IPPs) and Build, Own,
Operate and Transfer (BOOT) in an attempt to liberate (commercialize) the
electricity market. There are three long-term (20-year) BOOT contracts with the
Egyptian Electricity Transmission Company, 12 private electricity producers, and
15 licensed private electricity distributors (El-Salmawy n.d.). Acknowledging the
importance of the private sector in expanding generational capacity and improving the quality of supply, Egypt, under a new market reform in 2000, sought to
'establish a fully competitive electricity market, where electricity generation,
trading, transmission, distribution and supply activities are fully unbundled'
(OECD 2010). The reform ought to have commenced in 2000, the Egyptian
Electricity Holding Company (EEHC) having separate subsidiaries for the production, transmission, and distribution of electricity. The subsidiaries were to be
privatized under the economic reform, but this plan was never implemented. Thus
private sector participation is limited to the BOOT schemes (MENA/OECD
2010), with the state having excessive control over electricity production, transmission, and distribution. Nevertheless, Egypt offers a very low electricity rate of
between EUR 0.02 and 0.04 per kWh while many emerging countries' rates are
between EUR 0.06 and 0.18 per kWh. In addition, the Egyptian government has
encouraged local engineering companies to participate in the energy market
through a local content policy. The UPDEA (2009) concludes that the local
content policy has proven itself to be effective in increasing the share of local
components in the project and in reducing the cost per kWh.

14.2.2. The Ghanaian electrification experience

In 1989, a decade before the NEPAD's target, the government of Ghana embarked
on an ambitious programme—the National Electrification Scheme (NES)—to
electrify the whole country by the year 2020 (Opam 1995; Abavana 2004). The
electricity access rate prior to the launching of the NES was estimated at 15 per
cent, with rural access about 5 per cent. The NES comprised the District Capitals
Electrification Programme (DCEP) and the Self-Help Electrification Programme
(SHEP). The purpose of the DCEP was to extend the grid to all the district capitals

(Opam 1995) in line with the growth pole concept. The multiplier effects of the development of the district capitals were to be experienced in nearby communities. The DCEP successfully connected all the district capitals to the grid.

The SHEP, on the other hand, sought to connect communities within 20 km of an existing 33 kilovolt (KV) or 11 KV sub-transmission line to the national grid (Abavana 2004). Communities may apply for electrification projects provided they procure all the poles required for the low voltage (LV) network and have a minimum of 30 per cent of the houses within the community wired. The programme has a two-pronged objective: ensuring local ownership and ensuring that demand from such communities renders the connection viable. Once these conditions are satisfied, the government bears the cost of providing the conductors, pole-top arrangements, transformers, and other installation (Abavana 2004).

In 2001, the Ministry of Energy formulated an energy policy dubbed 'Energy for Poverty Alleviation and Economic Growth: Policy Framework, Programmes and Projects'. The government of Ghana underscored its commitment to universal access to electricity. The ministry thus continued to support rural electrification, but this time through both grid extension and other decentralized options—mainly solar PV. Like the Egyptian rural electrification programme, the government of Ghana introduced the decentralized systems to enable access to electricity for remote off-grid communities. The total installed capacity for PV systems was estimated at 1 MW peak (MWp) as of 2002 (Edjekumhene et al. 2006), and over 4,000 solar PV systems had been installed by 2004 (Abavana 2004). The Government's target is to increase the contribution of modern renewable energy (i.e. renewable energy sources excluding hydropower) from 0.01 per cent to 10 per cent of the electricity generation mix by 2020.

By 2011, 73 per cent of Ghanaians had access to electricity (Energy Commission 2011; Miller et al. 2011), which represents about a five-fold increase from the 1989 access rate of 15 per cent. Clearly, the government of Ghana's electricity programmes have enhanced access to electricity, but Ghana is not the only nation in the region to make progress. For instance, the Programme pour l'Electrification Rurale Global (PERG) in Morocco increased the rural electricity access rate in that country from 18 per cent in 1995 to 96.5 per cent by the end of 2009.

The private sector in Ghana, like that of Egypt, has been encouraged to support electricity generation. In 2000, Ghana's first IPP, the Takoradi International Company (TICO), commissioned an electricity-generation thermal plant. TICO has a 100 per cent capacity utilization and feeds the national grid with 1,040 GWh of electricity.

Energy efficiency has been a major area of concern, not only for the government of Ghana but for the international community as encapsulated in the UN General Assembly Sustainable Energy for All (SE4All) initiative. The government of Ghana replaced about 6 million incandescent lamps with energy-saving compact fluorescent lamps under the 'Efficient Lighting Retrofit Initiative Programme'. According to the Ministry of Energy (2012), the programme has led to savings of over 124 MW and energy cost savings of over USD33 million per annum. With assistance from the UNDP and the Global Environment Facility (GEF), the government of Ghana introduced the Comprehensive Refrigerating Appliance Efficiency Project, which involves the extension of standards and labels to refrigerating appliances and the replacement of obsolete refrigerators with more efficient versions under a turn-in and rebate scheme (Ministry of Energy 2012). Ghana has successfully banned the importation of second-hand refrigerators, which may impact energy consumption and efficiency should the law be effectively enforced.

14.2.3. Access to modern cooking fuels in Senegal

The government of Senegal in 1974 launched the national butanization programme with the ultimate aim to replace 50 per cent of charcoal consumption with LPG in key urban areas. Over 60 per cent of total energy needs and 90 per cent of household energy needs were derived from wood fuel, which was extremely detrimental to the country's natural forest cover. LPG consumption in Senegal subsequently expanded rapidly during the 1990s. According to the World Bank (2001), LPG consumption witnessed a triple jump from 33,500 tonnes per annum in 1990 to over 100,000 tonnes (Figure 14.2) by 1999 (World Bank 2001) and further to 124,000 tonnes in 2009 (World Bank 2011). With this impressive rise in LPG demand, Senegal has emerged as the top major consumer of LPG in West Africa on an absolute and on a per capita basis (World Bank 2001).

A number of policy measures were implemented in Senegal's efforts towards substituting biomass consumption with LPG. The government established the LPG subsidy in 1987 to promote LPG use, increased wood cutting licensing fees, and created a land allocation system for charcoal production. The other policies include a progressive increment in the official sale price of charcoal and the construction of new refilling centres in remote regions to encourage the utilization of LPG among the inhabitants of remote rural communities. We observe that enabling policies play important roles in increasing access to and utilization of LPG. This conclusion was reached after an analysis of the impact that government subsidies made on LPG use in Ghana. LPG emerged as a major fuel used by commercial drivers owing to the subsidies they enjoyed (Ahiataku-Togobo 2013).

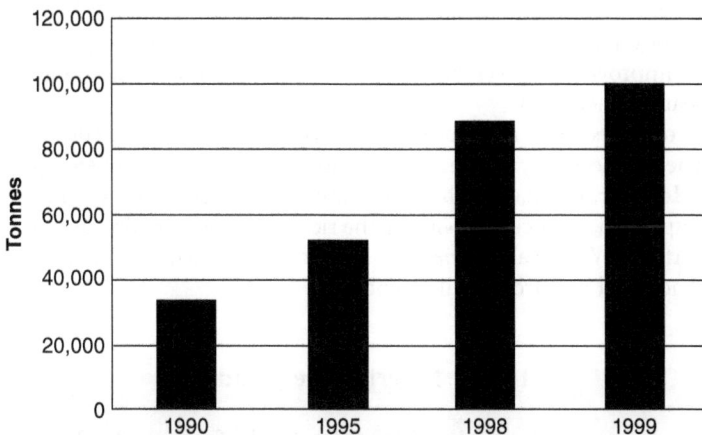

Figure 14.2. Senegal's LPG consumption from 1990 to 1999
Source: World Bank 2001

Ahiataku-Togobo argued that the benefits of fuel subsidies are hardly enjoyed by the target group. However, what remains as a consensus is that the LPG subsidies in Ghana promoted the domestic and commercial application of LPG.

The successful scale-up of LPG use in Senegal has not avoided practical problems—the first being gas stoves that were not well suited for Senegalese households and their cooking habits. Von Moltke et al. (2004) explain that the LPG cookstoves in Senegal were ill-suited to the needs of Senegalese households because the standard LPG stove was attached to a 12 kg cylinder by a flexible tube and a metal exhaust valve, which could not support the cooking habits of the households. Cooking utensils could not be used easily on the cooking stoves. The prices of cookstoves were also beyond the means of many households. The government thus introduced numerous measures to curtail the challenges, measures that are worthy of emulation by African governments. Stoves were redesigned to cater for the needs of Senegalese households with the introduction of the cooker model, which was equipped with a 2.75 kg gas bottle. The government also removed taxes levied on imported equipment and introduced a price structure taking into account different income groups. In 2002, around 71 per cent of urban households in Senegal and even 88 per cent of households in Dakar used LPG as primary cooking fuel (ANSD 2006). The Senegal Ministry of Energy estimated the annual saving of firewood and charcoal to be 70,000 tons and 90,000 tons respectively, which represents 15 per cent of current demand for those fuels (Sokona et al. 2003, cited in Schlag and Zuzarte 2008; Von Moltke et al. 2004).

14.3. WHERE WE WENT WRONG

As stated in the introductory section, Africa's failure in providing modern energy to the populace has not been due a lack of resources. We observe from Tables 14.2 and 14.3 that the largest ten hydrocarbon producing nations in Sub-Saharan Africa experience staggering levels of energy poverty which are not synonymous with their energy resource wealth. These countries also experience high levels of economic poverty as there is not enough access to modern energy services for people to improve their lives in terms of health, education, entertainment, and income-earning activities.

These countries could make use of their fossil resources or the currently flared gas to generate electricity or use it directly to turn the situation of energy poverty around. However, this has not been the case. In an attempt to find out why this has not happened, we focus on Nigeria, the richest nation in hydrocarbons in Sub-Saharan Africa. We will also review the case of Uganda, which is energy-poor but has not had fossil resources until recently.

14.3.1. Nigeria—the paradox

People usually express shock when they realize for the first time that Nigeria is one of the countries with the least access to modern energy services; what they usually know is that the country is very rich in fossil resources. Someone who recently

Table 14.2. Production and reserves in assessed Sub-Saharan African countries (ranked by oil reserves)

Country	Oil			Gas		
	Reserves (billion barrels)	Production (mb/d)	Exports (mb/d)	Reserves (bcm)	Production (bcm/yr)	Exports (bcm/yr)
Nigeria	36.2	2.35	2.03	5,207	29.3	18.9
Angola	9	1.70	1.64	270	0.8	—
Sudan	5	0.47	0.39	85	—	—
Gabon	2	0.23	0.22	28	0.1	
DR Congo	1.6	0.21	0.21	91	—	—
Chad	1.5	0.14	0.14	—	—	
Equatorial Guinea	1.1	0.36	0.36	37	1.3	—**
Cameroon	0.2	0.09	0.06	135	—	
Côte d'Ivoire	0.1	0.06	0.03	28	1.7	—
Mozambique	—	—	—	127	2.7	2.7
Total	56.8	5.61	5.09	6,008	35.9	21.6
% in world	4.30%	7.00%	12.10%	3.40%	1.20%	5.20%

** LNG exports commenced in 2007 from Equatorial Guinea.
Source: IEA 2008

Table 14.3. Number of people without access to electricity and relying on fuelwood and charcoal for cooking in the 10 richest hydrocarbon-producing countries in Sub-Saharan Africa

Country	Total population, 2006 (millions)	Number of people without electricity access	%	Number of people relying on fuelwood and charcoal for cooking	%
Angola	16.6	14.6	88	15.7	95
Cameroon	18.2	14.2	78	14.2	78
Chad	10.5	10.1	97	10.2	97
DR Congo	3.7	2.9	78	2.9	80
Côte d'Ivoire	18.9	11.6	61	14.7	78
Equatorial Guinea	0.5	0.4	73	0.3	59
Gabon	1.3	0.9	70	0.4	33
Mozambique	21	18.6	89	16.9	80
Nigeria	144.7	76.6	53	93.8	65
Sudan	37.7	26.9	71	35.2	93
Total	273.1	176.9	65	204	75

Source: IEA 2008

visited the country and saw first-hand the energy access situation in the country exclaimed; 'I was shocked when I went to the market, everybody had a generator and the whole place was so noisy' (Awafo 2013). Nigeria represents a classic paradox: abundant energy resources and widespread energy poverty. With 28 power cuts per day (see Table 14.4), it is not surprising everyone feels the need to get a generator to ensure constant supply of electricity in order to carry out their businesses.

Table 14.4. Nigeria socio-economic indicators

Population living in poverty (1980)	17 million
Population living in poverty (2010)	112 million
Access to electricity	47%
Number of households without electricity	15.3 million
Number of blackouts per day	28
Electricity consumption per capita	150 kWh
% of population dependent on biomass	72%

Source: Eleri et al. 2012

Only 47 per cent of Nigerians are connected to the national grid, with as many as 28 blackouts per day (Eleri et al. 2012). One writer refers to this level of power outages as embarrassing, where even the State House is not spared and has had to be powered with generators to ensure a 24-hour supply of electricity (Oluwole et al. 2012). One underlying cause of these power outages is the country's low generation capacity of about 3,600 megawatts (MW), leading to a demand deficit of about 7,500 megawatts (MW) (Oluwole et al. 2012; Aderibigbe 2010).

The country's electricity is generated using 72.9 per cent fossil fuel and 27.1 per cent hydropower (World Bank 2011). Data from the IEA also places access to electricity in Nigeria in 2010 at 50 per cent of the total population (OECD/IEA 2013).

Nigeria has proven oil reserves of 37.20 billion barrels and was ranked tenth in the world in 2012. It also had 187 trillion cubic feet of proven natural gas reserves in 2011 and was ranked eighth in the world. The country has some coal as well (EIA 2012). It is even reported that the gas Nigeria flares from oil production alone could sufficiently meet a substantial portion of the needs of the entire continent of Africa (Ayoola 2011).

The situation is no different on the cooking fuel scene. Again, in spite of these resources available to the country, 74 per cent of Nigerians rely on traditional biomass for cooking (OECD/IEA 2013). The use of firewood for cooking is not only prevalent in the rural areas but also in urban and semi urban areas. Whereas rural populations usually gather their own firewood, the urban communities buy theirs, with as many as 38 per cent of households in the country buying firewood from the market (Eleri et al. 2012). So why does a country with so many resources suffer such extreme energy poverty?

Explaining energy poverty in Nigeria

Eleri et al. (2012) found a number of factors that contribute to the high incidence of energy poverty in Nigeria. These included a weak political will to promote rural electrification. The country in 1981 embarked on a rural electrification pro-gramme which was intended to connect all local government headquarters to the national grid, allowing onward connection to all parts of the country. The Rural Electrification Agency (REA) was set up to carry out this objective, but political support to pursue these objectives has waned over the years. In countries where electricity access has improved tremendously, the key driving force, in addition to other factors, has been the government's will and commitment to providing energy access to the people. Thus if this key ingredient is missing, then one expects Nigeria, in spite of her abundant resources, to be energy-poor as there is no commitment from the key stakeholder to harness these energy resources.

The same authors report that there has been no budgetary attempt by any government, without exception, to deal with cooking fuel challenges faced by the poor, and the poor unfortunately constitute 69 per cent of the entire population. Once again, a conscious effort must be made to promote access to clean cooking fuels. In the examples given in the success stories of Africa, governments designed programmes with incentives to promote the use of clean fuels for cooking. The Senegal's LPG programme is a classic example of the role that deliberate government efforts can play in increasing access to modern fuels. Similarly, in Ghana, the government subsidized LPG as an incentive to get households to switch from the use of biomass. This raised the LPG usage rate to 18.2 per cent, according to the results of the country's 2010 population and housing census. Although 18.2 per cent is not necessarily an achievement to write home about, the point remains that deliberate efforts must be made by the government if households are to switch or at least reduce the use of biomass for cooking. Other factors that contribute to the low level of energy access in Nigeria are discussed in the following paragraphs.

Non-implementation of agreed policy, legal, and regulatory frameworks: like most developing countries, Nigeria has fine policies and programmes that sit on shelves and gather dust without being implemented to bring about desired changes.

Weak institutional structures: institutions are critical to expanding energy access to the energy-poor. Strong institutions that are tailor-made for the purpose of energy access provision are critical. These institutions must also work with other relevant agencies, as energy provision is broad and involves many actors. Such strong institutions are lacking in Nigeria. In actual fact, the board and management of the REA were removed in 2009 due to what is described as 'serious governance issues' (Eleri et al. 2012).

Inadequate access to finance: financing is, and will be, at the heart of every project or programme. When good plans are made to expand energy access, institutions are strong, and everything else is in place, there will still be no expansion of energy access to those who need it if there is no funding to finance the programmes. For most governments, the existence of different national issues which require government funding usually makes it difficult to expand energy access, especially to the rural poor. These poor people who are usually unable to pay for power and for clean and efficient cooking fuels are therefore left to their fate. Such is the case in Nigeria, according to Eleri et al. (2012).

Energy poverty is reportedly highest in LDC countries (UNDP; WHO 2009). Although Nigeria is not on the list of LDCs, the country nonetheless experiences a high level of poverty. The country has been unable to develop economically at the same rate as the population growth, thus resulting in the high incidence of poverty. In Table 14.4, it can be observed that there has been an exponential increase in the poverty level in Nigeria, from 17 million people to 112 million in 30 years, representing a rise from 27.2 per cent to 69 per cent of the total population.

We know that access to energy and economic development are related, but which comes first? Must a nation provide its populace with access to modern energy services for economic development to occur, or must that nation and its populace have the economic ability before they can afford the cost of modern energy access provision? This chapter does not seek to answer that question except to reinforce the fact that there is a link between economic poverty and energy poverty. Thus the high level of economic poverty in the country correlates with the high incidence of energy poverty.

Using what you have to get what you need

According to the IEA 2008 World Energy Outlook, providing access to modern energy services for the energy-poor in oil- and gas-producing countries is well within the means of such countries. The agency estimated that the capital cost of providing electricity and LPG stoves and cylinders to the households lacking access to these services in the top ten hydrocarbon-producing countries in Sub-Saharan Africa between 2006 and 2030 would be approximately equivalent to just 0.4 per cent of the governments' income from oil and gas exports over the said period. For Nigeria in particular, it would cost only 0.3 per cent of the government's revenue from the export of oil and gas products to provide universal access to electricity and LPG.

From the above illustrations it can be concluded that Nigeria is certainly able to provide universal access to modern energy services for the people; the question remains whether there is enough political will and commitment to advance such a cause. The IEA calls for efficient revenue allocation and government accountability in utilizing public funds (IEA 2008). If this happens, then revenue from oil and gas could be used to improve the energy access situation in the country, using locally available materials which are the abundant gas and oil resources. This should be done in a manner that reduces negative impacts on the environment by adhering to international standards of clean production and pollution control.

Resources should also be committed to harnessing the abundant solar resources available in the country, as well as wind and hydropower and other renewable resources. This will not only be good for the environment but also create jobs for sections of the population. Exploiting off-grid renewable energy solutions is especially useful in providing electricity access to remote areas where extending the national grid is cost-prohibitive. Biogas production from both solid and liquid wastes could also be promoted as a means of providing clean cooking fuels, with the added benefit of waste treatment.

Since affordability is also a barrier to access among the poor, especially the rural poor, the government should deal with energy poverty in a manner that links energy access provision with the larger development agenda and policies targeted at raising incomes and promoting economic development (IEA 2008).

14.3.2. Uganda

Uganda, unlike Nigeria, could not boast of fossil resources until recently; even now, these resources are not yet being produced locally. The country therefore imports all of its petroleum products from neighbouring Kenya and Tanzania (EIA 2012; Saundry 2009). In 2006, the first discovery of oil was made in the Lake

Albert Rift Basin. At present the country is reported to have 2.5 billion barrels of proven crude oil reserves and natural gas reserves of 500 billion cubic feet. This would place Uganda as one of the leading oil-rich nations in the sub-region. Actually, according to the EIA of the US (2012), Uganda and Madagascar are predicted to be the next big names in terms of oil production on the African continent. Although the oil discovery in the country is new, it goes to further consolidate the fact that Africa has abundant energy resources and should thus not experience the present levels of energy poverty.

Although Uganda has been poor in terms of fossil fuels until lately, the country has not been poor in terms of other energy resources and thus has no excuse for its inability to provide access to modern energy services for the populace.

Energy resources and energy poverty in Uganda

IRENA (2013) estimates the geothermal and solar potentials of Uganda as high and the wind and biomass potentials as medium. Also citing the German Technical Cooperation's (GTZ)[2] work in 2007, Energypedia reports the energy resource potential of Uganda to be an estimated 2,000 MW of hydropower, 450 MW of geothermal, 460 million tonnes of biomass standing stock with a sustainable annual yield of 50 million tons, an average of 5.1 kWh/m^2 per day of solar energy, and about 250 million TOE of peat. So again the question arises of why there is such energy poverty in a country rich with energy resources.

According to the Uganda Bureau of Statistics (UBOS), in 2002 only 8 per cent of households in the country had access to electricity, with a 39 per cent and 3 per cent urban and rural coverage respectively. Over 97 per cent of the nation relied on charcoal and firewood for cooking, with charcoal usage being high in the urban areas while firewood was largely used in the rural areas (UBOS 2006) The country's Total Electricity Net Generation as of 2010 was 2.406 billion kWh, of which 1.585 kWh was from renewable sources, mainly hydro sources. The remaining amount is generated from fossil resources imported from neighbouring countries (EIA 2012).

One obvious reason for the low access to energy in Uganda is the wars and conflicts the nation has experienced. The northern part of the country has experienced unrest since the 1980s and this continues to an extent, mainly between government forces and the rebel group called The Lord's Resistance Army (GAIHLHR 2011) Not only do these conflicts have the potential to destroy the physical infrastructure provided for energy access, but they also divert attention and resources which could be used to expand access to energy. Oil exploration in Uganda is reported to have begun as early as the 1920s but stalled due to political instability until about six decades later (EIA 2012). Thus, assuming that oil revenue would have been equitably managed and priorities set right, the country most probably would have a brighter energy outlook than what persists now were it not for these conflicts. Across Africa, conflicts and wars have been central to the ongoing under-development problems. Countries such as Libya, Egypt, and Tunisia, which recently experienced conflicts, lost much infrastructure

---

[2] Now German International Cooperation (GIZ).

as a result. Post-conflict rebuilding takes the continent many steps backwards, and this delays energy access provision even further.

Over the years, Uganda has formulated policies aimed at promoting energy access. These included the Electricity Act of 1999 whose implementation saw the separation of the Uganda Electricity Board (UEB) (which enjoyed monopoly) into three entities, namely generation, transmission, and distribution. The Act also saw the creation of the Electricity Regulatory Authority (ERA), the Rural Electrification Fund (REF), and the Electricity Dispute Tribunal (EDT) (MEMD 2012). The Energy Policy for Uganda, 2002, Renewable Energy Policy for Uganda, 2007, and The Atomic Energy Act, 2008, National Biomass Energy Demand Strategy (BEDS), 2001–2010, are the other energy policies formulated by the country (MEMD 2012). It has therefore not been long since Uganda started designing deliberate policies to target energy access. We can also observe that of all these policies, there is no specific policy targeting the cooking fuel sector. These (i.e. lack of any policy to promote clean cooking fuels and the absence of energy policies until the near past) are also reasons that can help account for the low level of modern energy access in the country. Furthermore, in spite of these energy policies being relatively recent, their effective implementation should have seen the country rising out of energy poverty. However, as in the case of Nigeria, many are the fine policies that are gathering dust on shelves without being implemented.

Extending grid electricity to all of Uganda is hampered by limited government resources, like in most developing countries. Increasingly off-grid solutions become the best option for rural areas which tend to be sparsely populated. However, households and rural communities are unable to pay for the cost of solar home systems or mini-hydro plants. In assessing the impact of off-grid electricity on microenterprises in rural Uganda, Muhoro (2010), suggested that electricity access could only rise above the established levels if subsidies and credit-based sales are pursued in addition to other favourable alternative financing. Thus the high incidence of poverty in Uganda hampers the country's efforts at providing universal access.

14.4. LESSONS AND CONCLUSIONS

The case studies have highlighted a number of issues that are significant for expanding energy access.

14.4.1. Lessons

The experiences in all the case study countries reveal that a strong political will is essential for expanding modern energy services to all. In the countries where considerable access to modern energy services has been achieved, as in the cases of Ghana and Egypt, there was a strong political will and commitment shown by the government. Once that critical ingredient was in place, programmes and policies were created to achieve success. On the other hand, countries such as Nigeria still experience high levels of energy poverty in spite of the abundant resources available to the country as a result of a lack of will and leadership shown by the government.

Target-setting is also significant in facilitating energy access. Egypt and Ghana set electrification targets that predated those set by regional bodies such as NEPAD; coupled with the political will to bridge the rural–urban access gap, the setting of targets has enhanced electricity access. Both countries have introduced lifeline tariffs to facilitate the poor's access to electricity. The government's hope to ensure effective access to electricity led to the introduction of affordable tariff regimes to encourage access to electricity and TO reduce poverty. The lifeline tariffs have not witnessed upward adjustment, though the other tariff bands have been increased several times. Consumers whose monthly consumption is not above 50 kWh are described as the lifeline consumers and thus enjoy the subsidies. The question worthy of answer is: what economic activities can 50 kWh support? The implication therefore is that pro-poor electricity should be planned to support income-generating activities instead of only household use. Ghana's LPG subsidy has been instrumental in increasing LPG usage not only among households but also among other commercial entities such as taxi drivers.

The private sector's involvement in the forms of IPPs and BOOT is playing key roles in expanding energy access in the well-performing countries. The lesson is that with the enabling policies, the potentials of the private sector can be harnessed to expand energy access. The case study has, however, revealed limited private sector participation owing to the uncompetitive tariff regimes in Egypt and Ghana.

Another significant lesson from the experiences of the case study countries is the need to ensure efficiency in the use of electricity. The review has indicated that Ghana was able to save about 124 MW, equivalent to 6.05 per cent of Ghana's total installed capacity, through the distribution of compact fluorescent lamps (CFL) under the 'Efficient Lighting Retrofit Initiative Programme'. Decentralized environment-friendly technologies have also been introduced to electrify the remote off-grid communities. The important issue is the extent to which households can make productive uses of these decentralized systems whose prices are high.

One important lesson to learn from the LPG access expansion is the commitment to increasing access to LPG. The redesign of LPG stoves with the introduction of the 2.75 kg gas bottle to respond to local use has been an important element in Senegal's LPG success story. Wood fuel use became a disincentive given the subsidy for LPG use, increased licensing fees for wood cutting, and charcoal price upward adjustments.

Designing the right policies that target specific aspects of energy access provision is important. In Uganda, the Electricity Act saw the separation of the UEB into three different entities of generation, transmission, and distribution, as well as the creation of other agencies as described earlier, to tackle the different aspects of providing energy access. We also see specific targets in the case of Ghana and Egypt, designed to create an enabling environment for private sector participation. However well designed, policies can only make impact if they are implemented. That is the only way success stories like those in Egypt and Ghana can be achieved. These policies were targeted at strengthening the institutional structures to promote expanding energy access, which is also critical. As was seen in the

Nigerian case study, part of the problem has been due to weak institutions that are plagued with 'serious' issues of governance.

14.4.2. Recommendations

This chapter has observed that target setting should be backed by a strong political will to drive the energy access goal in the worse-performing countries. To ensure universal access to energy, governments should highlight productive uses and thus introduce systems that can support enterprise development towards sustainable poverty reduction. Thus the lifeline tariff bands should encourage productive uses but not electrification for lighting and other domestic purposes that have marginal effects on poverty reduction. Energy access provision is multifaceted; the relevant institutions should work in concert to link energy access expansion to the broader goal of development. Creating wealth through energy access provision will reduce poverty and more people will be able to pay for the cost of energy.

Energy efficiency should be at the top of the agenda. Ghana's 'Efficient Lighting Retrofit Initiative Programme' could be adopted as a model for other developing countries in Africa while they work to increase generation capacity.

The fossil resources available should be exploited to provide energy access for the energy-poor. This should however be done in line with sustainability standards in order to reduce the negative impacts of these resources on the environment. The countries without fossil resources should exploit renewable resources, which are abundant, since there is appropriate technology to utilize these resources. Considering the finite nature of fossil resources, hydrocarbon-rich countries should explore their renewable energy options even as they utilize their fossil resources. This will not only guarantee energy security but also promote low-carbon development.

Safer cooking fuels could be promoted in Africa through customized designs to satisfy local needs. LPG stoves should be redesigned to enable householders to use them for the preparation of traditional cuisines and more importantly for operation in commercial enterprises. LPG could also be retailed in smaller bottles to curtail the high upfront cost, which deters people from buying. Wood fuel is going to be part of Africa's energy resources for a long time to come. Thus to use the fuel in an efficient manner, there should be deliberate policies and programmes to penetrate the market with improved cookstoves, something which is currently being done in Ghana.

14.4.3. Conclusions

A number of factors are responsible for the dismal levels of energy access in Sub-Saharan Africa in spite of the abundant energy resources available on the continent. This ranges from a lack of political will to weak institutional structures and funding challenges. It is certainly possible, however, for Sub-Saharan Africa to emerge out of energy poverty using both the renewable and non-renewable resources available. In order to do this, governments should show willingness and leadership. It is this willingness which will translate into funds being allocated as well as policy and institutional reforms that will promote the expansion of energy access.

Conflicts have also been responsible to a large extent for poverty and under-development in Africa. If Africa is to have progressive and uninterrupted development, nations would need to find diplomatic channels for solving arising issues instead of picking up arms every time and thus destroying and stalling development.

REFERENCES

Abavana, C. (2004). 'Ghana: Energy and Poverty Reduction Strategy', paper presented by Government of Ghana at the Facilitation Workshop and Policy Dialogue, Ouagadougou, Burkina Faso, 26–29 October.

Aderibigbe, D. (2010). 'Power Supply to Industries—Pros and Cons of Available Options', presentation at Conference of Nigerian Society of Chemical Engineers, Lagos, 10 July.

Ahiataku-Togobo, W. (2013). 'National Cooking Energy Strategies—Lessons from Ghana', presentation at the WACCA Workshop, Ouagadougou, April.

ANSD (Agence Nationale de la Statistique et de la Demographie) (2006). *Resultats du troisième recensement général de la population et de l'habitat (2002): Rapport National de presentation.*

Awafo, E. (2013). 'Nigeria and Generators', interview with G. Mensah, 5 March.

Ayoola, T. (2011). 'Gas Flaring and its Implication for Environmental Accounting in Nigeria', *Sustainable Development* 4(5), 244–50.

Brew-Hammond, A. (2007). 'Challenges to Increasing Access to Modern Energy Services in Africa', background paper for Forum of Energy Ministers of Africa Conference on Energy Security and Sustainability, Maputo, Mozambique, March.

Edjekumhene, I., Amaka-Otchere, A., and Amissah-Arthur, H. (2006). 'Ghana: Sector Reform and the Pattern of the Poor', paper presented at the International Bank for Reconstruction and Development Meeting on Energy Sector Management Assistance Program, Washington, DC, March.

EIA (2012). 'International Energy Statistics: Africa' [online], <http://www.eia.gov/cfapps/ipdbproject/iedindex3.cfm?tid=6&pid=29&aid=12&cid=r6,&syid=2007&eyid=2011&unit=BKWH> (accessed 23 June 2013).

Eleri, E., Ugwu, O., and Onuvae, P. (2012). *Expanding Access to Pro-Poor Energy Sevices in Nigeria.* Abuja: International Centre for Energy, Environment and Development.

El-Salmawy, H. (n.d.). *Egyptian Power Sector Reform and New Electricity Law* [online], <http://www.ecrc.org.eg/uploads/documents/Dr.Hafez%20El-Salmawy%20-%20EgyptEra.pdf> (accessed 9 March 2013).

Energy Commission, Ghana (2011). *2011 Energy (Supply and Demand) Outlook for Ghana,* Accra: Energy Commission.

GAIHLHR (Geneva Academy of International Humanitarian Law and Human Rights) (2011). 'Uganda' [online], <http://www.geneva-academy.ch/RULAC/current_conflict.php?id_state=163> (accessed 25 June 2013).

Hailu, Y. (2012). 'Measuring and Monitoring Energy Access: Decision-support Tools for Policymakers in Africa', *Energy Policy* 47: 56–63.

IEA (International Energy Agency) (2008). *World Energy Outlook 2008* [online], <http://www.worldenergyoutlook.org/publications/2008-1994/> (accessed 1 August 2013).

MEMD (Ministry of Energy and Minerals Development, Uganda) (2012). *The Electricity Act, 1999* [online], <http://energyandminerals.go.ug/policy> (accessed 23 June 2013).

MENA-OECD (Middle East and North Africa Organisation for Economic Co-operation and Development) (2010). 'Business Climate Development Strategy: Phase 1 Policy Assessment—Egypt' [online], <http://www.oecd.org/countries/egypt/egypt-busi nessclimatedevelopmentstrategy-oecd.htm> (accessed 28 June 2013).

Miller, V., Ramde, E., Gradoville, Jr., R., and Schaefer, L. (2011). 'Hydrokinetic Power for Energy Access in Rural Ghana', *Renewable Energy* 36(2): 671–5.

Muhoro, P. (2010). 'Off-Grid Electricity Access and its Impact on Micro-Enterprises: Evidence from Rural Uganda', PhD dissertation, University of Michigan, Ann Arbor.

OECD (Organisation for Economic Co-operation and Development) (2010). *Competitiveness and Private Sector Development: Egypt 2010*. Paris: OECD.

OECD/IEA (Organisation for Economic Co-operation and Development/International Energy Agency) (2011). 'Energy for All: Financing Energy Access for the Poor', *Special Early Report on the World Energy Outlook 2011* [online], <http://www. worldenergyoutlook.org/resources/energydevelopment/> (accessed June 2014).

OECD/IEA (2013). 'Global Status of Modern Energy Access' [website], <http://www. worldenergyoutlook.org/resources/energydevelopment/ globalstatusofmodernenergyaccess/> (accessed 10 March 2013).

Oluwole, A., Samuel, O., Festus, O., and Olatunji, O. (2012). 'Electrical Power Outage in Nigeria: History, Causes, and Possible Solutions', *Energy Technologies and Policy* 2(6): 18–23.

Opam, M. (1995). 'Institution Building in the Energy Sector of Africa: A Case Study of the Ghana Power Sector Reform Programme', presentation at the 6th Session of the Africa Regional Conference on Mineral and Energy Resources Development and Utilisation, Accra, 14–23 November.

Practical Action (2010). *Poor People's Energy Outlook 2010*. Rugby, UK: Practical Action.

Saundry, P. (2009). 'Energy Profile of Uganda', *The Encyclopedia of the Earth* [online], <http://www.eoearth.org/view/article/51cbef1e7896bb431f69c81d/> (accessed 12 March 2013).

UBOS (Ugandan Bureau of Statistics) (2006). *2002 Uganda Population and Housing Census: Analytical Report* [online], <http://www.ubos.org/?st=pagerelations2&id=16& p=related%20pages%202:2002%20Census%20Results> (accessed 1 August 2013).

UNCTAD (United Nations Conference on Trade and Development) (2012). *The Least Developed Country Report 2012: Harnessing Remittances and Diaspora Knowledge to Build Productive Capacities* [online], <http://unctad.org/en/pages/PublicationWebflyer. aspx?publicationid=249> (accessed 1 August 2013).

UNDP (United Nations Development Programme) (2010). *The Real Wealth of Nations: Pathways to Human Development. Human Development Report 2010* [online], <http:// hdr.undp.org/en/reports/global/hdr2010/> (accessed 1 August 2013).

UNDP and WHO (World Health Organization) (2009). *The Energy Access Situation in Developing Countries: A Review Focusing on the Least Developed Countries and Sub-Saharan Africa* [online], <http://www.who.int/indoorair/publications/energyaccesssituation/en/> (accessed 1 August 2013).

UNIDO (United Nations Industrial Development Organization) (2009). *Scaling Up Renewable Energy in Africa*. 12th Ordinary Session of Heads of State and Governments of the African Union [online], <http://www.un-energy.org/publications/185-scaling-up-re newable-energy-in-africa> (accessed 1 August 2013).

United Nations (2011). *Sustainable Energy For All: A Vision Statement by Ban Ki-Moon, UN Secretary-General* [online], <http://sustainableenergyforall.org/resources> (accessed 1 August 2013).

UPDEA (2009). *Comparative Study of Electricity Tariffs Used in Africa* [online], <http://www.updea-africa.org/updea/DocWord/TarifAng2010.pdf> (accessed June 2014).

Von Moltke, A., McKee, C., and Morgan, T. (2004). *Energy Subsidies: Lessons Learned in Assessing their Impact and Designing Policy*. Sheffield: UNEP and Greenleaf Publishing.

World Bank (2011). *The Little Green Data Book 2011*. Washington, DC: International Bank for Reconstruction and Development/World Bank.

15

Global Energy Subsidies

Scale, Opportunity Costs, and Barriers to Reform

Doug Koplow

Government subsidies to energy producers, transporters, and consumers are widespread throughout the world and represent a large public investment in the energy sector. In theory, this investment could be funding a variety of social goals such as providing the poor with access to basic energy services and addressing common environmental problems linked to energy extraction and consumption. Although some subsidies do address these types of concerns, most either do not, or do not do so effectively.

Far from helping to alleviate energy poverty, subsidies distort the relative prices of energy options, resulting in over-exploitation of fossil fuels and exacerbating associated environmental costs. Below-market prices to industrial, commercial, and retail customers mute incentives to conserve energy, and can contribute to 'subsidy clusters', pockets of industries that become reliant on subsidized energy inputs in order to be competitive (Koplow 1996). Capital investment into real estate infrastructure may undervalue energy efficiency as well, locking in a region or country to excess consumption for many decades.[1] Efforts to suppress domestic energy prices below world market levels often give rise to smuggling and black market operations as people try to profit from the pricing differentials. Finally, the fiscal cost of subsidies can absorb such a large portion of available government revenues that it crowds out spending in critical areas focused on improving population welfare or transitioning the country to a cleaner energy path.

While there is no exact global estimate, financial subsidies to energy are measured in many hundreds of billions of dollars per year. External costs of energy fuel cycles are relevant as well. They include a wide variety of negative impacts on human health and environmental quality from energy extraction, conversion, and consumption, and have been estimated to exceed a trillion dollars

---

[1] For both industrial and commercial consumption, looking at trends in energy consumed per unit of GDP can provide insights into how seriously countries are integrating global price signals on energy into capital investment patterns. It is notable, for example, that '[b]etween the oil price hikes of the 1970s and the global financial crisis in 2008–09, GDP per unit of energy increased in the oil-importing countries, but declined or stayed level in the oil-exporting ones: Saudi Arabia, Iran, Malaysia and Nigeria' (Lahn and Stevens 2011: 8). See also Fattouh and El-Katiri (2012).

per year globally. Though not directly funded by government budgets as financial subsidies are, external costs nonetheless exacerbate the pricing distortions caused by financial subsidies, further skewing energy investment and conservation patterns.

Despite the clear benefits of subsidy removal, political impediments have greatly slowed the pace of reform around the world. Once governments begin subsidizing particular fuels, they are often 'trapped' in policies that make little fiscal, developmental, or environmental sense but that are protected and defended by subsidy recipients. In some cases, reforms are successfully implemented but are then rolled back due to subsequent changes in political or market conditions.[2]

Subsidies to electric or natural gas distribution and generation infrastructure represent a slightly different twist on this same problem. Pricing of energy services at levels below break-even is common, as are utilities reliant on state subsidies so that they can continue to operate despite high rates of non-payment or theft. Both result in inadequate revenues to maintain and grow the enterprise. Existing customers may benefit from artificially low power rates and therefore resist price increases. Over time, however, the utility is starved of needed capital to maintain its existing system. The low or negative returns also preclude network expansion to the very customers and service regions it needs to reach in order to ameliorate energy poverty and improve the quality of life for the billions of people without access to modern energy services.

Energy subsidies are thus not an effective policy to alleviate energy poverty. As currently structured, they tend to be part of the problem, not the solution. Only recently has the international community begun to come to terms with just how big a problem they are—remarkably, the cost of energy subsidies far exceeds the estimated cost of effective energy access policies. This chapter reviews current estimates on the magnitude of energy subsidies globally, including what remains missing from the tallies; inefficiencies with subsidy targeting; the growing opportunity costs of not reforming; and the impediments to making subsidy reform a reality.

15.1. QUANTIFIED FINANCIAL SUBSIDIES TO ENERGY EXCEED $750 BILLION ANNUALLY

Subsidy measurement has been improving in recent years. The International Energy Agency (IEA), for example, has been producing estimates of consumer subsidies to fossil fuels (i.e. where local prices are below world prices) annually in its World Energy Outlook, rather than intermittently as was done in the past. Beginning in 2010, the Organization for Economic Cooperation and Development (OECD) began systematically inventorying fossil fuel subsidies that target producers within OECD member states (OECD 2013).[3] The OECD's effort is one of the

---

[2] Kojima (2009, 2013) and IMF (2013) summarize many past attempts at pricing reform around the world.

[3] OECD's review includes producer subsidies to extraction, transport, refining, and processing, as well as a granular review of consumer subsidies.

first to recognize the importance of capturing state and provincial subsidies to energy, rather than just federal or nationwide supports. Both the International Monetary Fund (IMF) and the World Bank have been assessing fossil fuel pricing regimes throughout the world, and the degree to which shifts in world prices show up in domestic markets. They have benefited from detailed pricing surveys of key petroleum transport fuels around the world undertaken every two years by the German Society for International Cooperation (GIZ 2012). More granular subsidy reviews, focusing on a specific fuel or a handful of countries, have also been produced by the United Nations Environment Programme and by non-governmental organizations such as Earth Track and the Global Subsidies Initiative.

Increased reporting is due in part to a growing recognition that the scale of subsidies is so large that competent fiscal planning requires that it is addressed, and that subsidy reform must be integrated into any logical response to global greenhouse gas emissions. Formal approval among the 20 largest economies in the world (the G-20) to phase out environmentally harmful subsidies to fossil fuels has also provided political support for action, though the success of that commitment remains far from certain.[4] Finally, a growing number of researchers around the world are focused on the issue and continue to produce important new analysis of the subject.

This progress is extremely valuable. Still, substantial gaps remain and a full accounting of global energy subsidies has never been done (Box 15.1). Global estimates of subsidy magnitude are likely well below actual levels of support. Further, and perhaps more important, available estimates are primarily broad national averages that miss subsidy 'hot spots'—specific regions or types of activities that are disproportionately supported by subsidies. Fossil fuel extraction in environmentally sensitive regions, and efforts to spur production of lower-quality deposits where such development would otherwise be uneconomic, are two examples.

A full subsidy review entails a systematic examination of a wide variety of policy instruments at multiple levels of government used to transfer value from the public to the private sector. Subsidies at the sub-national level can be surprisingly large (Koplow and Lin 2012). In addition to relatively transparent direct cash transfers, subsidies through credit markets, tax breaks, caps on private liability from spills or accidents, reduced royalty payments on publicly-owned minerals, and purchase mandates requiring market purchase of higher-cost resources are all common in many countries. Direct government ownership of energy infrastructure or service enterprises is also widespread globally and tends to be rife with subsidies.

The IEA's price gap measures do not detail specific instruments but rather capture subsidies only if they result in drops in domestic energy pricing (Koplow 2009). The OECD's current work is capturing some, though not all, of these instruments. An earlier review of fossil fuel subsidies in four countries (Koplow et al. 2010) found that basic data on many of these policy types was extremely difficult to obtain and sometimes nonexistent, particularly in countries without a

---

[4] As documented in Koplow (2012) and Koplow and Kretzmann (2010), tangible progress towards reform attributed to the G-20 commitment has thus far been fairly limited.

Box 15.1. **Gaps in global estimates of energy subsidies**

Geographic. Subsidies to producers in developing countries are systematically missing from global estimates, though coverage of consumer subsidies in these regions is improving. Outside of a handful of OECD countries, subsidies at the state or provincial levels are rarely captured, though they can be substantial (Koplow and Lin 2012; OECD 2012; IEA 2012).

Policy type. There is growing coverage of grants and many types of tax breaks (OECD 2013). Substantial coverage gaps remain for producer support via subsidized credit or insurance, regulatory oversight and site remediation, energy security (shipping lanes, stockpiling) and bulk transport costs, and tax-exempt corporate forms. Capture of subsidies through government-owned energy infrastructure or service organizations also remains low.

Non-payment. Price gap metrics capture underpricing, but do not capture power theft and non-payment. These 'hidden' costs of power were larger than underpricing in some regions (Joint Report 2010: 17).

User fees. Many countries levy a variety of fees or taxes on fuels that are earmarked (hypothecated) for specific uses closely linked to particular fuels—for example, building and maintaining transit infrastructure or cleaning up oil spills or abandoned sites. These fees are sometimes improperly deducted from subsidy estimates, or shortfalls in actuarially-based fee collections are not incorporated into subsidy tallies (Koplow 2009, 2010).

---

strong tradition of government transparency and public accountability. Box 15.1 identifies a number of core gaps among current numbers, not only in policy types but also in geographic coverage and in the calculation of price gap values.

Broader coverage would provide important insights into the real scale and distribution of subsidies, particularly in environmentally sensitive regions of the world. However, even currently available data indicates the scale of the problem is staggering. Subsidies to fossil fuel consumers alone were $523 billion in 2011 (IEA 2012). Adding available data on subsidies to the producer side, as well as subsidies to renewables and nuclear, drives the total up to about $840 billion annually (roughly 1 per cent of global GDP). Over the 2007–11 period, this amounts to more than $3.5 trillion (see Table 15.1).

Table 15.1. Total quantified financial subsidies to energy

	Billions of USD					
	2011	2010	2009	2008	2007	2007–11
Fossil fuels[1]	589	475	361	622	404	2,451
Renewables[2]	88	66	60	48	44	306
Nuclear[3]	162	159	157	156	152	787
All	839	700	579	825	600	3,544

Notes

[1] OECD consumer subsidies to South Korea and Mexico deducted to avoid double counting. IEA price gap subsidies to fossil-fuel electric allocated back to source fuels based on country-level data on the fuel mix of power generation. IEA (2011a, 2012, 2013); OECD (2012); and Sauvage (2013). [2] IEA (2011a and 2012). [3] Kitson, Wooders, and Moerenhout (2011) midpoint value. Single year annual value for 2009, adjusted for inflation, was applied to other years in the series. No adjustments made to incorporate the taxpayer costs of the Fukushima nuclear accident.

Subsidies are often defended on the grounds that they help transition economies towards more sustainable energy systems or that they alleviate poverty. However, their efficacy in achieving these ends is inadequate.

15.1.1. Subsidies delay achieving core environmental goals

Of the $3.5 trillion in quantified financial subsidies to energy between 2007 and 2011, nearly 70 per cent supported fossil fuels, versus only 8 per cent for generally cleaner renewables. In turn, nearly two-thirds of quantified subsidies to fossil fuels supported oil and coal rather than cleaner natural gas (Figure 15.1). It is estimated that removal of consumer subsidies would reduce global greenhouse gas emissions by 8 per cent by 2050, or nearly 10 per cent if OECD countries cap their carbon emissions at the same time (Burniaux and Chateau 2011: 16).

The external costs of energy extraction and consumption on human health, environmental quality, and the global climate are widespread and are additive to financial subsidies. They are properly included when measuring the under-pricing of particular energy resources in the marketplace, and they tend to exacerbate distortions created by financial subsidies. A literature review conducted by Geneva-based Global Subsidies Initiative found a very wide range of externality estimates (Figure 15.2), indicative of differing methodologies, time periods of analysis, and estimation challenges. However, even using a midpoint of the estimate range indicates a scale of external costs on the order of $1.5 trillion *per year* globally, the vast majority associated with fossil fuels.[5]

Thus, current information suggests that total financial subsidies and uncontrolled externalities top $2 trillion per year, equivalent to more than 3 per cent of global gross domestic product.[6]

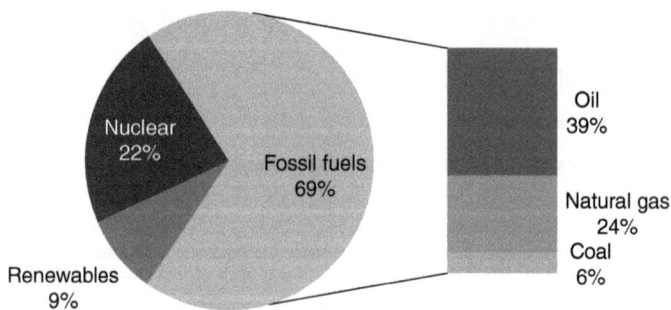

Figure 15.1. Fossil fuels receive most subsidies

Source: Earth Track calculations, OECD 2013, IEA 2011b, IEA 2012, Sauvage 2013, Bromhead 2013

---

[5] The IMF (2013: 9) estimated fossil fuel externalities based on damage estimates associated with greenhouse gas emissions. Though the approach was different, the resultant value was quite close to the $1.5 trillion per year midpoint shown in Figure 15.2.

[6] Based on World Bank GDP figures for 2009 and 2010 (World Bank 2013a).

Global external costs of energy fuel cycles
Estimate range and midpoint

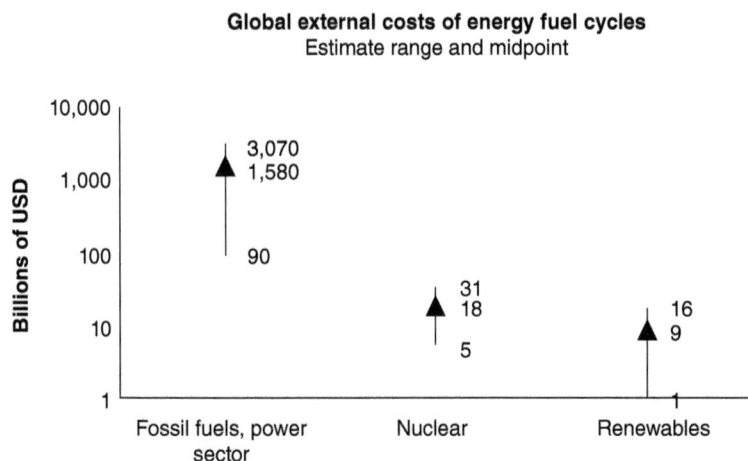

Figure 15.2. Global external costs of energy fuel cycles, estimate range and midpoint.
Source: Kitson, Wooders, and Moerenhout (2011)

The relative costs of energy options drive which fuels attract the most research and development spending as well as levels of capital investment. These factors, in turn, influence the market uptake of particular energy technologies. The misallocation of investment driven by combined financial subsidies and externalities is large, as are the resultant damages to human health and the environment.

15.1.2. Energy poverty

The IEA estimates that 1.3 billion people have no access to electricity, 85 per cent of them located in rural areas. This grouping includes two-thirds or more of the population in developing Africa (IEA 2012: 535). Roughly 2.6 billion people have no access to clean cooking fuels. They rely on traditional biomass instead—fuels that are time-consuming, and often dangerous, to collect. Notable pockets of underservice for cooking fuels include half the population of developing Asia and roughly 80 per cent of Sub-Saharan Africa. As with access to electricity, more than 80 per cent of those lacking access to clean cooking fuels live in rural areas (IEA 2012: 529–34).

Access to modern energy services has been clearly demonstrated to improve health, productivity, and welfare for recipient families. Access has important gender benefits as well. The World Bank notes that electrification allows women 'increased scope for evening activities, greater flexibility in organizing household activities as daylight is no longer a constraint, enhanced security, the potential for undertaking income-producing activities such as handicrafts, and reduction in time required for collecting water if electrification improves water supply' (World Bank 2010: 19, 45).

Much of the money spent subsidizing fossil fuels has aimed to keep prices on transport and cooking fuels below market levels. Although this would seem supportive of expanded energy access, in practice the policies have not been very effective in helping the poorest sectors of society. Higher energy consumption by the wealthy, political influences on subsidy awards, diversion of subsidized fuels for resale on the black market at higher prices, and gaps in infrastructure necessary for the poor even to access subsidized flows of natural gas or electricity have all contributed to a high 'leakage' rate of subsidy dollars away from those most in need of support.

Surveys of developing countries found that only 8 per cent of fossil-fuel subsidies reached consumers in the poorest 20 per cent of the population (IEA 2011a) and less than 25 per cent reached the poorest two quintiles (Joint Report 2010: 24).[7] Leakage rates for gasoline have been particularly high according to IMF analysis, with an estimated $33 in subsidies to gasoline required for each $1 that actually reached the poorest 20 per cent of society (del Granado, Coady and Gillingham 2010: 13).

15.2. REASONS TO REFORM

A combination of immense magnitude, high leakage rates to wealthier consumers, and support to more polluting fuel cycles all create powerful economic, social, and environmental pressures for reform. The poorer the country and the larger the energy subsidies, the more the policies constrain government operating flexibility and crowd out social welfare needs.

The fiscal burdens of these subsidies to developing countries in particular can be severe, absorbing a large portion of available government revenues. As shown in Table 15.2, even countries with relatively small subsidies in terms of absolute funding levels may be crowding out other public spending. In 15 of the 38 countries for which the IEA tallies consumer subsidies, fossil fuel subsidies exceeded 5 per cent of 2011 gross domestic product. As of the end of 2011, according to IMF analysis, half of the countries in the Middle East and Central Asia had fuel subsidies exceeding 2.3 per cent of GDP and half of the countries in Sub-Saharan Africa had fuel subsidies greater than 1.3 per cent of GDP (Coady, Flamini and Antonio, 2012: 48).

Government revenues provide a better proxy than GDP for the opportunity cost of squandering limited public resources on fuel subsidies. Whereas GDP picks up all actors in an economy, government revenues tie much more closely to the non-deficit budget constraint facing the very government with the power to set and modify the subsidy policy. The figures are striking: 30 of the 38 countries tracked by the IEA had subsidies in excess of 5 per cent of federal revenues, and nearly one-quarter of the sample was spending more than 20 per cent of federal revenues.

---

[7] Countries surveyed were Angola, Bangladesh, China, India, Indonesia, Pakistan, Philippines, South Africa, Sri Lanka, Thailand, and Vietnam.

Table 15.2. Subsidies to fossil fuel consumers crowd out other spending priorities

Country	Annual subsidy amount to consumers (billions of USD)[1]	Fossil fuel subsidy amount as percentage of:		
		GDP[2]	Federal revenues[3]	Public spending on health care[4]
Algeria	13.4	7.0%	16.9%	144.9%
Angola	1.3	1.3%	2.3%	47.7%
Argentina	10.0	2.2%	8.5%	34.9%
Azerbaijan	2.0	3.1%	8.6%	186.5%
Bangladesh	5.8	5.1%	41.5%	202.8%
Brunei	0.5	n/a	6.1%	100.1%
Chinese Taipei	1.6	0.3%	2.0%	n/a
China	31.0	0.4%	1.7%	11.4%
Colombia	0.7	0.2%	0.7%	2.9%
Ecuador	5.6	8.4%	17.8%	157.1%
Egypt	24.5	10.4%	43.3%	274.8%
El Salvador	0.6	0.0%	12.4%	34.7%
India	39.7	2.4%	23.1%	82.0%
Indonesia	21.3	2.5%	15.3%	159.8%
Iraq	22.2	19.3%	21.3%	244.0%
Iran	82.2	17.0%	62.7%	327.1%
Kazakhstan	5.8	3.3%	13.5%	109.1%
Korea	0.2	0.0%	0.1%	0.3%
Kuwait	11.1	6.3%	10.4%	432.8%
Libya	3.1	8.5%	5.5%	98.5%
Malaysia	7.2	2.6%	12.2%	70.1%
Mexico	15.9	1.4%	5.9%	29.5%
Nigeria	4.4	1.8%	18.7%	58.9%
Pakistan	11.1	5.3%	37.6%	273.1%
Peru	0.3	0.2%	0.5%	3.9%
Philippines	1.5	0.7%	4.3%	31.6%
Qatar	6.0	3.4%	9.6%	255.2%
Russia	40.2	2.2%	9.7%	45.7%
Saudi Arabia	60.9	10.6%	19.4%	356.7%
South Africa	1.4	0.3%	1.5%	6.7%
Sri Lanka	1.1	1.9%	13.4%	79.9%
Thailand	10.3	0.3%	15.1%	60.0%
Turkmenistan	5.8	22.8%	22.0%	962.8%
Ukraine	9.3	5.7%	17.5%	69.2%
UAE	21.8	6.3%	16.7%	240.7%
Uzbekistan	12.7	22.8%	77.5%	497.5%
Venezuela	27.1	8.6%	23.3%	449.3%
Vietnam	4.1	3.4%	9.7%	57.6%
Country counts				
Total countries		37	38	37
Subsidies > 100% of metric		0	0	18
Subsidies > 50% of metric		0	2	26
Subsidies > 25% of metric		0	5	32
Subsidies > 10% of metric		6	22	33

Notes
[1] Price gap subsidies to consumers in 2011 from IEA (2012). [2] 2011 GDP data from World Bank (2013a). [3] Federal estimated revenues for 2012 from CIA (2013). [4] Health-care spending based on World Health Organization data compiled by the *Guardian* newspaper (2012). Population data used to scale per capita to national figures from World Bank (2013b).

The scale of subsidies relative to other social objectives (Tables 15.2 and 15.3) is of equal concern.

- *Universal access to modern energy.* Overall subsidies to fossil fuels are more than 15 times the $34 billion per year in incremental funding that the IEA estimates would be sufficient to achieve universal access to clean cooking fuels and electricity by 2030. Many country-specific values are even worse. Unlike actual subsidies for universal access, most of the fuel subsidies benefit upper-income quintiles. Targeted funding for expanding energy access is paltry in comparison, even when supplemental support from international agencies and lending institutions is combined with funding by national governments. Total spending for universal access amounted to only $15.5 billion for the 2005–2010 period, or less than $3 billion per year (Piebalgs 2012: 82).[8]

- *Public spending on health care.* Inadequate availability and access to health care is an endemic problem throughout the developing world. Yet governments in fully half of the countries evaluated by the IEA spent more on fossil

Table 15.3. Fossil fuels subsidies are five times funding for climate mitigation (millions of USD)

Country	Fast start climate pledge (average, 2010–12)	Subsidies to fossil fuel consumers, 2011	Ratio of subsidies to mitigation
Australia	206	8,362	40.5 ×
Austria	—	509	0 ×
Belgium	63	2,770	44 ×
Canada	406	3,178	7.8 ×
Denmark	68	1,277	18.9 ×
Finland	46	2,323	50.5 ×
France	528	3,569	6.8 ×
Germany	528	6,603	12.5 ×
Greece	—	270	n/a
Iceland	—	—	0 ×
Ireland	53	101	1.9 ×
Italy	—	2,752	n/a
Japan	5,000	439	0.1 ×
Netherlands	130	440	3.4 ×
New Zealand	24	43	1.8 ×
Norway	333	698	2.1 ×
Spain	157	2,417	15.4 ×
Sweden	336	2,762	8.2 ×
Switzerland	49	269	5.5 ×
United Kingdom	793	6,606	8.3 ×
United States	2,500	13,146	5.3 ×
Total	11,220	58,534	5.2 ×

Source: Oil Change International (2012)

[8] The United Nations Sustainable Energy for All initiative may lead to some uptick in baseline spending. More than 100 commitments for a variety of appliance purchases, financing, and extended energy access will result in $320 billion in direct investments, 10 per cent of which is earmarked for modern energy access for the poor. However, overall spending remains below what is needed to provide universal access (IEA 2012: 531).

fuel subsidies to consumers than they spent on health care. Seventy per cent had subsidy levels equal to half or more their public spending on health care.

- *Commitments to fast start climate finance.* While fossil fuel subsidies within most OECD countries are not high relative to overall government revenues, they nonetheless have important social opportunity costs. The quantified subsidies to fossil fuels within the OECD (but a portion of the actual total) are more than five times the level of financial commitments that these very same governments have made to ameliorate climate change around the globe (Oil Change International 2012).

15.3. SUBSIDY 'TRAPS': WHY REFORM DOESN'T HAPPEN

Despite the compelling logic of eliminating fossil fuel subsidies, many attempted reform efforts have not been successful. The persistence of subsidies can be seen in the degree to which increases in world fossil fuel prices 'pass through' to end-consumers in countries around the world. An IMF assessment found that most low- and middle-income countries passed through less than 70 per cent of the sharp increases in global fossil fuel price increases that occurred through mid-2008, with similarly low levels of price adjustments for the 2008–2011 period. They observed particularly low pass-through in Sub-Saharan Africa and in the Middle East and Central Asian regions (Coady, Flamini, and Antonio 2012: 48).

World Bank data on the same issue provides more resolution by fuel type, but shows comparable results. Median pass-through rates in lower-income countries lag higher-income countries for all fuels but liquefied petroleum gas (LPG). Median pass-through for net oil exporters was barely more than one third of the change in world prices for gasoline, and less than 10 per cent for both kerosene and LPG (Kojima 2012: 21, 22). During times of rising oil prices, many countries stop or even reverse price reforms that they had previously begun to implement (Kojima 2013).

In contrast, full pass-through has been the norm in advanced economies. Energy prices are set to a much greater degree by market forces; adjustments happen automatically and are expected by key market participants. A variety of strategies, from the hedging of fuel supply costs by energy-intensive industries to lifeline rates or grants to let low-income residents afford basic energy, have been implemented.

15.3.1. Economic and political drivers of subsidy traps

For the many countries with government intervention in fuel prices, reform is challenging. Efforts to protect domestic consumers or industries often become entrenched and difficult to end. Governments end up 'trapped' into continuing these policies over a long period of time despite high fiscal, social, and environmental costs. The economic and political constraints to subsidy reform tend to feed on each other. Economic factors drive increased political activity, while political activity protects and expands the financial transfers.

Initial subsidy creation, and subsequent retention, is often driven by a mix of pure interest-based politics and 'legitimate' purposes of government such as poverty reduction or addressing other social ills (Victor 2009: 14). However, it is common to tie symbolic objectives with social qualities in order to bolster public support for subsidies that mostly transfer significant financial resources to concentrated economic interests (Koplow 2007). Indeed, this is not difficult to do, as with 'so many goals, there is rarely a shortage of inspiration for government to invest a subsidy to serve some purpose' (Victor 2009: 14).

With large financial flows at stake, groups organize to capture the economic rents that changes in government funding, market rules, or tax policies can send their way. These returns may be directly pecuniary, such as grants or artificially low consumer prices. However, they may also come in the form of greater market power or reduced market risks. In the case of declining or globally-inefficient industries, the primary impact of subsidies may be enabling globally uncompetitive firms to remain in the market, avoiding either closure or expensive restructuring. Even subsidies that initially start primarily to support social policies can, over time, be altered such that a greater share of the total support flows to more powerful segments of society.

Incumbents block reform. Though modifications do happen, the political process can make them challenging to accomplish. Even existing subsidy beneficiaries are often unable to optimize subsidy capture by modifying policies in light of new market entrants or changing market conditions, and thus they focus lobbying support on protecting what they have. This process 'locks in' political support for particular fuels, sectors, or technologies, slowing technical progress (Victor 2009: 19).[9]

The longer artificially low energy prices persist in an economy, the more difficult it becomes to escape the trap (Lahn and Stevens 2011: 20). The portion of a country's capital base that was procured assuming cheap energy grows over time, and this installed base drives up the expected economic dislocations from allowing prices to reach world market levels. Political factions benefiting from the established subsidy policies become increasingly entrenched and more sophisticated as well, compounding the challenge. While government revenues pay much of the cost of consumer subsidies, a portion is normally extracted from energy market participants as well—through regulation of prices or tax levels. These domestic energy firms face low returns, which discourage both new entrants and new investment by incumbents. The lack of new investments further worsens the stagnation.[10]

In practice, the economic gains from some types of subsidy policies may be short-lived. Economist Gordon Tullock noted a 'transitional gains trap' where

---

[9] Victor (2009: 19) notes that blocking new entrants from tapping into a particular subsidy is an important part of protecting current gains, referencing the example of US import tariffs on Brazilian ethanol that, for many years, helped to ensure the economic benefit of US tax credits flowed primarily to US producers.

[10] Steenblik (2007) describes additional categories of subsidies as *sympathetic support* (policies that influence the direction of technological development to support domestic producers) and *compensatory support* (policies that drive up input prices for downstream consumers, requiring related consumption subsidies to ensure that the higher-cost domestic products can find a market).

windfalls would accrue to market participants at subsidy initiation but would be quickly capitalized into the value of assets linked to subsidy eligibility (e.g. subsidy-eligible farm land, taxi operating medallions, legacy water rights, subsidized mineral leases). New entrants would pay more for these assets, bringing down returns on the subsidized activities to normal levels (Tullock 1975). End the subsidy, though, and asset prices immediately fall again, with the then-owner bearing the full cost. Thus, transitory gains or not, market incumbents all have a strong interest in defending the subsidy.

Poverty reduction. Most energy subsidies are ineffective policy tools to extend energy access to the poor, with high leakage rates to industry and wealthier citizens. Particularly in countries with corrupt or ineffective governance and few safety nets, however, even the small portion that does reach the poor can be important. As a result, poorly planned and executed subsidy removal schemes can disproportionately harm the lowest-income quintiles. Sudden increases in the cost of basic energy or energy-intensive goods and services (often food and public transit) can make them unaffordable, worsening energy poverty. A compilation of impacts from subsidy reform in nine developing countries, for example, found a more severe percentage decline in income or increase in expenditures for the bottom income quintile than for the top (Joint Report 2010: 80).

Where fuel price reforms inadequately addressed the basic needs of the poor prior to implementation, political unrest and sometimes violence has ensued. Many attempted energy subsidy reforms have subsequently been rolled back or weakened.[11]

Black markets. By definition, domestic price subsidies create two-tiered pricing for what is essentially a fungible, commodity product. Intermediaries diverting subsidized supplies away from their intended recipients to sell on the black market at a higher price are common in most countries subsidizing petroleum. Because subsidy reform would eliminate the pricing disparity, people involved with the black market (sometimes including government officials) will oppose reform.

All of these economic interests play into political strategies to expand subsidies, or at least to protect existing programmes from elimination or dilution by new claimants. The concentrated benefits to subsidy recipients provide both salience and funding for organization and rent sharing with politicians or other officials who rely on such contributions to fund electoral campaigns or remain in power. In contrast, the taxpayers who ultimately fund the subsidies are a diffuse group. Any single taxpayer will not see much financial gain from beating back a particular subsidy, and absent a crisis will not invest the necessary time or money to do so.

---

[11] The Joint Report (2010: 37) notes six examples where violence and protests followed fuel price increases that were required as lending conditions by the World Bank or IMF between 1977 and 1996. In half of these (Tunisia 1983, Egypt 1977, and Morocco 1981), the price increases were rolled back. While the external requirement may have made these reforms particularly unpopular, more recent reforms initiated internally have met a similar fate, such as in Pakistan and Nigeria. Kojima (2009, 2013), GSI (2013), and IMF (2013) also provide useful reviews of past reform efforts.

15.3.2. Beating the trap: what has worked?

There have been enough attempts to reform energy prices over the years that some common guidelines have been developed (Joint Report 2010; Bacon and Kojima 2006; Coady et al. 2010; Laan, Beaton, and Presta 2010; IEG 2008; TERI 2011; GSI 2012, 2013; IMF 2013). These fall into the general categories of using broader changes within the economy to also fix energy pricing; acknowledging and addressing from the outset whatever dislocations may result from reform; communicating clearly about the costs of current policies, and both the benefits and challenges of reform; and instituting reforms that are not reliant merely on political goodwill to remain effective.

- **Macro conditions can leverage reform.** Reforms are far more likely to be successful in times of crisis, such as when the fiscal costs of the subsidy are so high that some action must be taken (Joint Report 2010: 36).[12] These periods create the political will to make larger policy shifts, despite transition costs. Crises may also involve elevated assistance from the IMF or World Bank, providing an opportunity to thoughtfully link fiscal support to structural price reforms in energy markets.[13] National goals to join groups such as the European Union can provide similar (non-crisis) leverage. These factors were important in price reforms in the power sector within Eastern Europe (IEG 2008). Similarly, to the extent ancillary economic conditions (e.g. rising incomes, declining inflation, or the ability to boost public spending in other areas) can mute the impact of price shifts, resistance to subsidy removal will be reduced.

- **Mitigating measures should be built in from the outset.** Subsidy reforms will create some losers, and they may include concentrated and powerful interest groups. Mitigating measures can reduce resistance to change by allowing a transition period or by providing cash or in-kind compensation to the most vulnerable recipients of the subsidies. Transitional support may also sometimes be needed to achieve buy-in from powerful groups even if they don't face increased poverty from reforms. More careful targeting of the subsidy can reduce leakage rates while still protecting the needy.[14] However, political pressures to expand transitional payments and derail reforms are common, and phased changes must be structured so that they are very hard to roll

---

[12] The World Bank notes, for example, that it has been 'difficult to engage in price reforms in petroleum- or gas- producing countries not under fiscal stress' (IEG 2008: 55).

[13] Caution is needed on the linkage. When energy price increases were linked to accessing international assistance in the past, there was insufficient advance warning or explanation on the logic of the linkage. Political unrest resulted (Joint Report 2010: 37).

[14] A World Bank survey of cash transfer programmes across multiple sectors (not just energy) found they had a much lower leakage rate than universal fuel subsidies (Joint Report 2010: 39). Reforms can also target energy resources less central to the very poor first, such as premium gasoline rather than basic kerosene (IMF 2013). Kojima (2013) notes the variety of strategies that governments have deployed to mitigate energy price impacts more broadly across the economy. These include hedging, bulk purchases, improved infrastructure and storage to reduce logistics costs, and promoting more effective price competition.

back.[15] If transitional assistance is provided, the rationale for doing so must be logical and clearly stated, and the assistance must be strictly limited in scope and duration (Joint Report 2010: 38).

- **Transparency on existing subsidies and reform plans increases chances of success.** Data on subsidy programmes, conveyed through carefully designed communications, can help highlight programme inefficiencies and inequities. Too often subsidies don't even show up in budget documents, a data gap that has harmed may past reform efforts (IMF 2013: 23). Specifics on the total transfers and key beneficiaries can be important in silencing political resistance by well-organized current beneficiaries. For the general populace, education and communications on how the needs of the poorest citizens will be protected post-reform is critical. It is important that the negative effects of reform also be discussed openly (GIZ 2012). A review of scores of reform efforts by the IMF indicated that 'strong political support and proactive public communications' almost tripled the changes of subsidy reform success (IMF 2011: 47).

- **Government competence and reputation affects confidence in reforms.** Transitions require confidence that the government will have the will and the capability to make good on transitional support or other promises it made to achieve buy-in on the reforms, and to prevent backsliding that will undermine the positive aspects of the shift. Pairing subsidy reform with other reforms or actions that address long-running concerns on corruption, property rights, or welfare can help build confidence. Nonetheless, implementing reform is likely to be more challenging in countries with a history of poor governance.

- **Reforming the price mechanism is necessary to prevent backsliding.** Political support for subsidy reform ebbs and flows depending on local politics, broader economic conditions, and global energy prices. To prevent a reversion to subsidies when oil prices rise, for example, a shift from ad hoc (politically determined) energy prices to market prices is important. Where this isn't politically possible, shifting to automatic price adjustments based on internationally-measurable benchmarks is a second-best strategy to reduce (though not eliminate) the risk of backsliding (Coady et al. 2010; IMF 2013: 32).

15.3.3. Segmenting subsidy traps can help identify promising reform strategies

Individual country circumstances vary, and past experience indicates that not all attempted subsidy reforms succeed. However, there are a number of variants on the subsidy trap problem rather than a single one, and tailoring reform strategies to the type of subsidy in place can be helpful in boosting the success rate. Table 15.4 details the main policy issues and provides examples of appropriate reform strategies for each.

---

[15] Transitional payments to accompany fuel price reforms in Iran, for example, quickly expanded to cover a larger and larger portion of the population and ended up costing more than the original subsidies did (Kojima 2013: 30).

Table 15.4. Escaping the subsidy trap varies by policy type

Policy area and related reform trap	Potential solutions
Subsidized extraction or market rights	
Inexpensive access to raw materials or allocated market rights (e.g. operating permits or export licenses) provide windfall to initial recipient. Rights may be politically allocated rather than merit- or auction-based. Recipients often include state-owned enterprises.	Offset transitional losses via grants, tax write-offs, or slow phase-in. Political risks during transitional phase as incumbents seek to obtain new subsidies and also derail reform of old ones.
Asset prices often rise to reflect the enhanced value of the opportunity, though resale of rights at higher prices means subsequent owners earn only normal returns.	Publicizing the full costs of subsidies and their beneficiaries can be useful. Expanding beneficiaries of reform to include the populace (as with royalty trust funds that are required to distribute dividends each year) can also alter the political dynamics of reform. Privatization can be useful for nationalized assets or firms, forcing a shift to more market-based operations.
Incumbents fight reform or new entrants. National owners may be focused on politically-based wealth redistribution rather than efficient management.	
Socialization of high-risk portions of fuel cycle	
Advocated to 'jump start' risky technologies, key technical, environmental, or financial risks of fuel cycles are capped or shifted entirely to the public. Common in both developed and developing countries; examples include transport support to access Arctic oil, government investment into oil sands, socializing nuclear waste management and accident risks, and capturing carbon emissions from coal.	Key points of leverage are at policy inception. Force competition between all higher-cost marginal supply options rather than looking at a fuel or field in isolation. Re-price the government services giving rise to the subsidy so that fees adequately compensate taxpayers for the risks they are taking on, rather than targeting break-even at best. Finally, tightly restrict the subsidy duration and eligibility to ensure rapid phase-out as conditions change (including rising energy prices) and to prevent 'subsidy creep' to an ever-broader set of recipients.
Policies mask critical price signals, accelerating development of resources that may have elevated environmental problems or public risks at the expense of alternatives. Government involvement becomes an unquestioned part of fuel cycle economic baseline.	
Consumer subsidies	
To general populace. Despite high leakage rates, consumer subsidies remain important to many of the poor in the developing world. Because reform can cause undesired hardships for this group of citizens, inadequate planning and communication related to energy price increases, or a failure to build sustainable support for reform and credible substitute safety nets for the poor, have often resulted in popular unrest and violence.	Political support for reform is an absolute prerequisite. A credible transition plan drives success here: phased price adjustments; clear communication and education to affected sectors on reasons and impacts; and replacement of fuel subsidies to the poor with more efficient instruments such as cash transfers or vouchers.
	Challenges include a large black market or corruption; lack of a competent government to deliver replacement support; or an inability to limit transitional support only to the groups that need it. Even if somewhat inefficient, replacement policies may nonetheless improve on the prior subsidies.

To industrial and commercial sectors. Subsidized energy prices for industrial and commercial infrastructure skew investment patterns and over time result in 'subsidy clusters' filled with energy-inefficient plants, equipment, and buildings. Export-based commodity industries are particularly at risk from these policies over time, as they have limited ability to pass through higher fuel prices to customers. They are likely to resist reform more strongly than other sectors.

Phase-in of reforms can signal firms they need to re-price energy costs in all new capital decisions while reducing premature retirement of existing capital.

Direct subsidies to capital replacement may also be useful in particular sectors, such as when both the energy suppliers and the industries are government-owned. However, they are politically difficult to target efficiently and so should be used sparingly.

Market adjustments to privatize parts of the fuel cycle or allow in new competitors can also spur upgrades more quickly. For inefficient real estate infrastructure, mandated disclosure of heating and cooling costs can establish building operating costs as a competitive attribute of leases or sales.

Black-market intermediaries. Dual-tier pricing to consumers supports smuggling and black markets as well. These groups will seek to block reform to protect their market, sometime violently.

Full elimination of dual pricing structures can eliminate the black markets but can also cause other unrest. Using electronic debit cards to ration subsidized fuel or deliver lump sum cash payments in lieu of fuels can protect the poor, better target subsidies, and reduce or eliminate black market diversions. These approaches still require basic competence within the government, and thus sometimes break down. However, they can be a useful transition to market-based pricing, or to deliver substitute benefits.

Quantifying the costs of the market distortions, particularly to business and the poor, may help undermine any popular support the illegal operators have.

Pricing networked energy services (power, natural gas) at below long-term break-even
Existing customers may mobilize to protect favourable pricing, or to prevent extensive enforcement against resource theft or non-payment of bills.[16] Groups most hurt by these conditions tend to be the very poor who are not currently serviced by transmission or pipeline networks. They are not politically mobilized or powerful, and tend to be ignored.

Evaluation and disclosure of the scale of subsidies, and the groups that benefit most from them, can alter the political dynamics of reform. More clearly delineating the costs of unreliable and low-quality power can also help achieve buy-in for better maintenance, system upgrades, and enforcing non-payment and power theft. Surveys indicate that many customers are willing to pay more for higher-quality power (Komives et al. 2005).

Inadequate revenues over time lead to decay in existing infrastructure and returns too low to justify enhanced service quality or expanded service area.

More targeted subsidies to overcome barriers to access for key user groups such as the poor are far more efficient than maintaining low prices to all users. Obtaining separate funding for this group as a welfare transfer can protect the utility against

continued

---

[16] The World Bank (Komives et al. 2005: 1) notes that '[s]ubsidies to utility customers are widely popular among policymakers, utility managers, and residential customers alike, and yet subsidies remain the subject of much controversy'.

Table 15.4. Continued

Policy area and related reform trap	Potential solutions
	the revenue erosion that cross-subsidies can sometimes cause. Support payments predicated on adequate power efficiency, reliability, and quality can further spur better operations.
	Any reform will be more difficult if existing managers or utilities are viewed as corrupt or inept. In some cases, privatization can help align management interests with service efficiency. This is more difficult with utility-like services than with other sectors, as some natural monopoly aspects of the networks remain; solid regulatory oversight and enforcement would also be needed.

Cross-subsidies in power distribution, bulk energy transport

Gross revenue targets may support basic network operations but only by large cross-subsidies among customer classes that create inaccurate price signals for consumers and for utility planning. Favoured classes may also organize politically to protect their status.	Cost transparency is needed to highlight the real economics of different customer classes; this is helpful whether or not the information is immediately used to alter tariffs. Particularly in regions with sparse distribution infrastructure currently in place, proper costing should make decentralized power options more viable in many areas.[17]
Urban areas often face elevated charges to fund the higher cost of serving remote, low-density loads. Efficient break-points for grid extension versus decentralized power resources or suppliers are therefore lost.[18] IEA estimated that 93 per cent of the energy supply gap for mini-grid or off-grid supplies would be renewable; in that context, the cross-subsidies appear a significant market impediment for renewable technologies (IEA 2011b: 89). Cross-subsidies between higher-income and lower-income customers are also common, though these can be challenging in utility systems where overall revenues barely suffice to meet operational needs.	With accurate data, better decisions can be made on which groups require cross-subsidies and how best to fund those needs.

Although consumption may be subsidized for poor customers, connection fees and fixed monthly charges may remain high, resulting in many citizens not connecting to the grid. Targeted funding to cover the connection costs can be important in making expanded access work. Development aid is one funding source; redirecting savings from subsidy reforms (particularly provisions with high leakage rates) is another.[19] |
| Challenges differ somewhat depending on whether expanded access involves boosting connection rates to an existing grid or requires grid extension. | |

[17] Grid extension, separate mini-grids, or decentralized power resources can all be appropriate solutions to extended energy access, both individually and sometimes in combination. Policy frameworks will perform better if they don't bias decisions on the best service delivery model (World Bank 2010: 14).

[18] Rising power distribution and transportation costs in rural areas make off-grid sources of supply, including mini-grids or home-based solar systems, more economic. High-cost supplies that need to be distributed into areas with a low population density remain a challenge, but the off-grid solutions are nonetheless cheaper (World Bank 2010: 25).

[19] High fixed charges for connections can undermine lifeline rate structures for low-consuming customers (Komives et al. 2005: 87).

15.4. SUMMARY

Energy subsidy reform has attracted increasing attention and research in recent years as a useful lever to help countries reign in fiscal deficits, redirect public spending to areas with higher social benefit, and avoid undermining efforts to address global climate change. Financial subsidies totalled nearly $840 billion in 2011, and more than $3.5 trillion over the 2007–11 period. External costs of energy systems, primarily fossil fuels, are estimated to be an additional $1.6 trillion per year, and are additive to the financial subsidies in distorting the energy prices on which energy investment and deployment decisions are made. Combined financial subsidies to energy plus external costs amount to more than 3 per cent of global GDP. For many developing countries, spending on fossil fuel subsidies exceeds what governments spend on health care for these populations and absorbs an unsustainable amount of federal budgets.

Spending on fuel subsidies mostly leaks to higher income quintiles rather than helping improve the lives of the world's poorest citizens. The lost opportunity is huge: annual subsidies are more than 30 times the incremental funding needed to achieve universal access to modern energy services, a transition the IEA notes would bring with it large improvements in public health and quality of life. OECD countries, on average, are subsidizing fossil fuels at a rate five times the level they are willing to commit to addressing climate change around the world. The potential to achieve a variety of important social goals by redirecting current subsidy flows is clear. Yet despite the strong logic of reforming subsidies, many countries are trapped into continuing existing policies because ending them would cause political unrest among current beneficiaries or because they are unable to credibly provide an alternative safety net for the poor.

Successful reform efforts have involved a number of common themes. Leveraging macro-economic changes to incorporate price reforms can help governments implement reforms during periods that will cause less dislocation. Advance planning is needed, however, so as to be ready to implement changes when conditions are good.

Assessing which groups are likely to lose under reform and building in appropriate mitigation measures from the outset, particularly to protect the poor, has been critical in avoiding popular unrest as subsidies are phased out. Integrating subsidy reform more directly with universal energy access targets is also important. Many existing subsidies have been justified based on claims that they helped the poor; it is only fair to ensure that a portion of the savings is deployed to help achieve that goal. However, just as improperly-targeted government energy subsidies bleed budget capacity away from higher-impact social spending, so too does underpricing of grid-based power or gas erode the ability of utilities to remain viable and expand. Accurately measuring both utility subsidies and cross-subsidies is a first step in fixing the problem. Even if tariffs do not immediately change to target only those who need them, better decisions amongst core options can be made, such as whether to extend grids, to subsidize connection and fixed costs to existing grids, or to reach new areas via decentralized power resources rather than line extensions.

Particularly for very low income customers, subsidizing the tariff per kWh may be less important than reducing the connection fees and monthly fixed costs of services so that small initial increments of power or natural gas become affordable. Subsidies to clean cooking facilities have focused on technologies that can be supported at a small scale by an indigenous industry, rather than being reliant on outside support and providers. It would be useful to deploy additional funds freed up by subsidy reform to extend this type of approach.

The most difficult challenges of subsidy reform and effective redeployment of the savings to achieve goals such as universal energy access are political. A consistent finding from reform case studies is the importance of gathering much more detailed information on subsidy costs and core beneficiaries than is routinely collected, using that information to overcome more powerful vested interests that will try to block reform, and communicating transparently about both the benefits and the risks of planned changes. To ensure that successes are not rolled back in short order, the reform strategies—particularly how energy prices are set—need to be incorporated into the legal framework of the country rather than remaining within the decision-making domain of policymakers.

REFERENCES

Bacon, R. and Kojima, M. (2006). 'Viewpoint: Phasing Out Subsidies—Recent Experiences with Fuel in Developing Countries', *Public Policy for the Private Sector Note 310* [online], <https://openknowledge.worldbank.org/handle/10986/11178> (accessed 16 October 2013).

Burniaux, J. and Chateau, J. (2011). 'Mitigation Potential of Removing Fossil Fuel Subsidies: A General Equilibrium Assessment', *OECD Economics Department Working Papers*, No. 853, OECD [online], <http://www.oecd-ilibrary.org/economics/mitigation-potential-of-removing-fossil-fuel-subsidies_5kgdx1jr2plp-en> (accessed 16 October 2013).

CIA (United States Central Intelligence Agency) (2013). 'Government Revenues,' *The World Factbook* [website], <https://www.cia.gov/library/publications/the-world-factbook/fields/2056.html> (accessed 6 March 2013).

Coady, D., Flamini, V., and Antonio, M. (2012). 'Fueling Risk: Energy Subsidies to Low- and Middle-income Countries can Take a Big Toll on their Fiscal Health', *Finance and Development*, September [online], <http://www.imf.org/external/pubs/ft/fandd/2012/09/coady.htm> (accessed 16 October 2013).

Coady et al. (2010). *Petroleum Product Subsidies: Costly, Inequitable, and Rising*. Washington, DC: International Monetary Fund [online], <http://www.imf.org/external/pubs/ft/spn/2010/spn1005.pdf> (accessed 16 October 2013).

Del Granado, J.A., Coady D., and Gillingham, R. (2010). 'Working Paper: The Unequal Benefits of Fuel Subsidies: A Review of Evidence for Developing Countries', International Monetary Fund, September, IMF WP/10/202 [online], <http://www.imf.org/external/pubs/ft/wp/2010/wp10202.pdf> (accessed 16 October 2013).

Fattouh, B. and El-Katiri, L. (2012). 'Energy Subsidies in the Arab World', Arab Human Development Report Research Paper Series, United Nations Development Programme [online], <http://www.undp.org/content/dam/undp/library/Environment%20and%20Energy/UNDP-EE-AHDR-Energy-Subsidies-2012-Final.pdf> (accessed 16 October 2013).

GIZ (Deutsche Gesellschaft für Internationale Zusammenarbeit GmbH) (2012). *International Fuel Prices 2010/2011, 7th Edition*, April [online], <http://www.giz.de/Themen/en/dokumente/giz-en-IFP2010.pdf>

Global Subsidies Initiative (2012). 'Reforming Fossil Fuel Subsidies to Reduce Waste and Limit CO$_2$ Emissions while Protecting the Poor', prepared for the APEC Energy Working Group, September [online], <http://www.iisd.org/gsi/sites/default/files/ffs_apec.pdf> (accessed 16 October 2013).

Global Subsidies Initiative (2013). 'A Guidebook to Fossil-Fuel Subsidy Reform for Policy-Makers in Southeast Asia' [online], <http://www.iisd.org/gsi/sites/default/files/ffs_guide book.pdf> (accessed 16 October 2013).

Guardian (2012). 'Datablog: Health care Spending Around the World, Country by Country', 30 June. Data extract of health care spending data collected by the World Health Organization [online], <http://www.guardian.co.uk/news/datablog/2012/jun/30/ healthcare-spending-world-country> (accessed 16 October 2013).

IEA (International Energy Agency) (2011a). *World Energy Outlook 2011* [online], <http:// www.worldenergyoutlook.org/publications/weo-2011/> (accessed 16 October 2013).

IEA (2011b). *World Energy Outlook 2011* slide library.

IEA (2012). *World Energy Outlook 2012* [online], <http://www.worldenergyoutlook.org/ publications/weo-2012/> (accessed 16 October 2013).

IEA (2013). Statistics division, 'Electricity/Heat by Country/Region' [online], <http://www. iea.org/stats/prodresult.asp?PRODUCT=Electricity/Heat> (accessed 5 March 2013).

IMF (International Monetary Fund) (2011). 'Regional Economic Outlook: Middle East and Central Asia'. Washington, DC [online], <http://www.imf.org/external/pubs/ft/reo/ 2011/mcd/eng/pdf/mreo0411.pdf>.

IMF (2013). 'Energy Subsidy Reform: Lessons and Implications', 28 January [online document], <http://www.imf.org/external/np/pp/eng/2013/012813.pdf> (accessed 16 October 2013).

Independent Evaluation Group (IEG) (2008). 'Climate Change and the World Bank Group, Phase 1 – An Evaluation of World Bank Win-Win Energy Policy Reforms'. Washington, DC: World Bank [online], <http://siteresources.worldbank.org/EXTCLICHA/Re sources/Climate_ESweb.pdf> (accessed 16 October 2013).

Joint Report (2010). International Energy Agency, Organisation for Economic Co-Operation and Development, Organization of the Petroleum Exporting Countries, and World Bank. 'Analysis of the Scope of Energy Subsidies and Suggestions for the G-20 Initiative', prepared for submission to the G-20 Summit Meeting, Toronto, 26–27 June 2010, 16 June [online], <http://www.oecd.org/env/45575666.pdf> (accessed 16 October 2013).

Kitson, L., Wooders, P., and Moerenhout, T. (2011). *Subsidies and External Costs in Electric Power Generation: A Comparative Review of Estimates*, Geneva: Global Subsidies Initiative, September [online], <http://www.iisd.org/gsi/sites/default/files/power_gen_subsid ies.pdf> (accessed 16 October 2013).

Kojima, M. (2009). 'Changes in End-User Petroleum Product Prices: A Comparison of 48 Countries'. Washington DC: World Bank [online], <http://siteresources.worldbank.org/ INTOGMC/Resources/ei_for_development_2.pdf> (accessed 16 October 2013).

Kojima, M. (2012). 'Oil Price Risks and Pump Price Adjustments', *World Bank Policy Research Working Paper 6227*, October [online], <http://elibrary.worldbank.org/content/ workingpaper/10.1596/1813-9450-6227> (accessed 16 October 2013).

Kojima, M. (2013). 'Petroleum Product Pricing and Complementary Policies: Experience of 65 Developing Countries Since 2009', *World Bank Policy Research Working Paper 6396*, April [online], <http://econ.worldbank.org/external/default/main?pagePK= 64165259&theSitePK=469372&piPK=64165421&menuPK=64166322&entityID=000158349_ 20130401160010> (accessed 16 October 2013).

Komives, K., et al. (2005). 'Water, Electricity, and the Poor: Who Benefits from Utility Subsidies?' Washington, DC: World Bank [online], <https://openknowledge.worldbank. org/handle/10986/11745> (accessed 16 October 2013).

Koplow, D. (1996). 'Energy Subsidies and the Environment,' in *Subsidies and Environment: Exploring the Linkages*. Paris: OECD.

Koplow, D. (2007). 'Chapter 4: Energy,' in *Subsidy Reform and Sustainable Development: Political Economy Aspects*. Paris: OECD.

Koplow, D. (2009). *Measuring Energy Subsidies Using the Price-Gap Approach: What does it Leave Out?* Geneva: Global Subsidies Initiative of the International Institute for Sustainable Development, August [online], <http://www.iisd.org/publications/pub.aspx?pno=1165> (accessed 16 October 2013).

Koplow, D. (2010). 'EIA Energy Subsidy Estimates: A Review of Assumptions and Omissions'. Cambridge, MA: Earth Track, Inc., March [online], <http://earthtrack.net/documents/eia-energy-subsidy-estimates-review-assumptions-and-omissions> (accessed 16 October 2013).

Koplow, D. (2011). *Nuclear Power: Still Not Viable without Subsidies*. Cambridge, MA: Union of Concerned Scientists, February [online], <http://www.ucsusa.org/assets/documents/nuclear_power/nuclear_subsidies_report.pdf,> (accessed 16 October 2013).

Koplow, D. (2012). *Phasing Out Fossil-Fuel Subsidies in the G20: A Progress Update*, Earth Track Inc. and Oil Change International, June [online], <http://priceofoil.org/content/uploads/2012/06/FIN.OCI_Phasing_out_fossil-fuel_g20.pdf> (accessed 16 October 2013).

Koplow, D. and Kretzmann, S. (2010). *G20 Fossil-Fuel Subsidy Phase Out: A Review of Current Gaps and Needed Changes to Achieve Success*, Earth Track Inc. and Oil Change International, November [online], <http://www.earthtrack.net/documents/g20-fossil-fuel-subsidy-phase-out-review-current-gaps-and-needed-changes-achieve-success> (accessed 16 October 2013).

Koplow, D. and Lin, C. (2012). *A Review of Fossil Fuel Subsidies in Colorado, Kentucky, Louisiana, Oklahoma, and Wyoming*. Cambridge, MA: Earth Track, Inc., December [online], <http://www.earthtrack.net/documents/review-fossil-fuel-subsidies-colorado-kentucky-louisiana-oklahoma-and-wyoming> (accessed 16 October 2013).

Koplow, D., et al. (2010). *Mapping the Characteristics of Producer Subsidies: A Review of Pilot Country Studies*. Geneva: Global Subsidies Initiative, August [online], <http://www.iisd.org/gsi/sites/default/files/mapping_ffs.pdf> (accessed 16 October 2013).

Laan, T., Beaton, C., and Presta, B. (2010). *Strategies for Reforming Fossil-Fuel Subsidies: Practical Lessons from Ghana, France and Senegal*. Geneva: Global Subsidies Initiative, April [online], <http://www.iisd.org/publications/pub.aspx?id=1268> (accessed 16 October 2013).

Lahn, G. and Stevens, P. (2011). *Burning Oil to Keep Cool: The Hidden Energy Crisis in Saudi Arabia*. London: Chatham House, December [online], <http://www.chathamhouse.org/publications/papers/view/180825> (accessed 16 October 2013).

OECD (Organisation for Economic Cooperation and Development) (2013). *Inventory of Estimated Budgetary Support and Tax Expenditures for Fossil Fuels 2013*. Paris: OECD [online], <http://www.oecd-ilibrary.org/environment/inventory-of-estimated-budgetary-support-and-tax-expenditures-for-fossil-fuels-2013_9789264187610-en> (accessed 16 October 2013).

Oil Change International (2012). 'Fossil Fuel Subsidies vs. Fast Start Climate Finance in Annex 2 Countries', 12 December [website], <http://priceofoil.org/content/uploads/2012/05/Fossil-Fuel-Subsidies-vs-Climate-Finance-2-pg.pdf> (accessed 16 October 2013).

Piebalgs, A. (2012). 'Delivering Sustainable Energy for All', in OECD, *Development Cooperation Report 2012: Lessons in Linking Sustainability and Development* [online], <http://www.oecd-ilibrary.org/sites/dcr-2012-en/02/02/index.html;jsessionid=roso0lffmiej.x-oecd-live-01?contentType=/ns/Chapter&itemId=/content/chapter/dcr-2012-12-en&containerItemId=/content/serial/20747721&accessItemIds=&mimeType=text/html> (accessed 16 October 2013).

Sauvage, J. (2013). 'OECD', e-mail communication with Doug Koplow, 14 February.

Steenblik, R. (2007). *A Subsidy Primer*. Geneva: Global Subsidies Initiative.

TERI (The Energy and Resources Institute) (2011). 'Fossil-Fuel Subsidy Reform in India: Cash Transfers for PDS Kerosene and Domestic LPG'. Geneva: Global Subsidies Initiative, August [online], <http://www.iisd.org/gsi/sites/default/files/ffs_india_teri_rev.pdf> (accessed 16 October 2013).

Tullock, G. (1975). 'The Transitional Grains Trap', *The Bell Journal of Economics*, 6: 671–8.

Victor, D. (2009). *The Politics of Fossil Fuel Subsidies*. Geneva: The Global Subsidies Initiative of the international Institute for Sustainable Development, October [online], <http://www.iisd.org/publications/pub.aspx?id=1254> (accessed 16 October 2013).

World Bank (2010). 'Addressing the Electricity Access Gap: Background Paper for the World Bank Group Energy Sector Strategy', June [online], <http://siteresources.wor ldbank.org/EXTESC/Resources/Addressing_the_Electricity_Access_Gap.pdf> (accessed 16 October 2013).

World Bank (2013a). Databank data series, 'GDP (current US$)' [online], <http://data. worldbank.org/indicator/NY.GDP.MKTP.CD> (accessed 16 October 2013).

World Bank (2013b). Databank data series, 'Population, total' [online], <http://data.wor ldbank.org/indicator/SP.POP.TOTL> (accessed 7 March 2013).

16

Is Small Beautiful?

Phil LaRocco

16.1. INTRODUCTION

Originally this chapter bore the title 'Small is Beautiful'. Fortunately the editors allowed these three words to be turned into a question. This avoided significant problems. First, no one should presume to write under the banner of what the *Times* (*TLS* 1995) considered one of the most influential books since World War II, E.F. Schumacher's *Small is Beautiful: Economics as if People Mattered* (London: Blond & Briggs 1973). Second, by presenting a question instead of a statement this chapter can serve as an inquiry about how Schumacher's principles might apply to the more earthy matters of energy for development—40 years after his work was published. Third, though *Small is Beautiful* is a serious and diverse work by a complex and credible author, the title itself creates an impression of 'over-the-hill hippie' manifestos. People hearing 'small is beautiful' make assumptions not actually borne out by a critical analysis of the book.

Anyone who knows this work realizes that the title would have been far more descriptive had it been *Appropriate is Effective* or *Intermediate is Somewhere between Drudgery and Robotic*. As I don't think I would still have a 1973 Perennial Library edition of this work had it carried either of those titles (full disclosure: I am one of those over-the-hill-hippies), I am glad that *Small is Beautiful* made the cut as the published title. For those who have not read the book, however, *Small is Beautiful* is about much more than just the scale of technology—the work defies simple classification when read cover to cover. The title, however, is just a 'title', perhaps even a 'cliché' and certainly good marketing, but it is not a literal summation of Schumacher's philosophy.

Small is Beautiful emphasizes a number of key principles concerning what Schumacher believed would be lasting, people-centred development. The energy-for-development space is as likely a test-bed for his principles as any, given the close—perhaps inextricable—relationship between the availability of basic energy services and the beginning of a person's climb up the human development ladder.

This chapter summarizes some of Schumacher's principles and introduces a sample of energy-for-development experiences, asking how these stack up alongside Schumacher's principles and whether or not these principles still apply given the passage of 40 years. Along the way it tries to relate what Schumacher thought in 1973 to the state of play and significant issues of today.

16.2. SOME OF SCHUMACHER'S GENERAL PRINCIPLES AND ISSUES

Intermediate technology: technology that is neither traditional (in the sense of crude and non-productive) nor advanced (in the sense of expensive and automated)—a technology choice that gives priority to workspace creation, ease of use with local skills, and use of local materials.

Goals of development: development should emphasize self-reliance, the creation of meaningful work and quality of life improvements, and the avoidance of 'giantism' (a unique Schumacher term). Entities directly engaged in development should exhibit active owner engagement and resource efficiency. These activities should strive to be profitable and productive while optimized for employment opportunity rather than unit cost or mass production.

Successful development: because it is local and grounded in culture, successful development is organic and evolutionary, leading to financially sound, sustainable outcomes, using local resources—human and natural—whenever practical.

Industrial agriculture and urban misery: Schumacher critiqued large-scale agriculture as too resource-intensive (especially energy-intensive), leading to job and workplace elimination and a flight of rural workers and their families to cities where their skill sets prove marginally useful and their employment potential is low. He found cost-benefit analysis, with a laser focus on input–output per hectare, to be a short-sighted, one-dimensional (economic) analysis.

Economics: economics should be only one dimension of development because it is limited by its emphasis on problems readily reduced to quantities (convergent problems). Economics as a leading (dominating) discipline lacks the capacity to integrate meaningfully qualitative (meta-economic) values. Meta-economic problems cannot be solved by reduction; these divergent problems are the kind that need to be lived, thus the importance of evolution and adaptation.

'Religion of economics': Schumacher feared a decision-making process obsessed with unit cost, consumption, and automation. He saw this paradigm as dehumanizing to workers and underpinned by John Maynard Keynes' instruction that greed and usury need to rule ('foul must be fair') for a century before the march to peace through prosperity can begin. Schumacher wondered if there was precedent that rich people were any more peaceful than poor ones as a counter to the 'foul is fair' temporary injunction.

Stages of an idea: Schumacher believed that four stages occur. The first is sneering and laughter by the forces of conventional wisdom. The second is grudging acceptance that perhaps there is something to the idea but it really is not significant. The third is when serious work by serious people begins. The fourth occurs when the idea transforms into a practical and routine application, when the policy becomes implementation.

16.3. APPLICATION OF SCHUMACHER'S PRINCIPLES TO ENERGY FOR DEVELOPMENT: A THOUGHT EXPERIMENT

One of the Schumacher principles difficult to judge today is the trade-off between employment intensity and output cost and efficiency. Projects, programmes, and companies deal with dual and often distinct goals of, on the one hand, eliminating drudgery and dirty and dangerous conditions for, say, 5 thousand households and, on the other, creating profitable work space and employment opportunity for one local enterprise with perhaps 10 to 30 employees. Are these goals necessarily in conflict? Of course not, but if faced with the choice between a locally produced, labour-intensive improved cookstove with a price point of A and an efficiency of B versus an imported cookstove with higher efficiency and/or lower price point (but little productive, local employment or materials used), how does one decide? Putting aside the canard that we can just let the market decide—if all the sheet metal for local production is being exported for stove manufacture elsewhere, that is not an invisible hand but a fistful of purchasing power—I think Schumacher would want us to look closely at that imported stove and bring its production closer to the customer, even perhaps at the sacrifice of a few points of efficiency. There is no cost-benefit analysis to making this decision; this is one of Schumacher's 'divergent' problems that need to be lived rather than a 'convergent' one that can be simplified and reduced. You cannot reach a fact-bound conclusion. You need to make a *decision* with imperfect information and ambiguous circumstances.

Another Schumacher principle that might cause a problem today is the 'big push' or commodity manufacture approach. The location of the 'champions', the owners or organizers of the 'big push' or product production, would be important to Schumacher. Are they in the capital city, hours or days from the actual customer? Are they in a different country? Are they some NGO in Washington or international organization in Rome?

There would likely be a related issue of forced versus organic, planned versus evolutionary. 'Plan your work and work your plan' is an excellent slogan, but in practice nothing goes exactly as planned. Adaptations need to be made by the people on the scene. This comes to Schumacher's point of change being organic and evolutionary. It also introduces the element of *time* as less malleable than we like to think. It often said that time is money, but that is one of the great falsehoods of programme, project, or enterprise development. Time is an infinitely more valuable commodity than money, and all the money in the world cannot buy a minute of time (ask a dying billionaire).

Translating 1973 Schumacher into today's energy-for-development world in a hundred words might look something like this:

- Where choice exists, emphasize local resources and local talents.

- Nothing goes as planned. Time (evolution) and flexibility (organic growth) are required.

- Self-reliance and quality-of-life gains are both important goals. These represent challenges that cannot be reduced and solved but must be lived.

- In a reasonable trade-off between performance efficiency and employment creation, go with jobs.

- Owner-managers and decision-makers must be close to implementation.

(To further reduce the hundred words to ten: local resources, organic change, divergent problems, job intensity, and close ownership.)

Within the confines of this chapter, it is not possible to examine a sample of 10 to 20 examples involving different peoples from different parts of the world to consider Schumacher's principles in the twenty-first century. Instead, we look closely at three energy-for-development experiences to come to this understanding.

16.4. ENERGY-FOR-DEVELOPMENT EXPERIENCES: A SAMPLE

- Anagi Stove, Sri Lanka: a 40-year venture of numerous institutions and activities that resulted in the production of about 3 million more efficient cookstoves—and a self-sustaining annual production of 300,000—by a few hundred enterprises (Practical Action 2010).

- Household Biogas Programme (BSP), Nepal: a two decade public–private, domestic and international programme for a few hundred small contractors to install more than 200,000 household scale biodigesters using cow and sometimes human waste to produce methane (Ashden Awards 2006; Bajgain et al. 2005).

- Grameen Shakti, Bangladesh: a two-decades-old social enterprise that has delivered small solar solutions to more than 1 million households and that has reached an annual throughput approaching 200,000 units, leveraging the experience and lending techniques of its parent, Grameen Bank (IFC 2012; Aron et al. 2009).

16.4.1. A question of maturity

The preceding are but three of a long list of diverse energy development experiences gathered these last 20-plus years. This experience base shows a few things:

- Acknowledgement by major international institutions such as the World Bank Group, IEA, UNDP, Practical Action, and UNEP.

- Independent recognition by Ashden Trust, Zayed Future Energy Prize, *Financial Times*, Energy Globe, and the Ramon Magsaysay Award.

- Dedicated research and development and assessment of 'bottom of the pyramid' products (d-Light, Envirofit, SEF, Lighting Africa).

Energy for development (E4D) has emerged as a topic of importance (Rio+20, Year of Sustainable Energy for All). This acknowledgement suggests that the *idea* of energy for development may fall within Schumacher's second and third stages: begrudging acceptance that there may be something interesting going on, caution as to the scale of the 'something', and work being undertaken by serious people.

But the question remains: is it (are we) a recognizable industry? When did/does/will E4D reach what Schumacher called 'The Policy of Implementation'?

Some years ago, Bill Drayton, founder of Ashoka, spoke to a group assembled in Geneva on the subject of 'blended value' and what it meant. He framed a question using the analogy of feeding versus teaching:

> In the development business, we seem to know how to fish and we seem to also know how to teach others to fish. What I want to know is when are we going to organize fishing fleets and a fishing industry?

Given the width and depth of the E4D experience to date and the number of people both fishing and teaching others to fish, Bill Drayton's question is particularly apt: are we an industry? Are we even fleets of boats? Or are we lots of small and medium-sized fishing boats each doing its own thing? Do we have (or even need) Schumacher-like principles or do we just keep building and fishing?

With that in mind, let us explore our three examples of E4D experience through a Schumacher lens.

16.4.2. Anagi

Frustrate simplification . . . define small . . . define beautiful . . . define success . . . The first *Poor People's Energy Outlook* in 2010 (*PPEO 2010*) summarized Sri Lanka's experience with the Anagi improved cookstove: 3 million stoves had been sold and private sector production was reported to be 300,000 units a year. The more important point concerning the Anagi experience, however, is within the *PPEO 2010* narrative:

> The evolution of this sector was not due to any one project or body but was the result of a series of linked initiatives forming rough 'phases' of evolution . . . different actors, activities, types of finance, stimuli and policy and regulatory actions. Various entry points and approaches have led to a final market-based outcome in this product sector—although it did not—and it is argued, could not—start that way (Practical Action 2010).

To make the point that this success had many parents, *PPEO 2010* lists five different phases beginning in 1972 and names well over 20 significant institutional actors (*not* including the hundreds of private-sector artisans and enterprises involved or all the channels of distribution resulting). The narrative ends with two important cautions: that these 'phases . . . are only visible in retrospect' and while it is tempting to ascribe one actor or action with the 'key role', the truth is that this example 'frustrates such simplification'.

I believe Schumacher would want to know and confirm a few things about Anagi (even though it contains much of his DNA through the role of Practical Action): how many work places have been created at about what cost each? What improvements (in the stove, in its production, in its distribution) have been made over what period of time? How are such improvements made *now*? Where does this stove model fit along the spectrum of affordability and appropriateness for the customer? What other stove choices exist? (Schumacher was not afraid of either competition or profit.) I think he would also want some independent verification that the things reported in *PPEO 2010* were *essentially* true (meaning, in his calculus, that although they surely went through many false starts, in hindsight this fairly describes the current state of play; Schumacher seemed very comfortable with ambiguity as well as evolution as a principle).

How does this help us understand the state of the art today? First, the end-user (customer) focus is clearly evident in Anagi. Second, the metrics are driven by quality of life, not simply by productivity and income (Schumacher would like that—he thought measures such as GDP and averages were incredibly short-sighted). Third, the process has been evolutionary, learned through experience (not central planning), and eventually has become self-replicating and self-supporting. Fourth, the improved cookstove has created numerous jobs. Finally, the producers and the resources are close to the customers.

Would Schumacher care that the Anagi cookstove reached 3 million units distributed and 300,000 per year? Initially, I do not think those numbers would register as more or less significant to Schumacher without further explanation. Certainly he would object if you explained that all that you did was create a single job, pushing a button to produce that result. He would have trouble with the scale discussion so dominant today; Schumacher would not dismiss the discussion but simply relegate it to a secondary consideration.

What would be Schumacher's primary interests? As noted, job intensity was crucial to Schumacher. 'Intermediate' technology for him was optimal—neither traditional drudgery nor advanced technology, as measured by the efficiency of output per unit of capital input. Schumacher believed that somewhere between the $10 capital investment per traditional job and the $1,000 investment per hi-tech job will be found a choice of technology of $100 per job that creates ten jobs for the same $1,000 and perhaps *slightly less* economic output or efficiency. If you could do this a thousand times and create millions of stoves and many, many jobs, Schumacher would have concluded that 'this is development'. The fact that you also build 'human infrastructure'—a term I would never associate with Schumacher—would be excellent: *education* (broadly defined) was Schumacher's bedrock. All of our euphemisms for education would be distinctions that really do not reflect differences, just layers of complexity.

We, in the second decade of the third millennium, experiencing order-of-magnitude increases in the pressures that Schumacher foresaw, we with our iThings, are left to determine:

- Was Anagi beautiful while small?
- If not, when did Anagi become beautiful?
- Would Anagi be beautiful if it were only 10 per cent the size reported?
- Is Anagi still beautiful while no longer small?
- When did Anagi become successful and reach scale by today's standards?
- Does any of this matter?

Turned on their head, these questions ask: is Anagi a good example of small *becoming or being* beautiful? Quantitatively, surely. But taking 40 or 50 *years* to get there might rankle with some. Schumacher, fortunately, helps with this question. He would look at the eclectic, decentralized, organic nature of *Anagi* not as a single thing but as *many*—perhaps thousands—of appropriate-sized experiences that may be considered as a single entity (but are not). Separated from the broader experiences of energy for development (somewhere in a combination of the second and third stages: gathering momentum), Anagi separately might qualify as an example of Schumacher's fourth stage of an idea: the Policy of Implementation. That is, it is happening, full stop.

Scorecard to complete

- Emphasize local resources and local talents.
- Nothing goes as planned. Time and flexibility are required.
- Self-reliance and quality-of-life gains are both important goals. These represent challenges that cannot be reduced and solved but must be lived.
- In a reasonable trade-off between performance efficiency and employment creation, go with jobs.
- Owner-managers and decision-makers must be close to implementation.

16.4.3. Heading north: Nepal Biogas

A second example appears very different from the Anagi experience. The Biogas Support Programme in Nepal (BSP) is generally presented as a tightly designed, well integrated, multi-actor programme to bring household biogas digesters to different ecosystems within rural Nepal. Unlike the eclectic, evolutionary Anagi, the BSP seems an exercise in complexity with government oversight, collaboration, and coordination; detailed technical assistance and engagement of NGO expertise; training of private sector contractors, including control of materials and processes; fixed subsidies and end-user finance built in from the beginning; and tight organization, management, and documentation during specific phases that included monitoring, evaluation, and course correction. The time series for BSP is hardly short but pales compared to Anagi: a mere 10 to 15 years. Even in video form (Dev Part Consult Nepal n.d.) it has command-and-control features that a central planner would admire (and Schumacher would fret over).

But BSP and Anagi share some very important similarities. Both use local technology adaptations. Both employ many and different-sized contractors. Both strive for quality assurance, durability, and appropriateness of the product. Both seek affordable price points. Both exhibit all the signs of being a fourth-stage (mature) idea. While these two examples clearly differ in the formal versus the informal, the organized versus the organic, and the planned versus the improvised, would that matter to Schumacher? I do not know. He would probably worry over the tendencies that a command-and-control approach implied in a complex public–private partnership, but if the results stood the test of inspection (that certain principles were active) he would probably get over those concerns and celebrate the different ways of solving common problems.

Schumacher would probably ask serious questions along the same lines as those regarding Anagi: how many work places have been created at about what cost each? What improvements (in the digester, in its production, in its distribution) have been made over what period of time? How are such improvements made now? Where are we along the spectrum of affordability and appropriateness, subsidy to and benefits for the customer? What other choices exist? What are the qualitative benefits? For example, the provision of latrines has been known to reduce the incidence of snake-bites by eliminating modesty-driven toilet visits to weeded areas preferred by snakes.

Within the context of the question, 'Is small beautiful?', BSP begs us to ask a different question: *What if* the programme had *never* taken off in the way it did (hundreds of thousands of units, many dozens of contractors) but had reached a plateau at, say, a few thousand units per year and a handful of firms? Would these 'small' results in that case become 'ugly'? I think not (a personal bias), because we should fear, as Schumacher feared, that *scale* becomes the enemy of *success* if the answer is otherwise—or worse, that scale becomes the *measure* of success. Schumacher cautioned about worshipping what he called 'giantism' in anything (nations, cities, technologies) but he wisely recognized that appropriateness to the task was the measure—*not* bigness *or* smallness. Did it get the job done? Did it create individual opportunity (jobs)?

Abbreviated scorecard to complete: local resources, organic change, divergent problems, job intensity, and close ownership

Before considering the third example, let us remind ourselves that there is a difference between the goal of eradicating energy poverty (analogous to the elimination of a disease) and the goal of development (creation of human opportunity). While these two goals are linked in many ways, they are nevertheless different. One seeks to eliminate the drudgery of traditional energy and to empower (literally) the first steps on the development ladder. The other seeks to cement the foundation for that ladder—jobs and all the dignity implied. Can one goal be reached without the other? Probably not: whether you seek relief from drudgery, escape from poverty, or the creation of opportunity, *modern energy* is a key input. It is in the provision of this input that these multiple goals are linked. It is in the *choice among choices* that requires attention and screening via principles such as Schumacher's.

16.4.4. Grameen Shakti

Grameen Shakti (GS), in Bangladesh, is different from both Anagi and BSP. It originated as a social enterprise spin-off—one of many—from Grameen Bank. It is cited as the largest doorstep-delivery solar enterprise in the developing world. It is highly monolithic from a technical perspective (GS spent years getting the product features and price point engineered to its satisfaction). It is decentralized but highly structured from an implementation perspective, and its customer base overlaps but is not identical to that of its parent. In one (fairly loose) sense it grew from the Anagi-like evolution that characterizes Grameen Bank and all its spin-offs (with perhaps the difference of one significant and continuous 'actor', Professor Mohammed Yunus). In other ways it was very BSP-like in its design and execution (underpinned by the multi-decade partnership of Professor Yunus and his deputy, Dipal Barua). Was it ever small? Of course it was. Is it still small? No. Is it beautiful, appropriate etcetera? Not to belabour the point: does it matter?

16.4.5. Scale as a metric: a significant or incidental problem?

Looking at Anagi, BSP, and Grameen Shakti illustrates an easy path to victory in answering our question 'Is small beautiful?' Yes, as long as we are dealing with something that is today large *and* today deemed successful, *and* if we can demonstrate that it indeed started small (and what organization does not start small?).

With that simple decision screen, it requires no real synapse exertion to conclude that when a venture or organization was small, of course it was beautiful, and now that it is large and it is impactful, it remains beautiful—judging by the number of times the word 'scale' appears in the literature of E4D, size equals beauty to many of us.

Schumacher, I believe, would consider this whole conversation about scale as being beside the point in any discussion of practical, successful, people-focused development. He would also probably put it in the same category as using only profit as a measure: easy, mechanical, simplistic, and short-sighted. Unfortunately, he would probably find himself out of step with much of the discussion today. It sometimes sounds as if we are saying 'if not a million people, why bother?' This is doubly unfortunate, because life—whether in 1973 or 2013—is not that simple. Energy access does not lend itself to being one of those convergent problems (problems that can be reduced to their parts and solved accordingly). For every Anagi, BSP, and Grameen Shakti, how many failures can you find that had similar origins and stages to these successes?

It is because the E4D space is so eclectic, with myriad technologies, business models, delivery approaches, and (most of all) market segments, that I suspect that Schumacher's principles are helpful. He was on to something significant here, not just in the sense of local development emphasis but in the sense of the business principle 'be close to your customer.' Schumacher can help us supplement our quantitative approaches and our silo creation: for example, distribution of lique-fied petroleum gas (LPG) in urban centres to reduce charcoal use is an extremely close relation to improved cookstoves in rural locations. Perhaps Schumacher can get us off the quantitative treadmill; does it really matter if one stove or lantern is 5 per cent more efficient than another if each represents substantial improvement? And perhaps Schumacher can even offer some unifying ideas through which we can study our ragtag collection of fishing boats and assess the support systems we need to be a real energy-for-development *industry*. Otherwise we truly risk being an appendage to renewable energy or climate change mitigation, or a silent partner throughout Millennium Development Goals. Perhaps Schumacher's prin-ciples can move us beyond the constraints of trying to fit into a 'fund' model or 'private equity' or this or that technology silo, and make the sector intelligible to investors and donors *in general.*

If there is something in common across E4D (and something that differentiates it from renewable energy and climate change mitigation as well as many other topics) it is the requirement that we look at the 'last mile' or 'last kilometre' transaction. In renewable energy broadly or climate change mitigation or myriad other topics, the 'last mile' transaction is a sidebar or a piece of a more complex whole (e.g. climate change, vaccine development). In energy for development it is the centre of gravity.

- The 'last mile' transaction is the essential transaction. Without it, all the upstream planning, organizing, designing, and manufacturing is for nothing.
- It is an effectively neutral and 'pure' moment. No matter what has happened up to that point—carbon monetization, subsidy, imports from China, give-aways, false starts—it does not really matter to the immediate reality of the transaction between last-mile delivery entity and the customer. This last interaction is at arm's length. It either works as a lesson to take further or it does not (consider for a moment the half-life assigned to absurdly subsidized transactions).
- If we want to know if 'small is beautiful', it is pretty hard to get any 'smaller' than the moment where delivery enterprise and customer exchange value, goods, and services. This is either energy access or development or it is something else.
- Enormous reduction, aggregation, and simplification efforts take place upstream. We are a 'big ticket', 'big push' society. Few things unite industrialized and developing economies so solidly. We base our big pushes on something we call 'planning' and we base that on something we and Schumacher call 'assumptions'. One chapter in *Small is Beautiful* politely attacks any faith that we might hold in either assumptions or planning, especially in large projects and programmes. More to the point, however, practitioner experience shows that the actual delivery moment in energy access has all too often been considered as an afterthought or foregone conclusion to the mobilizing of upstream resources, policy, and regulatory support and scale considerations. Yet *all* the development impact we desire is captured in the effectiveness of this last step, not in the PPP organization matrix, the logical framework exercises, the leveraging of resources or project management or M&E protocols. The centre of gravity for energy access is where the delivery entity interacts with the customer, not in the capital city or on the factory floor.

16.5. IS SMALL BEAUTIFUL? SOME OBSERVATIONS

- 'Small' is not a designation of size—but the smaller an enterprise, the more closely it seems to interact with its customers and staff. Even big enterprises need to manage themselves with small units that have the authority and opportunity to innovate.
- 'Beautiful' is in the eye of the beholder—for Schumacher beauty encompassed at least an emphasis on local resources and local talents, self-reliance,

and quality of life; evolution and organic adaptations; employment creation and 'champions' close to implementation.

- Our experience base in energy for development is rich and at the same time confusing—not surprisingly, because it is an agglomeration of many technologies; diverse business models; varied cultural, market, and regulatory regimes; etcetera. There is a need for organizing principles other than technologies and business models if we wish our fishing fleets and fishing industry to be distinct from outwardly similar but quite different kinds of vessels.

- There are common features within this experience base: customer orientation, product appropriateness; we should accept and learn from the differences instead of trying to 'tame the beast' with one-size-fits-many policy, technology and enterprise solutions. Accept the ambiguity: *let knowledge frustrate over-simplification.*

- We should size activities in the smallest effective human units. Big is not better if it dehumanizes. The market, production, least cost are but some of the slices of the energy-for-development pie. Everything starts small. Only some things succeed. Only some things increase in volume. The measures of energy for development should be qualitative as well as quantitative matters. Job intensity, local resource use, sustainability and proximity to customer may be far more important in the long run, not size, ambition, or lack of ambition for something new.

- The current within energy for development is pushing against production job intensity and local natural resources, including labour. This is especially true in the electrical device category; while less so in the cooking and fuel categories, it is also in evidence. Our understanding of the consequences of these currents from an overall development perspective is meagre. Put bluntly, because the problem is so large we are rushing.

16.6. CONCLUSION

The size and scale of the energy poverty challenge are daunting, but its solutions should not be limited to what experts perceive as 'scalable' solutions because scale is realized only by looking in the rear-view mirror. These solutions are not in evidence at the outset. The height of the mountain that can be scaled successfully is a function of the climbers: their knowledge, training, equipment, and support: their technology ('any and all means to achieve an end'), not their ambitions or plans. Perhaps today's 3,000-metre climber will develop knowledge or equipment that enables others to climb 5,000-metre mountains. Is that 3,000-metre climber too small to warrant our attention? I think E.F. 'Fritz' Schumacher would hope not.

Box 16.1. Development in action: the experience of E+Co

Few have been able to translate the 'small is beautiful' paradigm into the world of impact investing like E+Co. Created in 1994 as an outgrowth of a Rockefeller Foundation program, E+Co's vision was to provide clean, affordable energy to the world's poor via locally developed and adapted solutions ('Energy through Enterprise'). The founders understood early on that rural energy access did not have a one-size-fits-all solution and oftentimes failed when solely dependent on mega-scale power infrastructure projects. It was also clear that alternative technologies such as hydro, solar, and wind were difficult to scale up in the field for various regulatory, market, and economic reasons. E+Co's response was to focus on empowering local small and medium-sized enterprises by providing financing and business development support to develop new products and services which were suitable to the energy needs in their communities. Working directly with local entrepreneurs, E+Co first identified the appropriate business model and market strategy and then provided the tools, training, and funding to grow their venture into sustainable businesses. The company's 'triple bottom line' (people, plant, profit) ensured that the investments made brought significant returns to the people and the environment in local communities. It should be noted that unlike microfinancing, which provides small-scale loans and other banking services for low-income individuals, E+Co provided debt and equity financing (averaging $110,000) to help entrepreneurs build their energy businesses to the point where they can access financing from commercial institutions.

Between 1994 and 2010, E+Co made over 280 investments totalling more than $45 million across 36 countries around the world. These efforts provided energy access to over 7 million people while supporting over 5,400 new jobs. E+Co's investments also resulted in over 4.8 million tons in CO_2 offsets across the globe, or the equivalent of offsetting emissions from 700,000 cars in the United States (figures from 1994–2010). Notable projects include:

Clean Thai: E+Co provided early equity of $150,000 to support the construction of Clean Thai's first biogas plant within the country's largest cassava-processing facility. The plant uses anaerobic digesters to convert organic waste byproducts from cassava processing into methane, which is then used to power the facility's production boilers and to provide electricity. Thanks to the start-up funding, Clean Thai was able to operate successfully and attract second-stage funding. The plant itself displaces $2.3 million in electricity and $2.2 million in heavy fuel costs per year. Clean Thai later went into an agreement with the Thai Biogas Energy Company to construct and operate a series of biogas power plants.

Anasset, Ghana: E+Co's seed capital and business development services helped Anasset create a LPG distribution service which provides the cleaner fuel to neighbouring house-holds, businesses, hospitals, and schools, replacing traditional firewood and charcoal. E+Co and its partner, African Rural Energy Enterprise Development, provided Anasset's owner with a four-year 7.5 per cent loan of $38,000 which helped increase Anasset's sales by 57 per cent in two years. The loan also enabled Anasset to obtain additional working capital for a second plant in Ghana.

Tecnosol, Nicaragua: Tecnosol was a small start-up renewable energy systems distributor offering pre-packaged systems for different ranges of affordability to the rural poor in Nicaragua. The company partnered with E+Co to carry out a market study to determine suitable markets and products to be offered. E+Co later invested a total of $1.3 million to help grow the company.

Note: Since 2012, E+Co has evolved into a for-profit entity, Persistent Energy Partners. Original co-founder Philip LaRocco has gone on to create Embark Energy (embarkenergy. com), a web-based portal centred on experiential learning, business coaching, links for entrepreneurs to financing, and access to a supply chain of clean energy products.

REFERENCES

Aron, J.E., et al. (2009). 'Access to Energy for the Base of the Pyramid'. Hystra/Ashoka [online], <https://www.ashoka.org/story/6072> (accessed 18 October 2013).

Ashden Awards (2006). 'Case Study Summary: Biogas Sector Partnership (BSP), Nepal'.

Bajgain, S., Shakya, I., and Mendis, M. (2005). 'The Nepal Biogas Support Program: A Successful Model of Public/Private Partnership for Rural Household Energy Supply.' Nepal: Biogas Sector Partnership.

Dev Part Consult Nepal (n.d.). Video training manual prepared for Biogas Support Programme on correct construction method for GGC 2047 Mode biogas plant and quality standards.

IFC (International Finance Corporation) (2012). 'Lighting Asia: Solar Off-Grid Lighting, Market Analysis of India, Bangladesh, Nepal, Pakistan, Indonesia, Cambodia and Philippines'.

Practical Action (2010). *Poor People's Energy Outlook 2010*. Rugby, UK: Practical Action Publishing.

Schumacher, E.F. (1973). *Small is Beautiful: Economics as if People Mattered*. London: Blond & Briggs.

TLS (The Times Literary Supplement) (1995). 'The Hundred Most Influential Books since the War', 4827(39), 6 October.

Part III

Challenges and Policy Options

17

United States' Approaches to Expanding Energy Access

Jason E. Bordoff

Access to affordable and reliable energy is a necessary condition to reduce poverty, improve public health, and increase economic growth. Unfortunately, the staggering and saddening scale of the energy poverty challenge is known all too well: 1.3 billion people do not have access to electricity, and 2.6 billion do not have access to clean cooking facilities. These people are mainly in developing Asia and Sub-Saharan Africa (Birol 2013). Although analysis of energy access often centres on electricity and clean cooking facilities, energy access is about much more, such as delivery of telecommunications, heating, and transportation services.

The challenge of expanding energy access is only going to grow, with the global population projected to increase by 1.7 billion people by 2035 (Birol 2013). Even as global GDP more than doubles in real terms over the same period (Birol 2013), the risk is that a large swathe of the world's population—the 'bottom billion'—will be left behind from this growth and prosperity without access to electricity and other energy services (Collier 2008).

Some progress has been achieved in recent years: according to the IEA, the number of people without access to electricity dropped by 4.5 per cent from 2009 to 2011, thanks to a mix of economic growth, urbanization, and energy access programmes. It was a welcome development that the United Nations designated 2012 the Year of Sustainable Energy for All, and that the UN Secretary-General included universal access to modern energy within his Sustainable Energy for All initiative (SE4ALL). But advances are still far too slow: in Sub-Saharan Africa, 68 per cent of the population still lacks access to electricity, compared to 17 per cent in developing Asia (although the total number in Asia is still greater) (Birol 2013). In Congo, Kenya, Uganda, and Tanzania, more than 80 per cent of the population lacked electricity in 2010 (Birol 2012).

In the race against energy poverty, Africa risks being left behind. While the total number of people without electricity is projected to drop over the next two decades, with developing Asia leading the gains, in Sub-Saharan Africa it is actually projected to increase (Birol 2012). Similarly, the number of people without access to clean cooking facilities is projected to remain unchanged worldwide over the next 20 years, but to increase in Africa (Birol 2012). Given these trends, IEA projects that, of the $1 trillion in spending needed to achieve universal energy access by 2030, two-thirds will need to be invested in Africa (Birol 2012).

Even this $1 trillion cost to provide universal energy access is probably too low if it is assumed that ending extreme poverty will require more than the miniscule amount of electricity needed to run light bulbs. Consider that the IEA 'Energy Access for All' scenario defines 'modern energy access' as 50–100 kWh per capita per year—compared to residential per capita consumption of around 4,700 kWh in the US and 1,700 kWh in Germany (Birol 2012; OECD 2013). For scale, 50 kWh would be enough to power one light bulb for one and a half hours a day. Based on World Bank and OECD data, Sub-Saharan Africa's electricity consumption would need to grow 16-fold if it were to enjoy the same per capita consumption levels as Germany by 2035. Moreover, as economist Catherine Wolfram and colleagues have observed, electrification results in sharp, non-linear increases in energy demand because energy-intensive assets are key to poverty alleviation (Wolfram et al. 2012).

For these reasons, it makes good sense for energy-access policy to focus on Sub-Saharan Africa, and it was thus encouraging to see President Obama give this issue high priority when he announced his administration's 'Power Africa' initiative during his 2012 visit to the region and again during the US–Africa Leaders Summit in August 2014.

17.1. EXPANDING ENERGY ACCESS THROUGH ECONOMIC GROWTH

Countless reports from international organizations have proposed a litany of assistance programmes to help expand energy access—several of which have indeed helped reduce the number of people without modern energy services. Moving forward, the focus should be on ramping up the small number of programmes that evidence and experience demonstrate yield the greatest results, rather than launching yet more initiatives.

Successful energy-access policies must first and foremost aim to create an enabling environment for trade and investment, so as to drive broad-based economic growth and provide the prosperity and infrastructure necessary to expand energy access. To that end, energy-access policies must rest on several core principles. These principles are not unique to energy access, but apply broadly to development policy in general. The Obama administration operationalized them in the autumn of 2010 through a Presidential Policy Directive on US global development policy. Among other things, this directive prioritized investments and policies to increase broad-based economic growth; focused resources on areas with the right conditions to sustain progress; held recipients of US assistance accountable for achieving results; sought to leverage private investment; and promised to let rigorous analysis of impacts drive programme and policy decisions.

This approach to development recognizes that growth is ultimately what provides countries with the resources they need to expand energy access.

Hundreds of millions of people have attained modern energy access over the last two decades, particularly in China and India, in large part due to the rapid rates of economic growth these countries have enjoyed.

Over the past several decades, aid has helped to manage poverty, but it is not the prescription for sustained improvement. As Paul Collier writes, 'Aid has been a holding operation preventing things from falling apart' (Collier 2008). And aid has diminishing returns, so just ramping up aid is not the answer. Rather, expanding growth through trade and investment can provide the resources needed to build out infrastructure for modern energy services.

Expanding energy access through faster growth is also self-reinforcing because more energy access itself can also help boost growth, which further expands access. Access to modern energy services and economic development are highly correlated. For example, a weak electricity infrastructure has been shown to be correlated with lower levels of productivity (Escribano et al. 2008). Electricity production of 100 kWh per person is associated with average income of $800 per person, while electricity production of 10,000 kWh per person is associated with income of $31,000 (Morris and Pizer 2013). According to an IMF analysis, long-term per capita growth rates for Sub-Saharan African countries would receive a boost of two percentage points if the quality and quantity of the electricity infrastructure were improved to the level of a country such as Mauritius (Clements 2013).

Never have the benefits of energy access been so great as today. In the past, energy access was often limited to building wires and generation capacity to bring lights and other basic services to rural and poor areas. In today's information economy, energy access is a prerequisite to build businesses, procure services, obtain education, and participate in the global economy. Electricity is no longer just a way for students to study longer into the night, but a necessity to access information in the first place. Technological innovations, such as liquefied natural gas (LNG) transportation or distributed renewable generation, mean there are far more potential options to bring electricity to rural communities than just building utility-scale oil or coal plants, as traditionally was the case.

This chapter explores five priorities, and describes US policy on each, that collectively can help expand energy access through a relentless focus on promoting economic growth: (1) encouraging private investment; (2) improving governance and transparency; (3) leveraging resource wealth; (4) improving access to technical information; and (5) eliminating inefficient energy subsidies.

17.2. ENCOURAGING PRIVATE INVESTMENT

Using public spending to leverage private investment is key to expanding energy access. The scale of investment needed to achieve universal modern energy access is simply too massive to be reliant solely on public funds. According to the IEA, the costs for achieving universal access are estimated at $50 billion per year through 2030, almost six times the current level of investment (Birol 2014). Foreign aid cannot achieve this level. For example, by the time of the Rio+20 Summit in June 2012, the UN Sustainability for All initiative, one the world's highest-profile attempts at raising awareness of this issue, managed to raise only a paltry 3 per cent in funding commitments of the $1 trillion that the IEA estimates will be needed to achieve energy access for all by 2030 (Birol 2012).

Expanding energy infrastructure in energy-poor countries will require investments by private-sector firms not as philanthropic or corporate social responsibility (CSR) initiatives, but based on market-oriented, commercial considerations. Return-oriented investments, not aid assistance, will be necessary to scale up investment in order to build out large-scale energy infrastructure.

Expanding energy access will require private investment in electricity generation and transmission—both fossil fuels and renewable sources. In 2014, the IEA estimated that global investment in renewable energy will constitute 61 per cent of the investment in new power plants from 2014 to 2035 (Birol 2014).

Private investment will also be needed in other areas, such as communications, technology, extraction, and refining infrastructure. Investing in infrastructure to capture flared gas, for example, would increase both access to energy for power generation and much-needed revenue. Nigeria has the second highest rate of flaring in the world, and lost revenues are estimated at $2.5 billion annually (Adeoye 2008; World Bank 2012). Many other countries flare even more per barrel of oil produced. Access to transportation fuel is also inhibited because governments have been unable to attract adequate private investment in refining capacity, and thus they export crude while importing higher-value refined products.

Scarce government dollars should be used to scale up private investment and increase its effectiveness. For example, public aid should focus on building tools to reduce policy and regulatory risks, conduct feasibility studies to break potential bottlenecks, and build the capacity of local actors financing and managing energy projects.

More specifically, public aid must identify and target the key barriers to large-scale private investment. These may differ depending on the types of projects. For large-scale, on-grid projects, which are mostly generation and distribution, a key barrier to private investment is the structure of Power Purchase Agreements between projects and utilities. To attract capital, these projects need secure, viable contracts that can effectively address technical challenges such as the intermittency of wind and solar power. In systems with poor cost recovery, private investment can be spurred by partial risk guarantees and insurance of PPAs, offered by OPIC or the World Bank, for example. The impact of on-grid projects in reducing energy poverty is limited by the small share of homes that are currently electrified, although a much larger share may be within very close proximity to the grid and thus could be connected relatively inexpensively (Lee et al. 2014).

As important as on-grid solutions may be, achieving universal energy access will also require leveraging private investment on small, decentralized projects, such as micro-hydro, small solar systems for battery charging, water pumps, and agricultural process machinery. These relatively low-cost, off-grid technologies can better reach remote communities than cost-intensive, on-grid projects. The key is the ability to replicate these investments many times over on a small scale. Scaling up off-grid investments can be challenging because the viability of the project depends on the quality of the technology being employed, and yet banks (especially smaller local ones) may not have the technical capacity to evaluate projects or accurately gauge repayment risk. For these reasons, local lending, in particular, often requires high collateral and interest terms. Financial institutions need the technical capacity to evaluate off-grid projects and the capability to scale up project investment, while project developers need standardized finance terms to lower transaction costs.

One high-impact way to lower transaction costs for energy projects is for public financing institutions to create easily replicable templates for energy finance contracts, particularly for clean energy and small-scale projects for which contracts are not as readily available as they are for larger-scale, fossil-fuel projects. Doing so obviates the need for parties to incur the costs of drafting individual contracts from scratch for each project, which can impede access to capital. Columbia University's Center for Climate Change Law, for example, developed such contract templates with leading US and Indian attorneys for clean energy projects in India (Columbia Law School 2013).

More broadly, public assistance spending should be designed to leverage private commitments by encouraging regulatory reforms that reduce corruption, provide more legal protections, and encourage trade. The US, for instance, is using the Overseas Private Investment Corporation and the Export-Import Bank to scale private investment and increase exports in this way. And the US has played a leadership role in the 'Connecting the Americas' initiative to achieve universal electricity access in the region by 2022, by creating economies of scale, attracting more private investment, and lowering capital costs. The Millennial Challenge Corporation has also negotiated compacts with several countries to help them undertake wholesale, systemic energy reforms that build private-sector confidence by complying with international standards of good governance, transparency, and responsible financial management.

As another example, the recent Power Africa initiative aims to bring a wide range of US capacity to support policy and regulatory best practices, pre-feasibility support and capacity building, and long-term financing, insurance, guarantees, credit enhancements, and technical assistance. More important than the $26 billion in investments the White House reports has already been committed, the initiative aims to create an enabling environment for private investment going forward by building host-government capacity to develop, approve, and finance energy projects and ultimately bring them on line. It is too early to tell how effective this effort will ultimately be, but the focus on increasing technical skills and improving market structures, along with the significant private and public sector commitments, is a welcome step in the right direction.

To be clear, investments to expand energy infrastructure and access are not solely the responsibility of the private sector. Governments need to make large-scale public-sector investments as well. This is particularly true in resource-rich countries. Too often the policy push is for resource-rich countries to save for future generations through sovereign wealth funds, as in Norway. But developing countries are in a very different position. Their priority is not to save, but to invest in infrastructure that will yield benefits in the long run and create the enabling environment for private investment and thus increased economic growth that benefits future generations many times over. Of course, the worst outcome would be if those resource rents were squandered. That is why public institutions should target their investments to help promote transparency and build capacity to increase the likelihood that resource-rich nations will spend their windfalls wisely.

One final point about public support for investments to expand energy access is that such support should take environmental and climate considerations into

account but should not categorically exclude coal-fired power plants. This issue
has attracted growing interest following the World Bank's decision in 2010 to
support a coal-fired power plant in South Africa and its current consideration of a
project in Kosovo (Yukhananov and Volcovici 2013). President Obama brought
greater attention to the issue in a speech laying out his Climate Action Plan in
mid-2013 by calling for an end to public financing of coal plants 'unless they
deploy carbon-capture technologies, or there is no other viable way for the poorest
countries to generate electricity' (Executive Office of the President 2013). US
Treasury Department guidance in October 2013 defined the poorest countries,
which includes most of Sub-Saharan Africa (US Department of the Treasury
2013). This approach strikes the right balance. Evaluation of proposed projects
should include social costs and encourage clean energy projects when feasible, but
it would be a mistake to pit climate change goals against development objectives in
the poorest countries. In those countries, if cleaner alternatives cannot find
funding and would lead to higher energy costs, organizations such as the World
Bank should not impose an additional burden on the poor or limit access to
energy. Rather, limiting support for coal plants to the narrow set of cases identified by the US and others can allow multilateral development banks to promote
both clean energy and the access to energy that is key to economic growth.

17.3. IMPROVING GOVERNANCE AND TRANSPARENCY

Achieving economic growth to expand energy access, particularly through the use
of natural resource revenues, requires strong governance and transparency structures. Empirical analyses have found a clear correlation between good governance
and commodity revenues and increasing economic growth (Collier and Goderis
2007; Iimi 2007). Although good governance is a vague concept, at its core is some
measure of democratic accountability, checks and balances, and the absence of
corruption.

Moreover, it is clear that leaders matter, as shown through groundbreaking
research by economists Benjamin Jones and Benjamin Olken (Jones and Olken
2005). Change of leaders can lead to sharp changes in economic performance.
Resource-rich countries that have fallen into the resource trap have lacked such
leadership. Nigeria, for example, has consistently been ruled by corrupt leaders—
General Sani Abacha shocked even those used to plundering by stealing around $4
billion during his five years in office in the mid-1990s—and currently ranks 152
out of 187 countries on the Human Development Index despite its massive oil
wealth (UNDP 2014). By contrast, even though Botswana had conditions that
might have been expected to lead to the resource curse, it avoided this fate in large
measure due to strong and visionary leaders committed to national rather than
personal gain. Botswana also formed an unusually strong and trusted working
relationship with DeBeers, whereby each party viewed the success of the other as
being in its interest. As more and more international oil and gas companies enter
the African market, governments should seek to replicate this model.

Promoting good governance is far from easy, of course. Recognizing that
challenge, the Obama administration's PPD promises to focus resources on

those countries where the assistance and investment is most likely to lead to improved outcomes, with good governance being a key metric. Although it is difficult to make such choices, in a world of scarce resources there is little reason to deploy resources where they are likely to have little impact or be squandered. Moreover, evidence suggests that countries respond to such incentives, seeking to improve scores on various governance indices when they perceive there to be a link between those scores and the level of aid they receive (David-Barrett and Okamura 2013; Öhler et al. 2012).

In the context of energy access, the Obama administration has promoted good governance by encouraging transparency and accountability in oil and gas mining through the Extractive Industry Transparency Initiative (EITI). These codes of conduct are developed by companies, governments, and NGOs to set compliance standards for improved oversight, monitoring, and control over the activities of resource extraction companies. By requiring publication of all payments from participating companies to governments, the programmes are intended to allow monitoring of the influence that energy companies have on host governments, and thus to reduce the potential for corruption. The initiative has proved extremely popular, with 29 countries now rated as EITI compliant, 16 having attained candidate status, and many others having signalled their commitment to implementing EITI.

Although experience with EITI, started in 2002, remains limited, it is nonetheless encouraging. Scholars at Oxford's Said Business School, for example, published data in November 2013 finding that implementation of EITI is associated with improved scores on a widely used corruption index (David-Barrett and Okamura 2013). In an effort to expand the use of EITI, the US, through the Cardin-Lugar Amendment, became the first country in the world to require that its extractive industry companies disclose any payments they make to any government worldwide. The EU also has recently enacted a new directive dealing with financial reporting that includes requirements for certain large natural resource extractive industry companies to report payments to governments (European Commission 2013).

17.4. LEVERAGING RESOURCE WEALTH

Africa is rich in natural resources. North and West African nations are major oil and gas producers. West Africa in particular has seen a sharp increase in production activity since the giant Jubilee oilfield was discovered in 2007 off the coast of Ghana. New discoveries are likely to mean much higher production levels in the medium term in East Africa as well, such as offshore natural gas in Mozambique and Tanzania. South Africa is one of the world's major suppliers of coal. And much of Africa remains unexplored, which suggests that large new finds are yet to come. IEA forecasts that oil production in non-OPEC African countries will increase 26 per cent between 2012 and 2020, while total African natural gas production is expected to grow by 40 per cent from 2011 to 2020 (Birol 2013). African oil and gas production to increase 26 and 40 per cent, respectively, between 2011 and 2020 (IEA 2013).

Natural resources are the largest asset available to most African countries and have the potential (sadly, often unrealized) to lift many of them to greater

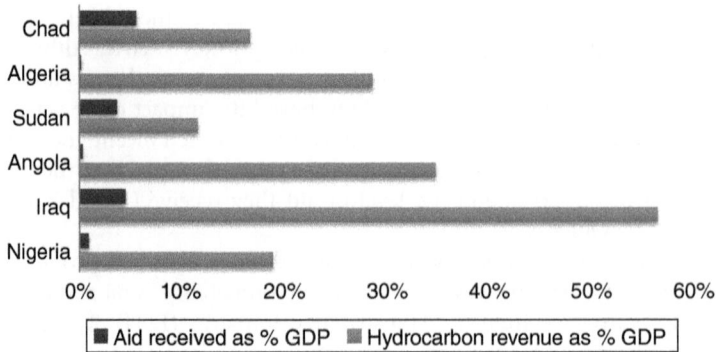

Figure 17.1. Average hydrocarbon revenue and aid received as percentage of GDP in key countries (2007–2011)

Source: IMF Country Reports, World Bank

prosperity. Oil and other natural resources accounted for one-third of Africa's GDP growth from 2000 to 2008 (Roxburgh et al. 2010). The revenues that they could get from natural assets are vast, dwarfing many times over any conceivable amounts of foreign aid (see Figure 17.1). If harnessed correctly—a big 'if'—the value of these assets could be transformative.

Moreover, developing more natural capital provides not only the prosperity needed to expand energy access and infrastructure but also the energy itself needed to deliver power. Africa is a major energy producer, yet today little of that energy is harnessed for its own domestic consumption. Despite its vast resources, more than half the people of Africa are without modern energy services while the continent exported more than half the fossil fuels it produced in 2011 (Birol 2012). In addition to fossil fuels that might be consumed in greater quantity domestically to expand energy access, renewable energy resources are also abundant—such as hydropower in Central and East Africa; geothermal resources in East Africa; wind power in North Africa and South Africa; solar power across the continent; and biomass in some areas.

According to the IEA, Africa's revenues from net energy exports are projected to increase from almost $280 billion to $415 billion per year in 2030. And the IEA estimates that achieving modern energy access for all of Africa by 2030 would require investments of $20 billion per year—or just 5.5 per cent of energy export revenue over that period (Birol 2012). Oil and gas exporters have strong growth prospects if they can use resource wealth to finance broader development of their economies. The experiences of resource-rich countries such as Indonesia, Botswana, and Malaysia show how important it is to invest in infrastructure, education, and other pillars of long-term growth.

Of course, it is far from certain that the benefits of increased natural resource wealth will accrue to a country's people. Too often, countries with high rates of natural resource extraction have fallen into the trap of the 'resource curse', leading to weaker long-run growth, higher rates of poverty, and higher inequality. There is a long-running debate in economic circles about the cause of the resource curse—or even its very existence—but several factors appear to play a role including corruption, poor governance, lack of democratic accountability, and exchange-rate depreciation (so-called 'Dutch Disease' because of the experience in the Nether-lands after finding natural gas in the 1950s). Such an outcome is not inevitable, however, as examples such as Botswana and Norway demonstrate. Rather than discourage resource extraction, which holds great promise for economic prosperity in these countries, the international community must help them avoid the resource trap by supporting the sort of governance, transparency, and structural reforms described above.

Increasing energy production also helps global energy markets and thus eco-nomic growth. Averaged over the last 3 years, oil prices in real terms are at their highest level ever and are projected to remain high, even with the explosive growth in US tight oil production and the potential in the longer term for similar uncon-ventional oil production elsewhere. The IEA projects continued growth this decade, but then increased tightness in the global market after 2020 in response to declining non-OPEC production and potential OPEC underinvestment in the interim—although several private forecasters are more optimistic about continued North American tight oil supply growth. Ironically, African nations may be among the largest losers if US tight oil and shale gas growth remains robust, as US LNG exports would displace other higher cost sources of global gas supply, with African offshore LNG projects, costing in the tens of billions of dollars, hit the hardest (Bordoff and Houser 2014).

Given the potential for resource development to support economic growth, as well as to supply and stabilize global energy markets, US policymakers have rightly prioritized engaging with developing countries to help them expand safe and responsible energy production. The US has actively engaged with countries to exchange lessons on developing unconventional energy resources, such as best practices for managing environmental risks, permitting, contracting, and pricing. The US is also working with such countries as Mozambique and Tanzania to help them establish responsible ways to develop and manage their newfound energy resources—although more needs to be done to help countries that will soon experience large energy windfalls to avoid the mistakes that have befallen others. Toward that end, the US administration's recent Power Africa initiative included a partnership with Uganda and Mozambique on responsible oil and gas resources management. Moreover, with the right governance reforms, legal regimes, and investment environment, a country such as Mozambique can use new offshore natural gas not only for export revenue but also to provide electricity for the people of Mozambique. Although transparency continues to need improvement, Mozambique has made notable progress building out its electricity grid to reach a much larger share of the nation's population.

Even if resource curse pitfalls can be avoided, which remains far from given, efforts to promote more oil and gas development in Africa also raise concerns about the impact on greenhouse gas emissions. Why seek to produce fossil fuels when the planet badly needs to start using less? While policies to encourage energy access must account for potential climate impacts, helping African nations increase energy production, even fossil fuel production, need not be inconsistent with making meaningful progress to reduce carbon emissions. Even if world leaders take the steps necessary to address climate change (also far from given), fossil fuels will still play a key role in the global energy mix for decades to come. According to the IEA, if we stabilize emissions at the globally agreed target of

450 parts per million, fossil fuels will still comprise 64 per cent of the global energy mix in 2035; oil demand in 2035 will decline by about 13 per cent from current levels, and natural gas demand will actually grow (Birol 2013). So there will remain a demand for Africa's oil and gas resources for decades to come—and that demand will be met by supply from elsewhere if production in Africa is constrained.

As for expanding energy access in Africa, the sources of Africa's energy consumption will need to come from both fossil fuels and renewable energy (both on-grid and off-grid). Limiting new energy projects to renewables would raise costs and reduce the scale of increased energy access, all with minimal climate benefits. The IEA estimates that if we achieve energy access for all, roughly half that energy will come from fossil fuels—and that will increase greenhouse gas (GHG) emissions only 0.6 per cent through 2030 (Birol 2012). (Of course, as noted earlier, this scenario assumes token levels of energy use to achieve universal energy access, so the actual numbers may be higher.) Developing countries use very little energy and emit very little GHG per capita. Africa emitted only 3 per cent of the world's CO_2 in 2010 (Birol 2012). The climate change problem is not due to energy consumption in the poorest parts of the world, and that is unlikely to change even if those regions enjoyed modern energy services powered, in part, by fossil fuels. Even as global leaders support more sustainable energy sources, it would be wrong to deny people in those regions low-cost energy from fossil fuels in the face of extreme poverty. As noted above, that is the approach the Obama administration recently adopted in the President's Climate Action Plan, in which President Obama called for an end to US government support for public financing of coal plants overseas, except in the world's poorest countries where other economic alternatives do not exist.

17.5. IMPROVING ACCESS TO TECHNICAL INFORMATION

In order for resource-rich countries to extract as much value as possible out of their natural assets, they need to know enough to structure lease sales and terms in such a way that does not give away valuable rents. Moreover, in order to develop these resources safely—particularly for new and unconventional sources of energy—they need to learn best practices from others for safe and responsible rules of development.

Unfortunately, developing countries often lack these tools. These countries usually have less information than private companies about how much of such resources they actually possess or what the results of past test wells have been. Improving access to that sort of geological information means that countries can set up better terms and not give away the scarcity rents to private companies. Recognizing this need (and also desiring to expand access for US companies), the US Geological Survey has undertaken resource assessments in numerous countries around the world and works in partnership with governments to build capacity. The US Energy Information Administration (EIA) has also commissioned a study of technically recoverable shale oil and shale gas reserves in 137 shale formations in 41 countries outside the US (EIA 2013).

Additionally, the US State Department set up an initiative focused on shale gas and tight oil to exchange lessons on developing unconventional energy resources (UGTEP 2010). Through this effort, the US government has shared best practices on issues such as water management, air quality, permitting, contract, and pricing. Sharing this expertise has allowed countries and companies to learn from the US experience that creating the right policy and investment environment is critical to successful development.

17.6. ELIMINATING INEFFICIENT ENERGY SUBSIDIES

Roughly half a trillion dollars is spent each year subsidizing the costs of fossil fuels: lowering market prices that consumers and businesses pay for oil, gas, and coal. Energy subsidies are concentrated in the Middle East/North Africa, Former Soviet Union, and emerging and developing Asia, and parts of Latin America (Birol 2013).

It may seem paradoxical at first glance to argue that subsidizing the prices consumers pay for energy is inconsistent with the goal of increased access to modern energy services. But the data overwhelmingly demonstrates that to be the case.

First of all, as already discussed, increasing economic growth is key to expanding energy access, yet energy subsidies undermine economic growth (Clements 2013). For example, energy subsidies lead to a less efficient allocation of capital and undermine incentives to invest in the energy sector. As the price of fossil fuels, particularly oil, has risen, these energy subsidies have placed greater strain on national budgets, reducing resources available for more productive purposes. A recent paper by the International Monetary Fund showed the fiscal weight of energy subsidies is growing so large in some countries that budget deficits are becoming unmanageable and threaten the stability of those economies. The IMF showed that 20 countries maintain pre-tax energy subsidies that exceed 5 per cent of GDP. In Sub-Saharan Africa, subsidies exceeded 2 per cent in 12 countries (Clements 2013).

Subsidies for energy consumption also reduce the incentives to invest in building out the energy infrastructure that would actually increase energy access. Low energy prices in the form of price controls reduce the returns to producers, and thus make it difficult for public and private energy companies to invest in production and infrastructure. In Sub-Saharan Africa, the losses incurred by energy providers, and their concomitant unwillingness to invest, help explain why installed per-capita generation capacity in the region is only one-tenth what it is in Latin America (Clements 2013).

Moreover, the evidence is clear that energy subsidies do not actually advantage the poor, who typically lack energy access; rather, the benefits of subsidies accrue far more to the wealthy. On average, the richest 20 per cent of households in low- and middle-income countries capture 43 per cent of fuel subsidies, according to the IMF (Clements 2013).

All of this does not mean simply that eliminating subsidies is desirable. Removing subsidies, and letting consumers face prevailing market prices for energy, could have severe adverse impacts on the poor, which is why subsidy reform needs to be accompanied by other measures to offset those impacts.

Recognizing the importance of this issue, the US led the effort in 2009 to get a commitment from the G-20 to phase out inefficient fossil fuel subsidies. Since then, 50 more countries have made the same pledge. Unfortunately, while some countries have taken promising steps, far too little overall progress has been made. Indeed, the total amount of global fossil fuel subsidies has risen sharply since 2009, largely tracking the rise in fossil fuel prices. The recent experiences of countries that have sought to curtail fuel subsidies, from Egypt to Indonesia to Nigeria, for example, demonstrates how challenging it can be to raise energy costs for poor consumers, as the moves were met with widespread public opposition and, in some cases, violent strikes and mass protest.

17.7. CONCLUSION

After years of talk of peak oil, the energy world these days is agog over the unconventional energy boom—from shale in the US to oil sands in Canada and perhaps Venezuela, to ultra-deepwater pre-salt in Brazil, to rapidly rising LNG trade. Amidst all this new supply, it is easy to forget that 1.3 billion people still cannot turn on the lights. While widespread reports of gas flaring are illustrated with nighttime satellite photos of the Bakken lit up at as brightly as Chicago, even more striking, though less widely reported, are photos showing the Niger Delta shining brightly in an otherwise unilluminated continent.

Providing access to modern energy services is essential to pull hundreds of millions of people out of extreme poverty in the decades to come. Doing so will require international support that goes beyond the holding pattern of foreign aid to foster broad-based economic growth through foreign investment and trade and that uses evidence of impact to target areas with the right conditions to make sustained progress. Recognizing the limits of public spending, US policy to expand energy access has sought to use limited government resources to leverage private investment; encourage governance and transparency reforms; help nations benefit from new resource wealth; provide resource-rich nations with technical informa- tion and assistance; and promote economic growth by phasing out fossil fuel subsidies.

Much more remains to be done in all these areas, of course. Enough increased funding has not been forthcoming to support these efforts, leaving the US administration to work largely with existing Ex-Im and OPIC funding to support Power Africa, for example. Too little progress has been made to phase out fossil fuel subsidies. Transparency, regulatory, and legal reform efforts still have a long way to go. And it is far from clear that African nations poised to reap large new resource windfalls from oil and gas have the policies or leaders necessary to avoid resource curse maladies. More generally, the challenge of providing universal energy access should be a higher priority for policymakers than has typically been the case.

To continue making progress, the United States should increasingly highlight the importance of expanding modern energy access, require accountability and transparency, and redouble efforts to create the enabling environment necessary for trade and investment to drive broad-based economic growth.

ACKNOWLEDGEMENTS

The author thanks Cherry Ding for helpful research assistance, and David Sandalow, Carlos Pascual, Paul Collier, and Benjamin Jones for helpful comments and discussion.

REFERENCES

Adeoye, Y. (2008). 'Nigeria: Country Loses 150 Billion Dollars to Gas Flare in 36 Years', *Vanguard* [online], <http://allafrica.com/stories/200807150018.html 15 July 2008> (accessed 9 December 2013).

Birol, F. (2011). 'Energy Access for All: Financing Access for the Poor'. *World Energy Outlook*. Paris: OECD/IEA.

Birol, F. (2012). 'Measuring Progress towards Energy for All.' *World Energy Outlook*. Paris: OECD/IEA, pp. 529–668.

Birol, F. (2013). 'World Energy Outlook 2013'. Paris: OECD/IEA.

Birol, F. (2014). 'World Energy Investment Report'. Paris: OECD/IEA.

Bordoff, J. and Houser, T. (2014). "American Gas to the Rescue? The Impact of US LNG exports on European Energy Security and Russian Foreign Policy," Columbia University Center on Global Energy Policy Working Paper Series (forthcoming).

Clements, B. (2013). 'Energy Subsidy Reform: Lessons and Implications', International Monetary Fund [online], <http://www.imf.org/external/np/pp/eng/2013/012813.pdf> (accessed 9 December 2013).

Collier, P. (2008). *The Bottom Billion: Why the Poorest Countries are Failing and What can be Done About It*. Oxford: Oxford University Press.

Collier, P. and Goderis, B. (2007). 'Commodity Prices, Growth, and the Natural Resource Curse: Reconciling a Conundrum,' CSAE Working Paper Series 2007-15, Centre for the Study of African Economies, University of Oxford.

Columbia Law School (2013). 'Clean Energy Investment US–India', Center for Climate Change Law [online], <http://web.law.columbia.edu/climate-change/resources/law-clean-energy-energy-efficiency/clean-energy-investment-us-india> (accessed 9 December 2013).

David-Barrett, L. and Okamura, K. (2013). 'The Transparency Paradox: Why do Corrupt Countries Join EITI?' [online], <http://eiti.org/files/The-Transparency-Paradox.-Why-do-Corrupt-Countries-Join-EITI1.pdf> (accessed 9 December 2013). Oxford: University of Oxford.

EIA (United States Energy Information Administration) (2013). 'Technically Recoverable Shale Oil and Shale Gas Resources: An Assessment of 137 Shale Formations in 41 Countries Outside the United States' [online], <http://www.eia.gov/analysis/studies/worldshalegas/> (accessed 9 December 2013).

Escribano, A., Guasch, L., and Pena, J. (2008). 'A Robust Assessment of the Impact of Infrastructure on African Firms' Productivity', Africa Infrastructure Country Diagnostic Working Paper. Washington, DC: World Bank.

Executive Office of the President (2013). 'Climate Action Plan', June [online], <http://www.whitehouse.gov/sites/default/files/image/president27sclimateactionplan.pdf> (accessed 9 December 2013).

European Commission (2013). MEMO/13/541 12/06/2013.

'IEA World Energy Statistics and Balances.' OECD library [online], <http://stats.oecd.org/BrandedView.aspx?oecd_bv_id=enestats-data-en&doi=data-00510-en> (accessed 9 December 2013).

Iimi, A. (2007). 'Escaping from the Resource Curse: Evidence from Botswana and the Rest of the World', *IMF Staff Papers* 54(4) [online], <http://www.imf.org/external/pubs/ft/staffp/2007/04/pdfs/Iimi.pdf> (accessed 9 December 2013).

Jones, B. and Olken, B. (2005). 'Do Leaders Matter? National Leadership and groWth since World War II', *The Quarterly Journal of Economics* 120(3): 835–64.

Lee, K. et al. (2014). Barriers to Electrification for "Under Grid" Households in Rural Kenya (National Bureau of Economic Research Working Paper No. 20327). Retrieved from <http://faculty.haas.berkeley.edu/wolfram/Papers/w20327.pdf>.

Morris, S. and Pizer, B. (2013). 'Thinking Through When the World Bank Should Fund Coal Projects.' Center for Global Development [online], <http://www.cgdev.org/sites/default/files/archive/doc/full_text/CGDEssays/3120619/world-bank-coal.html> (accessed 9 December 2013).

OECD (Organisation for Economic Cooperation and Development) (2013). 'Economic Outlook' [online], <http://dx.doi.org/10.1787/eco_outlook-v2013-1-en> (accessed 9 December 2013).

Öhler, H., Nunnenkamp, P., and Dreher, A. (2012). 'Does Conditionality Work? A Test for an Innovative US Aid Scheme', *European Economic Review* 56(1): 138–53.

Roxburgh, C., et al. (2010). *Lions on the Move: The Progress and Potential of African Economies*. New York: McKinsey Global Institute.

UGTEP (2010). 'Unconventional Gas Technical Engagement Program'. US Department of State [online], <http://www.state.gov/s/ciea/ugtep/index.htm> (accessed 9 December 2013).

UNDP (United Nations Development Programme) (2013). 'International Human Development Indicators' [online], <http://hdr.undp.org/en/data/trends/> (accessed 9 December 2013).

US Department of the Treasury (2013). 'Guidance for US Positions on MDBs Engaging with Developing Countries on Coal-Fired Power Generation' [online], <http://www.treasury.gov/resource-center/international/development-banks/Documents/CoalGuidance_2013.pdf> (accessed 9 December 2013).

Wolfram, C., Shelef, O., and Gertler, P. (2012). 'How Will Energy Demand Develop in the Developing World?' National Bureau of Economic Research [online], <http://www.nber.org/papers/w17747> (accessed 9 December 2013).

World Bank (2012). 'Estimated Flared Volumes from Satellite Data, 2007–2011' [online], <http://web.worldbank.org/WBSITE/EXTERNAL/TOPICS/EXTOGMC/EXTGGFR/0,,contentMDK:22137498~menuPK:3077311~pagePK:64168445~piPK:64168309~theSitePK:578069,00.html> (accessed 9 December 2013).

Yukhananov, A. and Volcovici, V. (2013). 'World Bank to Limit Financing of Coal-fired Plants', Reuters, 16 July [online], <http://www.reuters.com/article/2013/07/16/us-worldbank-climate-coal-idUSBRE96F19U20130716> (accessed 9 December 2013).

18

The Energy Access Practitioner Network

Richenda Van Leeuwen and Yasemin Erboy Ruff

18.1. THE PRACTITIONER CHALLENGE

Meeting the electricity needs of 800 million people—60 per cent of the total who now lack access and who, according to current projections, will not be reached by traditional utilities (IEA 2010)—is a daunting challenge. To date, this challenge has been met mostly by a group of practitioners, including small- and medium-scale enterprises (SMEs), social enterprises, non-governmental organizations (NGOs), and community-based organizations that struggle to provide energy products and services to households and communities sustainably under the most demanding conditions.

⇒ Technically, they may lack access to quality products and services, face supply chain challenges in delivering products to customers, or be without the market clout needed to obtain products in sufficient quantities at competitive prices in a timely manner.

⇒ Financially, these enterprises confront difficulties in attracting the right type and amount of capital and in ensuring a predictable flow of money to enable planned growth and to fund multiple generations of innovation. Investors prefer to finance more mature companies that have an established track record and proof of concept, require larger amounts of capital, bring an established investor base, and promise high commercial rates of return.

⇒ Organizationally, practitioners may lack the business and management skills needed to expand their business ventures to take advantage of existing opportunities in the market.

⇒ Potential customers—those most in need of electricity and the advantages it provides—need guidance to understand fully how these services can improve their quality of life, save them time and money, and be financed affordably.

Typically, energy entrepreneurs in these contexts operate in rural areas without being able to assess and benefit from the experience of their counterparts, either in their own country or in other parts of the developing world. They may lack information on best practices in providing energy services, business approaches to streamline their operations, and sources of assistance in the scale-up of energy delivery to those lacking modern energy services. They may not speak English, have Internet access, or come to the attention of international impact or other investors who provide financial support for social entrepreneurs. Thus they may be forced to bootstrap their own operations—a challenge in any context, and a particular obstacle for women seeking to become energy entrepreneurs in some parts of the world.

The number of players operating in the off-grid marketplace today, and the scale at which they are operating both as individual companies and as an aggregate, while increasingly impressive, is inadequate to reach universal electricity access in less than 20 years. Further, those entities that are already delivering decentralized electricity services in this market confront myriad obstacles that must be tackled in order for them to grow, expand, and replicate their approaches.

18.2. THE PRACTITIONER NETWORK

Following a year of consultations with practitioners working on decentralized energy solutions, who strongly desire access to peer networking and sharing of best practices, the United Nations Foundation established the Energy Access Practitioner Network[1] in 2011. The mission of the Practitioner Network is to support market-led decentralized energy activities towards achieving universal energy access by 2030. It catalyses energy service delivery by promoting new technologies, adoption of quality standards in technology and delivery, and innovative financial and business models. As such, the Practitioner Network's philosophy is supportive of and consistent with the UN Secretary-General's Sustainable Energy for All[2] initiative's objective of achieving universal access to modern energy services by 2030. It also operates as a sister initiative to the Global Alliance for Clean Cookstoves,[3] which addresses the other major component of energy access—the need for clean cooking and heating solutions for the 3 billion people who still lack them.

The Practitioner Network emphasizes community- and household-level electrification via mini- and micro-grids and remote applications (Figure 18.1), as well as small-scale products for switching out kerosene-based lighting as a first step towards energy access. Although humanitarian applications are included, the primary emphasis is on developing market-based solutions that can be scaled up or replicated to help achieve the universal energy access objective (Figure 18.2). The Practitioner Network is open to all organizations and individuals that are actively involved in the development, implementation, finance, investment, and other components of delivering electricity and increasing access to energy in developing countries and, where relevant, in OECD countries.

Since its inception, the Practitioner Network has grown rapidly to over 1,700 practitioners representing some 980 organizations from 191 countries (as of July 2014), including private-sector businesses ranging in size from SMEs to large corporations, equipment manufacturers, distributors, project developers, financial institutions, investors, and others involved or interested in scaling up the delivery of modern energy services (Figure 18.3).

Based on a survey conducted of Practitioner Network members in May 2013— to which about 20 per cent of organizations initially responded—they provided close

---

[1] <http://www.energyaccess.org>   [2] <http://www.sustainableenergyforall.org>
[3] <http://www.cleancookstoves.org>

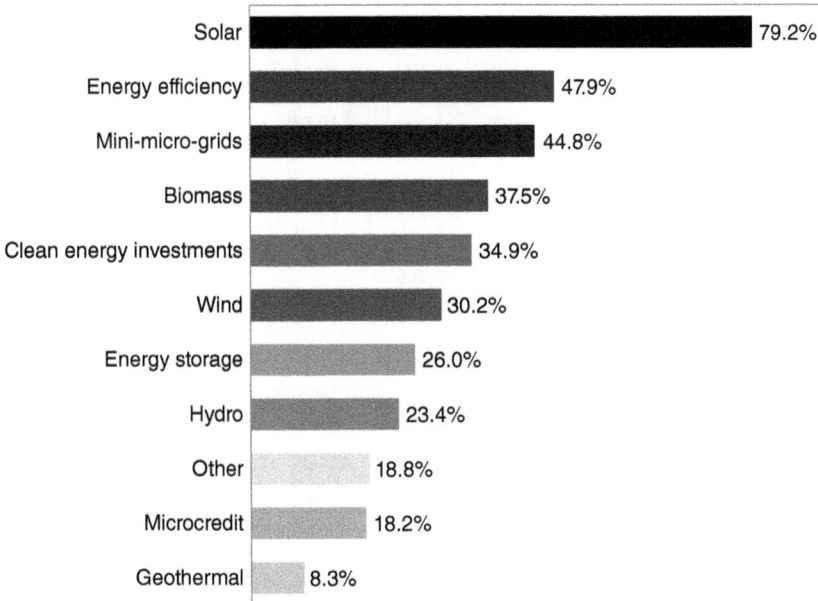

Figure 18.1. Breakdown of Practitioner Network membership based on the sustainable energy solutions they offer (based on 192 survey responses in May 2013)

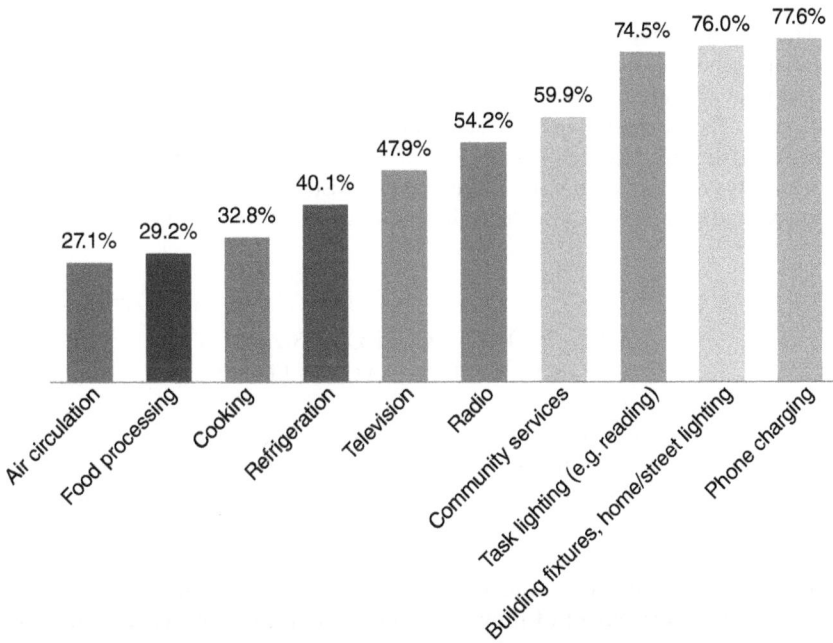

Figure 18.2. Breakdown of the primary use(s) of products/services by Practitioner Network members' customers

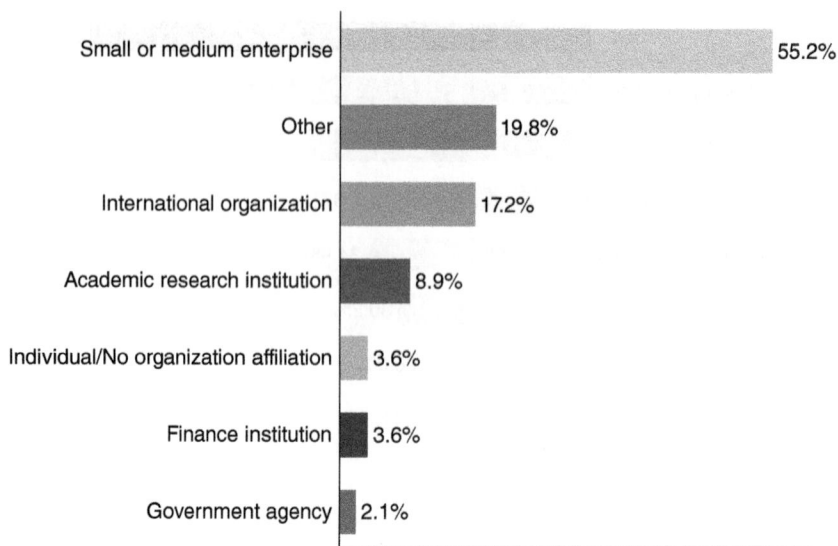

Figure 18.3. Breakdown of Practitioner Network membership based on organization type

to 19 million people with energy services access in 2012, with more than 17 million people having already been connected in only the first quarter of 2013. Cumulatively to date, the responding practitioners have provided electricity to a total of about 53 million people.

Though these numbers are significant and reflect the exponential increase in scale-up in the field, because of the response rate, the survey captures only a portion of the total activities and impact of such practitioners, and should be viewed as indicative of what practitioners are achieving on the ground and what they hope to achieve in the near future (Figure 18.4). The Practitioner Network conducts its annual survey in July each year, and updated statistics based on current membership are made available on the Practitioner Network's website in the fall.

18.3. WORKING GROUPS AND COUNTRY-LEVEL ACTIVITIES

Practitioner Network participants formed informal working groups on key topical areas of interest to them including energy and agriculture, energy and health, finance and investment, mini- and micro-grids, social protection and humanitarian issues, supply chains and entrepreneurship, and standards. Other areas may be added as additional issues and opportunities are identified.

The working groups were established to highlight existing and emerging issues hindering the achievement of universal energy access by 2030. Where possible, they offer specific recommendations for both immediate action and long-term solutions, including how each could potentially be realized, which are highlighted in Section 18.4. In addition, the working groups provide a mechanism to network, inspire, share information in a more focused manner, and motivate action. Two of

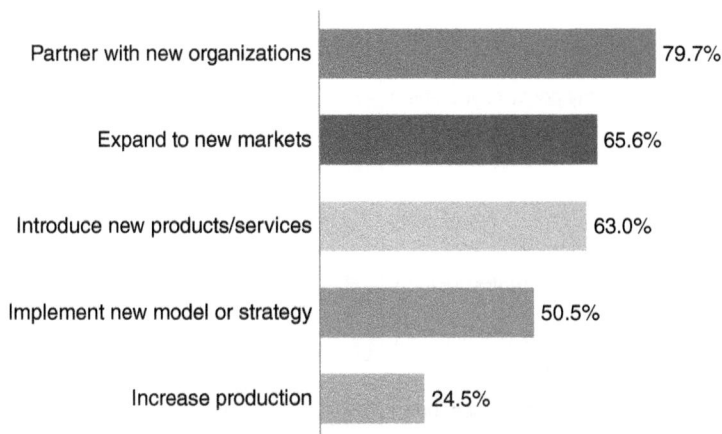

Figure 18.4. Breakdown of Practitioner Network member organizations' plans to scale up operations

the working groups have already been instrumental in the establishment of High Impact Opportunity Areas (HIO) under the Sustainable Energy for All initiative, namely Clean Mini-grids and Energy and Women's Health. Further work on these HIOs will benefit from continuous input from the relevant working groups.

The Practitioner Network also seeks to integrate existing regional initiatives and partnerships under one global umbrella by functioning as a 'network of networks', an integrating platform for action and partnerships among energy access practitioners globally. Where such partnerships do not yet exist, gaps will be addressed where possible through targeted activities at the regional and country level.

As the Sustainable Energy for All initiative moves forward at the country level, the Practitioner Network working with a variety of stakeholders from governments, UN agencies, the private sector, and civil society, increase its engagement on country-specific activities as well. The first country-level affiliate of the Practitioner Network, Sustainable Energy Network Ghana (SENG), was established in 2013 and has had its first annual meeting in May 2014. The UN Foundation has also led the creation of an off-grid country-wide alliance in India in collaboration with other partners. The Clean Energy Access Network (CLEAN) brings together 10 leading entities in the off-grid sector under one umbrella with the eventual goal of registering CLEAN as an independent entity. The UN Foundation is a Steering Committee member of CLEAN and plans to continue to be actively involved in its activities moving forward.[4]

18.4. PRACTITIONER PRIORITIES

Based on a series of consultations, the working groups identified a number of priority areas that are critical to the scale-up of off-grid energy service delivery. In addition, the need to address gender aspects across these priority areas was

---

[4] SENG: <http://www.senghana.com>.

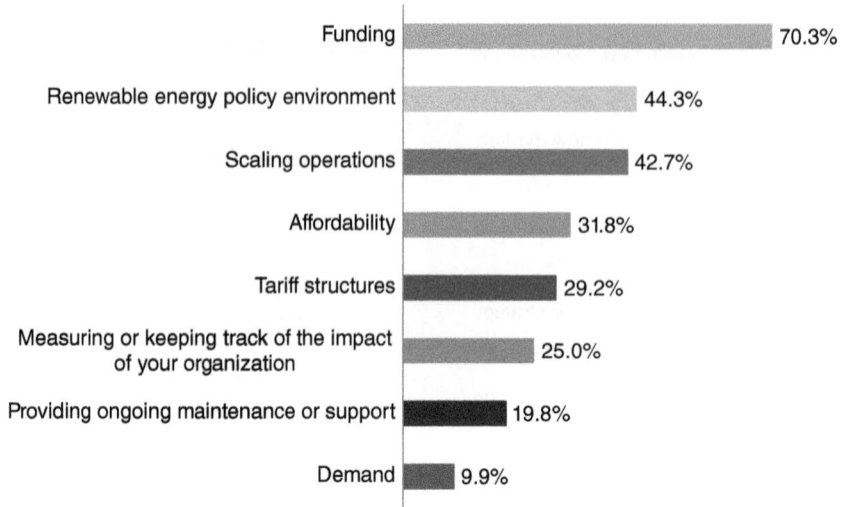

Figure 18.5. Breakdown of the biggest challenges Practitioner Network members currently report facing

stressed, and a recent survey of all members revealed more points that need to be addressed going forward (Figure 18.5).

In overcoming the challenges they report facing, members have identified the Practitioner Network's highest-value-added contributions as: introductions to potential partners, sharing best practices, advancing the energy access agenda at external high-level meetings, and providing a voice for Practitioner Network members within UN agencies.

18.4.1. Understanding the market

Practitioners have indicated the need for better access to business intelligence data, supply chain information, and capacity development to enhance their performance and scale-up activities. Even practitioners who have been active in rural energy service delivery for years identified these issues as an on-going priority given the dynamic, rapidly changing marketplace.

Business intelligence

Business information needs identified by practitioners include renewable energy resource data at both the country and the local levels (such as solar, wind, hydropower, and geothermal mapping) and renewable energy technology status, experience, costs, and suppliers. Additionally, information was requested on consumer attitudes and behaviour for off-grid energy products, to include energy usage and requirements, purchase patterns, price points, product preferences, and customer satisfaction. Collecting, collating, and reporting the information in easily accessible and under-standable formats would also be useful. Tools should be adapted for the specific use of the private sector and investors, as well as for practitioners on the ground.

At the national level, a compendium of information regarding policies in place around tariff structures has been identified as a need to help with initial macro assessments of market potential. To this end, the Practitioner Network has developed an online Import Tariff and Barriers to Entry Database[5] to provide a comprehensive overview of import duties and barriers to entry for sustainable energy products at the country level. The database addresses the common struggle of sustainable energy product importers to navigate initial entry into emerging markets due to their unfamiliarity with the tax regime, nuanced non-tariff barriers, and the local commercial climate, and serves as a starting reference for organizations interested in comparing and entering sustainable energy markets in a number of developing countries. Next steps will involve expanding the scope of the database to include more countries and products, provide in-depth data analysis, and incorporate direct experiences of conducting business on the ground.

Supply chain

The supply chain for sustainable energy products and services is very difficult to develop in these contexts, as most of the people living in energy poverty are in rural and remote places where last-mile distribution provides many challenges. These include little or no financial services for end-users, difficulties for energy enterprises to attract capital for growth, a lack of skilled technicians and employees, inability of distributors to have enough customers to buy wholesale in order to bring prices down, and a lack of efficient distribution channels, such as retail outlets, to make a business profitable.

It would be useful to leverage existing distribution channels that deliver products into rural markets, as well as the experiences of established companies such as microfinance institutions, large corporations in telecommunications/retail, and development and relief agencies. Timely dissemination of information on these networks and facilitation of market linkages would be valuable. Greater collaboration among the various actors and a series of interventions to support enterprise development and growth will help to build the significant scale required to achieve universal access by 2030.

Human resource capacity

To achieve significant increases in energy supply operations and delivery, there will need to be a corresponding growth of a trained and educated workforce. Training programmes have already been instituted by Practitioner Network participants that can assist in showcasing, replicating, and expanding energy access models in other countries and regions (see Table 18.1). Cataloguing training programmes to support workforce development in off-grid energy will be critical to sector expansion and ensuring the sustainability of the energy solutions installed in communities. In particular, incorporating rural electrification curricula into national-level vocational training programmes with a focus on mini-grid and off-grid approaches across a range of technologies should be seen as a priority focus area. This will help support needed systemic change and to ensure that suitably qualified local capacity exists to provide appropriate installation and maintenance services on an on-going basis.

---

[5] <http://www.energyaccess.org/resources/tariffs-database>.

Table 18.1. Practitioner priorities for understanding the market

Activity	Sample interventions identified	Key partners
Consumer and market research	Collect information on existing consumer and market research and make available to practitioners in a format that is easy to access and understand. This should focus on off-grid technologies, products, and resource areas. Conduct new consumer and market research where information is lacking. Organize consumer and market information into a database and other formats of most value to practitioners; provide mechanism for routine data update by practitioners and others.	International development organizations, private sector, trade groups
Supply chains	Catalogue information on existing supply chains delivering products and services into rural areas via practitioners and their networks. Create mechanisms for energy enterprises to develop and explore partnerships with organizations that have established distribution channels in rural areas.	Governments, international development organizations, local and international business associations, development/relief agencies, entrepreneurs, local financial institutions
	Exchange lessons learned about distribution models from those who have achieved success in energy and related areas.	Entrepreneurs and businesses working in key rural areas
	Conduct distribution channel mapping in target countries and regions that will benefit multiple partners.	International development organizations, industry associations
Human resource capacity	Invest in training and enhancing skills for technicians and entrepreneurs in the rural energy sector relevant to the needs of both businesses and communities.	Governments, international development organizations, industry associations
	Identify successful training programmes and showcase, replicate, and expand these.	Foundations, NGOs, SMEs
	Capitalize on available Internet learning and network tools that exist but are not easily located (e.g. databases, market research studies, business models, resources, etc.). Make these available where possible using other channels for entrepreneurs lacking Internet access. Develop a central information hub for tools addressing practitioner needs to connect with other professionals around the world.	Governments, NGOs, academia, specialized social networks, Facebook

18.4.2. Implementing policy and regulatory frameworks

Achieving universal access will necessitate political commitment to establish effective policy and regulatory frameworks for the off-grid energy sector, advance inter-sectoral coherence, and send appropriate market signals. Enforcing these policy and regulatory measures in a sustained, stable manner will also be important. Policy support is required to attract and keep investors at the global, regional, national, and local levels.

Over the last few years, progress has been made on the policy front, with 144 countries having put in place some type of policy target and/or support policy related to renewable energy—about half from developing countries, primarily focusing on grid connection (REN21 2014). Approximately 70 developing countries have established electricity access targets (IEA 2010). Several developing countries, including Bolivia, Bangladesh, Brazil, China, India, Pakistan, Tonga, South Africa, and Zambia, have adopted policies to provide access to energy services in rural areas (IPCC 2011). Kenya has declared its intention to be 'kerosene-free' in lighting before the end of the current decade.

Nonetheless, significant policy barriers persist in reaching universal energy access. Examples raised by the Practitioner Network include:

⇒ National energy strategies/policies that do not address rural energy, clean energy, or the 'base of the pyramid' (BoP) market.

⇒ Expensive and inefficient traditional approaches for grid expansion into rural areas.

⇒ Uneven subsidy regimes that distort markets.

⇒ Lack of incentives for off-grid energy sector services.

⇒ Unattractive business climates for clean energy enterprises in the off-grid market.

In establishing appropriate policy and regulatory frameworks, experiences of Practitioner Network members and others have helped identify a number of lessons to consider in addressing policy issues. For example, policymakers should send clear long-term signals about the expected future policy direction to ensure investor certainty and confidence, with minimum administrative procedures to minimize costs. Energy access policies need to be addressed at the national, state, and local levels and in many cases integrated into regional policy and regulatory activities.

Financial incentives can play an important role in levelling the playing field for renewable energy investments. These can help de-risk transactions, decrease upfront capital costs through subsidies favouring sustainable energy solutions, reduce capital/operating costs through tax credits, improve revenue streams with carbon credits, and provide financial support via loans and guarantees. However, incentives can also distort the market and encourage quantity without accounting for long-term quality of service if not structured appropriately. The elimination of incentives that hinder renewable energy products and technologies, such as fossil-fuel subsidies, is important. Fiscal policy interventions to reduce or eliminate high import duties and discriminatory taxes on renewable energy equipment are also valuable, and establishing an alignment among government ministries is important to help make this as effective as possible.

Policies should be open to renewable energy, not biased towards diesel-based systems. It is vital to integrate renewable energy into national development plans, rural electrification policies, and low-carbon development strategies. Similarly, tying renewable energy into cross-sector government agency programmes such as health, agriculture, and water is very important to support holistic approaches at the country level, as is working with those ministries directly to support the inclusion of renewable energy in planning for delivery of these services. Development of rural electrification master plans that include renewable energy options can help countries systematically prepare rural energy strategies and policy recommendations, outline a comprehensive, multi-year electrification programme, and delineate a financing plan.

Pro-poor policies and regulations can help expand access to this target group, diversify service quality (or offer different levels of services), and make prices more affordable (ADB 2010). For poor customers, subsidies will be required, except for the lowest-cost entry-level solar lighting solutions.

Box 18.1. Practitioner lessons learned in policy approaches for off-grid energy

For micro- and mini-grids, a clear, uniform, and well documented operating process is needed to ensure the reliability and safety of the system. Relevant government policies are needed to authorize micro-grids installations even when a franchised grid operator exists. Additionally, policies should ensure that micro-grids will not compete with larger utilities. Standardized power purchase agreements (PPAs) will help to simplify administrative procedures and enhance market transparency.

There is also a need for consistent, long-term policy to encourage open access and to incentivize the use of advanced technologies that increase capacity and enhance efficiency and reliability. Additional mechanisms such as pilot demonstrations must be created to bring together various industries, including power, information and communication technology, manufacturing, and government, towards advancing this agenda and producing a blueprint for smart grid implementation (see Table 18.2).

Table 18.2. Practitioner priorities for policy support

Activity	Sample interventions identified	Key partners
Policy promotion for clean energy	Tie mini-grid and stand-alone systems into broader rural electrification policies and define roles of key players. Ensure effective policies for private-sector and other service providers implementing rural electricity supply systems. Incorporate mini-grids and stand-alone systems into regulatory frameworks. Ensure provision of appropriate cost-recovery tariffs for rural energy service providers. Incorporate enforcement provisions into policy/regulatory development and implementation. Develop rural energy master plan to guide policy-making, strategy development, and financing.	Government, entrepreneurs
Level the playing field	Where possible, phase out fossil fuel subsidies to enhance competitiveness of cleaner energy solutions for off-grid applications. Simplify the regulatory environment for small-scale power generation (less than 25 kW).	Government, entrepreneurs
Pro-poor policies	Promote pro-poor policies such as direct government subsidies, connection fee support, grants in the form of equipment, technical assistance or cash, low-interest loans, and/or cross-subsidies.	Government, entrepreneurs
Fiscal incentives	Identify fiscal incentives to advance clean energy solutions for off-grid applications. Reduce/eliminate value added tax and import duties on renewable energy equipment and products, particularly those being used to deliver energy access.	Government, entrepreneurs
Dedicated funds	Establish dedicated funds to support rural energy projects. Funding sources can include a surcharge on electricity consumption to consumer electricity bills (e.g. Systems Benefit Charge), carbon taxes, and/or government or donor funds.	Government, entrepreneurs

Policy coordination	Coordinate off-grid energy policies at the national, local, and regional level.	Governments at national, local, regional levels, entrepreneurs
Cross-sector policies	Link renewable energy to cross-cutting sector policies and plans in relevant end-use sectors and across government ministries such as agriculture, health, water, and education, where they can address energy poverty and improve quality of life.	Government agencies, entrepreneurs
Capacity building	Provide policymakers and regulators with information to effectively incorporate mini-grids and stand-alone systems into energy planning, policy development, and implementation. Provide energy service providers with training and guidance in establishing tariff structures and administrative controls.	Governments, entrepreneurs

Mini-/micro-grids are operating in a number of countries throughout Asia, Latin America, and Africa. For customers at the 'base of the pyramid' (BoP), effective subsidy instruments that have been deployed for mini-grid applications include rural electrification funds, bulk power subsidies, lifeline rates and cross-subsidies, and subsidies to customers, thus involving a diverse customer base with wide variation in ability to pay.

For stand-alone systems, favourable government frameworks that support SMEs are important for delivery of rural energy services and clean energy. Interventions can reduce barriers to private participation in rural energy delivery, increase fair competition, advance supportive policies, and promote 'smart' subsidies that minimize distortions and target the poor.

For all the above, exemption from import duties on renewable energy products will reduce up-front costs, helping companies to get their operations off the ground as well as assisting follow-on operations.

18.4.3. Facilitating investment and finance

When projects are powered by renewable energy sources, as they often are in off-grid energy markets, barriers can exist such as high up-front capital costs and perceived technology risks due to a lack of familiarity with these options by financiers. Project developers may lack an established track record and work in a policy environment that favours and/or subsidizes conventional energy sources.

Most of the venture capital and private equity funds in the energy access space are interested in scaling up companies that already have a few years of revenue, leaving limited funding options for early-stage innovators. Small-scale entrepreneurs face additional challenges given their small project size, high capital costs of renewable energy systems relative to household incomes (despite low operation and maintenance costs), the level of due diligence required as a proportion of the deal size, uncertain legal and policy frameworks, issues related to currency risk where foreign capital is involved, and a lack of deep existing and proven retail and distribution channels that lend themselves to scale-up. As a result, some private investors have been reluctant to enter the sector except on a limited project-by-project basis.

Given this situation, development organizations and public finance agencies play an important role in addressing barriers and facilitating investment for these smaller-scale deals. A number of governments are incorporating off-grid renewable energy components into larger in-country grid-connected projects, providing funding to local private or public financing institutions that are committed to supporting rural and renewable energy projects. In addition, public agencies are critical in supporting energy access projects through local utilities as well as leveraging private sector finance for energy access projects. The latter includes provision of loan guarantees, partial loan guarantees, revolving credit lines, and carbon finance to reduce risks, increase returns, and encourage the private sector to engage in this market. Grants and other forms of technical assistance are also used to support rural solar home systems and sustainable access to other modern energy services (REN21 2011).

For energy entrepreneurs, the off-grid finance spectrum involves upstream support in obtaining start-up funding, operating capital, and project finance, as

---

Box 18.2. **Building financing capacities for energy access**

Capacity building needs to occur at all levels if efforts to rapidly expand financing for energy access are to be successful.

Philanthropists. Philanthropists need assistance on how to craft their support so that it encourages core development outcomes but at the same time promotes financial sustainability of the nascent BoP energy enterprises. Philanthropists need to help the energy enterprises anticipate the type of funding needed at later stages of growth and build capacity to attract that type of financing from the outset.

Investors. Both large and small investors need support in understanding the nature of BoP energy investment opportunities so that they can adapt their return expectations and due diligence processes to the realities of the typical BoP energy provider. Large investors, such as development banks, may be predisposed to supporting conventional energy projects because staff members are more familiar with large-scale, centralized energy sector development. Thus, staff need to be educated on the importance of finding ways to support smaller decentralized players that can reach remote off-grid populations.

Financing entities. Financing entities need training to understand the risks and opportunities in energy lending, the differences between the range of clean energy technologies on the market, and the appropriate financing package that should be provided for each type of technology. Banks and other financing entities also need to understand how to structure lending packages for end-user clients— including the amounts clients pay for traditional energy products on a monthly basis—and how to match new loan requirements to those realities. Banks need to identify the key information points required to evaluate a client's capacity to repay an energy loan and to streamline client documentation requirements accordingly.

Energy companies. Energy companies need to learn how to present themselves to potential financiers. They should be able to work out their current and long-term profitability, identify the right type of financing for their stage of development, and make a business case to investors, including anticipated break-even points and the real return expectations that can be anticipated. Energy companies need to know when to turn down funding that will undermine the company over the long term. Finally, energy enterprises need peer-learning opportunities about financing mechanisms that work and data on the most supportive investors.

Governments. Governments need to be educated about policy measures that promote financing for universal access to energy, as well as the unanticipated harmful outcomes and impacts of detrimental policies and how to address them.

well downstream financing for end-users. Funding and financing tools need to be highly focused and represent the most appropriate type of capital for the target stage of the clean energy sector.

Enterprise finance

⇒ *Start-up capital.* This is needed to initiate a new off-grid business or expand an existing business into the energy sector. Funding is required to understand appropriate technologies, business models, and service requirements and package these in a way that addresses market needs and conditions in the region(s) of interest. Energy service providers must build the infrastructure necessary to meet consumer requirements and through the time-consuming and costly process of raising capital for a business start-up or expansion, conducting market outreach and awareness to inform consumers about clean energy technologies and their attributes, and covering the risks associated with new product and service offerings. Assistance to train innovators in business development and planning is also essential. Additionally, in many countries, there are still significant barriers to women entrepreneurs in accessing start-up capital.

⇒ *Working capital.* Organizations involved in delivering energy access for all typically go through a challenging 'mid-life' phase in which they face three key issues: availability of the right type of capital, awareness of players providing these types of capital, and capital affordability. Between the promise of new solutions with evidence of success and the 'bankability' of well-established, profitable businesses is a large and difficult chasm for most mid-stage companies to overcome. Leaders of mid-stage for-profit manufacturers and distributors of energy-related products argue there is a limited range of working capital finance for companies that have 'proven success' on a small scale but that are not yet profitable and do not have a multi-year track record. Capital is also required for those companies serving as the import and distribution agencies for such vendors.

⇒ *Project financing.* Capital for longer-term investment in infrastructure is also required; however, few commercial lenders are currently financing off-grid energy entrepreneurs in developing countries. Accessing debt and equity finance is difficult due to the lack of operating and development track records that would provide faith in the management team's ability to deliver projects successfully to completion. Entrepreneurs require assistance in developing and documenting projects for financiers, in such areas as business plan development, risk identification and mitigation, competitive positioning, feasibility studies, and detailed project reports. Entrepreneurs speaking languages other than those in which business is generally conducted may be particularly disadvantaged.

⇒ *Institutional capacity building.* Capacity building is needed at all levels if efforts to rapidly expand financing for energy access are to be successful. Both large and small investors need support in understanding the nature of BoP energy investment opportunities so that they can adapt their return expectations and due diligence processes and avoid issues later in the investment process due to misalignment of expectations. Large investors, such as development banks that typically lend for larger scale conventional energy projects, must understand the role of smaller decentralized projects/ players in reaching off-grid populations, especially when presented as an aggregate pipeline.

End-user finance

One of the major challenges in scaling up the off-grid energy sector is providing end-users with access to appropriate financing to purchase sustainable energy goods and services, linked to their ability to pay. Microfinance institutions (MFIs) can be part of the solution, and examples exist where this has been successful: where MFIs have helped reduce first costs for consumers by providing micro-credit and down-payment financing, and have supported a variety of finance options such as micro-leasing and micro-rentals. More work, however, needs to be done for this sector to engage more effectively in energy lending.

In addition to MFIs, encouraging a range of financing options for end-users will help increase financing volumes dramatically. These include:

⇒ Provision of loans by local banks or entities that specialize in small-scale loans for solar and renewable energy systems, either at the household or village level.

⇒ Financing provided by credit unions, credit cooperatives, and self-help groups to energy end-users.

⇒ Channelling remittance flows for the purchase of energy products and payment for energy services.

⇒ Exploring innovations, such as mobile phone payments, pay-as-you-go systems, and linkages for renewable energy as they relate to consumer financing.

Another requirement is to create consumer outreach programmes on clean energy technologies, their benefits, life-cycle costs of products and services, and financing sources for end-users (including terms and conditions) in order to boost consumer confidence and increase credit uptake as it becomes increasingly available (see Table 18.3).

18.4.4. Advancing mini-/micro-grids

Mini-/micro-grids are projected to account for approximately 65 per cent of rural off-grid electricity requirements, or 36 per cent of overall additional generation required by the year 2030 (IEA 2011), as current grid deployments are unable to provide the energy services needed to keep pace in many developing countries.

The mini-/micro-grid is a scaled-down version of a large-scale utility grid that is localized to a community, village, or isolated collection of energy users. It can be connected to the main grid or islanded, and is comprised of locally based technologies that provide generation, storage, and load management—depending on priority. These grids are based on standard components and information that will be cost-effective and efficient, and, once deployed, adaptable to local market needs. Utilities can improve the cost-effectiveness of the services they provide

Table 18.3. Practitioner priorities for finance and investment

Activity	Sample interventions identified	Key partners
Enterprise financing	Train financial institutions on risks and rewards in off-grid energy enterprise lending, structuring lending packages for end-user clients, and training loan officers on energy loans. Work with these organizations to develop risk-management tools; co-invest with international financial institutions and others to diversify risk and increase financing effectiveness. Design and promote tailored products in cooperation with the clean energy sector. Develop mechanisms to bundle small-scale renewable energy projects.	Local financial institutions, local investors
	Conduct early-stage innovation funding that supports advances in technologies, business models, and financing approaches that enhance the capacity and experience of entrepreneurs in delivering off-grid energy for the poor. Interventions involve hard-to-get grants and soft funding to assist in addressing all aspects of rural energy access, as well as business plan competitions specifically focused on energy access solutions and the inclusion of women entrepreneurs in the sector.	International development organizations, foundations
	Increase working capital (or operating liquidity) for energy access enterprises that are capital-constrained in terms of growth capacity and/or unable to access regular private streams of capital due to high perceived risks in the environments in which they operate. Working capital loans provide companies with the liquidity to accept new business, grow international sales, and compete more effectively in the international marketplace; benefits are the ability to fulfil export sales orders, turn export-related inventory and accounts receivable into cash, and expand access to financing. Working capital ensures that a firm is able to cover operating expenses and has sufficient funds to satisfy maturing short-term debt. Managing working capital involves tracking inventories, accounts receivable and payable, and cash.	International development organizations, impact investors, governments
	Establish local business incubators to provide technical assistance and advisory facilities to off-grid energy service providers in project preparation capacity. Help create project portfolios and link viable projects to prospective investors.	Entrepreneurs, trade associations, donors
	Support off-grid entrepreneurs with pre-feasibility studies, feasibility studies, due diligence work, and business planning; create new capital approaches for enterprise development, such as support for early-stage seed capital funds; finance growth capital funds via blending arrangements that buy down risk	International development organizations, social investors

continued

Table 18.3. Continued

Activity	Sample interventions identified	Key partners
	and buy up returns for commercial investors; provide credit enhancements to share risks (guarantees) and buy down financing costs of commercial loans.	
	Explore opportunities for carbon finance in off-grid energy projects.	International development organizations
	Create information portals and events for energy enterprises, focusing on financing. This would involve developing a database of funders interested in supporting early stage innovators, peer-to-peer exchanges, webinars, and more.	International development organizations, entrepreneurs
End-user finance	Identify key end-user finance groups, develop training modules, and provide training and capacity building support to loan officers and management to promote off-grid energy lending.	MFIs, credit unions, credit cooperatives
	Work with interested MFIs to add energy lending to strategic priorities. Ensure appropriate incentive structures for MFIs and staff and strong processes/systems so that energy lending can be effective and grow to scale.	MFIs
	Create a regional/global funding facility(ies) from which MFIs, local banks, credit unions, and cooperatives could on-lend.	International development organizations, MFIs
	Create loan/risk guarantee funds to provide retailer/supplier credit and promote community-based financing solutions.	
	Identify and support remittance organizations—including goods remittance companies—as a source of financing.	Remittance organizations
	Conduct consumer awareness campaigns targeting the energy-poor.	International development organizations, MFIs

while expanding their service offering. Suppliers can integrate a broader portfolio of products and services. Customers receive services that are both more affordable and better aligned with their current and future requirements.

Electrical losses are significantly reduced in mini-/micro-grids, given their close proximity to the load, as are maintenance costs. Newer micro-grids include power electronics that are adaptable to the intermittent nature of some generators and can employ smart-grid and smart metering systems from their inception. These help developing countries secure utility-scale energy services and showcase the potential of rural supply companies. They also advance environmental steward-ship, help reduce dependence on often imported fossil fuels, and even stem the migration of rural populations to the cities.

Once income levels rise from the availability of local energy, other benefits accrue, such as improvements in health and education within the communities. Mini-/micro-grids contribute to local economic prosperity, not only through community level uses such as powering water pumps, providing energy to local businesses, and supporting needed services such as powering health clinics, but also via the local maintenance of the mini-/micro-grid. Maintenance revenue would be included in the operating costs for the system.

Practitioners recommend establishing new types of bonds, loan guarantees, and subsidies to reduce the high initial capital costs of mini-/micro-grids—which can be a major barrier to initial deployment—as well as combining energy production with other utility services such as Internet and telecommunications (see Table 18.4). Developing a template on micro-grid options for different types of communities is essential, to help practitioners standardize processes for these applications. Opportunities should be explored to permit mini-grid developers to take over in areas where a government or a large utility has installed rural distribution feeder lines that are not operational due to lack of generating capacity. Finally, in the planning and design of mini-/micro-grids, consideration

Table 18.4. Practitioner priorities for mini- and micro-grids

Activity	Sample interventions identified	Key partners
Planning and implementation support	Coordinate with central utility(ies) on grid expansion plans and implementation. Consider interoperability of mini-/micro-grid systems and the utility grid.	Conventional utilities, government agencies
	Collaborate with rural/renewable energy agencies, prevalent in Sub-Saharan Africa, on their energy expansion plans and activities.	REAs
	Institute a regional mini-grid deployment plan with associated monitoring and growth management.	Government agencies
	Identify and develop opportunities where rural distribution feeders are substantially in the dark.	Government agencies, community planners
	Develop standardized templates for mini-grids: PPAs; electricity collection and distribution, power quality, and storage; energy efficiency and generation options; needs assessments; feasibility studies; implementation and training plans; and record-keeping.	Government agencies, financiers
	Create universal design templates for energy efficiency and renewable energy generation options.	Practitioner Network
	Develop guidelines for system maintenance, operation, and training.	Practitioner Network
	When planning electricity requirements for the community consider range of potential needs, e.g. households, health facilities, clean water, etc.	
Policy support	Ensure that policies and regulations accommodate mini-grids and do not compete with larger utilities. Regulations should favour new projects, not be a burden; should incentivize development of micro-grids (including for smaller grid systems); and should protect rural consumers.	Government agencies, community planners
	Promote the use of local, clean energy in mini-/micro-grids. Provide exemptions from import duties on renewable energy products including micro-grid components.	Government agencies
	Create policy to allow micro-grid developer to 'take over' dark distribution lines.	Government agencies, community planners

should be given to potential interoperability of the micro-grid with the electrical grid, when appropriate.

18.4.5. Standards

Technologies that fail due to poor quality or execution create a negative association in the minds of consumers and lead to market spoilage. These products undermine the reputations of high-quality or better-executed technologies and products. Standards help to ensure product quality and operability, foster innovation, increase credibility, reduce costs, and enhance market share. Through product certification and labeling, organizations can boost their consumers' confidence in the products they are purchasing.

For renewable energy technologies—a fairly recent and rapidly advancing field—international standards can help ensure that products are comparable in quality across regional contexts. Without proper standards, energy products have the potential to be unsafe, perform poorly, and/or fail quickly in a fledgling market. Conversely, standards that are set too high can limit the scale and reach of the technology to only those few who can afford it. Therefore, a key guiding principle of the Practitioner Network is to strike an appropriate balance between better quality/performance and affordability, while utilizing recent advances in technology where possible.

As the Sustainable Energy for All initiative seeks to ramp up energy access, it is important to identify gaps, overlaps, and disharmony among current standards organizations as they relate to the off-grid electricity sector, close these gaps, and educate industry actors on harmonized standardization efforts. Progress is being made in the standards area as it relates to off-grid electricity: in the last few years, Lighting Africa has developed a comprehensive quality assurance programme for off-grid lighting. Global standards bodies, such as the International Organization for Standardization (ISO) and the International Electrotechnical Commission (IEC), bring widespread dissemination of information and credibility to markets across the world. They have well-established procedures for consensus review and input available to regional and national standards organizations.

Unfortunately, the cost of participation is prohibitive for many developing countries. As a consequence, for off-grid electricity services, these standards may lack the necessary inputs from the very markets they are intending to serve. In an effort to bring them into the process and facilitate accelerated adoption in those countries, the Practitioner Network has worked with IEC, the International Finance Corporation (IFC), and the World Bank to offer discounts as part of IEC's commitment to the Sustainable Energy for All initiative. IEC is offering three packages of standards pertaining to rural electrification, in particular the newly adopted global standard IEC 62257-9-5, 'Recommendations for small renewable energy and hybrid systems for rural electrification—Part 9-5: Integrated system—Selection of stand-alone lighting kits for rural electrification', concerning portable solar photovoltaic lanterns, at a reduced price for qualifying practitioners.

While standards are not always given the recognition of other focus areas such as financing and supply chains, they are equally important and highly interdependent. For example, where financial institutions are looking for reassurance that products and systems will work as advertised, standards can provide this level of comfort. Conversely, standard bodies need financial institutions and other organizations that are in positions of authority to demand quality products in their purchasing and investment decisions. An important next step would be to coordinate efforts on standards, financing, and supply chains in a target region or country to maximize leverage and results (see Table 18.5).

18.5. MOVING FORWARD

In addition to the working group priorities discussion in the prior section, a number of overarching activities will need to occur as the Practitioner Network

Table 18.5. Practitioner priorities for off-grid electricity standards

Activity	Sample Interventions Identified	Key Partners
Consumer market assessments	Obtain reliable data on end-user requirements for off-grid electricity access products and services to support the standards development process. Provide inputs to manufacturers, distributors, and energy service providers.	Consumer groups, United Nations agencies
Standards database	Identify/document applicable standards for distributed electricity equipment, products, and services, as well as training standards. Incorporate into a standards database for access by interested parties. Include assessments and case studies on how standards support affordability, quality, and availability of products and services in a given market.	Standards bodies (international, national), certification organizations
Harmonization of existing and pending standards	Harmonize global standards and associated certification and accreditation programmes, working through accredited international bodies. Harmonize test methods and evaluation metrics for a given technology type. Facilitate multi-level standards that allow for the use of different minimum requirements and performance targets in different contexts.	Standards bodies (international, national)
Standards promotion and application	Develop a business case outlining the importance of standards to help in securing funding for this activity. Involve manufacturers in meeting quality standards, distributors to stock and distribute the products, and customers to purchase standards-approved products. This should be linked with training programmes for assemblers, installers, and maintenance personnel. Incorporate standards into warranty and service terms offered by manufacturers and installers. Fund and facilitate participation of experts from developing countries to participate in international standards bodies. Select a region/country to pilot an activity that links finance, supply chain, and standards activities.	Trade associations, financial institutions, consumer groups

continues its efforts to support its members in delivering quality energy services and solutions in developing countries.

18.5.1. Partnering for success

A key focus of the Practitioner Network is to strengthen and integrate existing partnerships among practitioners under one global umbrella, while developing new alliances in strategic sectors to accelerate the uptake of off-grid energy services, contribute to economic and social development, help to achieve the Millennium Development Goals (and post-2015 development goals, once they are agreed by the United Nations), and foster collaborative and innovative solutions to modern energy access.

18.5.2. Strategic partnerships

At present, a number of regional partnerships focus on energy access issues; however, the global market remains extremely fragmented. The intent of the Practitioner Network is therefore to build on, rather than duplicate, the work of these existing networks, drawing them together under the auspices of the Sustainable Energy for All initiative in an action-oriented framework as a 'network of networks' focused on the common objective of achieving universal energy access by 2030.

18.5.3. Cross-sector linkages

Beyond powering households and small businesses—the cornerstone of the Practitioner Network members' efforts—modern energy services can support a range of social and productive applications for off-grid energy. In this context, the Practitioner Network has prioritized three sectors where the nexus with modern electricity services is particularly critical: health, agriculture, and telecommunications. However, this list is not exhaustive, as other important sectors have also been identified as important, including access to clean water and education.

Health. Safe, affordable, and effective health care is a priority for all. In many clinics and hospitals, the lack of availability of reliable electricity to power essential medical equipment prevents the saving of countless lives. Even in small, rural health outposts, electricity is needed for basic lighting, to power refrigeration for cold-chain vaccine storage, and other essential medical equipment. Yet a new analysis by the World Health Organization has found that in 11 major African countries reviewed, only 34 per cent of hospitals have reliable electricity (Adair-Rohani et al. 2013). Most facilities with electricity suffer chronic outages and shortages.

Energy plays an important role in addressing myriad health challenges:

⇒ Pumped water from clean sources and/or energy for purifying water reduces the spread of water-borne diseases—a leading cause of infant mortality.

⇒ With increased access to electricity, health clinics in rural areas can power lights, water pumps, fans, refrigerators for drugs and vaccines, sterilizers, and life-saving medical equipment. An increasing focus on the energy efficiency of these health-care appliances can also help to strengthen deployment in situations with highly constrained power supplies. Energy also powers computers to access medical information and data and to store and retrieve records. Energy services allow more effective community education regarding health care.

⇒ Energy-efficient systems and practices help hospitals reduce energy expenditures, allowing more investment in other critical supplies.

⇒ Energy services enhance amenities for health-care staff, improving staff retention rates.

In light of these considerations, the Sustainable Energy for All initiative has identified the nexus of energy and health services as they pertain to women's health as a key high-impact opportunity area for concerted action to provide sustainable, life-saving energy solutions.[6] The global community has rallied around the Millennium Development Goals of improving maternal health and reducing child mortality. Yet without electricity, mothers in childbirth are at particular risk; maternal and newborn mortality can be reduced significantly with the provision of electricity in health clinics and for health workers.

Agriculture. For agricultural production, processing, and distribution, electricity is indispensable. All aspects of agricultural production require some form of electricity, from irrigation, to production and application of fertilizers and pesticide, to crop cultivation, harvesting, storage, food processing, and transport. Expanding energy access in rural areas can therefore increase farm and labour productivity, raising returns across the value chain as well as helping to provide non-farm income through home-based businesses that can operate into the evening. Improvements in agricultural productivity can transform rural areas by raising incomes and reducing poverty.

It is important to note that agriculture and forestry are also primary drivers of climate change. Raising yields on existing agriculture lands in combination with best practices will reduce pressures to expand into forests and pastures that provide essential environmental services. Additional improvements in energy use and efficiency that lead to further greenhouse gas reductions, such as precise irrigation and fertilizer application, are also critical to the sustainability of the sector. Use of modern renewable energy technologies can displace traditional biomass as the primary source of energy for most rural and agricultural communities; otherwise, biomass harvesting and use can lead to significant environmental degradation, health impacts, and greenhouse gas emissions. In particular, bioenergy has an important role in providing energy access in remote areas and in transitioning rural communities to modern energy services. Moreover, bioenergy is unique in that it can be adapted to local situations, based on feedstocks and technologies that are locally available.

---

[6] <http://www.se4all.org/hio/energy-and-womens-health/>.

Recommended areas by the Practitioner Network focus on the entire value chain, not only on farm practices, and include:

⇒ Engaging farmers in knowledge-sharing networks to encourage best practices and allow farmers to provide feedback and inform actions higher up the value chain. The focus should be on disseminating information on the potential of renewable energy in improving crop quality and yields, agricultural processes, and income streams, as well as providing capacity and technical assistance on energy access.

⇒ Improving farmers' access to financing options pertaining to their work, and financing models based on income flows (e.g. during harvesting season) to increase access to renewable energy on the farm. Particular focus should be placed on microfinancing available to women and on weather-based crop insurance systems that enable farmers to reinvest in energy systems.

⇒ Expanding the integration of electricity with agriculture production for food processing and household use.

⇒ Removing barriers to adopting/implementing best practices in the agricultural sector across the supply chain, such as policies that promote fertilizer production and distribution, and elimination of tariffs on clean energy equipment and services in the agricultural sector.

Telecommunications. Electronic commerce and information and communications technologies (ICT) are essential tools for economic development. E-commerce helps farmers and local businesses to access timely market data, identify potential buyers and suppliers, reduce transaction costs, increase incomes, and facilitate trade. ICT enables industry and agriculture to make informed decisions about when and where to sell their products and to design and adapt products that suit customer needs. As international companies increasingly require their rural business partners to communicate electronically, enterprises in developing countries need to be able to quickly respond or be at a competitive disadvantage. For e-commerce and ICT, reliable electricity supplies are a necessity.

Additionally, many financial institutions operating throughout the developing world are now offering mobile banking services for account transactions, payments, and credit, including in rural areas. This provides an interesting model for the financing of energy services by off-grid customers that should be explored further. Both energy service providers and their customers are finding mobile phones to be instrumental in doing business in such places as rural Africa, using 'pay as you go' as well as providing needed data collection to support the delivery of health services.

18.5.4. Advocacy

To achieve the desired targets and impact of universal energy access requires a well-organized, structured, branded, and sustained advocacy effort over a number of years. The new global framework provided through the Sustainable Energy for All initiative led by the UN and the World Bank, in addition to the UN International

Decade of Sustainable Energy for All starting in 2014, should help to support the critical role that energy issues play in supporting sustainable development outcomes.

Expected outcomes from the initiative's focus as it relates to energy access are positive changes in government policies and regulations to support advancement of modern energy services; transformation of community attitudes towards clean energy and relative benefits of grid power; increases in the quality and number of entrepreneurs delivering modern energy services in rural areas; increases in the breadth and depth of financial organizations and programmes supporting clean energy in rural areas; a boost in consumer purchase of clean energy products and services; and accountability of energy providers and organizations for their actions in the area of clean energy service delivery.

18.6. CONCLUSION

Energy is critical to raising people out of poverty, advancing economic and social development, and achieving the Millennium Development Goals and their successors. Universal energy access is affordable, estimated at about \$48 billion annually (IEA 2011)—a small fraction of total energy infrastructure investments required by 2030, on par with the estimated current size of the kerosene lighting market. Current successful models can be expanded to the remaining 1.2 billion deprived of modern energy and the benefits it provides, using technologies that already exist today.

The Energy Access Practitioner Network was born out of the recognized need for electricity access for communities and households in the developing world, and emerging global recognition that renewable energy solutions can be particularly effective in reaching isolated and low-income communities. As meeting the energy needs of the rural energy-poor is beyond the capabilities of traditional utilities, the Practitioner Network is mobilizing a cadre of stakeholders to deliver electricity sustainably, even in remote, hard-to-access areas. Today over 1,700 participants have joined the Practitioner Network—with the number growing daily.

The Practitioner Network brings together practitioners and partners from around the world to accelerate universal energy access:

⇒ represents a cross-section of organizations involved in delivering rural energy services, from SMEs and large businesses to local banks and international development organizations;

⇒ mobilizes a range of skill sets—technical, financial, and business;

⇒ assembles knowledge and serves as a clearing-house for markets, technologies, and consumer needs;

⇒ provides doorstep services and tools to facilitate peer-to-peer learning;

⇒ shares information on the latest innovations in delivering modern energy services.

Practitioners have noted that a key benefit of the Practitioner Network is the provision of a platform for sharing experiences, streamlining activities, partnering on projects, and tackling common problems in the marketplace. The convening power of the United Nations Foundation, with its reach into the UN system, the private sector, the financial community, and other arenas, provides a collective access and clout that practitioners lack individually. Members identify, inter alia, the Network's role in catalysing the off-grid energy market in several areas:

⇒ forging new partnerships among practitioners, and between practitioners and investors, to scale up energy access projects;

⇒ facilitating practitioner participation in global fora to bring greater visibility to the challenges faced in delivering access to energy;

⇒ serving as an umbrella for existing efforts in the sector to encourage better coordination among key actors.

Practitioners are committed to scaling up their activities for poverty reduction in rural households and communities. The Practitioner Network is contributing to and amplifying their efforts by providing a range of tools, support, and know-how to help achieve this goal.

But practitioners cannot do it alone. The work of these on-the-ground entrepreneurs requires the continued and expanded partnership of national and local governments, international development organizations, NGOs, companies, foundations, communities, and others to join their resources in making the energy access vision for 2030 a reality.

ACKNOWLEDGEMENTS

Section 18.1 is adapted and updated from a report originally published in 2012 'Energy Access Practitioner Network – Towards Achieving Universal Energy Access by 2030'.

REFERENCES

Adair-Rohani, H., et al. (2013). 'Limited Electricity Access in Health Facilities of Sub-Saharan Africa: A Systematic Review of Data on Electricity Access, Sources and Reliability'. *Global Health: Science and Practice* 11(2).

ADB (Asian Development Bank) (2010). *Pro-Poor Policy and Regulatory Reform of Water and Energy Supply Services, Law and Policy Reform*, Brief No. 3, April.

IEA (International Energy Agency) (2010). *Energy Poverty—How to Make Modern Energy Access Universal? SPECIAL EARLY EXCERPT of the World Energy Outlook 2010 for the UN General Assembly on the Millennium Development Goals*. Paris: OECD/IEA.

IEA (2011). *Energy for All—Financing Access for the Poor—Special Early Excerpt of the World Energy Outlook 2011*. Paris: OECD/IEA.

IPCC (Intergovernmental Panel on Climate Change) (2011). *Summary for Policymakers*. Approved at the 11th Session of Working Group III of the IPCC, Abu Dhabi, 8 May.

REN21 (Renewable Energy Policy Network for the 21st Century) (2011). *Renewables 2011 Global Status Report*.

REN21 (2014). Global Status Report 2014: <http://www.ren21.net/ren21activities/global-statusreport.aspx>.

19

The Energy Plus Approach
Reducing Poverty with Productive Uses of Energy

Thiyagarajan Velumail, Chris Turton, and Nick Beresnev

19.1. INTRODUCTION

A single compact fluorescent bulb can light a home for over 50 hours using only one kilowatt of energy. By contrast, small agricultural water pumps can be switched on for less than an hour using the same amount of energy.[1] To many the difference may be insignificant, but to those without sustainable access to energy it can mean the difference between being able to irrigate a farm and having to choose not to sow a field. Energy access changes lives in many ways; it brings light to darkness, powers electronic devices, and makes cooking and heating cleaner and safer, but the way we define energy access and energy poverty has important implications for the standard of energy access that people have and how they can make use of that energy. If we understand energy access as only a single household electricity connection, we risk denying people the ability to use energy for productive purposes and to utilize that productivity to help them move out of poverty.

This chapter looks at a selection of case studies from around Asia where small-scale, off-grid energy solutions have been adopted by communities that have either previously lacked electricity access or depended on an unsustainable supply of biomass for thermal energy. Each of these examples provides a demonstration of how energy access can go beyond meeting simple domestic needs and be turned into a driver of productivity, income generation, and socio-economic transformation. These cases make a critical diversion from the way many traditional and ongoing government and international aid programmes approach energy access. Instead, they promote an alternative approach, referred to as 'Energy Plus'. Traditionally, many energy access programmes have adopted a minimalist approach, focusing almost entirely on subsidising or constructing energy infrastructure and overlooking other substantive requirements to make that infrastructure sustainable and able to contribute to human development. In contrast,

---

[1] Energy consumption of pumps varies, but see for example Ramachandra Murthy et al. 2012.

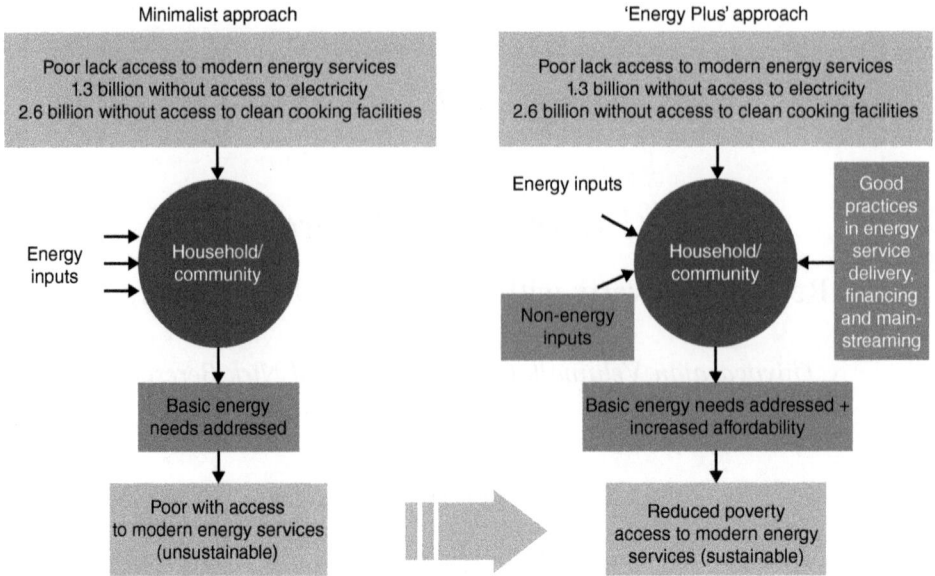

Figure 19.1. Transition to the 'Energy Plus' approach

Energy Plus can be defined as a method that incorporates productive uses of energy into energy access programmes in order to bring about positive impacts on human development, as depicted in Figure 19.1. This method recognizes that, in addition to promoting productive uses of energy, non-energy inputs such as technical skills and management capacity are needed to maintain energy infrastructure. It also recognizes that functional commercial and financing models need to be in place and energy concerns need to be mainstreamed into government policies, planning, and technical support in order to provide a comprehensive operational off-grid energy market.[2]

19.2. ENERGY ACCESS AND PRODUCTIVE USE

We see an ongoing discussion over the metrics used to define energy access and to determine how many people live in energy poverty around the world.[3] The broader definition, and one that we will use here, conceives of energy access by taking into account the extent to which it hinders or contributes to education, health, communication facilities, household lighting, transport, mechanical power, and enterprise activities.[4] In this sense, energy access includes electricity,

---

[2] This term was originally used in UNDP 2011f to refer to 'projects and programmes which combine the delivery of energy services with income-generating measures' (UNDP 2011f: 10). It has since been broadened to include other human development needs.

[3] For example, AGECC 2010, Nussbaumer et al. 2011, and Practical Action 2010.

[4] Practical Action 2010, pp. xi–x; Sovacool et al. 2012, p. 716.

modern fuels for cooking and heating, adequate power for industrial and non-industrial productive applications, and energy for transportation.[5] Such a perspective intentionally looks beyond the consumption of energy at a household level and extends to its use for both public services and commercial enterprise. Energy access then becomes a function of how the availability, affordability, and reliability of energy allows for a variety of uses of energy and how they impact the lifestyles of users.[6]

Without going into a detailed analysis of how energy poverty can be measured, it will suffice to highlight that there is no universal agreement on the minimum energy consumption levels used to define what energy poverty is.[7] Rather, acceptable levels will vary with geographic contexts, resources, and economic development and will sometimes be set on a country-by-country basis through national policies.[8] Such policies can have direct implications for the standards that energy service providers are required to adhere to, which in turn determine the quality of energy that users are ultimately able to access. If these standards are set too low they can become compounding as users are prohibited from using energy to expand income generation opportunities and are then unable to invest in continual improvements in energy services.

Many energy access projects instead focus solely on meeting household energy needs, that is residential lighting, cooking, and heating. While improving living standards, this approach does not make energy access projects effective at reducing poverty, as household energy services per se do not automatically translate into increased incomes.[9] Rather, 'they transform people from being "poor without energy access" to "poor with energy access"'.[10] Consequently, households often struggle to pay for energy services, and service delivery is abandoned in the absence of subsidies.[11]

There is an array of ways these restrictions can affect agriculture, industry, education, health, and social relations. For example, in agriculture and industry, energy poverty can result in an absence of motors and other machinery. The consequent dependence on manual labour leads to productive inefficiency and limits the range of income-generating activities that can be undertaken. On the other hand, introduction of machinery such as water pumps, grinders, and dryers increases yields per acre, improves cropping intensities, allows for better storage facilities, decreases labour time, increases areas for cultivation, and results in higher crop growth and the possibility of value-added processing and packaging.[12]

---

[5] UNDP and WHO 2009, p. 6; Practical Action 2010.

[6] For a discussion of the terms availability, affordability, and reliability see Rehman et al. 2012.

[7] A number of examples are given in Pachauri et al. 2004, such as where a household spends more that 10 per cent of its income on lighting, heating, and cooking energy, or consumption of less than 100 kWh of primary energy per capita per day.

[8] For example see the Chinese *National Programme for Rural Electrification Development (2001–2015)* which set an annual per capita electricity consumption target of 900 kWh for people living in less developed regions of the country (UNDP 2011c: 2).

[9] Brew-Hammond 2010, p. 2297; EUEI, PDF, and GIZ 2011.

[10] UNDP 2011f, p. 40.

[11] Even where physical access to networks is established, rural households often cannot afford to pay for energy unless it is heavily subsidized. UNDP 2011f, p. 16.

[12] Cabraal et al. 2005, p. 124.

Electric lighting also enables downstream productive activities to continue beyond daylight hours.

Additionally, schools without electricity are often poorly lit or inadequately heated, making it harder for children to study and read, and communities lack modern appliances that can be used for education—such as television, radio, and computers. On the other hand, access to modern energy in households reduces the amount of time spent by children collecting fuelwood, and thus frees time for education. Connecting schools and households to modern sources of energy can therefore improve school attendance and educational performance.

A lack of access to sustainable forms of energy also places a disproportional burden on women, who usually perform the bulk of household tasks, engage in subsistence agriculture, and work in home-based businesses. Conversely, providing access to modern energy services tends to have a distinct benefit for women by making these immediate tasks easier or more efficient and also bringing longer-term benefits such as improved health, education, economic opportunities, and involvement in community activities.[13]

The various ways that productive use of modern energy can improve income generation and contribute to poverty reduction and human development can be divided into five primary categories—areas within both economic and social spheres which are influenced by energy access. These would include:

1) Using energy to improve efficiency or productivity of existing economic activities, for example increasing agricultural productivity via mechanized irrigation.

2) Expanding operation of existing enterprises beyond daylight hours or into new services.

3) Establishing new energy-based enterprises and creating employment opportunities.

4) Improving operation of schools, health facilities, and other public services.[14]

5) Employing local people in the delivery of energy services—for example local masons building biogas digesters, local technicians servicing solar home systems (SHSs)—and in other productive activities.

These categories emphasize that productive use of energy not only shapes the way existing businesses and activities function, but also makes possible new opportunities and forms of economic interaction. Breaking these categories down further, an illustrative list, provided in Appendix A, shows the breadth of examples that have contributed to global community development in recent decades.

---

[13] A study of the impact of electricity on rural women in India showed that women from households with electricity had more time for leisure activities. In Mali, improvement in access to energy services led to an increase in the girl-to-boy ratio in primary schools, as well as in the number of prenatal visits to health clinics (Cabraal et al. 2005: 136–7).

[14] This report adopts a broad definition of 'productive use of energy', which includes direct and indirect impacts on income generation as well as on human development more broadly. For this reason educational and health outcomes are also able to be seen as productive uses of energy.

19.3. ENERGY PLUS IN THE ASIA-PACIFIC

Productive uses can lie at the crux of poverty reduction, but such uses do not emerge automatically simply because communities gain energy access, and in most cases these uses and opportunities need to be stimulated. As noted by the European Union Energy Initiative, 'in the absence of well-targeted support measures, productive uses of electricity often catch on much slower or to a much lesser extent than initially expected'.[15] This rationale has motivated the inclusion of strategies and activities which encourage productive uses of energy as components of energy access projects—the Energy Plus approach. There are a growing number of recent projects in Asia-Pacific, both government- and donor-led, which have used an Energy Plus approach, and a selection is given in this chapter to show the range of technologies and productive uses that can be included as well as to articulate the roles of enterprises and governments in leading and enabling energy access.

The case studies bring together several different technologies that are commonly adopted to solve the challenges of providing off-grid energy access. These solutions include distributed electricity generation technologies, such as solar, wind, biomass, and hydropower systems, as well as biogas production for use in domestic and commercial activities. In addition to these technologies, the Energy Plus approach commonly incorporates wider community support activities to enhance programme effectiveness. Examples of such activities include:

- Disseminating end-use technologies—energy-based machinery used in agricultural and industrial production (e.g. electric grinders and dryers), rather than focusing solely on residential energy access.

- Mobilizing communities to actively seek out and make use of public and private services available to them, such as microfinance or training on various productive uses of energy.

- Training potential entrepreneurs in business management.

- Improving access to finance for purchasing end-use technologies and raw materials and for business development, for example:
  - government subsidies for energy projects stipulating productive uses in eligibility criteria;
  - loans from commercial or development banks (against some sort of collateral) for meeting investments costs of starting an enterprise;
  - loans from microfinance institutions or local cooperatives;
  - revolving funds;[16]
  - improving households' awareness of available financing options;
  - improving the skills of financing institutions to assess the economic viability of energy projects.

- Establishing links between communities and rural and urban markets.

---

[15] EUEI, PDF, and GIZ 2011, p. 13.

[16] In revolving funds, reimbursements are returned for reuse in a manner that maintains the principal of the fund.

- Constructing infrastructure necessary for business operation (roads, transport, and telecommunication).

The application of many of these activities can be seen in the following case studies from around the Asia-Pacific region.

19.4. THE IMPROVED WATER MILL
PROGRAMME IN NEPAL (2003–PRESENT)

Traditionally, water mills have been used to transform the kinetic energy of running water into mechanical energy for a variety of uses, such as grinding grains and other basic agro-processing functions. More recently, Improved Water Mills (IWMs) have been developed by retrofitting traditional water mills with more efficient parts that increase processing efficiency and allow the mills to be adapted to other end-uses, including electricity generation of up to 3 kW. Retrofitting replaces the existing wooden components (rotor, shaft, and chute) with purpose-built metal components. These IWMs have been put to a wide range of innovative productive uses, and can be adapted for purposes such as paddy hulling and husking, rice polishing, saw-mill operation, oil expelling, and paper production. IWMs also provide energy services to households at low investment and maintenance costs and with construction that can be completed in a relatively short period of time.

In mountainous districts of Nepal, the IWM Programme has sought to expand the use of improved mills and adopt an Energy Plus approach by creating employment through the promotion of private microenterprises that provide fee-based services to communities (grain grinding, hulling, sawmilling, etc.). To encourage IWM uptake, the programme trains mill owners in various end-uses of IWMs and in business management and product marketing. The programme also researches and pilots additional end-uses of long-shaft mills (with 13 uses trialled to date), and organizes exchange visits for mill owners to highlight these end-uses. A government subsidy covers roughly 45 per cent of the IWM installation cost, and the programme is piloting a revolving fund which offers loans of up to 90 per cent of the remaining cost.

To date, the IWM programme has installed 6,349 IWMs, including 16 mills with electricity generation capacity. Consequently, services such as grinding and hulling have been provided to 330,148 rural households and lighting to 850 households. On average, the income of a mill owner rises from EUR 108 to EUR 461 per year after IWM installation (a fourfold increase).[17] Over 850 people are directly employed in the IWM sector as manufacturers, suppliers, and technicians.

However, the programme has encountered a number of hurdles. The uptake of the electrification option has been slow, primarily because of high costs. Remote districts lack financing for IWM installation due to the high cost of credit transfers

---

[17] UNDP 2011g, p. 10.

and risks of physical damage to IWM systems. Consequently, a fully viable market for IWMs does not yet exist, and reliance on donor subsidies continues.

19.5. COMMUNITY MICRO-HYDRO FOR SUSTAINABLE LIVELIHOODS IN BHUTAN (2005–2009)

Micro-hydro plants (MHPs) are defined by their energy capacity and are able to generate up to 100 kW of electricity. They can be distinguished from small and mini-hydro plants, which can generate up to 25 MW and 1 MW respectively.[18] Plants do not usually require dams, but instead include an intake channel which diverts water to a turbine before returning it to the original watercourse. MHPs generally have the advantage of benefiting from a consistent input of water and are thus able to produce a more stable output of electricity. While some systems are subject to seasonal fluctuations in water flows, most are designed to operate based on a calculated minimum flow of a river and can operate year-round. If adequate water is seasonally unavailable, MHPs may need to be backed up by diesel generation systems or other viable alternative energy systems.[19]

A MHP project in Bhutan sought to stimulate economic activity in a remote Sengor village by constructing a 100 kW plant to provide inhabitants with electricity.[20] The overall goals of the project were to improve the socio-economic status of the Sengor community and reduce greenhouse gas emissions from fossil-fuel-fired power generation.[21]

In this case, the Energy Plus approach focused mostly on providing financing and training for energy-based microenterprises. In order to encourage productive uses of the acquired electricity, the project sought to ensure villagers had access to finance through a Community Collateral Fund (CCF) and also provided a start-up sum for the CCF of USD 50,000. The CCF is managed by the Bhutan Development Finance Corporation Limited (BDFCL), a banking institution. In essence, the BDFCL provides loan facilities to villagers against the CCF as collateral. To encourage further enterprise development, the project conducted training programmes on poultry farming, vegetable production, tourism, and hotel management.[22]

In terms of outcomes, the MHP has been providing electricity to all 57 households of the wider Sengor community since May 2007. Access to electricity and the use of appliances such as rice cookers has drastically reduced the use of fuelwood and consequently also reduced time and effort spent collecting it. The village has also begun using electricity to boil drinking water, with 55 per cent of households now owning water boilers. Sengor's Basic Health Unit can now provide services at night and has a refrigerator for storing vaccines and medicines. With access to electricity, existing informal education classes provided by the

---

[18] See Purohit 2008, p. 2001, although classifications by size can vary between countries.
[19] Muhida et al. 2001.
[20] Sengor lies 370 km from the capital Thimpu at an altitude of almost 3,000 metres above sea level.
[21] UNDP 2011a, p. 3.   [22] UNDP 2011a, p. 5.

Ministry of Education to the adult population have been extended into late evenings.[23]

In terms of productive uses of electricity, only two CCF applications have been approved to date: one for establishing a poultry farm and another for starting a cable television business. Two reasons for such slow uptake have been identified:

- **Prevailing habits and preconceptions**: most Sengor households have had limited education and depend on livestock for their livelihoods. It has been difficult to convince people in the region to consider alternative income-generating activities.

- **Limitations of the small and medium enterprises (SME) sector**: enterprise development in Sengor is also affected by the overall challenges for SMEs in Bhutan. While the sector comprises 85 per cent of the country's business establishments, its development is hindered by a lack of entrepreneurial talent and business skills, limited access to financing, absence of clear sector policies, poor physical infrastructure, and concomitant high transportation costs.[24]

Nonetheless, Sengor has experienced an inflow of enterprises from non-electrified areas, including the franchise of a hotel from the nearby Gazamchu village. In addition, the Department of Livestock has established a community-owned milk processing unit, increasing income and employment in the village.[25]

19.6. THE RURAL ENERGY DEVELOPMENT PROGRAMME IN NEPAL (1996–2011)

In rural Nepal only 29 per cent of households have electricity, and 98 per cent of consumed energy comes from the traditional use of biomass. Causes of rural energy poverty include widespread poverty, geographic inaccessibility, and lack of technical staff to design, install, and operate energy equipment. While Nepal's economically viable hydropower resources are estimated at 42,000 MW, only 1 per cent of this potential has been developed to date.

To address these shortcomings, the Rural Energy Development Programme (REDP) promoted installation of community-managed MHPs, SHSs, toilet-attached biogas plants, and energy-efficient cookstoves in Nepal's remote rural communities. The programme was implemented between 1996 and 2011, with funding from the World Bank, UNDP, the Government of Nepal, and local communities.[26]

REDP's key element was the mobilization of micro-hydro functional groups (MHFGs), village-level organizations which are made responsible for MHP

---

[23] UNDP 2011a, pp. 6–7.   [24] UNDP 2011a, p. 8.   [25] UNDP 2011a, p. 6.
[26] Upon conclusion of REDP, the Government of Nepal and UNDP launched the Renewable Energy for Rural Livelihood (RERL) programme (UNDP Nepal 2012).

management. A community energy fund was established by each MHGF to manage REDP donor grants and government subsidies for construction of MHPs. MHFGs were trained in operation and management of MHPs; once a MHP had been running successfully for six months, MHFGs were usually registered with the local government as micro-hydro cooperatives.

Concepts of an Energy Plus approach were applied by ensuring that energy was allocated for use in community services, such as schools and clinics, and by setting up community institutions to support local businesses. REDP promoted productive uses of energy by encouraging each MHFG to establish Enterprise Development Funds and making an initial contribution of 10,000 Nepalese Rupees per kilowatt installed to each fund.[27] These funds provided soft loans for enterprise development to needy villagers.[28] In addition, collected electricity tariffs were deposited into a Community Energy Fund, which was managed by a MHFG and used (among other purposes) to provide additional loans for income-generating activities. The programme also provided potential entrepreneurs with enterprise development training, market information, and exposure visits. Entrepreneurs were also assisted in establishing links with city-based markets (e.g. linking rural craft producers with a Kathmandu-based company for the sale of traditional Nepalese crafts).[29]

Overall, REDP operated in 45 districts and installed 307 community-managed MHPs (5.5 MW of total capacity), 3,099 SHSs, 6,811 toilet-attached biogas plants, and 14,255 energy-efficient cookstoves; 550,000 people benefited from the programme, including 42,828 electrified households. Benefits of delivered energy included home lighting, powering enterprises, and household devices (refrigerators, televisions, etc.), energy for cooking, and more efficient, less polluting cooking facilities. Time spent on fuelwood collection and agro-processing was reduced by an average of three hours per day. Simultaneously, participation of men in household chores such as cleaning, agro-processing, and cooking increased, reflecting changing gender relations within households.

In terms of productive uses, 264 microenterprises were established in programme areas, including agro-processing mills, poultry farms, and carpentry workshops. Female entrepreneurs own 108 of these enterprises (41 per cent). Between 1996 and 2005, average household income in REDP communities increased by 52 per cent, compared to a national average of 46 per cent.[30] In addition, around three people gained direct employment from each MHP.[31]

REDP's successful mobilization of communities in managing funds and operating MHPs has been vital in ensuring sustainability of its activities. It should be noted that in later stages of the project, payment of electricity tariffs became more regular (an indication of increased household income and improved programme sustainability).

---

[27] Approximately USD 126 at the close of the project in 2011.

[28] REDP's motto 'one household, one enterprise' promoted the goal of having every household covered by the programme earn an additional monthly income of at least NPR 25, allowing it to pay its monthly electricity tariff.

[29] UNDP 2011b, p. 13. [30] UNDP 2011b, p. 11. [31] UNDP 2011b, p. 11.

19.7. PRODUCTIVE USES OF RENEWABLE
ENERGY IN CHITRAL DISTRICT, PAKISTAN
(PURE-CHITRAL) (2008–PRESENT)

Implemented since 2008, PURE-Chitral seeks to reduce the use of diesel for electricity production and kerosene for heating in the Chitral District of Pakistan. To this end, the project planned to install off-grid MHPs, with financing from the government and in-kind contributions from local communities.[32] MHPs provide electricity to nearby households and public facilities (schools, health-care centres) through mini-grid systems. Each MHP is owned by the community and managed by an MHP committee; the latter will set and collect electricity tariffs. The MHP will be operated and maintained by a local private leaseholder.[33]

To promote productive uses of acquired electricity, the project will design and implement micro-credit schemes. Specifically, local microfinance institutions will provide USD 100–10,000 loans to villagers to purchase equipment for agro-processing, animal husbandry, and skilled trades. Various financing options will be evaluated, including vendor financing, loan guarantees, and soft loans (14–18 per cent).[34] In addition, farming households will receive training in assessing different production options (agriculture, handicrafts, etc.) using electric equipment, business management, and financing schemes. The project will also conduct market studies to define the expected commercial value of local value-added products, and promote these products in regional and national markets.[35]

To date, the project has completed sustainability assessments of 10 MHP sites and assessed the feasibility of establishing repair and maintenance workshops. Training needs of local personnel to install, operate, and maintain MHPs have also been assessed.[36] Project targets include installing 15 MW of electrical capacity and a cumulative emission reduction of 461,465 tonnes of CO_2 equivalent during 2008–2014.[37]

19.8. BIOMASS UTILIZATION
IN SOUTH INDIA (1998–2009)

About 836 million people in India rely on fuelwood, animal dung, and agricultural residues for cooking and heating, and overall biomass energy sources account for about 33 per cent of the country's total primary energy requirements.[38] Rural populations continue to rely on biomass for most of their domestic and industrial energy needs and still use traditional low-efficiency wood-burning technologies in both the home and industry. In 2000, South India alone had about 8 million people directly or indirectly employed in small and artisanal industries which used fuelwood as a major source of energy.[39] The variety of industries fuelled by

[32] UNDP 2008, p. 3.   [33] UNDP 2008, pp. 15–17.   [34] UNDP 2008, p. 20.
[35] UNDP 2008, pp. 19–20.   [36] UNDP Pakistan 2012.   [37] UNDP 2008, p. 11.
[38] Pillai and Banerjee 2009, p. 970.   [39] Reddy 2000.

biomass combustion is extensive, ranging from sugar production to rubber vulcanization. However, the health impacts are equally substantial with workers suffering debilitating illnesses from chronic exposure to extreme heat and harmful pollutants, and with an ongoing loss of forest resources caused by fuelwood consumption. Taking an Energy Plus approach, this programme sought to make these enterprises, operated largely by poor and low-income communities, more profitable, sustainable, and efficient while also creating new opportunities for tradesmen to initiate energy-based service enterprises.

Efficiency improvements in stoves used in these informal industries have been minimal over the years because energy access policies and programmes have largely focused on household energy consumption and domestic stoves.[40] However, the programme was launched in South India to undertake a concerted effort to improve design efficiency of stoves, dryers, and kilns used in many of these existing industries.[41] Under this programme these industrial devices were redesigned to increase energy efficiency by reducing heat loss, improving combustion rates, and minimizing the need for modifications to be made by end-users. This often involved streamlining airflow, installing adequate ventilation, and improving insulation, and together these improvements were able to deliver at least a threefold increase in efficiency in all devices. In the latter stages of the programme, between 2007 and 2009, 11,840 stoves, dryers, and kilns were sold to informal industries across South India, and self-sustaining enterprises were established to continue construction and maintenance of improved devices over the following years.

Improvements in temperature stability and production control enabled end-users in many industries to increase the quality of their products, which subsequently allowed some industries to raise sales prices by as much as 10 per cent. Fuel consumption costs were also reduced by 30 to 60 per cent and some devices enabled producers to use alternative sources of fuel. For many of these industries, initial cost outlays were also recovered within one year through a reduction in fuel costs alone. These cost recovery rates meant that minimal subsidies or loans were necessary; where production costs were high, industry-specific government programmes were generally available to offset initial capital investment, and in many other cases entrepreneurs offered delayed repayment schemes to attract customers.

The programme was implemented through a participatory and locally embedded process. Appliances are all purpose-built, with exhaustive consultation from local manufactures and industries. Local entrepreneurs are engaged and provided with technical and business training, and pools of tradesmen are formed in each locality to offer ongoing construction and maintenance services after the programme ceases operation. The outcome of this approach was that entrepreneurs were often able to double their incomes in comparison with previous activities, and a host of new employment opportunities were created for skilled local masons and craftsmen.

---

[40] UNDP 2012, p. 2.   [41] UNDP 2012.

19.9. SUNLABOB SOLAR LIGHTING IN
LAO PDR (2006–PRESENT)

Approximately 45 per cent of people in Lao PDR do not have access to electricity, relying instead on energy sources such as fuelwood and petroleum products for cooking and lighting. Of those that do have access, about 8 per cent depend on small hydro plants, photovoltaic systems, and diesel generators.[42] While the hydropower resources are being increasingly exploited through large-scale projects, renewable energy otherwise remains largely underdeveloped. No single national government entity is responsible for small renewable energy development and financial incentives to encourage private sector activity are minimal.[43] Consequently, there are only a handful of local companies providing commercial renewable energy solutions.

However, in 2000 one company, Sunlabob Renewable Energy Ltd, began selling hardware and energy services, growing to a range that included solar water pumps and heaters, water purification systems, street lighting solutions, cooling units, and solar lanterns. After several attempts to refine a widely successful business model, the company began providing rechargeable solar lanterns in rural areas using a fee-for-service rental model in 2007. This model focused on the sale of light hours rather than the sale of equipment by offering lanterns which needed to be recharged at local solar charging stations. The service was successful largely because it removed the need for users to invest in prohibitive upfront equipment costs. It also veered away from the company's previous strategy of giving loans for purchasing equipment and making repayments on a monthly basis. Such an approach failed because many households had irregular, seasonal incomes and were not accustomed to putting money aside on a regular basis, which made collection of repayments difficult. Rather, the fee-for-service model worked because it meant that users could only use the service when they had money to pay for it, much as they were accustomed to doing when purchasing diesel for a generator.

Building village-level ownership and responsibility was also at the heart of Sunlabob's delivery mechanism. The company began forming and training village energy committees and making them responsible for purchasing lanterns in bulk and renting them to households, a method that would eventually begin generating an income for the committee. The committee would also oversee a trained local technician, who would be responsible for recharging and maintaining the lanterns, and manage a lantern maintenance fund. Public loans were sought with Sunlabob's help to support village start-up capital costs for purchasing lanterns if necessary. Once this local distribution system was established the company could undertake the installation, operation, and monitoring of village solar charging stations.

In line with an Energy Plus approach, access to lighting immediately opened up possibilities for consumers to extend productive activities, such as fabric weaving or children's education, beyond daylight hours. However, such changes were not limited only to existing productive activities. Once the lantern rental systems

---

[42] Messerli et al. 2008.   [43] UNDP 2011e.

began operating, local technicians started providing new goods and services: these included coolers for storage of perishable items, laptops with Internet connections, sterilized potable drinking water, and power for televisions and telephones. Sunlabob has been helping these ideas spread between sites by facilitating communication across its franchise network.

This model of energy service provision is not entirely unique to Lao PDR, but it does offer an important demonstration of how ownership and responsibility can be shifted away from individual consumers towards a collective governance structure, while still retaining local management and control. This distinction between individual and collective responsibility raises a number of important questions about how to conceive of energy access in the immediate and long term, and the role that an Energy Plus approach can have in promoting this access.

19.10. CONCLUSION

Each of these case studies demonstrates a different way of utilizing energy for productive purposes. While efforts to promote productive use are not always successful, such use can bring about fundamental social and economic transformation. These case studies also demonstrate that successful transformation requires more than just access to new technologies; the capacity of local governments, private enterprise, and financing organizations to support the introduction of energy services and the adoption of productive use activities is also critical.[44] Although programmes and initiatives such as those set out above are often aimed at bringing about positive impacts for communities, there are also many common considerations and challenges that affect the way such energy access programmes are designed and carried out. It is worth highlighting a selection of these.

- Productive uses need to be considered from the outset of a programme. This means that market opportunities for end-user enterprises and activities should be identified early on in order to build ownership and capacity for productive activities amongst communities and households. It also means that such productive use opportunities can inform forecasts of energy consumption when assessing the amount of energy required for a given area.

- Governments are crucial in shaping market environments to make it possible for energy service providers to operate where high capital costs and weak market demand would otherwise deter their investment. This is particularly true where the uncertainty of new markets places a high burden of risk on entrepreneurs and investors, and where the ability of the poor to afford energy services appears unlikely. Activities such as renewable energy resource assessments, clarity in local development planning, and regulation to encourage viable business models are among the roles that governments can play at a sub-national level to provide assurance to businesses.

[44] UNDP 2011c.

- Development benefits from energy access span multiple sectors, yet energy access programmes often only involve stakeholders from the energy sector and those directly related to it. Seeking out and managing clear engagement and inputs from other stakeholders, including across government ministries, can significantly increase impacts of a programme on other areas of human development.

- Sources of finance can be a determining factor in whether communities and energy service providers can invest in new energy systems, and in many cases funds can be available through government, aid organizations, or philanthropic bodies. Harmonizing these sources of finance can ensure that funding is more accessible to the poor and to those seeking to provide energy services to them. It can also help to ensure that finance is coordinated and available for all areas of need, such as energy infrastructure, training and capacity building, or microfinance for income generation.

- Moving from successful pilot projects and demonstration sites to national adoption is a challenge that can prevent programmes from meeting their full potential. Here, priorities lie in seeing that these practices are taken up as national goals and strategies, and also that this knowledge is shared across markets and across borders.

With these lessons in mind, the world is now looking towards reaching the goal of sustainable energy for all in the coming decades. Placing this Energy Plus approach at the centre of energy access programmes and building on the accomplishments so far will be crucial to ensuring that the poor will not merely be limited to a single, yet insufficient, household electricity connection, but that they will instead be able to fully participate in global sustainable development.

APPENDIX A—EXAMPLES OF PRODUCTIVE
USES OF ENERGY[45]

Agriculture and livestock

- mechanical power for water pumping, transportation, and crop processing
- electric water pumps for irrigation and for fish and shrimp farming
- agricultural power tools (fodder choppers, threshers, grinders, freezers, oil expellers, etc.)
- dryers for meat, fish, edible flowers, and spices
- grain milling (electric or hydropower)
- sugar cane processing
- coffee pulping
- milk collection centres
- electric fencing for grazing management
- shearing and carding equipment
- lighting and cooling for poultry factories
- beekeeping (honey and wax production)

---

[45] Cabraal et al. 2005, p. 123; White 2002; UNDP 2011c.

Industrial or commercial use

- operating general stores, hotels, restaurants, bakeries, etc. (lighting, cooking, space heating and cooling, refrigeration)
- workshops (electronic repair, gold and jewellery, car repair, etc.)
- rubber drying
- laundry shops
- soap production
- silk reeling (using electricity or heat produced from biogas)
- silkworm rearing
- cotton ginning (separating cotton fibres from their seeds)
- sawmills
- textile dyeing and dry-cleaning
- carpentry (saws, drills, rooters)
- tailoring (irons, sewing machines)
- kiln firing for pottery and brick-making
- welding and looming
- sewing and weaving
- handicraft production
- wood-working
- coconut fibre processing (for ropes, mats, etc.)
- processing of minerals and stones (marble and slate cutting, polishing of precious stones, etc.)
- cathodic protection

Water related

- water desalination and purification
- water pumps for potable water
- ice-making

Hospitals and clinics

- lighting for improved and longer operating hours
- space cooling and heating
- water pumping, heating, and sanitation
- refrigeration of medicines, including vaccines (e.g. measles and tetanus toxoid vaccine)
- powering of medical equipment (X-ray machines, electrocardiographs, etc.)
- sterilization of medical equipment
- radio and telecommunications equipment, essential for contacting doctors and obtaining emergency sources of medicines (single-side-band radios, two-way radios, mobile phones, satellite phones)
- basic living amenities to posted health workers (thus improving retention rates)

Education centres

- lighting for improved and longer operating hours
- powering of audio-visual equipment including computers
- space cooling and heating
- provision of potable water and meals
- basic living amenities to posted teachers (thus improving retention rates).

Energy production and conversion

- battery charging

Communication

- radio and television broadcasting
- Internet and telecommunication centres (e.g. used by farmers to check market prices, to order agricultural inputs, and to obtain information on good farming practices)
- photo and photocopying services
- distance education via satellite television
- navigational aids

Miscellaneous

- community street lighting
- community centre lighting
- veterinary clinics

APPENDIX B—PRODUCTIVE USES OF ENERGY AND INCOME GENERATION

Increasing agricultural productivity

- In India, the *e-choupal* initiative provides farmers with access to the Internet, allowing them to obtain current information on market prices and good farming practices and to order agricultural inputs.[46]
- Micro-hydro plants (MHPs) used in mechanical processing such as milling, hulling, and oil expelling can save 30 to 110 hours of labour per month.[47]

Extending working hours

- Rural businesses such as grocery shops, market stalls, cafés. And tailoring shops depend on adequate lighting in order to operate after sunset. An assessment of the SHS market in Bangladesh found that more than half of rural businesses that invest in solar lighting were able to extend opening hours and increase incomes.[48]
- A modelling study across four provinces in Luzon Island in the Philippines found that access to electricity increased operating hours of household businesses by two hours per day and increased incomes by approximately 50 per cent.[49]

Establishing new businesses

- A rural electrification project in Indonesia ran activities to promote the uptake of electricity services by rural business services. As a result of this project 66,000 rural

---

[46] Bhatnagar et al. 2003.   [47] Sovacool et al. 2012, p. 717.
[48] Kürschner et al. 2009, p. 35.   [49] ESMAP 2002, p. 72.

enterprises invested in electric equipment, and the subsequent growth of these businesses led to the direct creation of 22,000 new jobs.[50]

- A modelling study in the Philippines found that electrification increases the chances that a household will engage in a home business (primarily small general stores) by 10.7 per cent.[51]

- A 120-kW Cinta Mekar micro-hydro powerplant in rural West Java, Indonesia, sells excess electricity to the National Electricity Corporation. Monthly income is shared equally between the Hidropiranti Inti Bakti Swadaya company and a local rural community cooperative. Having formed as a social enterprise, the cooperative's income from 2004–2008 was used to fund construction of a village health clinic, radio station, and telephone line as well as educational scholarships for 156 poor students.[52]

Improving education facilities

- Basic electricity in schools can improve education by providing lighting, heating, or improved ventilation. A 1986 survey in Honduras showed that, in addition to education levels of teachers and student–teacher ratios, the provision of electricity in schools is one of the main factors correlated to students' future incomes.[53]

- A 1990–91 survey in Morocco showed that a 10 per cent increase in the presence of electricity in a village would lead to a 4.8 per cent increase in school participation for boys and 8.2 per cent for girls.[54]

- A modelling study in Philippines found that, combined with electricity, education leads to higher income, even for households with the same educational levels. As a result of a household's investment in education, wage earners in households with electricity could expect to earn 50–60 per cent more per month than their counterparts without electricity. In short, the presence of electricity in a household enhanced the returns to education beyond the effects of having electricity or having attained a certain level of education. This presents compelling evidence of the complementary nature of electricity in combination with schools and educational programmes.[55]

Improving medical facilities

- Provision of electricity, heat, liquefied petroleum gas, and kerosene to rural health clinics allows a cleaner and safer environment, power for operating lights, water pumping and heating, and sanitation and sterilization of medical equipment.[56] Many doctors and nurses will not work in health clinics that lack outdoor lighting to provide for their safety.[57]

- It is difficult, if not impossible, to establish a safe and efficient health clinic without electricity or more modern energy sources. Electrical connections to health clinics help to establish and maintain health-care facilities.[58]

Employing local people in service delivery

- Under the 2002–2011 Renewable Energy for Rural Economic Development Project in Sri Lanka, construction of a single MHP employed 8–11 members of the surrounding community for 18 months.[59]

---

[50] World Bank 2000.   [51] ESMAP 2002, p. 72   [52] IESR 2011; IBEKA 2007.
[53] Bedi and Edwards 2001.   [54] Brenneman and Kerf 2002, p. 19.
[55] ESMAP 2002, p. 70.   [56] Brenneman and Kerf 2002, pp. 24–30.
[57] Brenneman and Kerf 2002, p. 26.   [58] Brenneman and Kerf 2002, p. 26.
[59] UNDP 2011d, pp. 12–13.

- China's Township Electrification Programme constructed almost 700 solar PV power stations in remote areas, and trained local entrepreneurs in each town to run small businesses that maintain the PV systems and manage service delivery.[60]

REFERENCES

AGECC (The Secretary-General's Advisory Group on Energy and Climate Change). (2010). *Summary Report and Recommendations.* New York: United Nations [online], <http://www.un.org/wcm/webdav/site/climatechange/shared/Documents/AGECC%20summary%20report%5B1%5D.pdf> (accessed 16 October 2013).

Bedi, A. and Edwards, J. (2001). 'The Impact of School Quality on Earnings and Educational Returns—Evidence from a Low-Income Country', *Journal of Development Economics* 38: 157–85.

Bhatnagar, S., Dewan, A., Torres, M., and Kanungo, P. (2003). *E-Choupal: ITC's Rural Networking Project.* Washington, DC: World Bank Poverty Reduction Group [online], <http://siteresources.worldbank.org/INTEMPOWERMENT/Resources/14647_E-choupal-web.pdf> (accessed 16 October 2013).

Brenneman, A. and Kerf, M. (2002). *Infrastructure and Poverty Linkages: A Literature Review* [online], <http://www.ilo.org/wcmsp5/groups/public/—ed_emp/—emp_policy/—invest/documents/publication/wcms_asist_8281.pdf> (accessed 16 October 2013).

Brew-Hammond, A. 2010. 'Energy Access in Africa: Challenges Ahead'. *Energy Policy* 38: 2291–301.

Cabraal, R., Barnes, D., and Agarwal, S. (2005). 'Productive Uses of Energy for Rural Development', *Annual Review of Environment and Resources* 30: 117–44.

ESMAP (Joint UNDP/World Bank Energy Sector Management Assistance Programme). (2002). *Rural Electrification and Development in the Philippines: Measuring the Social and Economic Benefits.* Washington: World Bank [online], <http://siteresources.worldbank.org/INTPSIA/Resources/490023-1120845825946/philippines_rural_electrification.pdf> (accessed 16 October 2013).

EUEIPDF and GIZ (European Union Energy Initiative Partnership Dialogue Facility and Deutsche Gesellschaft für Internationale Zusammarbeit) (2011). *Productive Use of Energy (PRODUSE): A Manual for Electrification Practitioners.* Eschborn, Germany: EUEIPDF and GIZ [online], <http://www2.gtz.de/dokumente/bib-2011/giz2011-0462en-productive-use-energy.pdf> (accessed 16 October 2013).

IBEKA (2007). 'Community Private Partnership Pro-Poor Infrastructure: Cinta Mekar Microhydro Training Power Plant'. Presentation at the Seminar on Policy Options for the Expansion of Community-Driven Energy Service Provision, Beijing, 11–12 March 2007.

IESR (Institute for Essential Services Reform) (2011). *Cinta Mekar Micro-Hydro Power Plant Giving Power to the People.* IESR, Indonesia [online], <http://www.iesr.or.id/wp-content/uploads/2case-studies.pdf> (accessed 16 October 2013).

Kürschner, E., et al. (2009). *Impacts of Basic Rural Energy Services in Bangladesh: An Assessment of Solar Home System and Improved Cook Stove Interventions.* Berlin: Seminar für Ländliche Entwicklung [online], <http://edoc.hu-berlin.de/series/sle/238/PDF/238.pdf> (accessed 16 October 2013).

Messerli, P., et al., eds. (2008). *Socio-economic Atlas of the Lao PDR—An Analysis Based on the 2005 Population and Housing Census.* Swiss National Centre of Competence in Research (NCCR) North-South, University of Bern, Bern and Vientiane: Geographica Bernensia.

---

[60] Shyu 2012, p. 844.

Nussbaumer, P., Bazilian, M., Modi, V., and Yumkella, K. (2011). *Measuring Energy Poverty: Focusing on What Matters*. OPHI Working Paper no. 42. Oxford Poverty and Human Development Initiative (OPHI), University of Oxford [online], <http://www.unido.org/fileadmin/user_media/Services/Energy_and_Climate_Change/EPP/Publications/nussbaumer%20et%20al%202011%20measuring%20energy%20poverty%20focusing%20on%20what%20matters.pdf> (accessed 16 October 2013).

Pachauri, S., Mueller, A., Kemmler, A., and Spreng, D. (2004). 'On Measuring Energy Poverty in Indian Households', *World Development* 32(12): 2083–104.

Purohit, P. (2008). 'Small Hydro Power Projects under Clean Development Mechanism in India: A Preliminary Assessment', *Energy Policy* 36(6): 2000–15.

Pillai, I. and Banerjee, R. (2009). 'Renewable Energy in India: Status and Potential', *Energy* 34: 970–80.

Practical Action (2010). *Poor People's Energy Outlook 2010*. Rugby, UK: Practical Action [online], <http://practicalaction.org/docs/energy/poor-peoples-energy-outlook.pdf> (accessed 16 October 2013).

Ramachandra Murthy, K., Manikanta, K., and Phanindra, G. (2012). 'Analysis of Low Tension Agricultural Distribution Systems', *International Journal of Engineering and Technology* 2(3).

Reddy, A. (2000). 'Chapter 2: Energy and Social Issues', *World Energy Assessment: Energy and the Challenge of Sustainability*. New York: United Nations Development Programme, pp. 41–60.

Rehman, I., et al. (2012). 'Understanding the Political Economy and Key Drivers of Energy Access in Addressing National Energy Access Priorities and Policies', *Energy Policy* 47, Supplement 1: 27–37.

Muhida, R., et al. (2001). 'The 10 Years' Operation of a PV-micro-hydro Hybrid System in Taratak, Indonesia', *Solar Energy Materials and Solar Cells* 67: 621–7.

Shyu, C. (2012). 'Rural Electrification Program with Renewable Energy Sources: An Analysis of China's Township Electrification Program', *Energy Policy* 51: 842–53.

Sovacool, B.K., et al. (2012). 'What Moves and Works: Broadening the Consideration of Energy Poverty', *Energy Policy* 42: 715–19.

UNDP (United Nations Development Programme) (2008). 'Productive Uses of Renewable Energy in Chitral District (PURE-Chitral)', UNDP Project Document [online], <http://www.thegef.org/gef/sites/thegef.org/files/gef_prj_docs/GEFProjectDocuments/Climate%20Change/Pakistan%20-%20Productive%20Use%20of%20Energy%20in%20Northern%20Pakistan/CEO%20Approval%20Request%20PURE%20MSP%20April%202008.doc> (accessed 16 October 2013).

UNDP (United Nations Development Programme) (2011a). 'Augmenting Gross National Happiness in a Remote Bhutan Community', *Towards an 'Energy Plus' Approach for the Poor: A Review of Good Practices and Lessons Learned from Asia and the Pacific (Case Study 13)*. Bangkok: NDP Asia-Pacific Regional Centre [online], <http://www.snap-undp.org/elibrary/Publication.aspx?ID=614> (accessed 16 October 2013).

UNDP (2011b). 'Energy to Move Rural Nepal Out of Poverty: The Rural Energy Development Programme Model in Nepal', *Towards an 'Energy Plus' Approach for the Poor: A Review of Good Practices and Lessons Learned from Asia and the Pacific (Case Study 15)*. Bangkok: NDP Asia-Pacific Regional Centre [online], Bangkok: NDP Asia-Pacific Regional Centre [online], <http://www.snap-undp.org/elibrary/Publication.aspx?ID=612> (accessed 16 October 2013).

UNDP (2011c). 'Going Renewable: China's Success Story in Capacity Development', *Towards an 'Energy Plus' Approach for the Poor: A Review of Good Practices and Lessons Learned from Asia and the Pacific (Case Study 6)*. Bangkok: NDP Asia-Pacific Regional Centre [online], <http://www.snap-undp.org/elibrary/Publication.aspx?ID=608> (accessed 16 October 2013).

UNDP (2011d). 'Renewable Energy Sector Development: A Decade of Promoting Renewable Energy Technologies in Sri Lanka', *Towards an 'Energy Plus' approach for the poor: A review of good practices and lessons learned from Asia and the Pacific (Case Study 11)*. Bangkok: NDP Asia-Pacific Regional Centre [online], <http://www.snap-undp.org/elibrary/Publication.aspx?ID=613> (accessed 16 October 2013).

UNDP (2011e). 'Renting Lighting Services: Paying for the Service and Not the Hardware', *Towards an 'Energy Plus' Approach for the Poor: A Review of Good Practices and Lessons Learned from Asia and the Pacific (Case Study 14)*. Bangkok: NDP Asia-Pacific Regional Centre [online], <http://www.snap-undp.org/elibrary/Publication.aspx?ID=615> (accessed 16 October 2013).

UNDP (2011f). *Towards an 'Energy Plus' Approach for the Poor: A Review of Good Practices and Lessons Learned from Asia and the Pacific*. Bangkok: NDP Asia-Pacific Regional Centre [online], <http://www.snap-undp.org/elibrary/Publication.aspx?id=600> (accessed 16 October 2013).

UNDP (2011g). 'Turning Tradition to New Ends: Improving Water Mills in Nepal', *Towards an 'Energy Plus' Approach for the Poor: A Review of Good Practices and Lessons Learned from Asia and the Pacific (Case Study 15)*. Bangkok: NDP Asia-Pacific Regional Centre [online], <http://www.snap-undp.org/elibrary/Publication.aspx?ID=610> (accessed 16 October 2013).

UNDP (2012). United Nations Development Programme. 'Entrepreneurship at Grassroots: Diffusion of Biomass Devices in Informal Industries in India', *Towards an 'Energy Plus' approach for the poor: A Review of Good Practices and Lessons Learned from Asia and the Pacific (Case Study 15)*. Bangkok: NDP Asia-Pacific Regional Centre [online], <http://www.snap-undp.org/elibrary/Publication.aspx?ID=616> (accessed 16 October 2013).

UNDP Nepal (2012). 'Renewable Energy for Rural Livelihood (RERL)' [online], <http://www.undp.org/content/nepal/en/home/operations/projects/environment_and_energy/rerl/> (accessed 16 October 2013).

UNDP Pakistan. (2012). 'Productive Use of Renewable Energy (PURE)' [online], <http://undp.org.pk/productive-use-of-renewable-energy-pure.html> (accessed 16 October 2013).

UNDP and WHO (World Health Organization) (2009). *The Energy Access Situation in Developing Countries: A Review Focusing on the Least Developed Countries and Sub-Saharan Africa*. New York: UNDP.

White, R.D. (2002). 'GEF/FAO Workshop on Productive Uses of Renewable Energy: Experience, Strategies, and Project Development', workshop synthesis report [online], <http://siteresources.worldbank.org/EXTRENENERGYTK/Resources/5138246-1237906527727/5950705-1239294026748/GEF1FAO0Worksh1ent0Synthesis0Report.pdf> (accessed 16 October 2013).

World Bank (2000). *Implementation Completion Report: Indonesia Second Rural Electrification Project*. Report #20676. Washington, DC: World Bank [online], <http://documents.worldbank.org/curated/en/2000/09/729174/indonesia-second-rural-electrification-project> (accessed 16 October 2013).

20

Unlocking Financial Resources

Xia Zuzhang

In the 2011 edition of the *World Energy Outlook*, the International Energy Agency (IEA) estimated that about USD1 trillion investment would be needed to achieve universal access to modern energy services by 2030. In other words, an annual investment of $48 billion on average is needed from 2010 to 2030 to provide adequate energy services to billions of people without access to electricity and clean cooking facilities in developing countries, primarily in Sub-Saharan Africa and developing Asia. In terms of technical solutions, roughly 90 per cent of such investment would be needed for electricity generation, transmission, and distribution—including on-grid, mini-grid, and isolated off-grid solutions—and around 10 per cent for access to clean cooking facilities, including liquefied petroleum gas (LPG), biogas systems, and advanced biomass cookstoves (IEA 2011). Estimates from other sources indicate a similar scale of capital investment required for universal energy access (Bazilian 2010).

20.1. INVESTMENT NEEDS AND FINANCING SOURCES TO ADDRESS ENERGY POVERTY

The IEA has estimated that to provide the yearly $48 billion as required for universal access to modern energy services, $18 billion per year needs to be sourced from multilateral and bilateral development organizations, and $15 billion per year is required each from the governments of developing countries and from the private sector. In comparison, global investment in extending modern energy services is estimated at $9.1 billion, of which the largest portion of 48 per cent ($4.4 billion) was sourced from multilateral organizations and bilateral official development assistances, 30 per cent ($2.7 billion) from governments of developing countries, and 22 per cent ($2 billion) from private investors (IEA 2011). In the New Policies Scenario of the *World Energy Outlook 2011*, which assumes that recent government commitments are implemented in a cautious manner and primary energy demand increases by one-third between 2010 and 2035, the IEA has projected that an annual average of $14 billion would be invested in energy access worldwide during the period of 2010–2030, falling short of the required total (referred to as the 'Energy for All' case) by $34 billion per year. A breakdown of this required investment by region and technology can be found in Figure 20.1.

Investment in energy access projects involves a variety of public- and private-sector players, including multilateral development banks, bilateral development assistance agencies, export–import banks or guarantee agencies, governments of developing countries, state-owned utilities, national development banks, rural energy agencies or funds, foundations, microfinance institutions, local and inter-national commercial banks, investment funds, and private investors. Each of them uses different financial instruments, such as grants, subsidies, equity, loans, insurance, and guarantees, to invest in energy access projects. Table 20.1 shows a brief overview of various sources of financing and the traditional financial instruments relevant to energy access projects.

The total amount of $1 trillion to achieve universal access to modern energy services represents about 3 per cent of worldwide investment needed in the energy sector by 2030, or around 1.75 per cent of the projected amount of $57 trillion to be spent on roads, bridges, ports, power plants, water facilities, and other forms of infrastructure by 2030 (McKinsey 2013). The average annual investment of $48 billion only corresponds to 12 per cent of all global subsidies for fossil fuels consumption in 2010[1] (IEA 2011). However, it is widely considered to be challen-ging to secure such a scale of funding for universal energy access initiatives, which primarily target the population at the bottom of the economic pyramid featured by lower income levels, low ability to pay, low levels of consumption, or geographic dispersion. It is therefore critically important to create an enabling environment and remove various investment barriers to unlock financial resources, so as to facilitate universal access to clean, reliable, affordable, and sustainable energy services for the poor and remote populations in developing countries.

20.2. REMOVING BARRIERS TO INVESTMENT IN ENERGY ACCESS

All the players involved in tackling energy poverty worldwide have their own constraints and faces various barriers that limit their capacity and willingness to invest in energy access projects. The major barriers may fall into the categories of policy and regulatory, institutional, technological, and financial aspects, among others such as economic barriers, market barriers, social and cultural barriers, and so on.

20.2.1. Policy and regulatory barriers

Governments play a crucial role in the provision of an enabling environment for access to energy to the poor. Although the climate is growing more favourable in

---

[1] The IEA estimated that fossil-fuel consumption subsidies amounted to $409 billion in 2010, almost half of the total for oil products (IEA 2011).

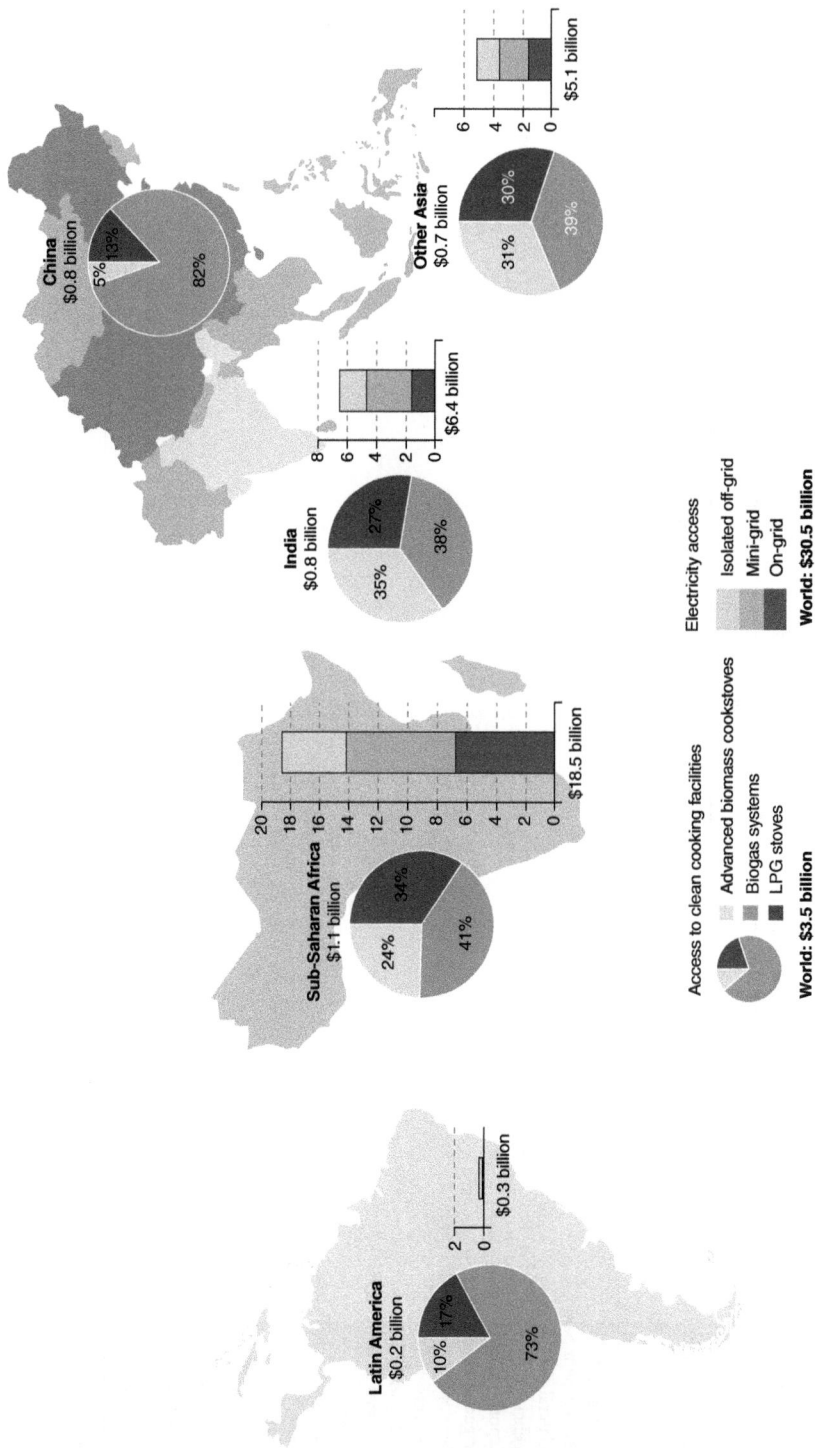

Figure 20.1. Average annual investment (in 2010 USD) required by region and technology in the Energy for All case, 2010–2030

Source: IEA 2011

Table 20.1. Sources of financing and financial instruments for energy access projects

Sources of financing	Financial instrument											
	Grants/ credits	Concessionary loans	Market-rate loans	Credit line for on-lending	Partial credit guarantees	Political risk insurance	Equity	Quasi-equity	Carbon financing	Subsidy/ cross-subsidy	Feed-in tariff	Technical assistance
Multilateral development banks	X	X	X	X	X	X	X	X	X			X
Bilateral development agencies	X	X	X	X					X			X
Export–import banks or guarantee agencies			X			X						X
Developing-country governments	X	X					X	X		X	X	
State-owned utilities							X			X	X	
National development banks		X	X	X	X							X
Rural energy agencies/ funds	X									X		X
Foundations	X						X		X			
Microfinance institutions			X									
Local banks			X									
International banks			X				X	X	X			
Investment funds							X		X			
Private investors							X		X			X

Source: IEA 2011

most countries, existing policies and regulatory frameworks are often a major barrier—they can be vague or uncertain, inconsistent or contradictory, or even hostile. Taxes and subsidies may undermine markets rather than encourage them. Supporting infrastructure may not be accessible or even exist. Certain players in the economy may be able to gain privileged access to subsidies that enable them to sell their products below cost (Chineyemba 2008). While government subsidies for particular energy sources or services help access to energy for the poor, the subsidies may also cause market distortions and disadvantages for some energy technologies or market players. The weak or unclear policy and regulatory environment creates uncertainty for project investors and developers.

In many countries where energy poverty is most acute, the nature of the regulatory and investment climate discourages business from investing, particularly in larger-scale infrastructure and capital equipment. This can be exacerbated by a lack of political will for reform and insufficient public investment in the energy system, especially the elements that support expansion of services to poor communities (WBCSD 2012).

To remove these kinds of barriers, governments of developing countries need to set up clear and consistent long-term policies and regulations on energy provisions to guarantee investment certainty and project viability. Without significant changes in such policies and in the regulatory environment, the flow of private-sector finance into energy access will be restricted.

20.2.2. Institutional barriers

Action to improve access to energy services requires adequate institutional capacity and effective coordination at national and local levels. Such capacity includes the formulation of effective policies, the establishment of incentive mechanisms, the provision of financial services, and project management. In this regard, institutional capacity within local governments in conjunction with the local private sector, civil society, and community-based organizations is especially important (PAC 2008).

In many cases, energy access services require decentralized solutions in which some or all of the major elements—resources exploration, engineering design, construction and installation, maintenance and operation, distribution and billing, and customer financing and governance—are handled at local levels (PAC 2008). This requires sound institutional capacities that may not be present in remote and poverty-stricken areas. To ensure the long-term availability of the service and encourage responsible behaviour and empowerment on the part of local users, the institutional aspects must be built into projects at the earliest possible stage to prevent any deterioration in the quality of the service and to avoid a hand-out situation. For example, the operation and maintenance of installations can be entrusted to clearly identified players with the capabilities to handle these roles. In some cases, the establishment of public–private partnerships can be a better, more feasible option to help public authorities fully assume their role and to put beneficiary communities in charge of organizing their own services.

20.2.3. Technological barriers

Extension of electricity grids to cover off-grid areas is often the most powerful and straightforward means to facilitate access to modern energy. However, it can be very costly to extend national power grids into remote and sparsely populated areas. In places where extension of centralized national grids is prohibitively expensive and not feasible, either financially or economically, small-scale electricity generation systems can be more effective and attractive. In this case, decentralized electrification solutions such as mini-grid or stand-alone power generation systems can be an option.

For off-grid electricity generation, small or micro-hydropower generators, small wind power generators, and solar PV can be found in many places. While providing an electricity solution for the isolated or remote sites, most of these technologies are only applicable at sites with adequate resources, and these resources are often available sparsely and intermittently. Some installations may not be at sufficient scale to provide certain required local energy services, such as shaft power; some other installations may need substantial up-front investment and regular maintenance by the users. This limitation imposes a major constraint on the electricity generating capacity of smaller and cheaper installations and makes electricity supply relatively vulnerable. Advances in technologies have, however, increased efficiency and reduced costs, opening up options for profit-oriented small companies.

For improved cooking solutions, the use of LPG can be clean and efficient compared to many traditional stoves that burn wood, charcoal, or agricultural wastes. However, the costs to establish delivery networks, not to mention up-front investment required for the purchase of LPG cylinders and gas cookers, can be significantly more expensive, and the running costs can be higher compared to burning traditional biomass fuels in areas where there is sufficient availability and accessibility of firewood and agricultural biomass.

Though technically not very complicated, advanced biomass cookstoves still need to be well adapted to local practices and social and cultural traditions, which may vary significantly across communities and regions. To ensure implementation of new technology, effectively involving local actors and supporting local solutions can be critically important. Many community-based non-governmental organizations have proved their value for engagement in such interventions.

Renewable energy sources are more sustainable and in general more environmentally friendly than fossil fuels in long run. They are better applicable as decentralized energy solutions to rural areas in developing countries with relatively lower energy-consumption intensity. The technologies, however, may not always be socially sustainable or pollution-free. For example, users of solar photovoltaic installations have often failed to renew batteries because the limited energy services the system provided, say for lighting and television only, do not well meet users' expectations at the cost of replacing batteries (PAC 2008), so in the end some small systems may be discarded inappropriately or not utilized at full capacity.

Technological solutions for energy access therefore have to be well tailored to the needs and contexts of specific communities in terms of technical, financial, natural, and cultural circumstances, even at the expense of reduced performance and less elaborate technical features. In prioritizing technology options, it is important to sufficiently consider how technologies could adapt to and take advantage of the contexts of the target areas.

20.2.4. Financial barriers

Creating access to modern energy services in developing countries would require contributions from different funding sources—governments, banks and other financial institutions, international donors, cooperation agencies, and also the actual users of the services.

Public finance has been playing a dominant role in addressing energy poverty. In many cases, extending sustainable modern energy services in rural and peri-urban areas is impossible without incentives such as grants, subsidies, or concessional finance to lower initial capital costs. However, public finance alone will not be able to deliver the investment required to cover most remote locations, particularly considering the limitations of available capital and budget constraints to the governments of developing countries and the competing needs for such capital to support infrastructure development for economic growth, social welfare, and political stability, as well as other national and local priorities. Legacy fiscal measures, such as heavy subsidies to fossil fuels, further restrain the capacity of governments in some countries to finance energy access initiatives. There are, however, encouraging signs with respect to achieving the necessary political commitment from the governments of developing countries. More than 50 governments of developing countries have joined the Sustainable Energy for All (SE4ALL) initiative, including seven countries with large populations lacking access to electricity and clean cooking facilities—Bangladesh, Indonesia, Nigeria, Ethiopia, DR Congo, Tanzania, and Kenya (SE4ALL 2013). This indicates a certain level of political commitment from the governments and would have a positive impact in mobilizing awareness to tackle this issue.

Multilateral financial institutions have been important players in supporting economic development and social advancement, including tackling energy poverty. Involvement of international institutions helps governments and the private sector to pursue development objectives. However, some of these institutions are generally structured to provide large-scale financing for multi-million-dollar projects, often focused on the construction of thermal or hydropower plants or the extension of electricity grids. This conventional approach may still leave large numbers of people, especially in rural areas far from transmission lines, without access to electricity.

The UN Secretary-General's Sustainable Energy for All initiative has been vital in raising public awareness of the urgent need to increase modern energy access for the poor worldwide. Following the UN declaration of 2012 as the International Year of Sustainable Energy for All, the political momentum is likely to continue over the coming years with the UN declaring 2014–2024 as the Decade of Sustainable Energy for All. This underscores the importance of energy issues for sustainable development, but the much-needed financing is not yet in place. The continuing economic downturn and financial difficulties in many developed countries in recent years have dimmed longer-term outlooks for providing funding to developing countries through official development assistance. For example, the energy access funding commitments received by the time of the Rio+20 Summit in 2012 only accounted for around 3 per cent of the needed $1 trillion in cumulative investment, and the portion earmarked for energy access projects only accounted for around 10 per cent of the committed fund (IEA 2012). Clearly, more international attention must be brought to improving modern energy access. The critical role of energy in the development process needs to be better recognized with stronger commitments and actions from the international community.

The involvement of the private sector directly in project development or through supplying goods and services has enabled millions of households to gain access to modern energy services. Globally, the poor spend $37 billion on poor-quality energy solutions to meet their lighting and cooking needs, which represents a largely untapped market opportunity for the private sector (IFC 2012). However, the return on investment in energy access for the poor is generally not competitive, so private players often face a trade-off between profitability and corporate social responsibility. Typically, multinational companies require business models to command the level of revenue and profit that meet both the financial bottom line and the triple bottom line (social, economic, and environmental). Such viable business models for 'base of the pyramid' customers are rarely available to reach the poorest people in remote communities who lack energy access. On the other hand, there is in general a lack of early-stage debt and equity financing for smaller energy companies seeking to commercialize their technology. Developers of small and medium-sized projects find it difficult to obtain capital from financial institutions at affordable rates. Partially due to the limited sound data available to investors, financiers, developers, and policymakers, developers of energy access projects have been experiencing challenging difficulties in financing their projects.

Some private financial institutions have begun to include social agendas in their lending criteria. However, many have not yet included access to energy as a means of promoting social and economic development and alleviating poverty. In the long run, private financial institutions may develop their policies in favour of addressing energy access. However, significant changes may not happen in the short term, as macro-level policy and regulatory barriers would not change overnight, and institutional and technical barriers may not be easily or effectively removed.

End-user financing is another impediment to increasing access to energy. Off-grid households usually spend a much higher proportion of their incomes—and often more in absolute terms—on energy than households connected to the electricity grids and/or public gas pipelines. Even if some poor households have the ability to pay for needed energy services, they may not be able to pay for grid connection fees or the upfront costs for installation of off-grid systems. In many cases, lack of end-user financing prevents some poor households within the range of the national grids from connecting to electricity or gas supply utilities. Micro-finance to the end-users has a role to play in this context and has proved effective with many good practices. There are, however, numerous identified barriers for scaling up energy lending through microfinance institutions, such as lack of coordination between the energy and microfinance sectors, inflexible microfinance lending models, lack of low-cost capital to lend for energy projects, insufficient business motivation to promote energy access products, and market-distorting subsidies. Very few commercial banks lend for small-scale energy investments, which are generally considered as not bankable, either because borrowers are not considered creditworthy or because the investments are not seen as productive. Without a significant breakthrough, end-user financing will continue to be a fundamental constraint to scaling up access to energy, especially for poor and remote populations. From a financial perspective, the cycle of energy poverty may only be broken if it is possible to combine investments to improve energy services with associated investments that generate income through productive activities—either by increasing productivity, by extending the range of outputs, or by improving output quality.

20.3. INNOVATIVE FINANCIAL MECHANISMS

Financing is a major stumbling block in achieving universal energy access. Traditional financial instruments, such as grants, subsidies, equity, loans, credits, insurance, and guarantees, are all important means of addressing energy poverty, though with mixed performance. As many traditional financial institutions are not geared to address the specific risks and relatively low returns of investment in energy access projects, innovative financial mechanisms need to be further developed through coordinated efforts from multiple stakeholders in order to supplement financing gaps and improve the effectiveness of conventional financing. A number of innovations have been developed to address financial difficulties thanks to the combined efforts of conventional market forces together with the public and private sectors and non-governmental organizations.

20.3.1. Smart subsidy

A subsidy is a financial mechanism to improve market outcomes by providing resources to the poor and underprivileged and/or by correcting market failures. Subsidies often take the form of direct payments, price cuts, or tax deductions to target social groups or the private sector in order to promote greater inclusion of the most vulnerable for social equity and/or to improve market efficiency. Subsidies are justifiable in an economic sense when the social benefits go beyond the private benefits associated with a given action. While having the potential to bring in a more equitable distribution of economic well-being than that generated by an unfettered free-market economy, subsidies are also under debate as many schemes miss their intended target group, lead to market distortions, or have to provide the cost-sharing to cover recurrent costs of certain market actors. This raises the question of how subsidies can be made smarter.

There is no unanimously agreed definition for 'smart subsidies' or as they are sometimes called 'least cost subsidies', though some general principles and rules may explain how to make subsidies more efficient and effective. Smart subsidies should be transparent, rule-bound, time-limited (Morduch 2005), and easily accountable. They are expected to catalyse systemic change that stimulates access to essential services; to accelerate technology adoption without distorting the behaviour of the private supply; to correct market distortions and create new markets; to create conditions to leverage additional investments from social, commercial, philanthropic, and civic sources; and to lead to the creation of sustainable financial solutions that will not require additional grants over time (Elisberg 2010). A subsidy can be considered effective if it may be withdrawn without the end-user noticing the impact either in price of products or the level of services. Smart subsidy schemes have been used by projects in various sectors so far, such as agriculture, water supply (WaterCredit 2013), public housing, and rural telecommunications (Silva and Tuladhar 2006). It may be further developed and adapted to energy access projects.

20.3.2. Output-based aid

Output-based aid (OBA), also known as result-based or performance-based subsidy, is a strategy that links the payment of aid to the delivery of specific services or outputs. The ultimate aim is to increase the effectiveness of scarce public resources for the provision of basic services. For addressing energy poverty, OBA is usually provided in the form of grant subsidy or a concessionary loan designed to address the gap between what poor users typically can afford for energy services and the real costs to provide quality energy services. The use of OBA in the energy sector is most widespread in individual systems for rural electrification. The predominant technology used in off-grid projects is solar home systems. Other technologies include small wind generators, micro- or pico-hydropower, and biogas (Mumssen 2010).

Under an OBA scheme, service delivery is contracted out to a third party— sometimes public or non-governmental though in many cases a private firm—as the service provider. The service provider is responsible for the pre-financing of the project, and only after service has been delivered and verified by an independent agent does the firm receive subsidies from the donor (GPOBA 2009). In such schemes, the service provider, rather than the aid donor, makes an initial investment and bears the risk of loss, while the target beneficiaries can choose between multiple service providers. This differs from traditional aid schemes that usually focus on the inputs to service providers rather than the outputs, and therefore it helps provide incentives for innovation in projects to mobilize expertise and finance from the private sector.

Some key factors have been identified to help make OBA projects successful. These include a sound regulatory environment, a reliable and motivated service provider, close links between the payment of subsidies and the outputs, recovery of operational and maintenance costs, and availability of funds to pre-finance the service delivery either from the operators' own resources or from banks (GPOBA 2009). OBA schemes are often limited by high administrative costs and the lack of well-established markets. So, depending on the region in question, additional efforts are frequently needed to develop this financial mechanism.

20.3.3. Guarantee and risk-sharing

Use of public funds as a guarantee is a risk-sharing mechanism to leverage private capital investment. Innovative approaches such as partial risk-sharing mechanisms, credit guarantees, and co-financing are particularly important in de-risking private investments (SGHG-SEFA 2012).

In case some energy access projects are too small for public financial institutions to finance directly, local microfinance institutions may act as the aggregator of these projects. Instead of providing direct financing or grant funding, public financial institutions may offer local financial institutions partial credit guarantees. The guarantee funds from the public sources may be used in a first-loss position to offer more attractive risk-sharing instruments, especially if these projects can be justified for subsidy. In this case, borrowers can obtain loans more easily, particularly medium- and long-term loans for business start-ups and business development. Local financial institutions can also reduce both their risks and their need for additional guarantees, which are often costly, uncertain, or not available at all.

20.3.4. Micro-credit to end-users

Affordability for improved energy services stands to be increased dramatically with credit facilities made available to end-users. Micro-credit is the provision of small amounts of credit to clients—generally those that lack assets as guarantee—who are under-served by traditional formal banking institutions. The credit to the low-income end-users may be provided by dealers themselves, by development banks, or by microfinance institutions. Local micro-credit institutions are usually closer to rural customers than commercial banks, enjoy lower transaction costs, and are more capable of dealing with the financing aspects than technology suppliers or dealers (ESCAP 2005).

An effective way to convert capital costs into affordable operating costs is through the investments of financial institutions and the private sector. One way to overcome this constraint is for villagers to form groups that are responsible for ensuring that each member of the group repays the loan. Another option is to lend to intermediary non-governmental organizations that are more likely to recover loan repayments than commercial organizations. A third option may be to encourage energy technology companies to lease basic equipment to rural households and communities (ESCAP 2005).

20.3.5. Carbon financing

The Clean Development Mechanism (CDM) under the Kyoto Protocol established carbon markets as a source of finance and investment to projects that reduce greenhouse gas emissions in developing countries. The verified emission reductions (carbon credits) associated with these projects can be traded in markets at regional and national levels to generate financial assets, which are mostly referred to as carbon finance or climate finance. A wide range of project types can qualify under the CDM, from wind and solar energy projects that earn credits by displacing electricity generation from the burning of fossil fuels, to projects that install more efficient cookstoves. The CDM Policy Dialogue Report released in late 2012 estimated that the CDM has mobilized $215 billion investments in developing countries over the last decade (CDM Policy Dialogue 2012). With more than 6,000 registered projects in 83 developing countries (UNFCCC 2013), the CDM has proven to be a powerful mechanism to deliver finance for emission-reduction projects.

By creating a commercial value for reducing greenhouse gas emissions, carbon markets provide an additional source of revenue for sustainable energy projects. This increases the commercial viability of a project, and therefore plays an important role in sustaining and growing the enterprise. Within this context, carbon finance can be an opportunity. However, there are significant barriers for small-scale energy access projects to obtain carbon finance, due to the cumbersome process and significant costs involved in the development, transaction, monitoring, and verification of CDM projects (Disch, Rai, and Maheswari 2010).

The closing of the phase-one commitment period of the CDM under the Kyoto protocol by the end of 2012, along with the economic and political environment in both developed and developing countries, has brought in negative signals to carbon finance in the coming years. However, carbon finance is expected to remain a key instrument for catalysing finance for low-emissions development, as more countries seek to establish domestic carbon markets and to use carbon pricing to achieve their climate-related objectives.

20.3.6. Public–private partnership

Public–private partnership (PPP) is an arrangement between a government entity and a private entity established for the purpose of providing essential services to the public. This arrangement can provide the services more efficiently and at a lower cost to the end-users than either entity could provide on its own.

PPP involves a contract between a public sector authority and a private party, in which the private party provides a public service and assumes substantial financial, technical, and operational risks in the project. It tries to allocate the risks of the venture fairly between the private and government entities, based on each entity's ability to manage these risks, and to provide rewards to each party based on the risks they have assumed. A key motivation for governments considering PPPs is the possibility of bringing in new sources of financing to fund public infrastructure and service needs. The operational models of PPPs include technical assistance, management contracts, lease contracts, concession contracts, build-operate-transfer, build-own-operate-transfer, and guarantees (Adenikinju 2008).

Public–private partnerships can combine the strengths of governments with the technical and financial strengths of the private sector. This mechanism has been employed as an innovative way to meet the challenges of widening access to energy services in rural areas, though numerous barriers exist to scaling up such a model, particularly the pro-poor elements with energy access projects.

Some of these aforementioned financial mechanisms may overlap each other or be applied simultaneously in a given project. In addition, many other mechanisms are available and applicable in certain contexts. New innovative financial mechanisms developed in other sectors may also be adapted to the needs of addressing energy poverty. It is noteworthy that while innovative financing mechanisms help address the constraints and risks that hinder both public- and private-sector investment in energy access initiatives, most of them need to be further explored and developed for more effective project financing. Each of these mechanisms has its own limitations and so can only be best applied under certain contexts. Also, these financial mechanisms are not geared to transform a poorly planned or managed project into a good one, so due diligence is always required to mitigate investment risks.

20.4. GOOD PRACTICES IN FINANCING ACCESS TO MODERN ENERGY

Many countries have successfully developed business models and financing mechanisms for the provision of energy products and services to the poor. Such good practices have hinged on market development driven by sound national commitments and multi-stakeholder engagement. These practices may be replicated or adapted into other countries to unlock the financing sources in support of energy access efforts. With many good practices worldwide in financing energy access, it would be helpful to pick up a few cases led by the public sector, the private sector, and NGOs and applied for different technologies.

20.4.1. Light for All programme in Brazil

The Light for All programme ('Luz para Todos' in Portuguese) was initiated by the government of Brazil in 2003 with an ambitious goal of providing universal access to electricity to the 12 million people who lived without it at that time, including 10 million people in rural areas. While the programme was running, new households without access to electricity were identified mainly in north and north-east regions and in areas of extreme poverty; the programme was therefore extended to 2014. By early 2013, this programme had benefited more than 14.7 million people throughout the country. The total investment so far has been about BRL20 billion, or roughly $10 billion, of which about 70 per cent has been provided by the federal government. The rest of the investment is shared between state governments and the companies that distribute electricity (Luz para Todos 2013).

This programme is coordinated by the Ministry of Mines and Energy, managed by Eletrobras and its subsidiaries, and implemented by electric utilities and rural electrification cooperatives in partnership with state governments. At the state level, the programme is administered by management committees whose members include representatives of federal and state governments, state regulatory

agencies, electricity distribution companies, city halls, and civil society. Programme agents play an important role in working with the target communities
to identify the specific requirements of each location. The construction work is
performed by energy distributors or rural electrification cooperatives.

The technical solutions of the programme include power grid extension, mini-
grids with decentralized power generation systems, stand-alone off-grid systems,
and hybrid systems. It was noted that more than 7.3 million power poles were
used, including 13,300 made of light-weight, floating fibreglass materials for easier
transportation by water—as roads are often impassable in the Amazon region of
the country. One of the major challenges of the programme is to serve isolated
communities, especially those located in remote and difficult-to-access areas in
the Amazon region, where extending the conventional electricity grid is not
feasible. To tackle this, the programme developed the Manual of Special Projects
to meet remote communities' particular needs with established technical and
financial criteria for the development of renewable energy sources. This includes
a subsidy of up to 85 per cent from the federal government. The technological
options for those remote areas cover mini- and micro-hydropower generators,
water mills, thermal power plants using biofuels or natural gas, solar PV systems,
wind turbines, and hybrid systems. An impact survey conducted in 2009 showed
that 79.3 per cent of the beneficiary households covered by this programme
acquired televisions, 73.3 per cent had a refrigerator in their homes, and 24.1
per cent bought water pumps (Luz para Todos 2013).

Differently from previous electrification initiatives, which demanded the beneficiary to share part of the installation and connection expenses, the federal
government made the electric installation and connection free to the beneficiaries
under the Light for All programme. The households have to pay only for what
they used based on power meters, like households in cities (Luz para Todos 2010).
The financial instruments used in this electrification programme are primarily
grant subsidies from the governments at federal and state levels as well as direct
investment from state-owned power utilities. With a total investment of
$10 billion, the programme helped 14.7 million people to access electricity. This
makes the electrification costs $680 per capita in average, which may reflect the
difficulties of electrifying the remote areas. In comparison, the electrification costs
were estimated at around $1,000 per connection in some studies (Bazilian 2010).
Also worth a note is the significant size of investment from the federal government at around $7 billion under this programme, compared to government
revenues of $911.4 billion in 2012 (CIA 2013).

20.4.2. Solar home system dissemination in Bangladesh

Around 60 per cent of the population in Bangladesh, or more than 90 million
people, did not have proper access to electricity as of 2009 (IEA 2011). Though the
situation has changed now with around 60 per cent of its population having access
to electricity, including power generated from solar photovoltaic systems, there
are still 40 per cent who do not have access (MPEMR 2013). Traditionally, most
rural dwellers rely on kerosene or candles for lighting, which are costly, give
negligible light, and emit fumes. The national electricity grid does not cover most

areas outside cities and is prone to frequent blackouts. Solar home systems (SHSs) were identified by the government a decade ago as an option for rural electrification in off-grid areas.

In 2003, the government initiated efforts to introduce SHS with funding support from the World Bank. Following the initial success, other development assistance organizations such as KfW, GTZ, the Asian Development Bank, and the Islamic Development Bank extended their support to this programme. By the end of March 2013, $400 million had been invested and 2.1 million SHSs had been installed with a total power generation capacity of around 100 MW. These SHSs benefit 10 million people or around 7 per cent of the total population, estimated as saving 30,000 tons of kerosene a year worth $110 million. The annual installations of SHSs and the grant subsidies during 2003–2012 are shown in Figure 20.2.

SHSs are stand-alone small systems usually with an electricity generating capacity of between 10 and 120 Wp, among which the 50 Wp module is the most popular. The system consists of a photovoltaic (PV) module, a rechargeable battery, a charge controller, fluorescent lamps or LED lights, and wiring and fixtures. A SHS with a PV module of 50 Wp can power four lamps or LEDs for four hours per day as well as a radio and a phone charger. Larger systems may also be able to support a TV for several hours a day. The cost of an average 50 Wp SHS is currently about $380, which is roughly equivalent to half the average annual income of a rural household in Bangladesh. The high up-front cost, often out of reach of the average villagers, and lack of household financing have been among the major barriers for its dissemination in the impoverished rural areas.

To tackle these challenges, the government of Bangladesh acquired grants, soft loans, and technical assistances from multilateral and bilateral development agencies, and channelled these resources through the government-owned financial institution Infrastructure Development Co. Ltd (IDCOL). IDCOL manages

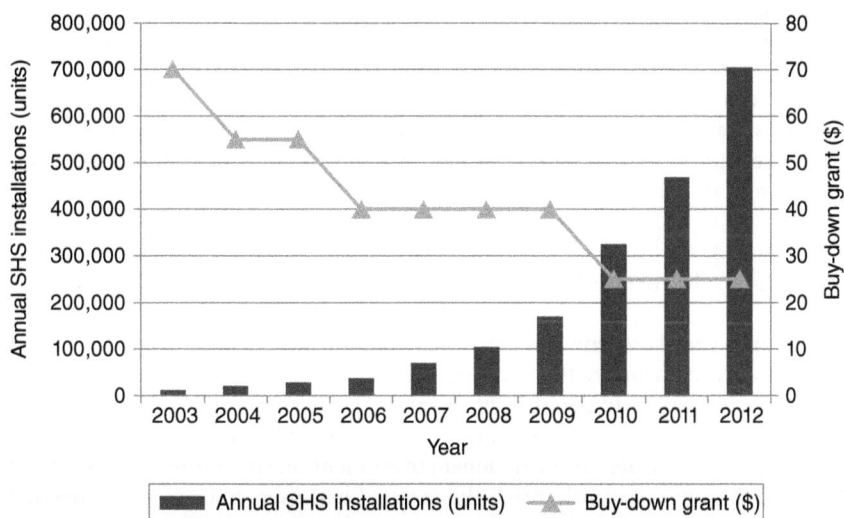

Figure 20.2. Annual installations and buy-down grants for SHS in Bangladesh, 2003–2012

Data sources: IDCOL 2011, Malik 2013, Haque 2012, Hamid 2013

the SHS programme and provides grants and soft loans to its partner organizations (POs). Upon completion of SHS installations, IDCOL claims from donors
the funds used to finance the installations in a phased approach. The POs are
responsible for project implementation, including selection of areas and customers, agreements with households and SHS suppliers, installation of SHS, extending
micro-credit to customers, and providing after-sales services. After the installation
approved by IDCOL, the POs can apply to IDCOL to refinance the customer
credit loans of up to 80 per cent. On receiving the funds from IDCOL, the POs pay
back the credit received from the suppliers (Kumar and Sadeque 2012). During
the journey of the programme, grant subsidies in the forms of buy-down grants to
end-users and institutional development grant to POs were gradually reduced
over time (see Figure 20.2).

To better illustrate the process, the installation of a typical 50 Wp SHS may be
used as an example. The cost of a 50 Wp SHS is currently about $380. The
government provides a grant of 10 per cent to buy down the cost to about $340.
The households may opt for a down-payment of 15 per cent at $50 and acquire a
customer credit loan for the remaining 85 per cent at $290. The customer credits
are made available by POs through their own capital or credits from SHS
suppliers. The households may pay back the loan to the PO in monthly instal-
ments of about $11 over three years at the market interest rate of 12 per cent per
year. A more rapid re-payment scheme is to make a 25 per cent down-payment
and repay the loan charged at a lower interest rate over two years. Alternatively, a
discount is offered for cash purchases (IDCOL 2011).

This Bangladesh SHS programme is widely considered to be the most successful
of its kind in the world and also the largest one in terms of dissemination scale.
Various financing mechanisms are applied, such as grant subsidies to buy down
the costs, concessional refinancing to operational partners, micro-credits to end-
users, and output-based aid from donors. The government plans to have 4 million
SHS units installed by 2015 to benefit 20 million people, or 12 per cent of the total
population of Bangladesh (Malik 2013). It was also planned to completely phase
out the grants in the future for the commercialization of SHS (IDCOL 2011).
Given the current pace of SHS dissemination, the satisfactory level of customers,
and the small portion of grant subsidies, it is very likely that this SHS programme
would be run on market basis in a more sustainable way, or less dependent on
grants from donor organizations.

20.4.3. Kerosene-to-LPG conversion programme in Indonesia

For many years kerosene has been heavily subsidized by the government of
Indonesia as the primary fuel for households and micro-businesses. The size of
subsidies on kerosene in the early 2000s ranged between 9 per cent and 18 per cent
of total government expenditures and reached around $4 billion in 2007, becom-
ing a significant burden to the national government. Furthermore, the subsidized
kerosene did not necessarily reach the targeted end-users but was often diverted
into areas of misuse, such as mixing with other non-subsidized fuels in the
industrial and commercial sectors (Pertamina 2012). Reducing the subsidy by

increasing the kerosene price, however, has long been a sensitive social issue with the potential of disturbing the country's stability.

In 2007 the Government of Indonesia initiated a massive programme with a target of switching the primary cooking fuel from kerosene to LPG for more than 40 million households and small and medium enterprises. The goals were to (1) greatly reduce the petroleum fuel subsidy; (2) divert kerosene into other more beneficial uses such as jet fuel for aviation use; and (3) provide a cleaner and healthier fuel to the end-users. This programme reached about 54 million households and micro-businesses by mid-2012 and is still expanding to reach more. The programme involves an initial investment of $1.4 billion from the government, leveraged $1.9 billion investment from the business sector, and is targeted to achieve a gross subsidy saving of $6.9 billion to the government, or a net savings of $5.5 billion (Pertamina 2012).

The key public sector players involved in the programme include the Ministry of Energy and Mineral Resources for overall coordination and the national oil company Pertamina for execution. Pertamina finances all aspects of the programme and then is reimbursed from the government. The programme initially planned to convert 42 million households and micro-businesses nationally, but adjusted the goal later to reach 58 million units across 23 provinces of the country by 2013. All communities meeting the programme requirements are eligible to receive the free initial LPG package, consisting of a 3 kg cylinder, a first gas-fill, and a one-burner stove, a hose, and a gas pressure regulator (see Figure 20.3). Priorities were given to areas with LPG-ready infrastructure and high

Figure 20.3. Distribution and delivery of LPG cylinders in West Java Province, Indonesia
Photo: Xia Zuzhang, 2011

consumption of kerosene. At the initial phase, the programme held consultations with local governments, local NGOs, and the communities to test the market on a small scale. Then the programme rolled out in full scale. Withdrawal of subsidized kerosene was done only in areas where the conversion packages had been distributed completely, by gradually cutting the allocation and kerosene supply. As the demand for LPG increased, the capacities of existing LPG terminals were increased and improved accordingly to ensure adequate reliability of LPG supply. With the geographic spreading of the programme, Pertamina established extensive business opportunities for the private sector, such as investment in building private LPG filling stations (Pertamina and WLPGA 2012).

Surveys by various parties indicate that a large majority of the recipients of LPG packages cook faster, have a cleaner kitchen, and reduce cooking expenditures. Over 94 per cent of the target beneficiaries are supportive of this government initiative. Thanks to this programme, the government subsidy on kerosene, which otherwise would have increased to over $5 billion, fell dramatically to below $3.1 billion in 2011 (Pertamina and WLPGA 2012). In the analysis and evaluation of the programme to date, Pertamina identified three key factors leading to the success of the programme: (1) strong governmental policies; (2) effective business and implementation models; and (3) clear benefits to all parties (Pertamina 2012).

In many emerging economies, government policies and programmes related to household fuels have been socially sensitive and financially difficult and risky. Using a conventional grant approach as the financial instrument, this programme successfully introduced an alternative cleaning cooking solution to poor households and avoided social objections to reducing kerosene subsidies. Though LPG is still under subsidy, the size of subsides is much smaller and this helped relieve the financial burden on the government. For countries struggling with fuel conversion issues, the experience and lessons from this programme can be very helpful.

20.4.4. Domestic biogas programme in China

China's domestic biogas programme underwent significant changes in the last decade due to increased investment from the government. With funding sourced from treasury bonds, the annual investment in the rural biogas programme reached a historic high of over 1 billion Chinese yuan in 2003, roughly corresponding to $124 million at that time. The investment scale kept growing until it reached CNY6 billion ($863 million) a year in 2008. During the period of 2003–2012, the accumulative investment from the national government alone was estimated at CNY31.5 billion ($4.5 billion). This pulled in CNY13.9 billion from local governments and CNY46.4 billion from rural households (MoA 2013). In addition to the funding sourced from treasury bonds and earmarked for rural biogas construction, some other government agencies also provided funding in support of domestic biogas development on a project basis, including those in charge of forestry conservation, public health, poverty alleviation, environmental protection, and ethnic minorities development.

Official figures from the Ministry of Agriculture indicate that biogas users in China reached 41.68 million households by the end of 2011, including 39.96

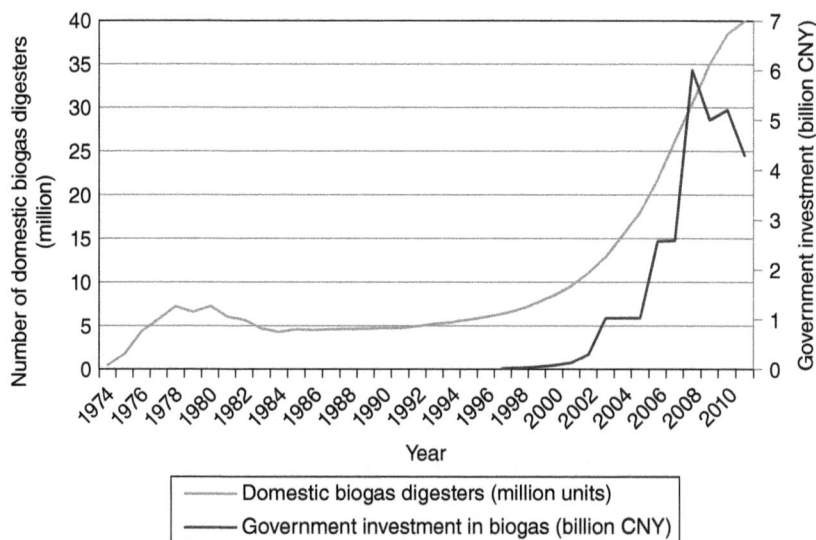

Figure 20.4. Number of domestic biogas digesters and government investment in biogas, China, 1974–2011

Data source: MOA 2012

million households with domestic biogas digesters. Biogas households accounted for 23 per cent of total households in rural China or about one-third of the rural households suitable for biogas installation (MoA 2012). The government plans to have 50 million biogas-user households and to increase biogas penetration rate to over 50 per cent of rural households suitable for biogas installation by 2015 (NDRC 2012). The total number of domestic biogas digesters and the annual government investment in biogas programmes are shown in Figure 20.4.

The justification for the government to promote biogas technology in China has changed over time, but may be attributed to one or a combination of the efforts in addressing the following problems: lack of clean cooking fuels in rural areas; water pollution and water-borne infection due to discharge of human and animal wastes; soil degradation due to wide application of inorganic fertilizers; and forest deterioration due to over-collection of firewood. The national biogas programme is considered effective in two major aspects: (1) mitigation of environmental pollution and ecological deterioration in rural areas through appropriate disposal and recycled utilization of animal wastes and agricultural residues, and (2) provision of an alternative clean cooking fuel to rural households (see Figures 20.5 and 20.6).

Multilateral and bilateral development organizations also played an important role in supporting biogas development in China. For example, in 2008 the World Bank provided a loan of $120 million under the Eco-Farming Project for rural biogas construction. The Asian Development Bank provided a loan of $33.1 million in 2003 for household biogas development, and another $66.08 million in 2010 for performance improvement of the biogas sector. In 2009, the French Development Agency provided a $50 million loan to China for the rehabilitation

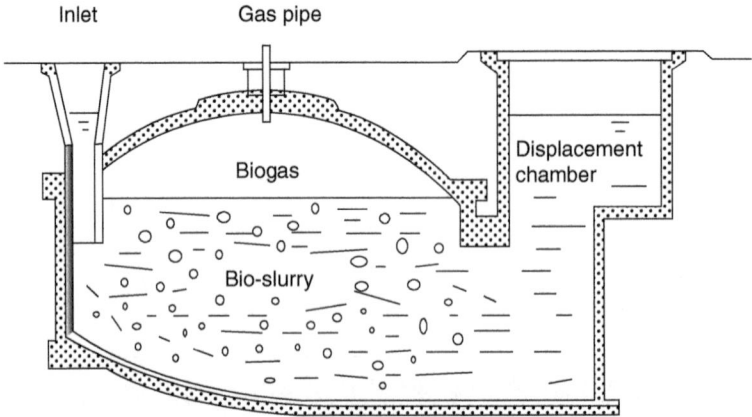

Figure 20.5. Schematic view of domestic biogas digester

Figure 20.6. Installation of domestic biogas digester
Photo: Xia Zuzhang, 2005

or reconstruction of rural biogas digesters in the earthquake-affected areas of Sichuan Province.

Private companies are also actively involved in biogas sector development through investment in manufacturing of biogas appliance and mass production of commercial biogas digesters. The aggregated annual production capacity of biogas cookers from more than 100 companies across China reached 10 million sets (Li and Xue 2010). One company in Sichuan Province invested about CNY300 million in the production facilities of fibreglass biogas digesters (Hongqi 2011).

Carbon financing also emerged as a source of funding to domestic biogas projects. Around 33,000 household biogas digesters in eight counties of Hubei Province were bundled together for carbon financing with the anticipated annual income of $0.82 million over 10 years, of which 60 per cent would be allocated to biogas households, 18 per cent to technical services, and 22 per cent to monitoring and project management (MoA 2009). The first half-yearly revenue from this CDM scheme was disbursed to the beneficiary households in August 2011. This additional income from carbon financing helps cut down the operational costs of domestic biogas digesters. During the period of 2011–2012, 38 household biogas CDM projects in eight Chinese provinces were registered (NDRC 2013).

In most cases, the beneficiary households contribute a significant portion of the installation costs and all the operational costs of biogas digesters in the form of cash and labour inputs. Reports from different sources indicate a wide range of varieties with the actual investment of beneficiary households from labour inputs only 50–70 per cent of cash inputs of the total installation costs (MoA 2007).

The financial instruments of government investment in rural biogas projects are primarily grant subsidies to households through provision of construction materials, biogas appliances, and technician services. Other financial instruments such as loans or guarantees are rarely used except for medium- and large-scale biogas plants. Subsidy or grant schemes from other funding sources vary from case to case. Most of them may end up with partial to full coverage of biogas construction costs, or add on government subsidy of biogas construction in remote places where construction costs are significantly higher, or filling the area gaps using the coverage of the national biogas programme.

The scale of the domestic biogas programme in China is much bigger than that of any other country, partially due to the large investment from the government. Though the quick increase in the number of domestic biogas digesters in China also brought growing pains, such as concerns over the utilization rate of biogas digesters, this programme has helped millions of households to access an alternative clean cooking fuel and the multiple benefits of biogas technology. A study of the World Bank indicated that the comprehensive economic benefits are much more attractive than the direct financial return of domestic biogas digesters (World Bank 2008). This may help explain the government's decision to invest in biogas technology through subsidy to end-users. In fact, subsidies from the governments have stimulated individual households to invest in biogas construction, leveraged investment from the private sector, and helped cost reduction through realization of economy of scale.

20.4.5. Commercialization of improved stoves in Cambodia

More than 90 per cent of the 14 million people in Cambodia use traditional biomass fuels for cooking, including wood and charcoal, which will probably be the primary cooking fuels for many years to come. Energy-related activities such as gathering wood, boiling water, and cooking take the poor as much as 3–4 hours a day (ASTAE 2010). A study in 2008 found that around 0.5 million households were using charcoal for cooking and that this number would increase to 1 million in 2015. Phnom Penh alone consumes 90,000 tons of charcoal every year, a market believed to be worth about $25 million (GERES 2010a). Cooking with charcoal is costly and uncontrolled wood consumption imposes adverse impacts on the ecological environment.

In the late 1990s, a non-governmental organization, GERES,[2] started promoting improved cookstoves in Cambodia. The initial results were not very encouraging—less than 10 per cent of the improved stoves are still around after two years. With technical improvements, the new stove consumes significantly less fuel than traditional stoves and last two to three times longer. However, the relatively high price and the lack of a wide distribution network kept its market penetration low (GERES 2010b; ASTAE 2010).

To facilitate commercialization of the new stoves, GERES managed to set up a micro-credit fund to finance entrepreneurs to buy equipment and raw materials in bulk for mass production of stoves; the fund also helped to establish a distribution network throughout the country. With financial and technical assistance from GERES, the stove producers, along with the distributors and retailers, formed the Improved Cookstove Producers and Distributors Association of Cambodia, which takes responsibility for quality control, price regulation, and market promotion. By the end of 2012, association membership covered 84 stove producers, 171 distributors, and 253 independent business owners across Cambodia with cumulative total sales of 2 million stoves since 2003 (see Figure 20.7). The improved stoves are sold on the market without any subsidies or credit facilities for the buyers. It was expected that the association members would soon be able to manage the association independently without the support of GERES (GERES 2013).

In addition to the microfinancing to stove producers, GERES also provides assistance in four interrelated areas—technology, marketing, forestry, and administration. The technical team carries out research and development on cooking stoves, provides training, monitors production, and helps quality assurance. The promotion and marketing team helps raise awareness by producing publicity material and facilitating business meetings between stove producers and other stakeholders. The forestry team works on sustainable charcoal production through community forests and plantations, and the administration team manages the credit facility to ensure its smooth operation (Wheldon 2006). All of these efforts have contributed to the success of the project, which not only helps to reduce the pressure on forestry resources but also creates income-generating activities for all the producers and retailers.

This project is considered as a milestone for all the NGOs around the world working on improving cooking devices. Through project intervention, a business model was created and large-scale dissemination of improved cookstoves on a commercial basis was accomplished. In early 2013, GERES Cambodia began

---

[2] GERES—Groupe Energies Renouvelables, Environnement et Solidarités (Renewable Energies, Environment and Solidarity Group).

Figure 20.7. Stove retailer in Kandal Province, Cambodia
Photo: Xia Zuzhang, 2010

collaboration with partners for a programme with a goal to disseminate over 100,000 improved cookstoves across five provinces in Lao PDR through a market-based multi-stakeholder approach.[3] With financial support from the French Global Environment Fund, GERES has also launched a Global Stove Programme and Partnership to support nine major stove programmes in South-east Asia and West Africa, targeted at creating a favourable environment for the dissemination of 2 million improved cookstoves during 2013–2016 (GERES 2013).

20.5. REGIONAL PROGRAMMES AND INTERNATIONAL INITIATIVES

20.5.1. Domestic biogas programme in Asia and Africa

The programme led by SNV Netherlands Development Organization had helped the installation of 0.5 million biogas digesters in 17 countries across Asia and Africa by the end of 2012. The financial contributions made available to SNV by various donors were mainly used to cover the costs to local partners of setting up and running national biogas programmes, to subsidize rural households to invest

---

[3] SNV Press Release (6 February 2013).

in biogas digesters, and to facilitate the engagement of private companies in providing financial and technical services. Rather than simply making grant subsidies available to have biogas digesters installed, the programme strengthens the institutional capacities of local and national partners towards a supportive public environment and the development of commercially viable biogas sectors in the target countries (SNV 2013).

In project operations, SNV made major efforts in building up partnerships with the private sector (construction companies, financial service providers) and the public sector (policy and regulations, standard-setting, quality control). Though various funding sources have been mobilized to support the development of the biogas sector in the target countries, the beneficiary households in almost all the cases still finance the major part (about 70 per cent) of the installation costs in cash or microcredit facilitated by the programme.

How to best use aid funding to support governments in addressing constraints in public service delivery has been a concern for a long time. The multi-stake-holder sector development approach as adopted by this programme can be inspirational. It aims to strengthen organizational and institutional capacities and to establish and optimize cooperation between all actors involved. In the long run, it is expected that a commercially viable biogas sector can be sustained without recurrent funding support from donor agencies.

20.5.2. Lighting Africa

This is a joint IFC and World Bank programme launched in September 2007 with the aim of catalysing and accelerating the development of sustainable markets for affordable, modern, off-grid lighting solutions for low-income households and microenterprises across Sub-Saharan Africa. The longer-term goal of the programme is to eliminate market barriers for the private sector and provide modern off-grid lighting products to 250 million people in Africa by 2030 (IFC and World Bank 2013).

The Programme addresses the challenges through a comprehensive set of tasks, including: (1) strengthening ties between the international lighting industry and local suppliers and service providers for manufacturing, marketing, and distribution of significantly lower-cost products; (2) facilitating consumer access to a range of affordable, reliable, and high-quality lighting products and services—for example, by providing consumer education services and consumer finance, and by executing a product quality assurance programme; (3) improving market conditions for the scale-up of modern lighting products by reducing existing technical, financial, policy, information, and institutional barriers; and (4) mobilizing the international community—governments, the private sector, international organizations, and NGOs—to promote aggressively the use of modern lighting services for the poor in Africa (IFC and World Bank 2013).

By the end of 2012, around 1.4 million solar lighting products were sold, benefiting nearly 7 million people. So far, 49 quality-assured solar lighting products have been made available in 20 African countries. The annual sales of the quality-assured solar lighting products have been doubling every year for the last three years, which shows a vast business opportunity for investors and entrepreneurs (IFC and World Bank 2013).

Different from many conventional development assistance programmes, this programme very much emphasizes market-oriented approaches and the roles of the private sector in delivering basic lighting services for the poor in off-grid areas. It is likely that these efforts would bring in major changes in serving the people at the bottom of the economic pyramid while providing business opportunities to private-sector players through the development of financially viable delivery models.

20.5.3. Mainstream energy access in bank operations at the Asian Development Bank

In response to the energy poverty prevailing throughout Asia and the Pacific Region, among other considerations, the Asian Development Bank (ADB) updated its energy policy in 2009 in congruence with its Strategy 2020. The energy policy emphasizes energy security, transition to a low-carbon economy, and universal access to energy. It aims to help its developing member countries to provide reliable, adequate, and affordable energy for inclusive growth in a socially, economically, and environmentally sustainable way (ADB 2009). As part of the efforts, ADB has launched the Energy for All Initiative to bring energy access into the operations of the bank and to engage regional partners working on energy poverty, aiming to help 100 million people to get access to modern energy by 2015. The initiative has four priorities of action: (1) increase ADB investments in energy access; (2) address financial, technical, and regulatory barriers to facilitate energy access; (3) capture and disseminate best practices and sustainable business models in the energy access sector; and (4) facilitate the replication and expansion of energy access enterprises in Asia and the Pacific Region.

Changes in the strategies and policies of a multilateral financial institution may have strong impacts on its investment priorities and its conventional ways of investment. In the case of ADB, significant changes are observed with its investment in energy access after the issuance of the energy policy. A total of over $3 billion was invested in energy access during the period of 2010–2012. This helps provide access to modern energy to an estimated 9.3 million households. In comparison, the annual investment of ADB in energy access was about $421 million in 2009, which benefited 258,000 households (Acharya 2013).

20.6. CONCLUDING REMARKS

There are many other good practices developed around the world at regional, national, and local levels. While all these good practices have different approaches and emphases, they have similarities in terms of resource mobilization, creation of enabling environments, making available technologies more affordable, exploring locally available resources, strengthening institutional capacities, establishing efficient and effective financial mechanisms and delivering networks, involvement of community-based organizations and targeting beneficiaries, and engagement of the private sector. All these efforts have helped millions of people toward getting out of energy poverty. Replication of these good practices and further adapting them to local contexts and market conditions would help scale up energy access under financial limitations.

To move towards energy poverty eradication requires the following:

1) Sound political commitments from the governments of both developing and developed countries. The Light for All programme in Brazil and China's domestic biogas programme show how such commitments can lead to large investments from governments and international development assistance. Countries at different stages of socio-economic development may face more or less challenging situations in public financing capacities, but sound political will and commitments to reform energy policies and regulations will help create a better enabling environment for private-sector finance to flow into energy access projects.

2) More effective technical solutions. Diversified technical solutions that better fit country-specific conditions should be explored to meet the energy needs of the poor. While extension of national power grid and decentralized power generation schemes under the Light for All programme provide more reliable access to electricity in Brazil, the solar home systems in Bangladesh and smaller solar lighting systems in Sub-Saharan Africa are probably more cost-effective solutions, given the overwhelming financial challenges to extending the national power grid to cover all the sparsely populated off-grid areas.

3) Robust pro-poor financial arrangements for productive activities and live-lihood improvement. Other than the capacities of the public and private sectors to supply modern energy services, the affordability of the poor to pay for needed energy services is an important factor in energy poverty eradication. Many case studies show that better access to modern energy services can significantly improve the livelihoods of the poor; introducing income-generating opportunities through productive use of modern energy enhances the ability and willingness of the poor, which in turn helps ensure the long-term market viability of providing modern energy services.

4) Large-scale replication of best practices developed so far around the world. Many good practices have shown cost-effectiveness in both financial and economic terms. Large-scale replication of such practices would help save time and effort for project interventions under similar contexts. For example, the SHS dissemination model of Bangladesh may have good potential to be replicated in other countries with similar situations to provide basic lighting to the poor in off-grid areas. The experience and lessons from the successful conversion of kerosene to LPG in Indonesia and from the commercialized dissemination of improved cookstoves in Cambodia can be helpful in addressing social and market barriers, though it may be necessary to further adapt these models to better fit in the unique market characteristics and socio-economic conditions of target countries.

5) Innovation in sustainable business development. Energy poverty has been a long-lasting problem affecting billions of people, particularly those at the bottom of economic pyramid. Many previous project interventions have failed to bring in lasting impacts at scale. Innovative business models and financial mechanisms are necessary to make a difference, so that much-needed financial resources flow more smoothly into energy access initiatives.

The public sector, the private sector, and civil society have made some headway in reducing energy poverty in the world, but collaboration and coordinated action between all three is essential in order to make serious inroads towards eradication of the problem.

REFERENCES

Acharya, J. (2013). *Energy for All—Increasing Energy Access in Asia and the Pacific.*

ADB. (2009). *Energy Policy* [online], <http://www.adb.org/sites/default/files/pub/2009/Energy-Policy-2009.pdf>.

Adenikinju, A. (2008). 'Promotion of Public Private Partnership to Improve Energy Access for Poverty Reduction and Growth in Sub-Saharan Africa', *Energy Poverty in Africa* 39: 185–208.

ASTAE (2010). *Cambodia—Supporting Self-Sustaining Commercial Markets for Improved Cookstoves and Household Biodigesters* [online], <http://www-wds.worldbank.org/external/default/WDSContentServer/WDSP/IB/2011/04/20/000356161_20110420051332/Rendered/PDF/611560WP0P12021usehold0Biodigesters.pdf>.

Bazilian, M. (2010). 'Understanding the Scale of Investment for Universal Energy Access', *Geopolitics of Energy* 32(October–November): 10–11.

CDM Policy Dialogue (2012). *Climate Change, Carbon Market and the CDM—A Call to Action* [online], <http://www.cdmpolicydialogue.org/report/rpt110912.pdf>.

Chineyemba, P. (2008). 'Energy Access in Rural Areas', *Energy Poverty in Africa.* The OPEC Fund for International Development, pp. 113–14.

CIA (Central Intelligence Agency, USA) (2013). *The World Factbook* [online], <https://www.cia.gov/library/publications/the-world-factbook/>.

Disch, D., Rai, K., and Maheswari, S. (2010). *Carbon Finance: A Guide for Sustainable Energy Enterprises and NGOs* [online], <http://www.gvepinternational.org/sites/default/files/carbon_finance_guide.pdf>.

Elisberg, J. (2010). *Smart Subsidies in Market Facilitation* [online], <http://www.aptenterprise.org.uk/downloads/Smart%20Subsidies%20in%20Market%20Facilitation%20with%20a%20Focus%20on%20Agricultural%20Knowledge%20and%20Information%20Systems.pdf>.

ESCAP (Economic and Social Commission for Asia and the Pacific, United Nations) (2005). *Energy Services for Sustainable Development in Rural Areas in Asia and The Pacific: Policy and Practice.* United Nations Publications.

GERES (Groupe Energies Renouvelable, Environnement et Solidarités) (2010a). 'Biomass Briquette Plant Opens in Cambodia', S. Marks, interviewer. *New York Times*, 4 January.

GERES (2010b). 'GERES Cambodia's Improved Cookstove Project Reaches 1 Million Stoves' [online], <http://www.nexus-c4d.org/news/news-updates/43-geres-cambodia-reaches-1million-stoves>.

GERES (2013). GERES press release, 28 January.

GPOBA (Global Partnership on Output-Based Aid) (2009). *Output-Based Aid—Fact Sheet* [online], <http://www.gpoba.org/sites/gpoba.org/files/GPOBA_fact_sheet_english_0. pdf>.

Hamid, M. (2013). 'Photovoltaic Based Solar Home Systems – Current State of Dissemination in Rural Areas of Bangladesh and Future Prospect', *International Journal of Advanced Research in Electrical, Electronics and Instrumentation Engineering* (2): 745–9.

Haque, N. (2012). *IDCOL Solar Home System Program in Bangladesh.* Infrastructure Development Company Limited.

Hongqi. (2011). Company brief [online], <http://www.hongqizhaoqi.com/about.asp> (accessed 16 October 2013).

IDCOL (Infrastructure Development Company Limited, Bangladesh) (2011). 'IDCOL Solar Home Systems Model—An Off-Grid Solution in Bangladesh' [online], <http://www.idcol.org/Download/IDCOL%20SHS%20Model_30%20Nov%27111.pdf> (accessed 16 October 2013).

IEA (International Energy Agency) (2011). *World Energy Outlook 2011* [online], <http://www.worldenergyoutlook.org/publications/weo-2011/#d.en.25173> (accessed 16 October 2013).

IEA (2012). *World Energy Outlook 2012* [online], <http://www.iea.org/W/bookshop/add.aspx?id=433> (accessed 16 October 2013).

IFC (International Finance Corporation) (2012). 'From Gap to Opportunity: Business Models for Scaling Up Energy Access' [online], <http://www.ifc.org/wps/wcm/connect/ca9c22004b5d0f098d82cfbbd578891b/EnergyAccessReport.pdf?MOD=AJPERES> (accessed 16 October 2013).

IFC and World Bank (2013). 'Lighting Africa—Catalysing Markets for Modern Lighting' [online], <http://www.awea.org/files/FileDownloads/pdfs/Judy-Siegel-Lighting-Africa. pdf> (accessed 16 October 2013).

Kumar, G. and Sadeque, Z. (2012). 'Output-Based Aid in Bangladesh—Solar Home Systems for Rural Households', *OBA Approaches* 42(April).

Li, J. and Xue, M. (2010). 'Review and Outlook of Biogas Industry Development in China', *Renewable Energy Resources* 28(3): 1–5.

Luz para Todos. (2010). *Light for All: A Historic Landmark—10 Million Brazilians out of the Darkness.* Ministry of Mines and Energy, Brazil.

Luz para Todos (2013). *Light for All Programme,* Ministry of Mines and Energy, Brazil [online], <http://luzparatodos.mme.gov.br/luzparatodos/asp/> (accessed 1 March 2013).

Malik, M. (2013). *IDCOL's Renewable Energy Initiatives.* Infrastructure Development Company, Bangladesh.

McKinsey Global Institute (2013). *Rethinking Infrastructure.* McKinsey Global Institute.

MoA (Ministry of Agriculture, China) (2007). 'National Rural Biogas Programme Development Plan 2006–2010' [online], <http://www.moa.gov.cn/zwllm/tzgg/tz/200704/P020070418570346665578.doc> (accessed 16 October 2013).

MoA (2009). 'Household Biogas CDM Project Successfully Registered', 23 February [online], <http://www.gov.cn/gzdt/2009-02/23/content_1240220.htm> (accessed 16 October 2013).

MoA (2012). 'Seeking for Five Major Shifts in Rural Biogas Development' [online], <http://www.moa.gov.cn/zwllm/zwdt/201212/t20121226_3116842.htm> (accessed 16 October 2013).

MoA (2013). 'Strive to Create New Situation for Rural Biogas Construction', *China Biogas,* 31(1): 3–4.

Morduch, J. (2005). 'Smart Subsidy for Sustainable Microfinance', *Finance for the Poor,* 6(4): 1–7.

MPEMR (Ministry of Power, Energy and Mineral Resources, Bangladesh (2013). *Power Division* [website], <http://www.powerdivision.gov.bd/user/brec1/30/1> (accessed May 5, 2013).

Mumssen, Y. (2010). *Output-Based Aid: Lessons Learned and Best Practices,* The World Bank [online], <http://www-wds.worldbank.org/external/default/WDSContentServer/WDSP/IB/2010/03/25/000333037_20100325013914/Rendered/PDF/536440PUB0outp101Official0Use0Only1.pdf> (accessed 16 October 2013).

NDRC (National Development and Reform Commission, China) (2012). *China Twelfth Five-Year Plan for Renewable Energy Development* [online], <http://cdm.ccchina.gov.cn> (accessed 16 October 2013).

NDRC (2013). *CDM Project Database* [website], <http://cdm.ccchina.gov.cn> (accessed 16 October 2013).

PAC (Practical Action Consulting) (2008). 'Access to Sustainable Energy Sources on a Local Level—Expert Analysis and Study on the Current Policy Issues' [online], <http://www.globalbioenergy.org/uploads/media/0812_Practical_Action_-_Access_to_sustainable_energy_sources_on_a_local_level.pdf> (accessed 16 October 2013).

Pertamina (2012). *Providing Cleaner Energy Access for Indonesia: Case Study from Kerosene to LPG Conversion* [online], <http://www.esmap.org/sites/esmap.org/files/8.%20Gusrizal%20Taib_120505%20-%20ESMAP%20clean%20energy%20initiatives_v4_RESIZED.pdf>.

Pertamina and WLPGA (2012). *Kerosene to LP Gas Conversion Programme in Indonesia* [online], <http://www.exceptionalenergy.com/uploads/Modules/Ressources/Kerosene%20to%20LP%20Gas%20Conversion%20Programme%20in%20Indonesia.pdf> (accessed 16 October 2013).

SE4ALL (Sustainable Energy for All, United Nations) (2013). *Country Actions* [website], <http://www.sustainableenergyforall.org/actions-commitments/country-level-actions> (accessed 16 October 2013).

SGHG-SEFA (Secretary General's High-level Group on Sustainable Energy for All, United Nations) (2012). *Sustainable Energy for All—A Framework for Action* [online], <http://www.un.org/wcm/webdav/site/sustainableenergyforall/shared/Documents/SE%20for%20All%20-%20Framework%20for%20Action%20FINAL.pdf> (accessed 16 October 2013).

Silva, H. and Tuladhar, R. (2006). *Smart Subsidies: Getting the Conditions Right* [online], <http://www.lirneasia.net/wp-content/uploads/2006/02/de%20Silva%20Tuladhar%202006%20Nepal%20final.pdf>.

SNV (Stichting Nederlandse Vrijwilligers [Foundation of Netherlands Volunteers]) (2013). 'Milestone 500,000 Biodigesters Reached' [online], <http://www.snvworld.org/en/regions/africa/news/milestone-500000-biodigesters-reached> (accessed 16 October 2013).

UNFCCC (United Nations Framework Convention on Climate Change) (2013). Press release, 30 January [online], <http://cdm.unfccc.int/press/releases/2013_01.pdf> (accessed 16 October 2013).

WaterCredit (2013). *Smart Subsidies* [website], <http://watercredit.org/philanthropy/smart-subsidies/> (accessed 16 October 2013).

WBCSD (World Business Council for Sustainable Development) (2012). 'Business Solutions to Enable Energy Access for All' [online], <http://www.wbcsd.org/pages/edocument/edocumentdetails.aspx?id=14165&nosearchcontextkey=true> (accessed 16 October 2013).

Wheldon, A. (2006). *Commercialisation of Efficient Charcoal Stoves in Cambodia.* Ashden Awards [online], <http://www.ashden.org/files/reports/GERES Cambodia2006 Technical report.pdf> (16 October 2013).

World Bank (2008). *Project Appraisal Document for China Eco-Farming Project* [online], <http://www-wds.worldbank.org/external/default/WDSContentServer/WDSP/IB/2008/11/13/000333038_20081113003334/Rendered/PDF/397810PAD0P0961E0ONLY10R20081023211.pdf> (accessed 16 October 2013).

21

Alleviating Energy Poverty in Africa

A Story of Leapfrogging, Localizing, and Fast-Tracking

Leila Benali and Andy Barrett

March 2013 in Houston, Texas—a group of consultants were invited to present to a world-leading philanthropist on solutions to address Africa's energy issues. During the dinner preceding the presentation, the business guru had waxed lyrical about the importance of mega-power grids, energy highways, and other capital-intensive infrastructure in front of a captivated audience of global energy leaders. Panic struck the consulting team: they had spent the last 48 hours preparing a presentation which championed micro-grids, localized solutions, and distributed energy. Brief text messages were exchanged among the members over dinner, and the team's final word was 'we'll stick to our guns!'.

At the same time, in south-east Guinea, a task force of a large mining company was finalizing a report, questioning the viability of an unprecedentedly large iron ore mining project for the region. The team was fully conscious of the strong sensitivities that their recommendations would arouse. The project was unprecedented both in absolute terms and relative to the country's economy. The company had already invested more than $1.1 billion in the past year in preliminary works and utilities alone. The senior management had widely communicated that the annual project revenue at full capacity would be more than double the country's GDP and would add significant added value. If the project were to be shelved, so would all the associated investments bringing electricity, fuel, and optical-fibre-based communication to local populations and businesses alongside the so-called 'National Interest Project corridor'. The 1.8 million population of this 47,000 km² area is extremely poor. The average annual revenue per capita does not exceed $200, well below the national average of around $500. The isolated population of the Nzérékoré region would be particularly disillusioned if the project were cancelled.

21.1. IS SMALL ALWAYS EXPENSIVE, INEFFICIENT, UN-SCALABLE?

These two stories happening simultaneously on two different continents illustrate the different dilemmas of energy access in many parts of the world, but most blatantly in electricity-poor Africa. They reflect the paradigms embedded in developed economies, misalignment between multiple interests, and the size of the complex issues at stake. The issue of energy access is all the more intricate as it is directly linked to income and wealth creation. If energy poverty is directly correlated with economic poverty, it looks difficult to solve the first without at least offering solutions to the latter. On the other hand, reliable energy access remains a prerequisite for economic growth. And finally, numerous case studies and theoretical analyses have demonstrated that economic growth does not necessarily eradicate poverty (e.g. Equatorial Guinea registers a GDP per capita of around $16,000, while more than 30 per cent of the population is unemployed and the country ranks 121st on the United Nations Human Development Index—out of 177 countries).

On a cumulative basis, the investment figures look big and the challenge overwhelming. More than two decades of failure to build new capacity leaves a gap of more than 250 GW needed by 2030 (or around $20 billion per annum of investment need), according to the International Energy Agency. Add to it the more than $20 billion per year required for universal electricity access during the same timeframe. Furthermore, the average per capita power consumption in Sub-Saharan Africa is 25 per cent that of India. More than 600 million Africans lack access to electricity. At first sight, the temptation is to embark on top-down approaches, for example by trying to mobilize large funds to invest in country-wide electrification programmes. That could be a valid approach, but here we argue that tailored technology deployment, both urban and off-grid, is required.

Nevertheless, sources of financing, particularly multilateral, are available and multiple and have been regularly used to finance various sorts of projects using diverse financing tools: the Clean Technology Fund, the European Investment Bank, national development agencies, the World Bank Group, Sovereign Funds, and so on. Today, Sub-Saharan Africa energy investments from the World Bank Group alone amount to around $10 billion: $2.27 billion for grid extension, $4.59 billion for grid-connected supply, $1.37 billion for off-grid renewable electricity, $1.07 billion for policy and regulation, and $0.76 billion for the efficient use of electricity (Monari 2011). More interestingly, the World Bank estimates that every $1 it spends leverages between $0.29 and $1.84 of additional private and client investments in each of the aforementioned areas. The emergence of China as a key energy and infrastructure investor in Africa has been widely documented (e.g. investing $700 million in Guinea's rural electricity system). The last decade also witnessed the development of alternative sources of investment in the energy sector, such as projects related to the Clean Development Mechanism (CDM), or charitable and not-for-profit approaches. However, there remains a large un-tapped potential: while accounting for more than 17 per cent of the population of the 'less developed world', Africa only registers 3 per cent of CDM projects currently in the pipeline (UNEP 2013).

For policymakers and energy planners, there are usually two traps that need to be avoided:

- *Missing the 'last mile'*: the macro-view often overshadows the more distrib-uted issues related to energy access, either because of lack of market data, lack of finance, institutional bottlenecks, or all of the above. Even in Morocco, a rural electrification success story (97 per cent rural electrification in 2010, up from 22 per cent in 1996), inhabitants of villages 50 km away from Casablanca are not connected to the grid and lack access to electricity.

- *Ignoring the existing*: there is a serious potential for improvement reliability and costs (including fuel). In addition to the often weak and fragmented transmission and distribution networks, electricity blackouts and reliance on expensive diesel can cost between 1 and 5 per cent of GDP annually—small diesel or kerosene generators in Sub-Saharan Africa account for up to 35 per cent of generation.

For private project developers and financiers, adaptability is one of the key requirements. For example, the average size of power projects in Africa is 30 MW, while the average size of power projects globally is 300 MW. Historically, the international model has led to a natural preference towards specific technology choices, widely used and tested but generally applicable to the large-scale archetypes (combined cycle gas turbines for instance) and financial models suitable for OECD best practice (i.e. large scale generation facilities connected to reliable centres of demand through reliable HV transmission grids). In theory, the economies of scale can be secured and the low-cost power supplies can be passed on to consumers. In reality, this model has broken down in most of Sub-Saharan Africa for various reasons. It has blocked a large share of the needed investments because of investors' concerns regarding reliability of infrastructure, fuel supply security, and creditworthiness of offtake, amongst others. Even in the case of Nigeria, which has invested billions of dollars in its power sector, dispatchable power capacity has actually fallen. Finally, passing through costs to consumers (or to governments) means that someone has to deal with the intricate issue of energy pricing. Industry players have called on a few occasions for the need to adapt the 'classic' IPP model to Africa's specificities and consumption patterns. The decrease of solar PV generation costs and the maturing of the wind industry might gradually deal, although only partially, with this concern in the years to come, catalysing secure nexuses of small-scale generation which can underpin local 'micro-grids'.

Then there is the disturbing 'trilemma' between energy–economy–environment. Most institutional investors face these trade-offs: access to energy should not occur at the expense of the environment, while access to energy is a necessity to induce economic and social development. It is true that inefficient appliances (cookstoves, lighting, water heaters) deteriorate local environments, and natural resources, including forests, are being depleted in some areas. However, access to modern and efficient forms of energy could in fact generate larger amounts of emissions. More importantly, projects mitigating major energy-sector-related greenhouse gas emissions are highly favoured by institutional investment bodies but difficult to implement. Gas flaring accounts for around 20 per cent of Africa's GHG emissions, and Nigeria alone could be foregoing more than $5.5 billion annually in export revenue and carbon offset credits. Even though the country has spent more than a decade trying to reduce flaring, it still records the second-highest level in the world after Russia. Flare-gas-to-power monetization schemes have been aggressively pursued but remain largely unrealized. Most gas-flaring reduction projects still face challenging requirements for the registration by the UN Framework Convention's CDM Executive Board and host countries' governments (at the end of 2010, only half of the submissions were successfully registered). A less stringent CDM scheme for African projects, and a tailored framework for 'Programmes of Activities' focused to match Africa's needs, might lead to a higher registration rate for the continent and should encourage project developers.

21.2. THE INTRICATE QUESTION OF PRICING: HOW TO CREATE CUSTOMERS?

The reliability of a network of ten 50 MW generating stations in providing 400 MW of capacity is far greater than a single 500 MW station. A distributed grid is much harder to disrupt than a single line. No individual community feels that it owns a 500 MW power station but local communities that are served by a 10 MW plant can see a direct link between power availability and the local facility. (Excerpt from 'Strong Generation, Transmission and Distribution Infrastructure to Support the Country's Rapid and Sustainable Economic Growth', a discussion paper prepared during the official visit to the United Kingdom of His Excellency Alhaji Umaru Musa Yar'Adua, President and Commander-in-Chief of the Armed Forces, Federal Republic of Nigeria, by Cambridge Energy Research Associates, 2012.)

Even resource-rich countries can struggle to guarantee access to reliable energy sources to a large share of their population. Here again, Nigeria, with a less than 50 per cent electrification rate, is one example where lack of political will has hampered energy programmes for several years, including gas monetization efforts and electrification programmes.

Strong political will and clear targets are often cited as two of the key requirements to advance universal access to modern forms of energy. These two requirements are certainly critical for any energy reform programme, but such programmes have generally shown clear failures when it comes to tackling energy poverty in Africa.

1) They have often presupposed that solutions can only be centralized and government-driven and have consequently proved difficult to implement in cases of low or weak governance.

2) They have linked access to energy (meant to be universal) either to regional political agendas or to national energy programmes, while the issues are often highly localized.

3) More importantly, they have generally postponed the discussions on energy pricing to a later stage in attempts to avoid the political sensitivity that surrounds the issues of subsidies.

This third point creates a vicious circle and is not unique to Africa: lack of price visibility discourages would-be investors and usually leads to non-existent or unreliable energy supplies; in this context, would-be consumers turn to local solutions to cover their basic energy needs—often distress solutions at high energy costs.

In all the parts of the world where energy or electricity supplies are unreliable, consumers rely on distributed generation of electricity or energy, be it their own diesel generators, local biomass, or renewable energy. These are residential, commercial, or even industrial consumers, who may never become customers, even if they are connected to the central system, unless they are provided the appropriate price incentive and guaranteed reliability. Furthermore, the so-called non-technical losses (theft, unpaid bills) on transmission and distribution networks can exceed 25 per cent in some areas of the developing world.

In areas where energy supplies are non-existent, but with a clear potential for pent-up demand, energy prices need to be set at the right level to serve two purposes: to attract investors and to turn would-be consumers into reliable paying customers. As mentioned in other chapters of this book, energy poverty is endemic in less- or least-developed countries. Therefore, energy supply will often be in competition with one of the first basic needs of the population: food.

21.2.1. Should customers first pay for food or for energy?

Governments have even considered imposing taxes on electricity to reduce food prices. The issue of subsidies and cross-subsidies becomes central in the debate, in terms of actual funds that governments would set aside to support energy consumption, and relative to other subsidy priorities such as food. The IMF concludes that energy subsidies in Sub-Saharan African countries are poorly targeted and often create a disincentive for maintenance and investment in the energy sector, perpetuating energy shortages and low levels of access (Alleyne and Hussain 2013). The study estimates that direct power subsidies average 0.4 per cent of GDP for Sub-Saharan Africa, and up to 0.8 per cent in Mali. Comparing these figures to food subsidies in Sub-Saharan Africa would be extremely helpful, particularly since food and fuel prices grew in tandem in recent years (pass-through of fuel price hikes into food and transportation, lower labour supply in agriculture, poor access to inputs including fertilizers, etc.). Data is limited, though (see Section 21.2.2. on the case of Mali).

By way of comparison, the IMF estimates energy subsidies to have reached 8.5 per cent of GDP in 2011 in MENA, compared to 0.7 per cent for food subsidies. To add some perspective, the pre-reform EU's Common Agriculture Programme was equivalent to around 0.5 per cent of GDP. It is worth noting that the IMF's methodology measures subsidies as the difference between the value of consumption at world and domestic prices. In view of the goal of facilitating energy access in Africa and setting attractive prices for both investors and would-be customers, a phased approach could be considered. The first step would be to focus on energy and food subsidies based on the difference between actual supply costs (including imports if any) and domestic prices. In terms of price support, allocating this foremost to food and agriculture should clarify that incurred for domestic energy prices. Thereafter the optimization of added value between international market opportunities and domestic economic growth can be considered. In reality, further trade-offs may be needed to attract foreign investors.

21.2.2. Reconciling the numbers to target subsidies—the case of Mali

In Mali, estimates of food subsidies are relatively well documented, particularly around the Government's Rice Initiative (extended to wheat, maize, and other products), launched in 2008. They now amount to around 0.1 per cent of GDP (FAO 2013). However, even basic data such as the share of population living below the poverty line is debatable. INSTAT (2011) and the UNDP (2012) estimate that 43.6 per cent and 50.4 per cent respectively of the population live below the UN poverty line of $1.25 in purchasing power parity (PPP) per day, while the World Bank indicates a relatively low poverty rate: 16.4 per cent below $1.25 PPP per day. The Malian rural population is spending 43 per cent of its income on food and the average income per capita in Mali stood at $1,030 PPP in 2010.

Access to energy, energy pricing, and energy reliability are therefore intimately linked. Greater emphasis will need to be placed on breaking the vicious circle of low incomes preventing access to modern energy services, which in turn puts severe limitations on the ability to generate higher incomes. There are business models which seem to reconcile this trilemma in many instances through communities' ownership and/or through productive use of energy. Local companies, including microenterprises and industrials, have played active roles in the marketing of petroleum products or, to a lesser extent, electricity generation services. In some countries, local community-based organizations have been mobilized to provide mechanical power, while larger municipalities have on occasion owned and operated electricity generation and distribution systems. These successful business models usually fall short of addressing the issue of scalability. We argue in Section 21.3 that government intervention could then be needed to scale up and to fast-track projects. Anchoring local electricity supply around the needs of large-scale energy-intensive industrial developments or social development of electricity supply as a condition of upstream access by oil companies are variants of this model.

21.3. THE CASE FOR FAST-TRACKING

Ideally, regulations and stakeholders' interests should be aligned to kick-start investments. Even in resource-rich countries, fuel availability and costs can be serious bottlenecks, as highlighted earlier. Therefore, regulatory frameworks are increasingly designed to induce local fuel monetization. However, the longer the debate on local monetization, the more often it leads to a stalemate: it can also become so sensitive that it jeopardizes energy exports and their associated revenue, much needed to underpin resources development. It also hampers local consumption of fuel in the absence of a proper regulatory framework. In all cases, securing fuel arrangements in an accelerated manner is vital, even when the fuel is produced domestically.

Fast-tracking a renewable investment programme or quickly implementing a local gas monetization plan is all the more relevant and easier to justify when there is a clear case for cost improvement, for example when targeting to substitute expensive diesel generation. Governments could then intervene to set achievable targets and to fast-track planning and permitting, particularly for small-scale projects.

Italy's case, albeit with its own characteristics, could be cited as a benchmark. In March 2011, the Italian government approved a Renewable Decree that transposes the EU Directive 2009/28/EC (17 per cent of its energy sourced from renewables by 2020) into law. The 2020 renewable target for the power sector specifically is 26.4 per cent of electricity demand. With the Renewable Decree accelerating permit approvals to a maximum of 90 days and simplifying the procedure for plants smaller than 1 MW, coupled with a set of phased subsidies for most renewable technologies (either through green certificates, FITs, or feed-in premiums), the country was able to install 9.5 GW of solar in 2011 alone. Italy's installed net capacity in 2011 was around 122 GW.

The viability of the fast-track approach, of related business models such as 'one-stop-shops' for approvals, permitting, land allocation, and grid connections, and of common project funds across multiple investments, will ultimately depend on guarantees on stability of supply and on the sustainability of price differentials. The quick development of small-scale land-based gas liquefaction plants (producing liquefied natural gas, LNG) in China is another example which could be particularly enlightening for Africa. Although small LNG plants were first built in China in the early 2000s, it is really only since 2008 that capacity has taken off (Yang 2013). Land-based liquefaction in China (typical plant sizes of 40,000 metric tons per year) started out as a way for developers to monetize stranded gas from fields with no access to infrastructure. Total installed capacity by the end of 2012 was 7.4 Bcm per year, with 72 per cent growth year on year. According to the China National Petroleum Corporation an additional 14.6 Bcm is under construction or planned to come online by 2015. Plant owners are arbitraging price differentials between price-controlled pipeline gas and international LNG prices available in coastal regions. They also take advantage of the increasing customer needs for flexible and reliable gas supplies, including more expensive truck supply of LNG to consumers, of the lack of local access to pipeline capacity, and of the emerging use of LNG in transportation (trucks, ships, and barges). This is a potential model for Africa which could address the needs for both local secure fuel supply and local monetization of flared gas. The prize of backing out expensive diesel-based generation should more than cover the costs involved.

The danger is that several African countries have a long history of not delivering against plans, or rushing ill-prepared programmes. This has often been due to poor governance, populist agendas, or lack of human capital. There are lessons to learn from Ghana's recent history. The Jubilee oilfield discovery, and its coming on-stream in 2010, transformed Ghana into an emerging oil and gas powerhouse in Sub-Saharan Africa. But before that, Ghana had experienced two major energy crises in the previous two decades, while an unused 126 MW barge-mounted gas turbine has been kept idle on the shore. During the last energy crisis, the government acquired a number of open-cycle turbines to run on diesel (one of which was reportedly a 60-Hz unit for a 50-Hz grid). Today, associated gas from the Jubilee complex and the TEN (Tweneboa, Enyenra, and Ntomme) discoveries are being developed to feed domestic users, and gas exploration programmes in the Jubilee–Côte d'Ivoire basin have been successful. The government has defined a clear prioritization order for gas allocations: electricity, fertilizers, and then petrochemicals. Nevertheless the country is considering building a floating LNG-importing terminal off its Atlantic Ocean coast, with the view of using it to fuel 1,500 MW of electricity by 2016. It is still early to assess the exact cost of indigenous offshore gas, but it remains to be seen whether a floating LNG import terminal could really supply gas at competitive prices compared to indigenous gas.

In summary, adapting and scaling business models will be key to successfully fast-tracking energy access or electrification programmes, for example by:

- adapting and scaling the Independent Power Producer model to foster investment to guarantee its scalability, with a focus on local/distributed generation;

- setting a fast-track supportive regulatory environment, with a focus on facilitating multiple small-scale projects, and on setting clear pricing expectations;

- encouraging small-scale flare gas capture projects, fuelling flexible or scalable electricity generation (small-scale gas liquefaction, developed alongside trucked LNG to facilitate the logistics, could enable additional development of stranded gas, reach new customers, and provide attractive economics for local generation investments);

- adapting financing requirements and other policies to allow for fuel and technology flexibility to facilitate project development, even if it means later retrofitting to combined-cycle plants in later scale-up stages;

- promoting cooperative projects with industry, including the oil and mining sectors—in this case, energy/fuel supplies to the local populations and businesses would ideally be scalable in parallel.

21.4. ADDITIONAL PERSPECTIVES ON SOLUTIONS: THE IMPLEMENTATION

When it comes to guaranteeing quasi-universal energy access in Africa, the parallel with the mobile phone industry has often been drawn—almost to exhaustion, but it still provides a useful framework. In short, Africa is today the fastest-growing market for the mobile phone industry. This sector's aggressive growth has enabled the penetration of modern means of communication without the investment in long fixed-lines grids. With more than 650 million subscribers, this market is larger than that of the US or the EU, according to the World Bank. Africa is directly exposed to mobile devices that offer first contact with the Internet at broadband speeds, while like the rest of the world, the continent is dependent for Internet access on a growing network of undersea fibre-optic cables. Can the energy industry do the same? The key seems to be enabling twenty-first-century technologies to penetrate nineteenth-century markets without using twentieth-century-type infrastructure.

Innovative funding and commercial approaches will be required, but the 'innovation bubble' should come from fostering local entrepreneurial ideas and locally-driven investments. In the first instance, these ideas generally try to leverage the existing: for example, leveraging biomass consumption (over 80 per cent of energy used in Sub-Saharan Africa excluding South Africa), recycling used oils, and so on. Then, public–private partnerships could indeed be useful in large-scale programmes or in the framework of master-plans. But in most of the cases which require flexibility and fast-tracking, we would foresee only a minimal role for governments: primarily as facilitators, probably as regulators, but not as shareholders.

Distributed off-grid power solutions look unavoidable in many instances—primarily because of the large distances. They would be enhanced by a focus on local resource utilization, including the huge photovoltaic potential of the continent—now available at significantly reduced costs. These solutions could be combined with new storage technologies and existing generators, with flexible 'pay-as-you-go' financing. However, successful case examples are yet to be proven. Azuri Technologies combines solar and mobile phone technologies to enable users in off-grid markets, primarily in rural Africa, to pay for their solar electricity as they use it and avoid the unaffordable upfront costs. The aim is to demonstrate the viability and the scalability of the business model and make the case for further commercial funding.

Grid planners in Africa could 'leapfrog' from existing large-scale focus to grids optimized for distributed generation, while seeking to build up reliability and minimize intermittency back-up needs. Often we see the rise of 'clusters' around industrial anchor off-takers with micro- or mini-grid connections. Local micro-grid reliability can be substantially improved with 'smart' software and demand-side management tools—new challenges for the IT sectors.

But let's face it: there are still deep-rooted preferences for macro-grid extension over decentralized options, and there are still strong opponents of distributed generation, more particularly of solar PV, because of some of the failed donor-funded schemes in Africa. Fast-track programmes can look highly risky if implemented under weak governance or regulation. The development of small-scale liquefaction and LNG trucking could only happen if a strong pent-up demand is demonstrated and if price arbitrage can be established. Subsidies and energy pricing are highly sensitive issues for any government, let alone in countries with extreme poverty levels or with fragile governments. As to our opening conundrums, the mining company is still waiting for a workable compromise with the government; and the philanthropist took on board the challenges that the consultants' ideas posed to traditional concepts of scalable energy solutions. However, there was no disguising that one paradigm shift might not be enough to alleviate energy poverty in Africa; many paradigms will need to drastically change.

REFERENCES

Alleyne, T. and Hussain, M. (2013). 'Energy Subsidy Reform in Sub-Saharan Africa: Experiences and Lessons'. International Monetary Fund [online], <http://www.imf.org/external/pubs/cat/longres.aspx?sk=40480> (accessed 24 October 2013).

FAO (Food and Agriculture Organization, United Nations) (2013). 'Review of Food and Agricultural Policies in Mali 2005–2011' [online], <http://www.fao.org/fileadmin/templates/mafap/documents/Mali/MALI_Country_Report_EN_Feb2013.pdf> (accessed 24 October 2013).

INSTAT (Institut National de la Statistique, Mali) (2011) Institut National de la Statistique. Republique du Mali [website], <http://instat.gov.ml/index.aspx> (accessed 24 October 2013).

Monari, L. (2011). 'Financing Access at the World Bank', IEA workshop, Paris, May [online], <http://www.iea.org/media/weowebsite/workshops/weopoverty/07_Dr_Monari.pdf> (accessed 24 October 2013).

UNDP (United Nations Development Programme) (2012). 'International Human Development Indicators' [website], <http://hdrstats.undp.org/en/countries/profiles/MLI.html> (accessed 24 October 2013).

UNEP (2013). United Nations Environment Programme, October [website], <http://www.cdmpipeline.org/cdm-projects-region.htm#> (accessed 24 October 2013).

Yang, J. (2013). 'Small Land-Based Liquefaction Plants: A Third Dimension for China's Gas Market' [online], <http://www.ihs.com/products/cera/energy-report.aspx?id=1065982850> (accessed 24 October 2013).

Index